Models of Matter

Principles and Perspectives of Chemistry

Models of Matter

Principles and Perspectives of Chemistry

Gayl H. Wiegand
Idaho State University

West Publishing Company
Minneapolis/St. Paul New York Los Angeles San Francisco

Composition The Clarinda Company
Design Hespenheide Design
Page Layout Hespenheide Design
Copyedit Yvonne L. Howell
Index Pam McMurray
Artwork Precision Graphics and Wiest Publications Management, Inc.
Cover Image Jan Cobb, The Image Bank
Photos Credits follow Index.

WEST'S COMMITMENT TO THE ENVIRONMENT

In 1906, West Publishing Company began recycling materials left over from the production of books. This began a tradition of efficient and responsible use of resources. Today, up to 95 percent of our legal books and 70 percent of our college and school texts are printed on recycled, acid-free stock. West also recycles nearly 22 million pounds of scrap paper annually—the equivalent of 181,717 trees. Since the 1960s, West has devised ways to capture and recycle waste inks, solvents, oils, and vapors created in the printing process. We also recycle plastics of all kinds, wood, glass, corrugated cardboard, and batteries, and have eliminated the use of Styrofoam book packaging. We at West are proud of the longevity and the scope of our commitment to the environment.

Production, Prepress, Printing and Binding by West Publishing Company.

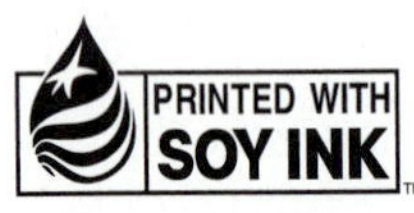

British Library Cataloguing-in-Publication Data. A catalogue record for this book is available from the British Library.

02 01 00 99 98 97 96 95 8 7 6 5 4 3 2 1 0

Library of Congress Cataloging-in-Publication Data

Wiegand, Gayl H.
Models of matter : principles and perspectives of chemistry / Gayl H. Wiegand.
p. cm.
Includes index.
ISBN 0-314-04573-2
1. Chemistry. I. Title.
QD33.W66 1995
540—dc20 94-34066
CIP

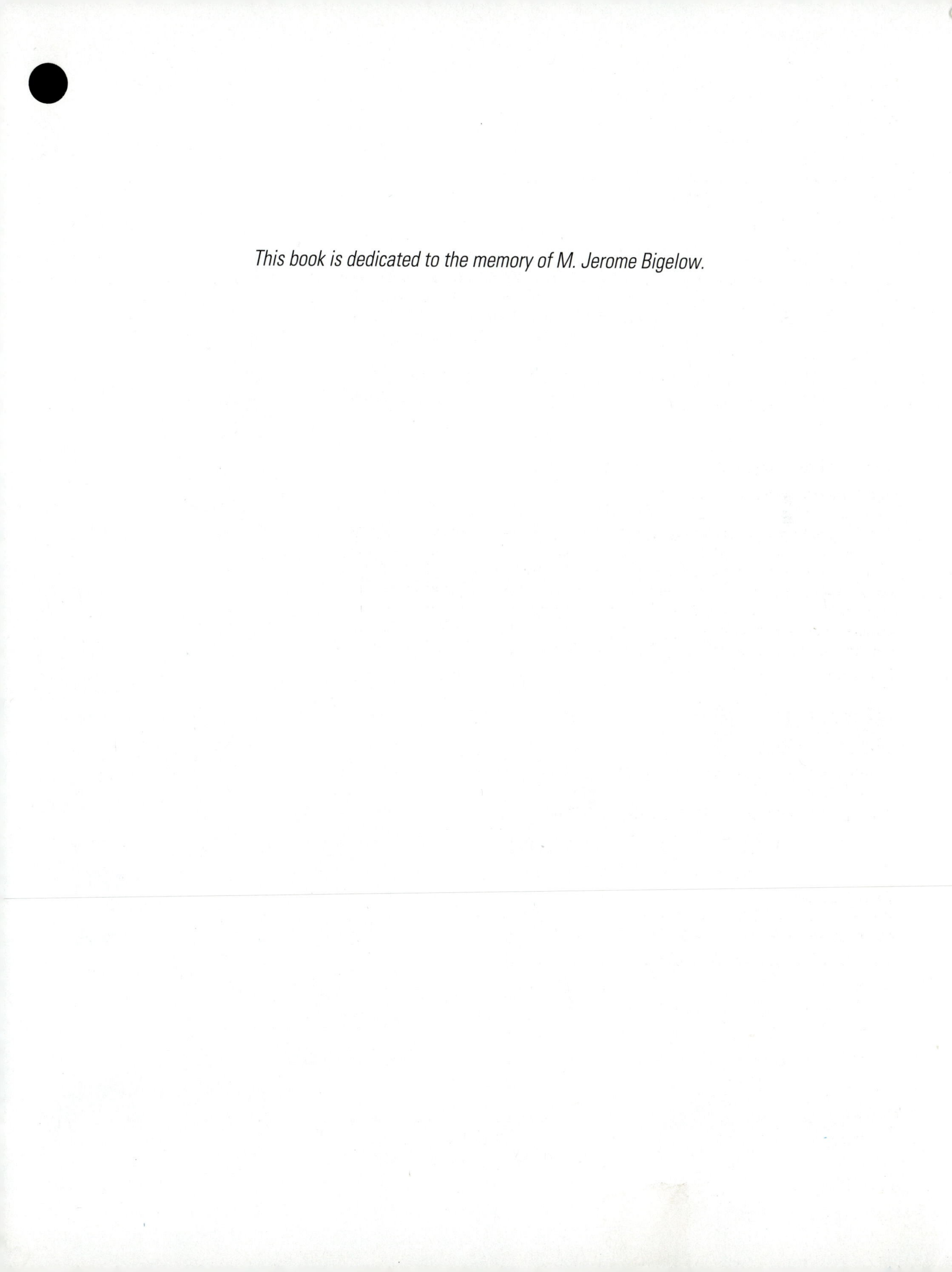

This book is dedicated to the memory of M. Jerome Bigelow.

Contents in Brief

Contents

Preface

Models of Matter is a text for a one-semester introductory college chemistry course for nonscientists, those students majoring in business, education, and the arts, humanities, and social science disciplines. Many have little previous knowledge of chemistry. Some harbor negative attitudes toward it, and a number fear having to learn it. *Models* offers a new approach that is sympathetic to the point of view of these students.

OVERVIEW

This new approach is based on the author's belief that besides introducing basic chemical principles and exploring how these influence our lives, a course of this kind should examine the processes by which scientific knowledge is gained. In fact, the development of scientific processes is used as the vehicle for introducing many chemical principles. Though many nonscientists are uncomfortable about learning science, they usually like to discuss its philosophical and social implications. *Models* takes advantage of this interest and the aptitudes of these students by initially engaging them in general discussions *about* science, something to which they can relate. Later a transition in viewpoint is gradually made, using the ideas of scientific model building and how the processes of science work, to give students the skills to be comfortably learning chemistry. No prior knowledge of chemistry is assumed, and little mathematics is required. The latter part of the book is more traditional in its approach, although the concepts presented there continue to draw on earlier discussions of scientific models. In addition, most of the chapters on chemical concepts and principles integrate a wide range of applications, rather than reserving them for separate discussions later.

OBJECTIVES

In a one-semester (or one-quarter) course for nonscientists, instructors have a limited opportunity–and possibly only one chance–to present chemistry in a meaningful way. In the early 1980s my colleague Ken Faler and I developed such a course with the following objectives in mind: namely, to provide our students with

1. an understanding of basic chemical principles and terminology,
2. an awareness that chemistry is part of their everyday lives,
3. a confidence in learning chemistry that hopefully would carry beyond the course,
4. an idea of how the process of science works, and
5. a more positive attitude toward chemistry (and science in general).

Based on the success of our approach, I began writing summaries of class discussions for distribution during the next class period. These notes formed the basis for the first draft of *Models,* which I wrote during a sabbatical leave in the fall of 1988. Since then, with many refinements, additions, and revisions, it has evolved into its present form.

GENERAL APPROACH

The general approach of *Models* is principles-driven and not applications-driven. My experience, and that of other teachers, shows that an associative conceptual framework that makes sense within student experience must be built, which can be used later, even after most of the facts are forgotten. To be sure, applications are very important in showing the relevance of principles to everyday life, but as facts themselves they are not easily incorporated into long-term memory or the associative reasoning of those who do not use chemical principles as a scientist does.

I chose to make this book relatively brief because there is usually little time to cover more. It is written with the belief that it is better to teach a coherent body of concepts and applications than to provide a disconnected assortment of topics. This approach gives nonscientists more opportunity to understand chemical principles, to carry them into later life, and to use them as tools for understanding. This approach also forces some hard decisions about which topics should be covered (and, of course, about those that cannot be covered). Not everyone will agree which topics are the most important, which skills may be most useful to students later, and which principles are the most applicable. The topics included here will meet the needs of most instructors, based on the comments of the many who reviewed this book.

ORGANIZATION

Models can be viewed as divided roughly into two parts. The first third develops the nature of the scientific process, model building, and the limits to scientific knowledge, while introducing a number of basic chemical ideas. The latter two-thirds of the text contains a more traditional development of chemical principles and applications.

Although the beginning chapters explore the nature of scientific knowledge, a surprising number of chemical principles are presented early. (It is not necessary for an instructor desiring a change of viewpoint to completely remake her/his course.) For example, scientific measurement is introduced in the second chapter; and in subsequent chapters, topics such as changes of state, chemical change, the "energy hill," Dalton's atomic theory, chemical periodicity, and writing chemical formulas and equations are presented. A considerable amount of chemistry is offered, but gradually and in small doses, so students do not feel overwhelmed by it. The presentation leads the students from idea to idea, a little at a time, pausing to give examples of each thought. It introduces points and then returns to them later for reinforcement and amplification. This, I have found, is what students unfamiliar with these ideas need. I also try to anticipate the limit to students' patience when learning chemical principles and give them a change of pace, by introducing difficult material in small "bundles" separated by breaks or problems for reflection, or by returning to the more philosophical narrative. Although the mathematical expectations in the course deliberately have been set low, the level of conceptual rigor has not been.

The structure of the atom itself is introduced in Chapter 5, whereupon the evolution of modern atomic theory is described. Here, the process of science is shown at its best, reinforcing the themes opened in the earlier chapters and leading naturally to the concepts of the nuclear atom, the Bohr model, and finally the wave model of the atom. Chemical periodicity is then restated in terms of the wave properties of electrons in atoms. The philosophical implications of wave/particle duality and the uncertainty principle are discussed at the end. Though this section can be skipped, it seems to interest students a great deal, probably because absolute certainty in science is called into question and the models themselves lose much of their commonsense relationship to "reality."

At this point, the students have sufficient chemical skills and confidence to handle chemical bonding. This is introduced in Chapter 6 and begins the more traditional approach to presenting chemistry. There may be more emphasis on skills here (and a few places later) than some instructors would prefer. I have written these sections so that the development of skills can be followed as far as desired and the rest can be omitted. (I sometimes do so myself.) Chapters 7, 8, and 9 present solutions, acids and bases (including equilibrium), and redox, respectively. Chapter 10, on energy and natural resources, offers a serious qualitative discussion of thermodynamics and the relationship of the second law to the depletion of nonrenewable natural resources. I have found that students can gain a real appreciation of thermodynamic principles from this presentation. Chapter 11 presents a survey of organic chemistry, and Chapter 12 follows with an introduction to polymers. Chapter 13 on environmental chemistry presents an overview of the major environmental problems facing us today and the insights provided by an understanding of chemistry.

Originally, a chapter on "household chemistry" had been planned. As later chapters were written, however, many household applications fell so naturally into them that a separate chapter seemed unnecessary and artificial. Environmental applications appear in various chapters, as well as in Chapter 13. The mole concept is introduced in Chapter 7 on solutions, but it can be skipped or used elsewhere if desired; stoichiometry and kinetics have not been included.

A qualitative conceptual approach is used in all of these chapters, and little quantitative problem solving is attempted. This is consistent with my view that in a course of this kind it is more important, for example, that music majors understand why the creatures in a pond are in big trouble if the pH of their environment is 6 and going down than for those students to be able to calculate the pH of the pond itself, with little idea of what it means.

PEDOGOGICAL FEATURES

- **You Try It** problems are interspersed at appropriate points throughout the chapters to reinforce ideas and give students the opportunity to test their understanding of new material.
- **Chemistry in Your Day** boxes present the experiences of a group of students as they encounter, discuss, and understand various aspects of chemistry in their everyday experiences.
- **Guest Essays** by individuals with various backgrounds and occupations give students perspectives on the importance of chemistry in different professions. Guest essayists include an art conservator, a solar engineer, a writer, a technical illustrator, a pharmacist, a politician, and others.
- **Did You Know?** boxes present high interest applications that augment discussions in the text: for example, Perpetual Motion Machines (Wouldn't It Be Great?), The Option of Nuclear Power, Easter Island, and Acid Deposition (Acid Rain).
- **Thinking It Through** exercises provide critical thinking challenges. For example, students are asked to assess the thermodynamic and social implications of using electric automobiles as a means to "control pollution at the source," or to decide what to take with them in a backpack in order to survive on a desert island to which they have been exiled. These questions immediately follow the regular end-of-chapter exercise sets. Note that answers to exercises with a blue number are given in Appendix D.

GENERAL FEATURES

- Emphasis is on understanding the relationship between ideas. A qualitative conceptual approach is used that requires little quantitative problem solving.
- Emphasis is on better teaching of fewer concepts. The book is designed as a "kit" of basic concepts for students to take with them after the course.
- A gentle introduction to chemical principles allows nonscientists to become comfortable doing chemistry gradually.
- Themes of the book are science as a human process, the nature of scientific knowledge, how science works, and how scientific models are made and used to explain and control our natural world.

- Chemical principles and their applications are clearly integrated. Applications illustrate principles, reinforce them, and show their relevance to everyday experience. This principles-driven approach provides easy and effective opportunities for instructors to use their own favorite applications or to integrate examples that are timely or of local interest.
- Ideas and principles are introduced in one chapter, then revisited and built on several times throughout the book. Students' comprehension increases because they return to principles for review, reinforcement, and amplification.
- A serious, qualitative treatment of thermodynamics, not found in other texts of this kind, leads to a discussion of how the second law relates to our use of nonrenewable natural resources.
- Instructors do not have to remake their existing courses to gain the advantages of this approach.

ACKNOWLEDGEMENTS

I owe much in the early chapters of *Models* to Ken Faler, who with his scientific insight, intellectual breadth, and sometimes warped wit inspired many of the thoughts I have presented here. I would not have conceived this kind of book by myself. Thank you, my friend.

I also wish to thank my colleagues for their help and suggestions, in particular Dennis Strommen, my department chair, who has given me his enthusiastic support, Jeffrey Rosentreter who read the chapter on environmental chemistry and offered suggestions for its improvement, and Paul Tate who reviewed the early chapters for philosophical correctness and balance. I also wish to thank Dean Victor Hjelm for his continued encouragement, Paul Link and Michael McMurray for helpful discussions about geology, and Richard Bowmer for assistance with questions of biology.

I thank Walter Manch, who with an editor's eye read the entire manuscript for consistency and clarity. His suggestions and encouragement are very much appreciated. Also, I want to thank the many students at Idaho State University who used earlier versions of this manuscript and provided me with valuable feedback.

My wife, Jeanette, has given me much support and encouragement. Though she will deny herself the attribute of patience, I thank her for it, as well as her good humor and eagerness to help with the task of putting the book together. Most of all, I thank her for being with me.

I wish to thank my editor, Richard Mixter, who provided the resources and encouragement for this project, and who skillfully guided it from my rough manuscript to the book in its ultimate form. I thank him particularly for the freedom to develop my own model for the text and to retain much of the character I originally wanted it to have. I thank developmental editor Keith Dodson for his contributions to this project, particularly in assembling the supplemental materials and working with accuracy checkers. I wish to thank production editor Christine Hurney, not only for her outstanding work in producing this book, but also for the enjoyment of working with her. Thanks also to Mary Steiner for her efforts in marketing.

Finally, I wish to express my gratitude to all of the reviewers who offered their various perspectives and ideas to make this book much better:

John Amend, Montana State University
Lamar Anderson, Utah State University
Ronald Backus, American River College
Robert Belloli, California State University–Fullerton
James Bills, Brigham Young University
Donna Bogner, Wichita State University
Kathryn M. Borgelt, Tennessee Technological University
Ronald E. DiStefano, Northampton County Area Community College

Jerry Driscoll, University of Utah
Seth Elsheimer, University of Central Florida
Stephen Foster, Florida State University
Lynne Hardin, Tarrant County Junior College
Robert Harris, University of Nebraska
Alton Hassell, Baylor University
LaRhee Henderson, Drake University
Richard Hoffmann, Illinois Central College
Colin Hubbard, University of New Hampshire
Toney Keeney, Southwest Texas Junior College
Keith Kennedy, St. Cloud State University
Christine Kerr, Montgomery College
Leslie N. Kinsland, University of Southwestern Louisiana
Jim Klent, Ohlone College
Donald E. Linn, Indiana University–Purdue University at Fort Wayne
Kenneth Loach, State University of New York–Plattsburgh
Steve K. Lower, Simon Fraser University
Glenn McElhattan, Clarion University–Venango Campus
David Newman, Bowling Green State University
Joseph P. Nunes, State University of New York–Cobleskill
Gordon Parker, University of Michigan–Dearborn
Robert Perrone, Community College of Philadelphia
Ralph Powell, Eastern Michigan University
Ronald Ragsdale, University of Utah
Edith Rand, East Carolina University
James Schreck, University of Northern Colorado
Sarah Selfe, University of Washington
Leo Spinar, South Dakota State University
Byron Strom, Des Moines Area Community College
Dan M. Sullivan, University of Nebraska at Omaha
Robert Swindell, Tennessee Technological University
John Thompson, Texas A&I University
George H. Wahl, Jr., North Carolina State University
Robert W. Wallace, Bentley College
Karen Weaver, University of Central Arkansas
Stanley Williamson, University of California–Santa Cruz
Robert Yolles, De Anza College

I also welcome your contributions to this book. If you should find any errors, I would appreciate knowing so that I can correct them. Your comments and suggestions about any part of the text or its supplements will be appreciated.

Gayl H. Wiegand
Department of Chemistry
Idaho State University
Pocatello, Idaho 83209

1 Science as Knowledge

The courts of science are always open, and every litigant has an unrestricted right of moving for a writ of error.

Sir William Ramsay, Nature, Feb. 1895.

At most American colleges and universities, students are required to take a physical science course to satisfy graduation requirements. A question frequently asked by those not majoring in some science discipline is, "Why should I need to take a chemistry (physics, geology) course when I'm going to be an artist (or musician, or sociologist)?" And it is an appropriate question. Just what is the purpose of taking a physical science course, anyway?

Simply put, to be literate in science is the duty of every citizen in a participatory democracy such as ours. Although counselors or bankers or English teachers may not directly use knowledge of physical science in performing their jobs, it is imperative that all of us, as responsible citizens, are aware of what physical science is, how it is done, and what it can and *cannot* do.

Science and technology are integral parts of today's world, and we cannot escape the effects they have upon our lives. To deal effectively with such problems as world hunger, air pollution, and the AIDS epidemic, and to assess the urgency of the greenhouse effect, we must possess at least a basic, general knowledge of scientific principles. Most of the politicians who formulate our laws have little scientific knowledge or training, and very few have been practicing scientists. When confronting these issues, they must turn to the advice of scientific "experts." However, these experts often disagree over what course of action should be taken. Most who have studied global warming agree that ocean levels will rise, for example; but there is considerable controversy about how rapidly they will rise and by how much, and how detrimental (or possibly beneficial) the impact will be.

How is the politician to know which expert is right? A scientifically literate public must insist on being directly involved in the process of making informed decisions based on the best and most complete evidence at hand. Otherwise, the choices will be made either in ignorance or by an elite few. The future of our society, both politically and economically, may well depend on the degree of scientific literacy our citizens have attained.

Science enables us to view our environment in a different way from that of philosophy or music or accounting, and the knowledge it gives us is of a very different kind. This does not necessarily mean that scientific knowledge is better than the other kinds of knowledge, but the perspective of science is unique, powerful, and very useful.

OPERATIONAL DEFINITIONS OF SCIENCE

But what is science? To distinguish clearly between science and what is not, we need an operational definition of science. An **operational definition** of a category of actions or things (such as science) is not a general statement of what it does or looks like but is a statement of some kind of operation or test of the category that any person can perform to determine experimentally which actions and things fit into the category.

There are actually three operations that enable us decide what leads to scientific knowledge. The first test we apply is that **science** is the body of knowledge that is based on repeatable, testable observation. Second, valid scientific theories based on these empirical data must be refutable, at least in principle. And finally, truly scientific theories must also lead to verifiable predictions.

OBSERVATION IN SCIENCE

Observations that lead to scientific knowledge are unique in that they must be repeatable and verifiable. The observation that a hot object cools when placed in contact with a cold one, while the cold one simultaneously warms, has been verified many times by observers in various laboratories in different countries all over the world. No one yet has seen the reverse of this process. Although this may be an extreme example of repeatability, it illustrates that scientific knowledge must be based on repeatable, testable observations made by different observers at different times and in different places.

Scientific **observations** must be verifiable.

Because scientific theories are based on reproducible observation, the more repetitions of the observation by different individuals (with the same results) the better. Some observations can be carried out only at limited intervals or particular places. An example is that Halley's comet could be observed close to Earth in 1986 for the first time since 1918 (and in 1834 before that; at 76-year intervals). Many observations made in 1918 were repeated in 1986, and a host of new ones were made that used methods based on the vast amount of new technology we had developed.

The criterion of repeatable, verifiable observation, then, is an operation that must be applied to distinguish science from that which is not science. Astrology and "ancient astronaut" theories do not pass this test, nor do unidentified flying objects (UFOs).

The nature of the subject matter being studied can influence whether or not scientific knowledge is obtained, either because some subjects do not easily lend themselves to scientific scrutiny or because suitable tools and methods are not yet available for observing them. For example, theories suggesting that people have a common psychological experience when they die cannot be tested. Because these near-death experiences are based on the testimony of persons who for all practical purposes have died and have then been "brought back to life," the experimental difficulties are obvious. Sometimes, nature "imposes ignorance" on us because we are unable to devise an experiment that will enable us to answer a specific question, even in the realm of science. We shall discuss this idea later.

Whether knowledge is scientific or not does not necessarily depend on who the observer is, scientist or nonscientist. Some terribly unscientific data have been obtained by trained scientists and passed off as being scientific. On the other hand, persons without formal credentials in science, for example Benjamin Franklin, have contributed enormously to the advancement of science.

DISCONFIRMATION OF SCIENTIFIC THEORIES

Scientific theories must be refutable.

Scientists go to considerable trouble to design and carry out experiments that could disprove a theory, to obtain even one small shred of evidence to the contrary, because any valid theory must survive all serious attempts to disprove it. We cannot prove that theories are true simply by amassing evidence to support them. We also must do as much as possible to disprove them if they are to have any scientific validity. The theory that UFOs exist is not refutable because no one has found a way to experimentally determine their identity. It is the nature of scientists to have a highly critical attitude toward untested theories, and even toward established ones. To the uninitiated this may seem rigid, closed-minded, and perhaps even somewhat arrogant, but it is a necessary attribute for doing science.

SCIENTIFIC FACTS, THEORIES, AND LAWS

SCIENTIFIC FACT

A **scientific fact** comes from observation.

A **scientific fact** is something that exists, discernible by observation, that has the consensus of most observers. (As we shall see later, scientific theories and laws are not considered to be scientific facts.) Most other facts differ from scientific facts in that they usually are based on the testimony of one or a few persons, and the event usually is not repeatable. A good example of this is an historical event. An event accepted as fact by historians depends a great deal on the credibility of the observer or observers who recorded it, as well as on other supporting evidence from artifacts, memorabilia, and records of other events parallel in time. It is impossible, however, to repeat the event to ascertain whether it really happened the way it was recorded. Even though the assassination of President Kennedy in 1963 was recorded on movie film, rewatching that event (the movies) has provided little evidence to show how he was killed or by whom.

In some special respects, scientific facts also have to be accepted on trust because it is physically impossible for an individual scientist (or even a group) to verify all the information that is accumulated by the scientific community as fact. Doing science is a communal effort, and the collective knowledge earned by its many participants, past and present, is held in trust in libraries and computer data banks for access by all. Being trained to observe and record carefully and accurately, scientists expect and trust one another to produce reliable data. For the most part this is the case, but if there is reason to doubt that a certain observation is a scientific fact, the observation can almost always be tested by repeating it. The body of knowledge in the scientific literature, then, is accepted as fact by the scientific community, but it is always open to question.

▶ Figure 1.1
Hot-air balloons rise because the warm air inside is lighter (less dense) than the cooler air outside. The density of a gas can be calculated using the ideal gas law.

SCIENTIFIC LAW

A **scientific law** summarizes patterns in nature.

If a scientific fact is an observation that is accepted by the scientific community, what is the nature of a scientific law? A **scientific law** is a description of a natural regularity or pattern that summarizes a large number of scientific facts. Generally speaking, a law is a universally accepted generalization of experience in the form of a verbal statement or mathematical relationship (or both) that has been shown always to be true (▶ Figure 1.1). For example,

HENRY'S LAW

Scientific Laws at Work

When you open a container of a carbonated beverage, you hear a *pfssst!* and bubbles form in the liquid inside (Figure 1). Carbon dioxide gas, stored under four times atmospheric pressure, comes out of solution when the pressure is released. Carbonated beverages are pressurized so that more carbon dioxide will dissolve in them, giving them more effervescence and better taste when opened. When carbonated beverages go "flat," they usually don't taste very good.

Henry's law is at work here. In 1803 William Henry, an English chemist, observed that the solubility of a gas in a liquid is directly proportional to its pressure above the solution. This means that twice as much carbon dioxide will dissolve in a beverage if the pressure is doubled to twice atmospheric pressure, three times if tripled, and so forth. Likewise, reducing the pressure decreases the solubility of a gas. Cold, high-mountain lakes contain less dissolved oxygen for fish to live on than cold lakes at lower elevations.

This effect is used in medicine to help persons with carbon monoxide poisoning and with gangrene infections. Carbon monoxide poisons by displacing oxygen from hemoglobin in the blood, thereby reducing the blood's ability to carry oxygen to the cells. The bacteria that cause gangrene are anaerobic. That is, they must grow in an environment that contains no oxygen, and one that contains oxygen kills them. Both of these conditions can be treated by high-pressure (hyperbaric) oxygen therapy, which increases the solubility of oxygen in blood.

Henry's law has its down side. Decompression illness, or the "bends," is caused when a person's body is subjected to a sudden reduction in pressure. Divers working at great depth under water are subjected to high pressure, which increases the solubility of the air they breathe in their blood. The divers' bodies use up the oxygen in this air, but the nitrogen remains dissolved in their blood. If the pressure is reduced too rapidly, some of the nitrogen comes out of solution and forms bubbles in the divers'

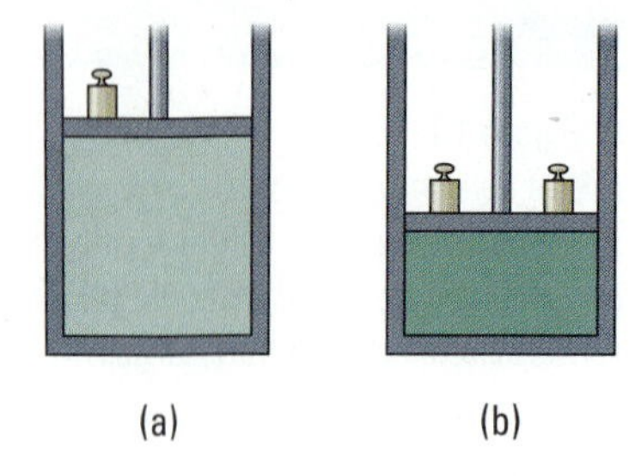

▶ Figure 1.2
Boyle's Law. *Gas chamber with piston and (a) one weight, (b) two weights. Doubling the weight on the piston doubles the pressure on the gas inside, causing its volume to be halved.*

Boyle's law states that the volume of a given amount of a gas is inversely proportional to the pressure under which it is measured, providing the temperature is held constant. That is, if the pressure on a gas such as nitrogen is doubled, its volume is halved (▶ Figure 1.2); and if the pressure is tripled, the volume becomes one-third. This observation, within certain limits, holds true for all kinds of gases. The law also enables us to predict the effect of halving the pressure on a gas that we have not tested before: Its volume will double. When we carry out the experiment using argon, for example, we find that this is indeed the case.

blood and body fluids. These bubbles cause extremely painful cramps and can even cause death. The bends can be prevented by decompressing a diver slowly or by using a breathing mixture that contains helium instead of nitrogen. Much less helium dissolves in the blood at high pressure.

Interestingly, fish can get the bends, too. When water falls from very high dams into deep pools below, large amounts of nitrogen from the air are dissolved on the way down. Fish tend to congregate at the bottoms of such pools because temperatures are often ideal and there is an abundant food supply. The water there can contain too much dissolved nitrogen, however. Then, the fishes' blood can also contain too much nitrogen, and the fish, like the divers, may die if they come to the surface too rapidly.

Figure 1
Carbonated beverages contain carbon dioxide gas dissolved in water under pressure. Henry's law states that the amount of a gas dissolved is directly proportional to the pressure under which the gas is stored. At four times atmospheric pressure, four times more carbon dioxide will dissolve in water than at ordinary atmospheric pressure. Opening the container releases the pressure, allowing bubbles of carbon dioxide to escape.

SCIENTIFIC THEORY

A **scientific theory** explains scientific facts.

Scientific facts and laws are explained by theories. (In the example, Boyle's law is explained by the *kinetic molecular theory*.) A **scientific theory** is an explanation that accounts for all of the facts currently known about a phenomenon and is an attempt to provide an understanding of its underlying structure. Because theories must account for all of the facts about some aspect of nature, they often must be changed to accommodate new observational evidence as it is gathered. Consider two theories: "the world is flat" and "the world is round." The flat-world theory, or model, was prevalent until new evidence was accumulated that the world could not be flat, and probably was round. The *theory of relativity,* as another example, is still being tested. So far all of the evidence suggests that it is valid, but let's just suppose we observe an exception to the theory of relativity. We have two options. If the exception is not too inconsistent with the theory, we may be able modify the theory to accommodate the new observation. If the exception is glaring, we must, as last resort, scrap the old theory and propose a new one in its place.

The new theory, of course, must also be tested by experiment. When the flat-world theory was in doubt, for instance, a cubical-world theory could have been proposed. An experiment to test this idea could then have been devised. Proving this theory to be wrong would have shown that the world is not cubical, but it would not necessarily have shown that the world is round. Although today we can "see" that the world is "round" by relying on information supplied from satellites in orbit high above the earth, this evidence does not prove conclusively that this theory is true. To date, however, no evidence has been found to disprove the round-Earth theory.

The development of scientific theories is a dynamic process.

The development of theories is thus a dynamic process, with theories being modified or replaced as necessitated by observational evidence. A current theory is said to be "accepted" or "established" by consensus: it is never proved. In this dynamic process of establishing theories, it may be that two or more explanations, or models, fit the existing experimental evidence. In this case, further experiments are needed to distinguish between (or among) the models until, by a process of elimination, one theory stands alone. This theory then becomes the accepted one until new evidence causes its modification or replacement (if this should indeed occur).

Sometimes an established theory persists in science even though it is disconfirmed by a number of observations. Usually, in this case, no better theory exists; and because the theory works for the most part, the scientific community is reluctant to throw it out until a better one is developed.

If theories must be modified to fit new experimental evidence, aren't they really mere hypotheses? Or if they are always being changed, in what way are they "truth," or what good are they, anyway?

HYPOTHESIS

A **hypothesis** is a scientific proposition that must be tested by experiment.

Hypotheses are much more "primitive" than theories in the sense that they are quite tentative and are based on limited observational data. They are propositions that are initially advanced to explain some observations and to provide a basis for further investigation. To become theories they must survive more experimental scrutiny, must usually be modified many times, and must be supported by a large observational data base.

There is some concern today, for example, that persons living near electrical transmission lines and transformers may be at a higher risk than other people of contracting cancer, possibly caused by the electromagnetic radiation emanating from them (▸ Figure 1.3). The validity of this hypothesis must be tested in at least two ways: it must be conclusively established that people living near power lines do indeed have a higher incidence of cancer, and then (because correlation by itself is not proof of cause) it must be shown that exposure to electromagnetic radiation causes cancer.

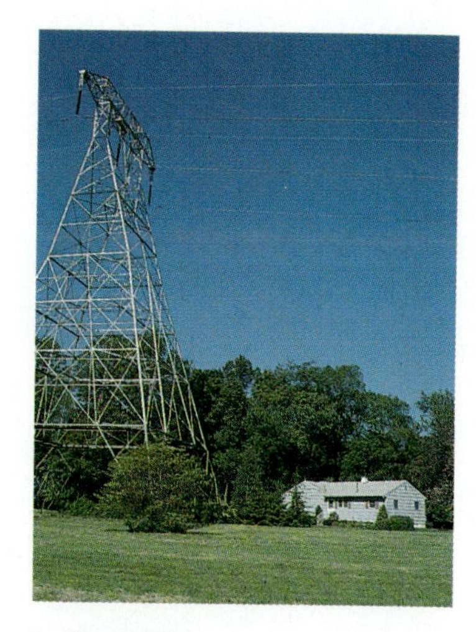

▸ Figure 1.3
Electromagnetic Fields and Health. *Do electromagnetic fields cause adverse health effects in people? This hypothesis is presently being tested for validity.*

The fact that theories are modified, changed, and established by agreement does not detract from their usefulness. Consider the "theory" of gravity as proposed by Newton. It explained the behavior of falling bodies and did a pretty good job of describing and predicting the motions of the planets around the sun. However, there are some peculiarities in the orbits of some of the planets (among other things) that could not be accounted for by Newton's theory. The introduction of the theory of relativity by Einstein made it possible to modify Newton's theory to explain all aspects of the orbits. Did this make Newton's theory wrong? No. It was incomplete and based in part on incorrect assumptions about space and time. In some respects, it was too simple a theory (to explain all the motions of the planets, for example), but it did (and does) work very well for many other explanations and predictions about gravitational phenomena. The power of Newton's theory of gravitation was that *it worked, as a single useful concept,* to explain and predict many gravitational phenomena not well understood before it was developed.

Valid scientific theories lead to reliable prediction.

In science, when more than one theory explains a particular phenomenon, one of the theories usually is an extension or elaboration of another. To explain and predict, we often just use the theory that does the job. At the heart of science is the fact that valid theories enable us to make verifiable predictions. Indeed, truly scientific theories *must* lead to reliable prediction. The philosophical and practical validation (soundness) of scientific theories is that they serve some useful purpose in explaining and predicting (and thus enabling us to control) natural phenomena. In other words, *they work.*

Although there is a common approach to gaining scientific knowledge, there is no special process that can be used to make scientific discoveries.

Notably absent from our discussion so far is any mention of the "scientific method." This is intentional because this author thinks that there is no such thing. There is a methodology or approach to doing science, to be sure, but there is no prescribed process or recipe for making scientific discoveries. It is not unheard of for scientific insights (products of inventive scientific minds) to precede experiment, and many times the creative synthesis of various perspectives to gain these insights defies any attempt at logic. Of course, any "discovery" must eventually be validated by experimental evidence or it fails at being science.

SCIENTIFIC TRUTH

But what about scientific theories as "truth"? One might be suspicious that theories, because they are part of a process of change, are questionable when

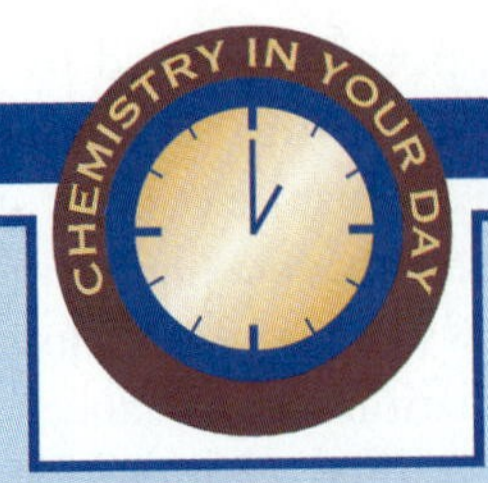

EMPHASIS ON EMFS

Tired of dormitory life, you and four friends decide to rent a house. You manage to find two houses large and cheap enough. The nicer one, more convenient to classes, is located near some high-voltage power lines, and Charley brings up the question of a possible link between electromagnetic fields (EMFs) and cancer. You all decide to investigate EMFs.

But that turns out not to be easy. You personally are assigned to discover the nature of EMFs, and two hours at the college library gives you a lot to consider. Yes, electricity flowing through wires does create EMFs, and EMFs pass through walls and human bodies without hindrance. You learn that some laboratory studies found that high-intensity EMFs can cause biological changes in rodents; for example, they promote existing breast tumors. But these studies used EMFs of much greater intensity than you'd receive if you lived right under the wires. Furthermore, researchers in other labs have been unable to reproduce these results, and reproducibility is the most basic test for scientific validity.

Jeanne finds the study that started the EMF debate: twice the number of children who lived in houses that *were likely* to have high-intensity EMFs had developed leukemia. But this study was based on EMF intensities *inferred* from wiring patterns in houses and neighborhoods by *people who knew where the children with leukemia lived*–it was a flawed study. Later studies couldn't correlate the incidence of leukemia with actual measurements of EMFs in children's houses. A more recent article reporting on six of the most careful studies of childhood leukemia and EMFs says that five of the studies showed some statistical association between elevated EMFs and childhood cancer.

But Rhonda, the science major in your group, points out that you're not children. She finds a report of studies on EMFs and adult cancer, but they're even cloudier than those of children because adults may have been exposed to EMFs earlier in life or at work. One study examined more than 4000 cases of cancer that occurred over 20 years among 220,000 male electric utility workers in Canada and France. That's one cancer in every 55 people from all causes over 20 years! Is that significant to five of you for a few years? Oddly, there was no indication of an overall statistical increase of cancer among those people exposed to the strongest EMFs, but there was an apparent twofold increase in development of one type of adult leukemia among the 50 percent of those workers exposed to comparatively weak fields of 1.6 milligauss (mG, the standard unit for measuring magnetic fields) or higher. Rhonda says these peculiarities could be artifacts, flaws in the statistics.

You learn that the strength of EMFs increases with current strength but falls off rapidly with distance from the source. The EMF of your 115,000 volt power line is 30 mG at the base of the 65 foot tower, but it falls off to 2.0 mG 100 feet away, and the house is about 150 feet from the line. You also learn that most homes have background EMFs of 0.1 to 2 mG; some ordinary appliances generate higher EMFs than transmission lines, though their strength falls off within a yard or so.

From your information, you think your preferred house is exposed to about 1 mG EMF on average from the power lines. To this you have to add the background EMF of the house, which includes that coming from the service lines into the house. To decide knowledgeably between the two houses you'd also have to know what the background EMF is at the other house.

You all make your reports, and the argument lasts into the night. In the end, there has to be a decision, even if it's just to stay in the dorms. But Charley, after starting all this and being too busy to share in the research, shrugs that idea off: who knows what the EMFs are in the dorms? In the wee hours everyone agrees that you really can't make a knowledgeable decision. To do so, you'd have to measure the average EMFs of both houses and the dorm over weeks, and then you still wouldn't know whether EMFs cause cancer. That knowledge would require a study of thousands of people over years.

Anyway, measuring EMFs is incompatible with lectures and quizzes, so the only way to decide is to give everyone a chance to register their anxieties and preferences by voting. Charley says that's one of the failures of democracy: we vote on things no one knows anything about . . . we might as well toss a coin. You feel he's wrong about that, but you can't put your finger on why. Then you realize that knowing no one knows about something *is* knowledge–the only knowledge available about whatever it is–and voting with this knowledge protects us from being led by charlatans who claim knowledge they don't have.

Speaking of voting: power lines could be buried at great cost, and that would eliminate the possibility of their being a threat. Appliances and electric-service connections would still generate EMFs, but then we would each be able to choose whether to do without electricity and therefore be free of possible threat.

If you had to contribute a significant part of your income to bury all power lines, would you vote for or against it because of the possibility that EMFs might cause cancer?

it comes to being "truth." First, scientific theories usually don't change very fast. It was about 215 years from the time that Newton's theory of gravitation was accepted until Einstein's relativity corrected it, and many of Newton's equations are useful (and are still used) today. Second, we need to decide what "truth" is, at least for the purpose of this discussion.

"Truth," as it is usually envisioned, can be divided roughly into two categories: absolute, universal, and unchanging truth, and relative truth. The former is sometimes represented as "TRUTH" and the latter as "truth." This author holds that scientific truth falls into the latter category.

Scientific truth is based on observation, and the theories and laws that emerge from observation are subject to change, albeit usually slowly. Certainly, not theories nor even laws can comfortably be called TRUTH. If one looks for truth in science, it is best found at the level of the actual reproducible observation itself.

If scientific truth is so nebulous, what does a scientist mean (or, more properly, what should she mean) when she says, "I can prove that the tire tracks on this pair of pants came from that blue Cadillac"? First, we must remember that science is based on observation, disconfirmation, and prediction, and that there are certain rules of the game for doing science. In this context evidence and arguments can be presented to "prove" that such is the case beyond a reasonable doubt, much as in a court of law. The weight of the evidence and the use of established theories (theories that work) give the scientist the confidence to say that she has the *knowledge* that the tire track on the pants came from the tires on the Cadillac. This is not a statement of absolute certainty. Instead, it is a statement that she knows, based on her experience as a scientist dealing with the way nature works, that it is true to a high degree of probability.

SCIENCE AND HUMAN VALUES

Another question that arises concerns the relationship of science to human values, usually in terms of religious and political beliefs. Religious philosophies usually are based on the premise that there is a God, or gods, or some supernatural being or force. Likewise, political philosophies are based on certain premises about how people should be governed. These kinds of assumptions are notably absent from science, and its working premises neither support nor disavow any kind of value system. One assumption of science is that the acquisition of scientific knowledge is a worthwhile human enterprise; but beyond this, science has little to say about human values.

The working assumptions of science are value neutral . . . .

To politicize science or inject religious belief into it destroys it as a unique way of gaining knowledge about the world. It is important to note that the very internationality of science requires that it have no bias in this regard.

A tragic example of political interference in science occurred in the Soviet Union, beginning in the Stalin era. Modern genetic theory tells us that organisms pass traits to their offspring through their genes and that such inheritance largely determines what offspring will look like. This view was held by biologists almost everywhere then, as it is today.

In the Soviet Union during the 1930s, however, Trofim Lysenko was advocating a different theory—that environmental conditions, not heredity, were the dominant influence. For example, he believed that by manipulating environmental conditions it was possible to convert, say, rye into barley or oats. Although his theory was based on questionable experimental evidence and could not be validated by the experiments of others, Lysenko gained considerable political support from Joseph Stalin because the idea fit communist ideology. Russian scientists who questioned this authority were systematically exiled, sent to prison, or executed, and for a period of about thirty years, Lysenko's theory of genetics became a part of Communist party doctrine. It was not until 1966, long after Stalin's death in 1954 and after several years of disastrous crop failures, that the Soviet government finally was forced to recognize that Lysenko's ideas were wrong.

Recent accounts from Russian scientists tell that Soviet science and technology, particularly in medicine, biology, and agriculture, have been severely crippled by this legacy. Not being self-sufficient in the production of food supplies, for example, the former Soviet Union has had to import vast quantities of grain at the expense of its meager supply of hard currency. Although collectivization of farms has been blamed, the knowledge necessary to develop a modern agricultural technology, such as in animal husbandry, pest and weed control, and the development of disease-resistant plants, draws heavily from modern genetics. The lack of skilled geneticists has tragically handicapped Russia in its efforts to feed itself.

. . .but scientists are not.

Although science itself is supposed to be neutral concerning religion, ethics, and politics, scientists are not. Like everyone else, they have a variety of personal views, and in spite of their efforts at objectivity, these beliefs sometimes affect their scientific work. For example, experts can differ about whether medical-waste incinerators pose a health hazard, even when they have the same pertinent facts at their disposal. Worse, people's egos, ambitions, and personalities, and the pressure to get results, sometimes get in the way of good science. Old theories sometimes are held too long and others are not carefully considered in the light of strong evidence. Sometimes experimental design is faulty, or the work is not carefully done. Sometimes data are interpreted to fit preconceptions. And sometimes data are even falsified, fabricated, sabotaged, or stolen in the race to be first with an important discovery.

Because everything done in science is subject to critical scrutiny by others, these transgressions of objectivity are eventually put right. (Indeed they must be put right if the process of science is to work properly.) In time a good theory is accepted, inaccurate experimental work is corrected, and perpetrators of fraud are discovered and discredited.

The nature of science reflects society as a whole. Much scientific work is carried out by industrial corporations hoping to develop new technologies for financial gain. Here, the marketplace points the direction for scientific endeavors. Other scientific investigations are sponsored by private institutions trying to alleviate a human problem, such as cancer, heart disease, or diabetes. Our government also directs the substance of science by funding projects considered to be of national importance; here, an informed public must be involved in making the decisions about where these resources are allocated.

SCIENCE AND TECHNOLOGY

Technology is a kind of knowledge that fits the operational definitions of science but is not, in the strict sense, science. Its difference from science lies in the intent of the observer.

If we compare a chemist doing basic research with one engaged in technology, we find that for the most part the former is interested in gaining new knowledge without being much concerned about the potential use of that knowledge. The technologist, on the other hand, is concerned with applying knowledge that the scientist has gained to a technique for making a better plastic, drug, or photographic film, or for finding a way to clean up oil spills. In this context, **technology** is a social use of science, often called applied science. Science seeks to understand nature, and technology seeks to control or modify it.

Technology is the social use of science.

The chemist engaged in technology holds to the same basic criteria of reproducible observation, disconfirmation, and prediction as the one doing basic research: it is only in intent that they differ. An overlap between science and technology always exists, and the distinction between them often is not sharp. For example, the practice of medicine is applied science (technology), and molecular biology is a science. Medical research to find a cure for cancer or AIDS involves a combination of the two fields; that is, more knowledge in molecular biology (basic science) is necessary before the cure in medicine (applied science, technology) is found. In this case the search for new knowledge is directed toward an application, so the line between science and technology becomes blurry. Although this distinction between science and technology may seem arbitrary, it is often necessary for the purpose of discussing ethical and moral issues concerning them.

DID YOU LEARN THIS?

For each of the statements below fill in the blank with the most appropriate word or phrase from the following list. Do not use any word or phrase more than once.

scientific law	relative	technology
hypothesis	observation	verifiability
scientific fact	science	absolute
theory		

a. _______ "Better things for better living through chemistry."
b. _______ That which is needed before an observation becomes scientific fact.
c. _______ The kind of truth that scientific truths are considered to be.
d. _______ A scientific idea about the nature of things that must be tested for validity by experiment.
e. _______ The basis of scientific knowledge.
f. _______ A description of a natural regularity or pattern that has been shown to always be true.

EXERCISES

1. What is an operational definition?
2. Provide an operational definition of science that includes all of the tests necessary to determine whether an activity is science or not.
3. How do the following differ in their operational definitions. (In other words, what test would you apply to tell the difference between them?) What similarities do they have?

   a. an electric refrigerator and an electric stove
   b. a motorcycle and an automobile
   c. ice and water
   d. a computer and a television set
   e. a piano and an organ
   f. astronomy and astrology
   g. biology and history
   h. sugar and salt
   i. music and sculpture
   j. geology and geography

4. Define a scientific

   a. fact
   b. theory
   c. law
   d. hypothesis

5. Distinguish between science and technology. What test would you use to tell the difference between them? In each of the following pairs, tell which is science and which is technology.

    a. certain minerals help plants grow
       crop yields are increased by fertilizing a garden
    b. parachutes are made of Nylon
       polyamide fibers are made strong by pulling them
    c. an extract from the American western yew tree retards the growth of some forms of cancer
       taxol (a substance from the yew extract) is being tested in clinical trials
    d. new refrigerants are being sought
       chlorofluorocarbons (CFCs) deplete the ozone layer
    e. hydrogen nuclei behave like miniature magnets
       magnetic resonance imaging (MRI) is useful for finding tumors in the body

6. Various individuals and groups holding beliefs that are not scientific seek approval of their views by the scientific community. A common complaint of many of these groups is that the scientific community is "closed" to their ideas and that in debate with scientists they are being "persecuted" for their opinions. What do you think there is about the nature of the scientific enterprise that could cause these groups to feel this way?
7. What do you think is the nature of scientific knowledge that makes it "unique and powerful"?
8. Decide which of the following is a fact, law, hypothesis, or theory.

    a. The atoms in elemental sulfur are arranged in eight-membered rings.
    b. Twelve grams of carbon always combines with 32 grams of oxygen to form 44 grams of carbon dioxide.
    c. The properties of the elements are a periodic function of their atomic numbers.
    d. Acids taste sour because their particles have little points on them that sting the tongue.
    e. When water pipes freeze in a house, the hot water lines often freeze and the cold water lines do not. Therefore, hot water must freeze faster than cold water.
    f. Like dissolves like.
    g. The volume of a gas at constant pressure varies directly with its temperature in kelvin.
    h. Methane contains 75% carbon and 25% hydrogen by weight.
    i. The sun is made up mostly of hydrogen.

9. What do you think are the limitations of scientific knowledge?
10. You have to explain "What is science?" to the following persons. What would you say to each?

    a. your cousin who is seven years old
    b. an automobile mechanic
    c. a reporter from your local newspaper
    d. an attorney
    e. a cattle rancher
    f. your dentist

11. Do you think that scientific knowledge itself is good, bad, or neither? Explain.
12. In your view, is technology good for society, bad for it, or neither good nor bad? Justify your answer.
13. How do scientific laws differ from civil laws? What similarities do they have, if any?

THINKING IT THROUGH

Reproducible Observations

Certain scientific observations cannot be repeated at will, for example observations of erupting volcanoes such as Mt. St. Helens or Mt. Pinatubo. Though not directly reproducible, the data collected from these events are usually considered scientific if they are obtained by scientists trained to observe these phenomena. How much scientific credibility would you give these kinds of data? How would you go about "reproducing" them?

Learning from History

It is often said that "history repeats itself" and that we can "learn from the mistakes of the past" to guide us in making decisions involving social or political action. Is this science?

Perceptions of Science

Ask some people of your everyday experience how they would define science. Try to get responses from people of different ages, education, and occupation, and compare them with what is presented in this chapter. What does this tell you about a "public perception" of science? What kinds of misconceptions about science do you see?

ANSWERS TO "DID YOU LEARN THIS?"

a. technology
b. verifiability
c. relative
d. hypothesis
e. observation
f. scientific law

SCIENCE AND PUBLIC POLICY

Ronald V. Dellums

Congressman

B. A. Psychology, San Francisco State College

M. A. Social Welfare, University of California, Berkeley

As a youngster growing up in west Oakland's African-American community, my dreams had little to do with science. For that matter, most boys and girls in my neighborhood did not think of science as a career option. Structurally, society discouraged young African Americans from going to college, much less pursuing a science career. Our neighborhood high school did not even offer many of the math and science classes needed to meet college admission requirements.

Despite my social ties to my community high school, my mother wanted to ensure that the avenues to college were open for me. She therefore transferred me from my neighborhood high school into one of the city's college preparatory high schools. Not only did this school offer the math, science, history, and language courses required for college admission, but the coach did not believe that Black youth could play baseball. My mother was pleased, but I saw the transfer as the end of my baseball career.

I approached my high school science classes with some consternation. Biology, chemistry, and physics were the standards at the time, and for the life of me I could not figure out how they would benefit me in the future.

After high school, a hitch in the Marine Corps, and two years at Oakland Community College, I found myself at San Francisco State College. Throughout college, I gravitated toward the social sciences, receiving a bachelor's degree in psychology and a master's degree in social welfare at the University of California, Berkeley.

At first, my indifference to science persisted in college. But, as I began to find my way, I became interested in what I regarded as the related sciences of the development of our planetary system and the evolution of life on Earth. These were personal interests, however; I did not think these subjects would have any direct application to my chosen career as a psychiatric social worker.

As the 1960s were ending, I decided to seek election to the United States Congress. By this time, I had worked as a social worker and manpower development and training specialist and had served on the Berkeley City Council. All of these endeavors convinced me that reordering our priorities at the federal level was essential to the betterment and well-being of our nation and its various communities.

My concerns were political, not scientific: I went to Washington to achieve civil rights for our citizens, end the war in Southeast Asia, and redirect funds from the military budget to our urgent social needs. Two years later, I wound up on the House Armed Services Committee—and ultimately became chair of the very technologically oriented Subcommittee on Research and Development.

I learned very quickly that the debate over the military budget was conducted in terms of technology, not just priorities. At first, I was overwhelmed by the testimony of the "rocket scientists," who appeared before the committee to discuss the "yields" of various nuclear weapons, the "circular error probability" of a missile landing near a target, "exchange" rates, computer "architecture," and all of the other esoterica of the Strangelovian world of nuclear war planning. I wanted to discuss whether or not we should build weapons per se; I found myself having to discuss whether or not these systems would work, alone or in combination with each other. To do that, I had to know the science behind them.

Physics, chemistry, and math all came into the dialogue. I found that my credibility in the debate over nuclear weapons systems was enhanced when I could discuss them in terms of scientific principles. I therefore forced myself to return to those lessons I had so painfully learned during my high school days. At first, summoning up that half-forgotten scientific knowledge was difficult. But once I had remembered it, I found that it played a significant role in my daily work as a public policy maker—work that, unfortunately, requires me almost constantly to contemplate systems that might one day destroy our planet.

No longer could my arguments be attacked as wistful liberalism or naïve pacifism. Now those who favored the development of more nuclear weapons system had to confront my challenge to the internal logic of the systems themselves. They were forced to defend the inconsistency of developing weapons that would, if used, destroy the very planet and society they were obstensibly designed to protect. They also had to acknowledge the instability created when massive destructive power is linked to computer technology and put on a short fuse. Time, distance, yield, instability, symbiosis, synergy, and other elements of weapons technology became my allies.

In my more reflective and philosophical moments, I also recall my exploration of planetary development and species evolution in college. I believe profoundly that the human spirit, given its current state of evolution, is capable of reaching beyond violence and war. The longer I am in public life, the more I hope this vision will be realized. Therefore, I not only apply the hard sciences to challenge these insane weapons systems, but I try to do my part to hasten our evolution toward a more peaceful and cooperative species that lives in harmony with the planet rather than as its principal threat and antagonist.

Our future as a species depends on our ability to understand the chemical and physical impact of our actions and the threats to the environment posed by our conduct. Accordingly, our public policies must be devoted to the enhancement of the quality of life on this planet. Without an understanding of science, I, as a public policy maker, would be unable to discharge my duty to the future generations who rely upon us to leave them a healthy and vibrant legacy—the Earth.

2 Scientific Reality

There was a faith healer of Deal,
Who said, "Although pain isn't real,
When I sit on a pin,
And it punctures my skin,
I dislike what I fancy I feel!"

What is the nature of reality? Is the "world" real? When we see an object or touch it, is it really there? Or is the object simply a product of our consciousness, of the fact that we are aware of our "world" and can interact with it? Interact with *what?* What about objects that we cannot see, such as atoms and distant stars? Are these real or are they just objects we think are there and have looked for with our instruments, which tell us we have "seen" something?

TWO VIEWS OF REALITY

According to the **classical view of reality,** we can be objective observers of nature.

Two differing views about the nature of reality are prevalent in science today. The **classical view of reality,** which is the legacy of Newtonian physics, holds that the objects we observe have an existence independent of our observation. In this view, we as observers can be detached from the objects we observe. The objects "are what they seem to be," so to speak, and if we are careful enough, we can determine their true nature. This leads to the notion of *scientific objectivity,* meaning that we can observe things or processes without ourselves affecting the outcome of the experiment in any way. Any theories that we base on these observations, then, should correspond to some element of actual truth, even though we may be unable to observe it directly.

The **Copenhagen interpretation** holds that the observer is a part of an experiment and affects its outcome.

A more recent perspective of reality arises from modern physics and is called the **Copenhagen interpretation.** This view does not assume an absolute reality outside that of our own experience. Theories merely correlate observations correctly and permit reliable prediction, rather than correspond to some facet of actual truth. Nor can we be objective observers. The observations we make are in some manner gathered by our senses, which are products of our consciousness. (When we are unconscious, we usually cannot observe what is going on around us.) Observation is an interactive process: what we observe ("out there") is connected, in each direction, to our consciousness ("in there"). Another way of expressing this idea is to say that reality is not outside of perception but is part of it; this view makes the observer an active participant in and part of any experiment.

The classical Newtonian view has been a part of Western culture for about 300 years. For this reason, it is more familiar to us and is more widely held than the Copenhagen interpretation, which is about six decades old. We are comfortable with the idea that scientists (and science itself) can be objective and that there is something special about scientific knowledge that makes it "closer to the truth" because of this impartiality. It is easy for us to accept scientific facts because we expect that what is observed "out there" is seen as being the same by all who look at it. On the other hand, it bothers us to think that, for example, the apparent orderliness observed in nature just happens to be the same to everyone, even though we cannot say for sure whether it actually is that way. There is evidence for this latter view, however, and we shall discuss it in more detail later.

THE ASSUMPTIONS OF SCIENCE

We pursue scientific knowledge with only *four basic assumptions* about the nature of reality:

1. The world exists. (In this context, the term *world* means all that is accessible to our senses, not just the earth.)
2. We can know the world through observations of nature.
3. There is an orderliness in nature.
4. The acquisition of this kind of knowledge is a worthwhile human enterprise.

The first assumption is an assertion that some kind of reality exists, without which it would be pointless to pursue scientific knowledge. Note that the Copenhagen interpretation does not deny reality but rather says it is a different kind than we usually envision.

Observation is the cornerstone of science, and science cannot be done without it. Knowing the world means we have some perceptual connection with it, either one-way (Newtonian) or reciprocal (Copenhagen). Implicit in this assumption is the belief that, somehow, it is possible for all observers to come to some consensus about what is being observed.

Order in nature seems to be all around us, and we as humans are order-seeking creatures. From this apparent order and patterning we make classifications. Classification helps simplify our world, and we get some kind of meaning out of the classifications that we make. Our systems of **nomenclature** (naming) usually imply relationships. For example, diamond and graphite are different forms of the element carbon, called allotropes (▶ Figure 2.1). Diamond is very hard and crystalline, whereas graphite is the grey, slippery material that makes up pencil "lead." Both substances are made only of carbon atoms; their properties differ because their atoms are arranged differ-

Classification helps simplify our lives.

(a)

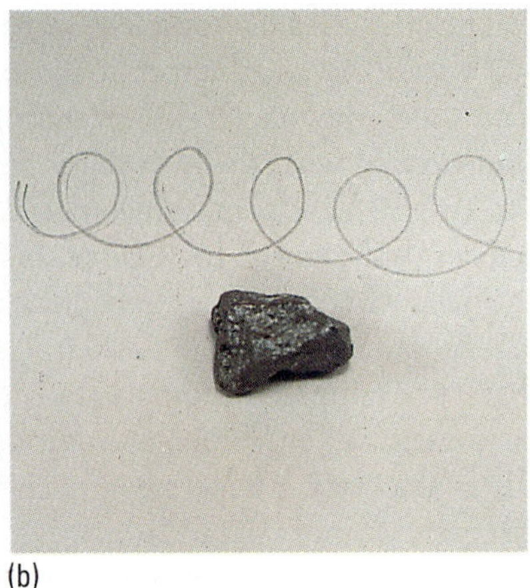
(b)

▶ Figure 2.1
Allotropes. *Diamond (a) and graphite (b) are classified as allotropes. Both are made only of carbon atoms, yet diamond is the hardest natural substance known, and graphite is used in making pencil leads.*

ently. Classification also leads to the recognition and identification of things originally not placed in a category but which belong there because they have similar attributes. Highly colored crystalline solids called buckminsterfullerenes have been discovered recently. They also are made entirely of carbon atoms, but the atoms are arranged in the general shape of a soccer ball (very different from the arrangements of the atoms in diamond or graphite); they have been classified as new allotropes of carbon.

The only value judgement in our short list of assumptions is that having scientific knowledge is worthwhile, but doing science without this assumption is almost impossible (and certainly no fun). Science makes no other philosophical assumptions, and the ultimate authority in science is observation.

TECHNOLOGICAL INNOVATION AND BASIC SCIENCE

This is a good place to digress for a moment and ask the question, "Why do science?" Why do we in the modern world spend so much of our money on it? Those who do pure science seek new knowledge, often because of a fascination with nature or to satisfy intellectual curiosity, but it is not obvious to many people why they should be expected to help finance this kind of activity. Much more visible is the impact technology has on our lives. Technology, the social use of science, is what enables us to predict the weather, clean up our environment, make a drug to treat heart disease, create a stronger plastic, and in general "make our lives better."

In the United States we spend a large sum of money on "science and technology," but most of it is spent on technology. A relatively small portion goes to pure science. Our position in the world as a technological power is being threatened in part because we have failed to understand the fact that technology usually *follows* scientific discovery, and we have not placed enough emphasis on acquiring basic knowledge.

Technology usually *follows* scientific discovery.

In addition, there is often a delay, called a **lag time** or **incubation period,** of 10 to 20 years (or more) between the time a basic discovery is made and the time its application is available for widespread use. Some examples are given in ■ Table 2.1. Also, no one can infallibly predict what new scientific knowledge will be good for, if anything. It could not have been known, for example, that the biochemical techniques developed in the early 1970s to study insect viruses would prove to be of crucial importance in our desperate battle with the AIDS virus, which was unknown then. And, although it sometimes seems that one area of scientific endeavor is more important than another, all are potentially valuable in the long run.

■ **TABLE 2.1 LAG TIMES FOR SOME TECHNOLOGICAL INNOVATIONS**

Technological Innovation	Date of Discovery	Date of General Availability	Time Elapsed
Ball bearings	A.D. 40	ca. A.D. 1900	ca. 1900 years
Light bulb	1879	1897	18 years
Kodachrome film	1924	1935	11 years
Artificial heart	1935	Not yet perfected	Unknown (at least 60 years)
Xerox copier	1938	1949	11 years
Nonstick Teflon pan coatings	1938	1960	22 years
Magnetic resonance imaging (MRI)	1973	1989 (clinical trials)	7 years

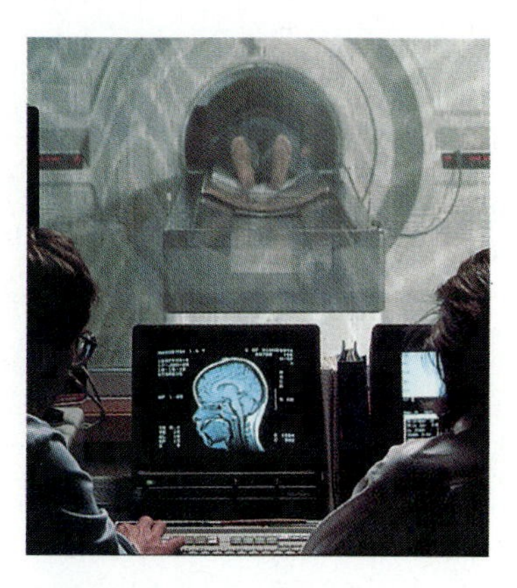

▶ Figure 2.2
Magnetic resonance imaging (MRI) uses magnetic fields and radiowaves to look inside the body without harming it.

EXTENDING OUR SENSES: USE OF INSTRUMENTS

Many of the objects and events in the world that we want to understand are beyond our realm of direct experience, so we must aid our senses with the use of instruments (▶ Figure 2.2). We can use a magnifying glass or a microscope to see small objects or bacteria, or a telescope to view distant stars. We can measure a human hair with a micrometer or the size of an atom with X-rays.

Instruments are devices that enable us to extend our senses. But how do we know that the reading on a dial or the squiggly line on a graph translates (somehow) into a real event or object that our instrument "saw" but we couldn't? Does the "beep" of a Geiger counter really tell us that radioactive particles are present? Does the gas company's "sniffer," when its needle swings upward and a light flashes, really tell us that there's a leak in our

Using **instruments** allows us to extend our senses.

basement? In both cases, we *believe* that the device is telling the truth, that it is sensing something we cannot perceive with our unaided senses. But what leads us to this belief?

Consider the cells in the leaf of a plant, which are too small to be seen with the unaided eye. We can use a strong magnifying glass, or better yet a simple microscope, to help us see their outlines and perhaps some of their details. We cannot clearly see what these details are, so we logically decide to use a microscope of greater power. This enables us to view finer structure with more clarity, and now we see the nucleus, mitochondria, and other parts of the cell. However, the nucleus seems to have some fine structure that we want to "enlarge," so we employ a microscope of even more power. Now we can see the poorly defined shapes of the chromosomes and note that they, too, appear to have a more detailed structure.

Images from instruments often must be visualized and interpreted.

At this point, however, we find that we have arrived at the limit (a magnification of about 1500×) to what a microscope using ordinary light can tell us. To make a more powerful microscope to further magnify the cell's genetic material, we need to use ultraviolet light. But the human eye is insensitive to this kind of light. To "see" the magnified image produced by our new microscope, we must have some means of *visualizing* it, for example, on photographic film. With this method we can see even more detail in the cell, but again there is more that it cannot tell us. Our next step is to use an electron microscope to resolve (see clearly) even smaller detail, and now the structure of the chromosomes becomes clearly visible. But the images we obtain from this instrument aren't really what it "saw" either, because again we must use some means of visualization to produce them. Still finer features, such as the structure of the DNA, that are not clearly resolved by the electron microscope (limited to a magnification of about 200,000×) can be viewed with X-rays. The X-rays give patterns of dark spots on photographic film, which we must *interpret* mathematically to get an idea of the helical structure of the substance. In principle, we can delve even deeper into the structure of the DNA. Using a technique called scanning tunneling microscopy we can obtain a computer-generated picture of the *atoms* on the DNA's surface! (▶ Figure 2.3 shows such a picture of the atoms of graphite.)

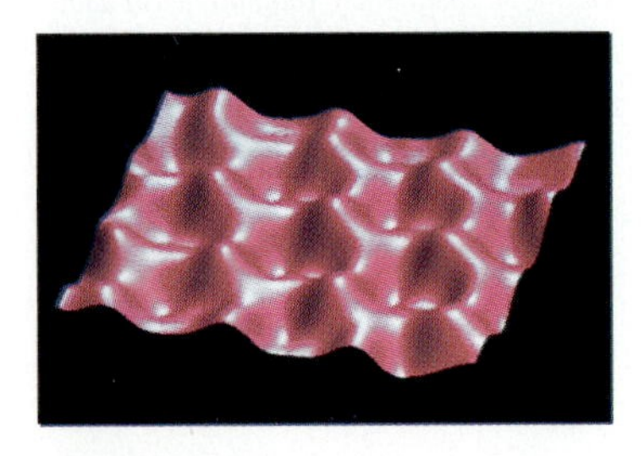

▶ Figure 2.3
A Scanning Tunneling Micrograph of Carbon Atoms. *The carbon atoms on the surface of a crystal of graphite appear as "bumps." The signals obtained by the scanning tunneling microscope were interpreted by a computer to produce this visual image.*

Are we seeing something that really exists? Do we really believe that DNA is coiled up like a miniature spring? That its atoms have surface shapes corresponding to those predicted by current theories? Yes, we do, because each view of our leaf leads us to predict what the successive level should look like. If what we expect to see matches what we observe with a more sensitive technique, we can be reasonably sure that the more detailed picture is reliable and useful. [What we are actually doing here is making a model (Chapter 4) of the next level of detail and then using that model to predict what the next more sensitive technique should tell us about the next level. If our observation is consistent with our prediction, then the new technique is considered valid.]

It usually is possible to observe the same feature with various techniques that have overlapping sensitivities, and thus to establish continuity among the various images. When the pictures obtained from all of the methods used to study the cell are laid out in order of their magnification, they form a sequence that links the least detailed view to the most detailed one, and each level of detail can be predicted from the preceding level and can be verified by a more sensitive experimental technique. This gives us confidence that the information we gather with instruments is indeed a reflection of reality, even though we have not observed it directly.

We also can be confident that a new technique that is applied in this way on one system is generally applicable for observing other systems at the same levels of detail. For example, if we trained our electron microscope on the cells of a plant stem, a seed, or a root, we would expect, based on what we observed in leaf cells, to see similar features in these other cells. And we do. After we use our instrument often enough, observing as many substances as possible, we eventually can say that the electron microscope is a reliable tool for observing many objects at magnifications of about 1500 to 200,000 times, and that the images it provides represent reality in a particular way.

VIEWING ALL OF REALITY

Our instruments enable us to view only a part of our world.

Couldn't we be missing something when we use instruments to extend our senses? Our instruments only let us look where we "point" them, so perhaps we are like the person who uses a flashlight to explore a dark room. He can see at one time only the part of the room that is illuminated by the flashlight; the rest of the room is dark.

This appears at first to be no problem. Our observer with the flashlight can systematically view overlapping portions of the room and combine all of his observations to make a composite picture. But what if our observer doesn't know the shape of the room (it might be spherical), or for that matter doesn't know where he is (say the room has no door)? He will then have trouble finding a *point of reference* from which to make measurements. In addition, he will be unsure whether he saw *all* of his environment. What if the objects in the room are moving? Can he ever be sure he sees them all and knows in what direction they really are moving?

This example illustrates just a few of the problems we face when we cannot see all of what we are trying to observe. Perhaps we cannot see everything "out there" with our instruments, and we are observing only a part of reality. Until a new technique is devised that enables us to see into a different part of reality, we have to make do with what we *can* see—and that is certainly better than nothing. The example also illustrates the point that there will always be

YOUR REARVIEW MIRROR: AN INSTRUMENT

The rearview mirror in your car is an instrument that lets you "see behind you" while driving forward (Figure 1). You are not actually seeing the scene behind you, you are looking at a piece of silvered glass in which you are observing a reflected, or "second hand," image. This image is reversed, left-to-right, so you have to interpret it. (This is why you see ƎƆИA⅃UᗺMA on the front of ambulances. Your mirror shows you AMBULANCE. There may be several mirrors in your car whose magnifications are different: one may make reflected objects appear closer, and another may make them look farther away. There may be blind spots in your view to the rear. Yet most of us use a rearview mirror without giving it a thought, trusting our lives to our interpretations of its reflections in spite of its limitations. How many of us check what the mirror is telling us by looking back over our shoulders?

Figure 1
The rearview mirror in your car is an instrument that allows you to "see" what is behind you without turning around.

something new for scientists (and science) to study because there will always be new "places" in reality to look. It is in the nature of science that a new discovery (or an answer to a question) invariably opens many new avenues to other discoveries (or new questions to be answered). The more we know in science, the more we discover there is to learn, and we find ourselves on a never-ending quest for more knowledge.

It has sometimes been said of science that when we use our instruments to look for something that we cannot actually see, we set up ourselves to "see what we are looking for." When we test theories, we design our experiments to look for the things the theory predicts, and often we *do* find them. Sometimes, however, we find something that we are not looking for. Also, we can use our instruments to ask the question, "What if ____?," not necessarily seeking a particular answer. Either of these situations can lead to exciting new discoveries. Were it not for the unexpected, science would hold little excitement for those of us who pursue it.

MEASUREMENT

Observation in science depends not only on being able to see something but also on measuring it. **Measurement** is the quantitative determination of the dimensions of things: what length, what weight (what mass), what duration of time, what temperature, and so forth. These dimensions are expressed in some kind of units (▶ Figure 2.4). For example, we once used units such as the cubit for length, the grain for weight, and the number of days for time.

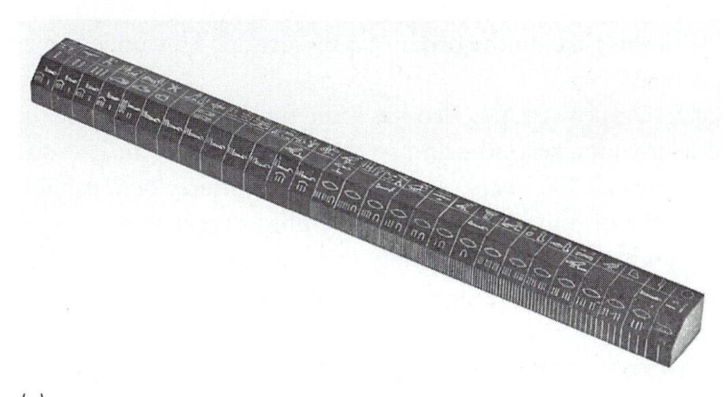

(a)

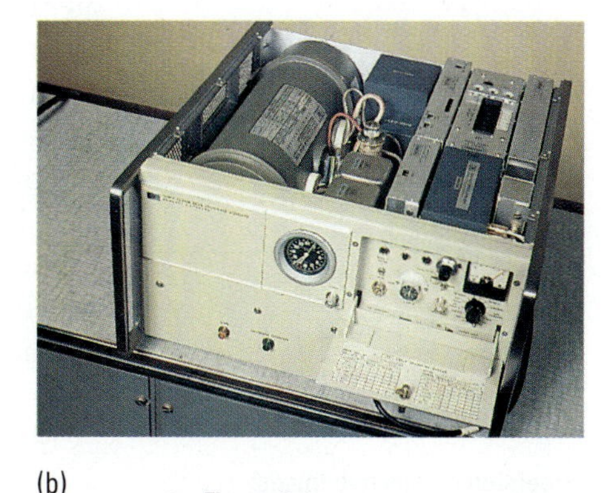

(b)

▶ Figure 2.4
Ancient and Modern Measuring Instruments. *(a) A replica of the Egyptian royal cubit, the standard unit of length for the builders of the Great Pyramid of Cheops (ca. 3000 B.C.). A cubit is about 52.5 cm. (b) Interior view of an atomic clock, the most accurate type of clock known to date. Time scales for atomic clocks are based on periodic processes that occur in atoms.*

THE NEED FOR STANDARDS

With few exceptions, measurements can only be accomplished by the use of instruments, whether they be simple or sophisticated. The cubit, a unit of length corresponding to the distance from the elbow to the fingertip, could be conveniently applied by almost anyone because the "instrument" was always at hand. A simple balance could be used to find the weight of an object in actual grains (of wheat, for example). However, measurements of this kind were crude because the length of a cubit depends on the size of the person doing the measurement, and the weight of a grain varies with the type and size of the seed used to balance the object being weighed. Counting days is not difficult, but estimating the length of a shorter period of time is. Eventually, a cubit became 18 "inches," a grain 1/7000th of a "pound," and a day 24 "hours." The only reason these newer units are considered better is that there is widespread agreement on what they mean; that is, they have become *standardized.* An inch on a ruler is **calibrated** by comparison with a bar of known standard length, made for just this purpose. Likewise, a pound is calibrated against a piece of metal of standard weight. Although these units of measurement (pounds, inches, and grains) are now also out of date, and the metric system of units is now in general use, they do illustrate that *any* measurement, to be accurate and meaningful to other observers, must in some way be calibrated by comparison with a known **standard**. In the United

Instruments must be **calibrated** against a **standard** if the measurements made with them are to be meaningful.

States the standards for weight (mass), length, temperature, and time are kept in the National Institute of Standards and Technology (▶ Figure 2.5).

PRECISION AND ACCURACY

The international definition of a second is the amount of time it takes a cesium atom to vibrate 9,192,631,770 times, meaning that we can divide a second into over nine billion parts! This is equivalent to a difference of one second in 300 years. We cannot measure this with a stopwatch, however, because any measurement of elapsed time made with such an instrument always has a built-in uncertainty, in this case on the order of 0.05 second. This uncertainty is often called **tolerance.**

All instruments have some built-in **tolerance.**

A cesium atomic clock, now in use, also has *some* uncertainty—on the order of about 10 billionths of a second—and so the best values it can give still vary by about 10 seconds in 300 years. An even better atomic clock, capable of measuring a difference of one second in over 3 million years is now being tested, but not even it will give us a completely error-free measurement. This is true of all instruments (measuring devices), no matter how they work or how well they are made.

Measurements with very small tolerance, such as those with atomic clocks, are said to be of high precision. When we talk of the **precision** of a measurement we mean two things: how fine a measurement is and how reproducible it is. *Fineness* is the number of digits in the measured value. A value of 1.537 inches is more precise than one of 1.5 inches. We would have confidence in a pharmacist who prepared three doses of a drug weighing 125, 126, and 124 milligrams. Such agreement of values *(reproducibility)* shows good precision. However, we would have second thoughts about taking a drug if we knew the doses weighed 122, 145, and 110 milligrams.

Precision means two things: fineness and reproducibility.

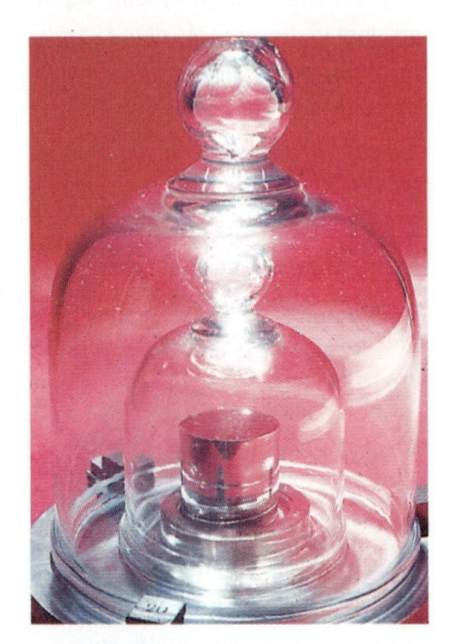

▶ Figure 2.5
The Standard Kilogram. *This duplicate of the international standard kilogram is kept at the United States National Institute of Standards and Technology (formerly the National Bureau of Standards) located near Washington, D.C.*

Even though a measurement is precise, it may not be accurate. The **accuracy** of a measurement is determined by how closely it corresponds to the actual dimensions of the object or event it has examined. The terms *precision* and *accuracy,* then, have very different scientific meanings.

An excellent example of this difference can be drawn from competitive rifle shooting (▶ Figure 2.6). One contest places emphasis on how close together a shooter can place ten shots at a certain distance; another contest involves the number of times one can hit the center of a bullseye at 1000 yards. In the first event the best competitors often place all ten shots in an area the size of a nickel at 300 yards. The smallest group, or cluster, has the greatest precision and determines the winner. No one cares whether the shots hit the bullseye, which serves only as an aiming point. On the other hand, the object of the 1000-yard event is to place all ten shots in the bullseye, and the competitor who shoots the smallest, most exactly centered group in the bullseye wins. It is not sufficient that the cluster of shots be small, they must also be in exactly the right place. It is perfectly possible for a person to shoot a very

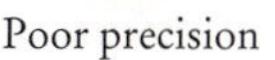

Poor precision

Good precision wins match

Good precision poor accuracy

Poor precision good accuracy better score

Better precision better accuracy wins match

(a) Accuracy Not Important

(b) Precision and Accuracy Important

▶ Figure 2.6

Precision and Accuracy. *(a) In the 300-yard contest, the smallest cluster of shots wins the match. (b) In the 1000-yard competition, bullet holes must be grouped tightly and centered in the bullseye to win.*

small group off the bullseye but lose the match to someone who shoots a larger group that is centered on the bullseye. Thus, in this contest, accuracy is the deciding factor. Winners of this event often place ten shots inside the 10-inch circle at the center of the bullseye.

Note that even though an activity may produce high precision, it may not necessarily produce a high degree of accuracy. As another example, measurements of 1/64th (.016) of an inch or so can be made with a good ruler that relies on visual acuity alone. A well-made micrometer, on the other hand, can measure to 0.0001 inch, 160 times smaller. Though the micrometer is thus capable of more precise measurement (smaller tolerances) than the ruler, it is possible for the ruler to yield a value that is closer to the actual size of the object being measured. In other words, the measurement with the ruler could be

2.1 Which of the following methods should give a more precise measurement of the quantity indicated?

a. pacing off 100 yards or finding the distance with a surveyor's chain
b. weighing a pound of fruit on a bathroom scale or on a postal scale
c. measuring 1.75 gallons of gasoline with a 1-gallon bucket or a 1-quart milk bottle
d. finding the length of an average room with a 3-foot length of rope or a 6-inch dollar bill
e. using a clock whose face is graduated by the hour to determine when it is 3:23 PM or one whose face is marked at 3, 6, 9, and 12
f. finding how fast you are driving a distance of exactly 10 miles by looking at your speedometer, or by timing to the nearest second how long it takes you to drive the distance
g. measuring the width of this page with a ruler that has thin lines or with one that has wide lines (same number of graduations)

more accurate even though it would be less precise. This might happen if a micrometer was improperly made, if its scale was not zeroed (that zero on the scale means zero inches), or if its operator used it sloppily or read its scale incorrectly. The result of such a measurement is a combination of errors, called **systematic errors,** which might be sufficiently large to overcome the advantage in precision of the micrometer over the ruler if the ruler was properly made and used with skill. The micrometer would show the precision it was made to deliver (assuming the errors were consistent), but the measurement would be inaccurate.

No measurement can be exact.

Thus, to obtain error-free measurements, or, more properly, *error-minimal* measurements, it is necessary not only to employ an instrument capable of high precision but also to use it properly so that the values obtained accurately reflect reality. Though it usually is possible to refine instruments and techniques to make measurements with precision and accuracy, no measurement can be exact. The added possibility that operator error can lead to more inaccuracy makes it clear that scientific measurements must be performed with utmost care and repeated many times if we are to believe they have anything to do with reality. In fact, scientific reality takes on new meaning: we now must view our observations as "reality within the limits of error, or tolerance"; and absolute reality, as far as we can know it, is suspect.

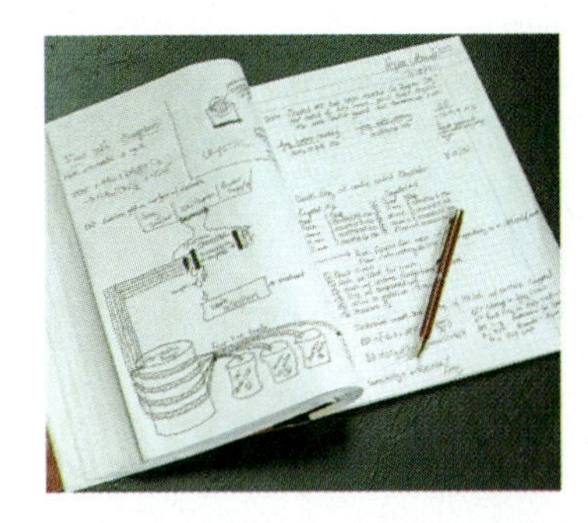

► Figure 2.7
Laboratory Notebook. *Quantities and their units are always recorded at the time that measurements are taken.*

UNITS OF MEASUREMENT

To be meaningful, all measurements must be expressed in some kind of units. Units qualify the measurement, or tell us what kind it is. For example, to say that a handball court is 16 × 16 × 32 is meaningless. We might guess that the units of the handball court must be units of length. This qualifies the measurement (length), but to say that the dimensions of the court are 16 high × 16 wide × 32 long still is not enough. Thirty-two what? Inches? Centimeters? Meters? Miles? Furlongs? Light years? Because the appropriate unit is feet, we get the perspective that a handball court is a moderately large room. Had the unit of length been inches, we would have visualized the court to be a medium-sized box. Playing handball in such confined quarters would be difficult.

All measurements must be expressed with their **units.**

Thus, **units** not only qualify a measurement, they quantify it (give it a size). It is essential that the appropriate units accompany the numbers generated by experiment if the numbers are to be meaningful. This means it is always necessary to record the units of measurements at the time they are taken in the laboratory (► Figure 2.7). It cannot be neglected: a measurement is not complete without its units and without units its numbers are useless. A few measurements, such as specific gravity and pH, are expressed as dimensionless numbers; but such numbers are always labeled, for example, sp. gr. = 1.37.

2.2 Which of the following operations would give or require greater accuracy?

a. weighing dirty potatoes or washed potatoes to get 10 pounds of potatoes
b. weighing yourself with your shoes on or in your stocking feet to get your body weight
c. using a barn door or a pie plate as an archery target
d. hitting a baseball or a basketball with a bat
e. kicking a field goal from 20 yards or from 40 yards
f. marking the position of this dot · on this page with a felt-tipped pen or with a ballpoint pen
g. pacing off a quarter mile or using the odometer on your car to find the distance
h. measuring the width of this page with a ruler that has thin lines or with one that has wide lines (same number of graduations)

Compound units permit us to make direct comparisons.

Some units are expressed as a *ratio* of two other units, which means that two measurements must be made to obtain them. We will call these **compound units** here. Some familiar compound units are miles per gallon (miles/gallons), dollars per pound (dollars/pounds), cents per dozen (cents/dozens), kilometers per second (kilometers/seconds), and calories per day (calories/days). Note that in each compound unit one measurement with its unit (in the numerator) is divided by another measurement with a different unit (in the denominator); thus, the numerical value of the unit in the denominator becomes one. This enables us to make direct comparisons without doing arithmetic. If chuck steak is $2.25/pound, hamburger is $1.19/pound, and filet mignon is $7.49/pound, we can easily decide which to buy if we must feed 20 hungry picnic guests on a small budget.

One unit that is commonly used in science is **density,** another compound unit (▸ Figure 2.8). It is imprecise to say that lead is heavier than water or wood because we haven't said how much lead is being compared with how much water or wood. A correct comparison must be based on measurement of the same volume of each substance, such as a cubic foot, a gallon, or one of the metric units (discussed shortly) such as the liter (L) or milliliter (mL). Using English measurements for now, we find that lead has a density of about 687 pounds/cubic foot, water 62.4 pounds/cubic foot, and wood (walnut) about 42 pounds/cubic foot. Now it is easy to see that lead is about 11 times more dense than water and walnut is about 2/3 (0.67) as dense as water because we're comparing the same volume of each material (1 cubic foot). Lead

▸ Figure 2.8
Ice floats on water because it is less dense than water. The gold ring is more dense than water and sinks.

sinks in water because it is more dense than water, and the wood floats on water because it is less dense.

DIMENSIONAL ANALYSIS (FACTOR-LABEL METHOD): WORKING SIMPLE PROBLEMS WITH UNITS

Although you will be asked to do only a small number of calculations in this course, it is a important for you to be comfortable when handling straightforward quantitative relationships. When you are comfortable, you will find that many scientific concepts are easier to understand and that laboratory experiments are more meaningful.

Dimensional analysis, sometimes called the factor-label method, is a useful method for solving problems. It uses the units associated with measurements or quantities as a guide to setting up calculations and as a means for checking answers.

"How many feet are there in 2.50 miles?" Before turning on our calculators, let's examine what this "story problem" is asking us to do. Notice that it gives us a distance, and it is 2.50 miles. This is called our *given quantity.* (We know it is a distance because its units, miles, are units of distance.) We are being asked to convert this given distance, expressed in one unit, the mile, to an equal distance in another unit, the foot. This distance in feet is our *unknown quantity.*

A *conversion factor* from an *equality* or an *equivalence* has a value of 1.

We first need to find the relationship between the units of feet and miles, and looking in Appendix A we find that 1 mile = 5280 feet. This relationship represents an *equality* of distance—namely, that a distance of exactly 1 mile is equal to a distance of exactly 5280 feet. Because 1 mile and 5280 feet are the same quantity, we can divide one by the other, and the quotient is exactly 1 (dividing a quantity by itself yields unity). Expressing this another way we obtain the following *conversion factors.*

$$\frac{1\text{ mile}}{5280\text{ feet}} = \frac{5280\text{ feet}}{1\text{ mile}} = \frac{\text{a certain quantity}}{\text{its exact equality}} = 1$$

Because our unknown quantity must be the distance in feet that is identical to our given distance of 2.50 miles, we cannot do any operation that changes the given distance. This means that we must multiply our given quantity only by unity, or any other quantity that is equal to 1. The conversion factor, 5280 feet/1 mile is exactly unity, so

$$2.50\,\cancel{\text{mile}} \times \frac{5280\text{ feet}}{1\,\cancel{\text{mile}}} = 13{,}200\text{ feet}$$

This equation is simply handled algebraically by multiplying 2.50 by 5280 feet to obtain 13,200 feet and multiplying the units of miles by feet/mile to give the units of feet on the right-hand side of the equation. Notice that the units of miles appear in both the numerator and the denominator on the left-hand side of the equation, so that they cancel algebraically and leave the units of feet in the answer. This is the big advantage to using dimensional analysis. If the conversion factor is set up and used properly, the unwanted units cancel out and only the desired units appear in the answer. When the units come out as desired, the setup for the conversion is probably correct (although this does not mean that the arithmetic used to obtain the numbers is necessarily right). If the units come out wrong, however, the setup is definitely wrong, and the problem needs to be rethought. For example,

$$2.50 \text{ miles} \times \frac{1 \text{ mile}}{5280 \text{ feet}} = 0.000474 \frac{\text{miles}^2}{\text{feet}}$$

produces a nonsense unit.

To illustrate the utility of this method, let's try another problem, this time with a compound unit. If an automobile gets 28 miles per gallon of gasoline, how many gallons of gasoline does it take to drive 560 miles? In this instance, our unknown quantity is the amount of gasoline, and its unit is the gallon. Our given quantity is 560 miles. What conversion factor do we use? It takes one gallon of gasoline to go 28 miles, but in this case we don't have an equality because gallons are units of volume and miles are units of distance. We get around this problem by noting that, although these quantities are not equal, they are an *equivalence*—namely, 1 gallon of gasoline is equivalent to a distance of 28 miles, and conversely, 28 miles is equivalent to 1 gallon of gasoline. Treated this way, the conversion factors are

An *equivalence:* hamburger is \$1.89/pound, or, \$1.89 gets you 1 pound of hamburger.

$$\frac{28 \text{ miles}}{1 \text{ gallon}} = \frac{1 \text{ gallon}}{28 \text{ miles}} = \frac{\text{a quantity}}{\text{its equivalence}} = 1$$

We can now choose a conversion factor that we can use to multiply our known quantity, 560 miles, to yield our unknown quantity, the number of gallons this distance is equivalent to. We want the units of miles to cancel, leaving gallons in our answer, so we use

$$560 \cancel{\text{miles}} \times \frac{1 \text{ gallon}}{28 \cancel{\text{miles}}} = 20 \text{ gallons}$$

Let's do one more for the road, this time using two conversion factors, which is about as many as we will need to learn for this course. Going back to our picnic, at which we wanted to feed 20 hungry guests, let's assume we've just received a large bonus from our boss and decide to celebrate with the filet mignon. The stuff is $7.49/pound and one-half pound per person should make for a real pig-out. How much do we have to pay for the meat?

Our unknown quantity is the dollars we have to pay, and our given quantity is 20 persons, meaning that we must find the amount of money (dollars) equivalent to 20 persons. Because we do not have a single equivalence that contains both dollars and persons, we must use two. The two conversion factors, based on the equivalences we have at our disposal, are

$$\frac{7.49 \text{ dollars}}{1 \text{ pound}} = \frac{1 \text{ pound}}{7.49 \text{ dollars}} = 1$$

and

$$\frac{0.5 \text{ pounds}}{1 \text{ person}} = \frac{1 \text{ person}}{0.5 \text{ pounds}} = 1$$

To get from the units of persons to the units of dollars, we choose the appropriate conversion factors from those just given and multiply, canceling the units of persons and pounds. This gives the dollars the filet mignon costs:

$$20 \cancel{\text{persons}} \times \frac{0.5 \cancel{\text{pounds}}}{1 \cancel{\text{person}}} \times \frac{7.49 \text{ dollars}}{1 \cancel{\text{pound}}} = 74.90 \text{ dollars}$$

Here are some general rules for using dimensional analysis:

1. Identify the known quantity and write its unit.
2. Identify the unknown quantity and write its unit. This will be the answer to the problem.
3. Plan a way to convert the unit of the known quantity to the unit of the unknown quantity (answer). Decide which conversion factors you need to multiply to cancel the units you don't want and retain the ones you do.
4. Set up your equation, including the numbers contained in the conversion factors you have chosen, and calculate your answer, making sure that unwanted units cancel and that your math is correct.

THE METRIC SYSTEM OF UNITS

The United States is the only major industrialized country that continues to use the English system of weights and measures as a national practice. With few exceptions, all other countries in the world use the **metric system of**

2.3 Use Appendix A to make conversion factors for working the following problems.

a. Convert 210 seconds to an equivalent number of minutes.
b. Find the number of quarts contained in a 55-gallon barrel.
c. What is your age (to your last birthday) in months? In days?
d. A particular city is laid out so that there are 16 city blocks to the mile. A student lives 72 blocks from city hall. How many miles does she have to walk to get there from her home?
e. How many miles can a person drive in 3.5 hours at 50 miles per hour?
f. A pronghorn antelope can run about 50 miles/hour. How many minutes does it take it to run 5 miles at that speed?

units. A new system of units, which is an extension of the metric system, was adopted by the international scientific community in 1960. It is called the International System of Units, or **SI system** (for the French *Systéme International*). For our purposes, however, the fundamental metric system will do nicely, and although it is likely that you are already familiar with it, a brief refresher is included here.

One of the main virtues of the metric system is its simplicity. For each fundamental kind of measurement there is a basic metric unit, such as the *meter* (abbreviation m, length), the *gram* (abbreviation g, mass, Chapter 3), the *liter* (abbreviation L, volume), the *second* (abbreviation s, time), and so forth. Larger or smaller metric values can be easily expressed by multiplying any of these basic units by ten or some multiple (or power) of ten.

Certain of these multiples of ten are given word prefixes which, when combined with the name of a basic metric unit, describe another metric value. For example, the prefix, *milli-*, means "one one-thousandth," or a multiplier of 0.001. Thus, a millimeter is one one-thousandth of a meter, or 0.001 meter. You should learn the common prefixes and their multipliers, which are listed in ■ Table 2.2. Others will be defined for you when it becomes necessary to use them. ■ Table 2.3 gives some useful approximations (▶ Figure 2.9).

Metric prefixes are multipliers of a basic metric unit.

EXPONENTIAL NOTATION

Some of the metric multipliers listed in Table 2.2, such as pico, milli, kilo, and mega, are either very small decimal numbers or very large multiples of 10. As you will notice in You Try It! 2.4, writing a large number of zeros to position the decimal point is tedious and time-consuming, and you may be wondering why we have the multipliers in the first place. The reason is that many

► Figure 2.9
Approximation. *A Golf club is about 1 m long, and 1 m is about 1.1 yards.*

TABLE 2.2 SOME COMMON METRIC PREFIXES AND THEIR MULTIPLIERS

Prefix	Multiply Basic Metric Unit by	Abbreviation	Meaning of Prefix
pico	0.000000000001 or 10^{-12}	p	one millionth of one millionth
nano	0.000000001 or 10^{-9}	n	one thousandth of one millionth
micro	0.000001 or 10^{-6}	μ	one millionth
milli	0.001 or 10^{-3}	m	one thousandth
centi	0.01 or 10^{-2}	c	one hundredth
deci	0.1 or 10^{-1}	d	one tenth
kilo	1000 or 10^{3}	k	one thousand
mega	1000000 or 10^{6}	M	one million

TABLE 2.3 SOME USEFUL APPROXIMATE METRIC-TO-ENGLISH UNIT CONVERSIONS

Length

1 in. is about 2.5 cm.

10 cm (1 dm) is about 4 in.

1 m is about 1.1 yd, and 100 m is about 110 yd.

1 mm is about 1/32 in., and 2 mm is about 1/16 in.

1 cm is about 3/8 in.

1 km is about 0.6 mi.

Volume

1 L is about 1 qt.

30 mL is about 1 fluid oz.

5 mL is about 1 tsp.

15 mL is about 1 tbsp.

250 mL is about 1 cup.

Mass

1 kg is about 2.2 lb.

450 g is about 1 lb.

30 g is about 1 oz and 1 g is about 1/30 oz.

Temperature

1 °C is about 2 °F.

0 °C = 32 °F (exactly).

100 °C = 212 °F (exactly).

2.4 Use the appropriate decimal multiplier from Table 2.2 for each of the following.

a. one milliliter = _____ liter
b. one kilogram = _____ gram
c. one decimeter = _____ meter
d. one picosecond = _____ second
e. one centimeter = _____ meter
f. one microgram = _____ gram

Supply both the decimal multiplier and the name of the basic metric unit for each of the following.

g. one mg = _____
h. one km = _____
i. one μs = _____
j. one nm = _____
k. one cg = _____
l. one ML = _____

measurements in science involve very large or very small numbers; the metric prefixes and their multipliers were actually devised to enable us to express these numbers more easily. Still, writing a bunch of zeros and doing the arithmetic with them each time we want to do a calculation is no fun and can lead to errors. In Table 2.2 a different kind of notation is listed for each multiplier, one that uses **exponential numbers** or powers of ten. An **exponent** is the power to which 10 is raised, such as 10^3, 10^{-6}, and 10^{-1}. A positive exponent simply represents the number of times 10 is multiplied by itself; so 10^3 is $10 \times 10 \times 10$, or 1000. A negative exponent represents the number of times the number 1 is divided by 10; so 10^{-1} is 1/10 or 0.1, and 10^{-6} is $1/(10 \times 10 \times 10 \times 10 \times 10 \times 10)$, or 0.000001. The value of 10^0 is 1. Using exponents may seem complicated at first glance, but it makes handling very large and very small numbers much easier. Thus, 4.6 μg can be expressed as 4.6×10^{-6} g, and 8.3 km as 8.3×10^3 m.

2.5 Express the following as an exponential multiple of the basic metric unit:

a. 7.73 cm
b. 1.2 mL
c. 8.27 kg
d. 3.56 μs
e. 5.36 dm
f. 4.44 mg
g. 9.3 ML

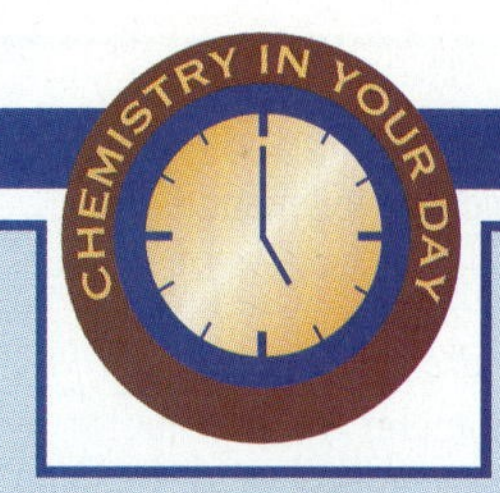

METRICKS

You and your housemates are fixing up the house, and Jeanne, who's French, is presenting a problem. You knew that scientists use the metric system, and you knew that some people use it for ordinary measurements, but you can't believe that anyone doesn't know the *real* way to measure, that is, the way *we* do it. Jeanne simply doesn't speak "English" in measurement, and you can't help but think she's only pretending confusion when you say something is three feet long or a yard wide.

"Just use the yard stick. Here, I'll show you."

"No," she says, "it is time for you to join the rest of the world! *You* learn metric and then I will help you to measure."

She walks away, leaving you talking to yourself. "It's just like a French person to want everything her way." But something about her statement bothers you for some reason.

So you get your chemistry book and find the section on metric, where you read that Liberia and the United States of America are the only countries in the world that don't use metric for everyday measurement. There are also some rules of thumb there for changing English units into metric units.

A meter is a yard plus about three and a third inches (1.1 yards). You can say "a yard's a little less than a meter" for translating a few yards into meters, but for 80 yards you'll do better to subtract 10% of the number of yards to get meters: $80 - 8 = 72$ m (the actual number of meters is 73.15), and if Jeanne gives you a measurement in meters just add 10% of the number of meters to get yards ($72 + 7.2 = 79.2$ yd). For feet, which we usually use in construction, multiply the meters by three and increase your answer by 10%. However, if you're replacing a window pane, get out your calculator. An inch is about 2½ centimeters, and so 10 cm is about 4 in., which is ⅓ ft, so 30 cm is about 1 ft, which is about ⅓ m.

A liter and a quart are close enough to be approximately interchangeable, but there's no equivalent in metric for a gallon. Such amounts are measured in liters, and some gas stations, even in the United States, sell gasoline in liters. If it takes about 12 gallons of gasoline to visit your folks, how many liters do you need? Are four liters close enough to a gallon? Multiply gallons by four and get 48 liters, which is actually twelve and two-thirds gallons: close enough, unless money is very short.

Jeanne wants to cook dinner for everyone tomorrow and you offer to shop for her. She says she needs 1½ kilos (short for kilogram, used by people who live with metric) of *bifteck* and 2 kilos of potatoes. How much should you buy? Well, a kilogram is about 2.2 pounds, so "about three pounds and some" is close enough for the steak. You can figure out the pounds of potatoes needed. She also wants 60 grams of baking chocolate. How much chocolate should you get? There are about 28 g in an ounce, so rounding off to 30 and dividing 60 by 30 gives you two ounces, and add a little bit to be safe. Actually, we don't buy chocolate that way here, so you'll probably have to buy an eight- or ten-ounce package and bear her objections to the way we do things.

A nickel weighs about 5 grams. A cup is about 250 mL. A tablespoon is 15 mL, and a teaspoon is about 5 mL.

You like to run four miles a day to keep fit, and Jeanne likes to run eight kilometers. Will you be happy running together? To find out, add 10% of the metric distance to one-half the metric distance. Then she wants to know whether to wear a sweat shirt for running, and you have only Fahrenheit thermometers in the house (at present . . . that will no doubt change). It's 58 °F outside this morning, so subtract 30° from 58° and divide by two for an approximation of Celsius. When you get Celsius thermometers you can reverse the process. Don't use this trick for baking though, get your calculator.

It's just a matter of getting used to metric, and if you use these approximations often enough, you'll soon get a feeling for it. But for accurate measurements we'll have to change all the wrenches and measuring tapes in the country, turn our yardsticks into meter sticks. What will we call our teaspoons and tablespoons? Milliliterspoons? Football fields will be 91.4 m long, home plate will be 27.4 m from third base, and you'll be ______ cm tall and weigh ______ kilos.

Exponential notation is also useful for converting between different metric units that have the same metric base. For example, if we want to convert a measurement of 36.5 cm to meters, we use a factor-label setup containing the proper exponents. Our given quantity is 36.5 cm and the units of our unknown quantity are meters, m. Since 1 cm = 10^{-2} m, we can use the conversion factor 10^{-2} m/1 cm, and

$$36.5\,\text{cm} \times \frac{10^{-2}\,\text{m}}{1\,\text{cm}} = 36.5 \times 10^{-2}\,\text{m}$$

Likewise, we can convert 0.572 L to mL. In this case, our conversion factor is 1 mL/10^{-3}L, and

$$0.572\,\text{L} \times \frac{1\,\text{mL}}{10^{-3}\,\text{L}} = \frac{0.572\,\text{mL}}{10^{-3}} = 0.572 \times 10^{3}\,\text{mL}$$

Note that in the first example the exponent, 10^{-2}, is in the numerator. In this case we simply multiply the number 36.5 by the exponent and obtain the product, 36.5×10^{-2}. In the second example, the exponent, 10^{-3}, is in the denominator, which means that we must divide the number 0.572 by 10^{-3} to get our answer. The simple way to do this is to move the exponent from the denominator to the numerator, change its sign (from – to + in this case), and multiply the number in the numerator by the new exponent; thus,

$$\frac{0.572}{10^{-3}} = 0.572 \times 10^{3}$$

Multiplying and dividing exponential numbers is an extension of this idea. Appendix B contains further discussion on this topic.

DID YOU LEARN THIS?

For each of these statements, fill in the blank with the most appropriate word or phrase from the list. Do not use any word or phrase more than once.

lag time	calibration
nomenclature	classification
Copenhagen interpretation	simple unit
Newtonian view of reality	compound unit
milliliter	measurement
liter	SI units
millimeter	National Institute of Standards and Technology (NIST)
meter	exponent
kilometer	metric units
precision	orderliness
accuracy	scale
standard	

a. ______ a system of labeling things that implies they fall into a particular order
b. ______ the most appropriate metric unit to use in expressing the amount of liquid contained in a small bottle of expensive perfume
c. ______ how closely the measurement made agrees with the actual size of the object
d. ______ an object or substance that is used in the calibration of an instrument
e. ______ the system of units that is now being used by the international scientific community
f. ______ a dimension that is expressed as a ratio of two measurements, such as density
g. ______ the process used to determine the length of a table using a meter stick
h. ______ what science assumes exists in the world because patterns seem to exist in nature
i. ______ an orderly arrangement of things based on certain properties they have in common
j. ______ the idea that the observer is an active participant in an experiment and affects its outcome
k. ______ always expected to be a part of the process when scientific discoveries are converted into practical application
l. ______ the most appropriate metric unit to use in expressing the size of the period at the end of a sentence in this book (.)
m. ______ the divisions on a meter stick, regardless of their size

n. ______ the idea that a scientist carrying out an experiment is a separate observer and is detached from what is being studied
o. ______ the amount of fineness in a measurement
p. ______ the power to which the number 10 is raised when it is used as a multiplier
q. ______ the place in the United States where standards for measurement are kept
r. ______ a process of comparison by which instruments are made to be accurate

EXERCISES

1. What attributes of the Newtonian view of reality lead to the notion that there can be "scientific objectivity"?
2. Compare and contrast the Newtonian and Copenhagen views of reality. What attributes of the Copenhagen interpretation might be difficult to reconcile with the requirement that scientific observations be repeatable, verifiable, and validated by a consensus of most observers?
3. List the four basic assumptions of science. What are the implications of each?
4. How do classification and nomenclature help us do science?
5. Distinguish between science and technology.
6. What is lag time? What implications does it have for developing a strong, competetive technological base in a country?
7. In this chapter the argument was presented that because, by using our instruments, we can see a composite sequence of detail from the crudest experimental view to the most detailed, we can say that what our instruments tell us is a reflection of reality, even though we cannot directly observe the reality. Using successively more powerful microscopes to view a leaf cell was used to illustrate this point. Another example of a field of science that affords this continuum of detail is common in your everyday experience, but instead of observing objects on an increasingly smaller scale, it consists of observations on an increasingly larger scale. What is this branch of science and what does it observe? What kinds of instruments does it use?
8. Our scientific instruments enable us to observe only a part of reality at a time. What limitations does this impose on the kind of knowledge we can obtain with them? Does this prevent us from gaining new scientific knowledge in the future? Explain.
9. Why is it necessary to standardize units of measurement? Why is it necessary to calibrate instruments using these standards?

10. Distinguish clearly between precision and accuracy in the scientific sense. Find the definitions of these two terms in a modern standard dictionary, read them carefully, and compare your answer with them. How can a measurement be precise but not be accurate? Under what circumstances could a less precise measurement be more accurate? From your everyday experience, give an example of the difference between precision and accuracy.
11. Why is it more appropriate to use the term *error-minimal measurements* rather than *error-free measurements?* Is a precise measurement an error-minimal measurement? To obtain error-minimal measurements, what conditions must be met in making them?
12. How could an error-minimal measurement you made in the laboratory become a scientific fact? Suppose that it did become one. How closely would it represent absolute reality?
13. What two attributes do units give to measurements? (In other words, how do units make measurements meaningful?)
14. What is a compound unit (as defined in this text)? Why are compound units useful?
15. How does the use of dimensional analysis make solving problems involving units easier? In the context of dimensional analysis, what is an exact equality? Give examples. What is an equivalence? Give examples. How are these used to set up and solve problems?
16. Using Appendix A and approximations from Table 2.3, set up the conversion factors for working the following problems.

    a. Convert 16 in. to an equivalent distance in cm.
    b. About how many pounds are in a metric ton (1000 kg)?
    c. About how many mL are there in 5 oz?
    d. About how long is a football field (100 yd) in meters?
    e. A pharmacist sold a woman 100 mL of medicine for her child, with instructions to give 2 tsp of it at each dose. About how many doses can the woman give her child?
    f. Mercury, the only metal that is liquid at room temperature, has a density of 13.6 g/mL. How many grams does 100 mL of mercury weigh? How many kilograms?
    g. Water has a density of 0.998 g/mL at room temperature. How many grams does 10.0 μL of water at this temperature weigh? How many micrograms?

17. Make the following metric conversions, expressing your answer in exponential notation.

    a. 65.2 cm = ____ m
    b. 4.7 km = ____ m
    c. 8.637 g = ____ kg

d. 1.20 ms = _____ s
e. 99.09 nm = _____ m
f. 22.414 L = _____ mL
g. 405 dL = _____ L
h. 56.4 mg = _____ g
i. 1.25 g = _____ μg
j. 0.18 cm = _____ m
k. 3.0 m/s = _____ cm/s
l. 4,500 mL = _____ L
m. 0.00004930 m = _____ nm = _____ km

THINKING IT THROUGH

Everyday Instruments

Develop a list of instruments that you use or rely on in your daily life. In what manner does each extend human senses and for what purposes is it used? What are the limitations of each? Do you really believe the information that is obtained? Why?

Precision and Range

Which has a more precise scale for measuring your body temperature, (a) an outdoor thermometer or a medical thermometer? Which probably would be more accurate? Which would enable you to obtain a greater range of measurement? (In this case a greater temperature range.) Now answer these same questions for (b) finding your weight on a bathroom scale or a highway scale, and (c) finding the length of a city block with a surveyor's tape or the odometer on your car. Does there appear to be a general relationship between the precision of a measurement and the magnitude of the range over which it can be made? What is it, and what problems does it pose for making very precise measurements spanning large dimensions?

ANSWERS TO "DID YOU LEARN THIS?"

a. nomenclature
b. milliliter
c. accuracy
d. standard
e. SI units
f. compound unit
g. measurement

h. orderliness
i. classification
j. Copenhagen interpretation
k. lag time
l. millimeter
m. scale
n. Newtonian view of reality
o. precision
p. exponent
q. National Institute of Standards and Technology (NIST)
r. calibration

THE PHARMACIST WHO HATED ARITHMETIC

Jeanette K. Wiegand

Pharmacist

B.S. Psychology, Idaho State University

Pharm. D., Idaho State University

As was typical with young girls in the 1950s and 1960s, I thought I hated math and science. In the 7th grade I finally had a real introduction to science—a genuine science teacher and a good-looking one at that. He had just graduated from college and was motivated (inspired) and made science fun and interesting. I had this same science teacher the next year, and he convinced me to take as much science as possible in high school. My small, rural high school in Nevada didn't offer much variety in the sciences, but chemistry and physics were options that were available. I am sorry to say that math remained uninteresting to me, and no similar mentors surfaced.

My first job was in a drugstore when I was sixteen. The pharmacist suggested I try pharmacy as a career goal. On graduation from high school I entered pharmacy school in Idaho, but other things got in my way of a pharmacy degree (marriage and a child to raise). After living in the East for several years, I returned to Idaho and earned a Bachelor of Science degree in psychology. An undergraduate liberal arts degree wasn't very helpful in the job market though, and I wasn't interested in pursuing a graduate level education at that time.

After many years as a bookkeeper (interpretation: doing arithmetic every day), I went back to college to finish my pharmacy education. It had been 13 years since I had graduated from college and almost 20 years since I had taken any chemistry, but I completed organic chemistry and all of the other classes I was required to take. So at age forty-five, and 28 years after starting pharmacy school, I received a Doctor of Pharmacy degree.

The curriculum in pharmacy is based on chemistry, biology, and math. Most of the classes use chemistry either directly or indirectly. My current job as a pharmacist in a pediatric setting also uses chemistry (and biology) either directly or indirectly. Much of my job also requires communication skills—dealing with parents and answering their questions. Many of these questions involve "chemicals." All drugs, even the natural ones, are chemicals. Drugs work in the body or on body invaders (bacteria, fungi, viruses, and cancer cells). The job of the health professional is to know the targets for the drug (such as receptor sites, organs, and bacteria) and to know and inform the patients of the effects of their medications, both desired and undesired (side effects). I must be aware of (or look up) the pH of drugs and of the body systems they target (e.g., acidic or basic). I have to know about protein binding (chemical bonding). I have to recognize "functional groups" to be able to predict what a particular drug's mechanism of action is.

Occasionally I have to "compound" a drug, that is, combine two or more ingredients to make a special medication or change the form of an existing medication. For example, I often make suspensions from tablets, prepare special medicated creams, or make small doses out of larger ones. Laboratory skills such as weighing and measuring are frequently used by pharmacists and pharmacy technicians. The most frightening aspect of my job is preparing the chemotherapy for cancer patients. There is a very fine line between killing the cancer and killing the patient. Obviously we want to do as much damage as possible to the cancer cells while causing the least amount of damage to the patient.

So, although to the average customer or outside observer it may appear as though pharmacists only transfer medications from larger bottles to smaller ones and put labels on them, there is much more involved. For the most part, I really enjoy my profession and the challenges that are presented to me every day.

3

Chemistry: The Study of Matter

I now mean by elements . . . . certain primitive and simple or perfectly unmingled bodies, which not being made of any other bodies or of one another, are the ingredients of which all those perfectly mixt bodies are immediately compounded and into which they are ultimately resolved.

Robert Boyle, "The Sceptical Chymist," 1661

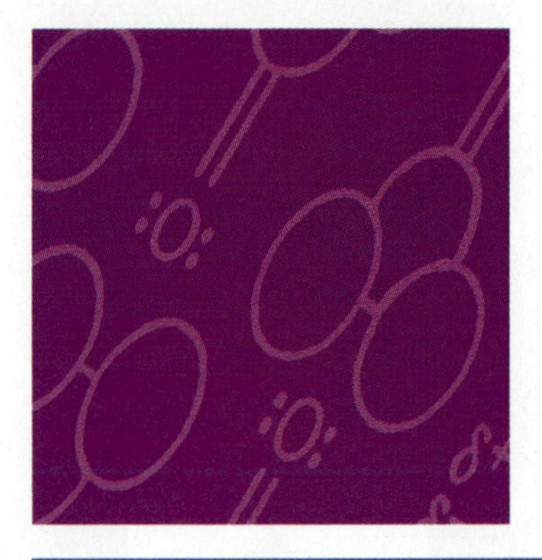

It is often said that "Chemistry is the central science," meaning that chemistry as a science lies about midway on a somewhat arbitrary spectrum between physics and the biological sciences (▶ Figure 3.1). *Physics,* the most abstract natural science, deals with matter and energy and their relationship in terms of motion and force. *Biology,* the natural science with the most variables, is concerned with life and living matter, its forms and processes of change. Between these extremes we find chemistry, which includes aspects of both. *Geology,* the earth science, lies between chemistry and biology, having more variables than chemistry but being less abstract.

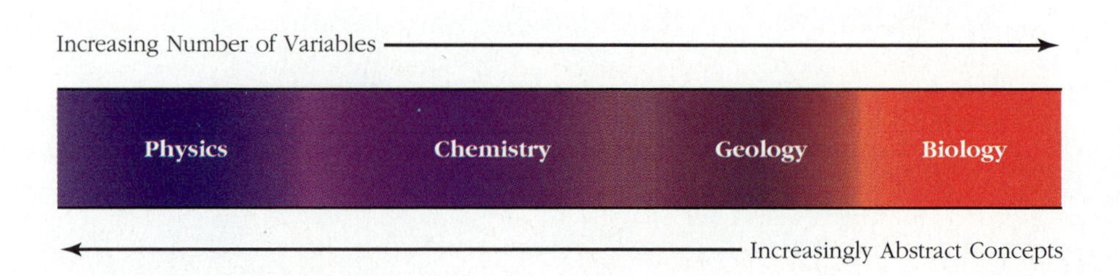

▶ Figure 3.1
Chemistry lies in the center of a spectrum of the natural sciences, between physics, the most abstract, and biology, which has the most variables.

A DEFINITION OF CHEMISTRY

Chemistry is concerned with the structure of matter and how this relates to its behavior.

It is chemistry's perspective of the world that makes it central. All natural sciences deal with matter in some way, but **chemistry** is concerned primarily with the structure, or architecture, of matter and how structure relates to behavior. It examines the categories, properties, structure, and transformations of matter. Chemistry often involves physics or biology. For example, many chemists apply quantum mechanics (theoretical physics) to the structure and dynamics of chemical systems; others study the chemistry of metabolism (biochemistry) or the transmission of impulses in nerves (biology). Some chemists use both ends of the spectrum to study the quantum mechanics of vision. The science we call genetic engineering is actually biochemistry and microbiology, both of which use a combination of chemistry and biology.

Knowing how matter is put together and how its structure can be changed helps us understand the properties of substances around us. We can explain why diamond is the hardest natural substance known and yet can be easily split by the stroke of a diamond cutter's tool, why chlorofluorocarbons (CFCs) are thought to be destructive to the ozone layer, and why bubbles form when baking soda is mixed with sour milk. Understanding how the properties of a substance are related to its structure also enables us to modify materials to make them more useful to us or to give them properties not found in nature. Nylon is far superior to silk in strength, and laboratory-made antibiotics are more specific and potent than natural penicillin. Graphite fishing rods are much stronger and more sensitive to a tug than bamboo, and Teflon has no peer in nature for slippery durability.

MATTER AND ENERGY

As far as we know, the universe consists of only two things: matter and energy. Matter is all around us, and we know intuitively what it is. It makes up this book, our bodies, the ocean, the moon, and bacteria. Some of it is invisible, like the air we breathe (▶ Figure 3.2).

▶ Figure 3.2
Invisible Matter. *Air is matter. Its resistance to the motion of the hang glider wing creates lift to keep the glider aloft.*

To sharpen our understanding of what matter is, we need to define it operationally. **Matter** is anything that has mass and occupies space. (The "occupies space" part is clear, but we have defined *matter* by using a new term,

Matter has mass and occupies space.

Mass is the measure of resistance to being moved by a force (push).

Weight is a measurement of the attraction of gravity for an object.

Energy makes things move.

mass, which isn't much help. However, the use of unusual terms is often necessary in science. When it happens in this book, we shall successively define each new term until we get to something we can relate to.) **Mass** is the measure of matter's resistance to being moved by a force (push). A large mass requires more force to make it accelerate than a smaller mass. It is easy, for example, for an average golfer to drive a golf ball 200 yards, but driving a lead ball of the same diameter that far would be impossible. Mass is different from weight in that mass does not change, regardless of where a piece of matter is located (► Figure 3.3). **Weight** depends on the action of gravity and hence is location dependent. Our golf and lead balls would weigh about one-sixth (1/6) as much on the moon as on Earth, but the mass of each would stay the same. Although we could hit the golf ball much farther on the moon, driving the lead ball for 200 yards would still be next to impossible. (Gravity on the moon is less, so the balls wouldn't fall back to the ground as rapidly; but we still would have to get them moving in the first place.)

Everything that is not matter is energy. It is fairly easy for us to comprehend what matter is by thinking about pushing against something or having it push against us, such as the wind. To get a mental picture of something that is *not* matter is more difficult. One way to describe **energy** is that it is the part of our natural world that makes things move. For example, heat from a gas stove causes water to boil, sunlight powers our solar calculators, and flowing water carves the Grand Canyon.

► Figure 3.3
Mass Versus Weight. *Although a satellite may be weightless in space, its mass is the same in space as it is on Earth.*

STATES OF MATTER

There are three common **states of matter:** solid, liquid, and gas. We need to define each of these terms operationally, as we did with *matter.* It does us no good, for example, to describe a gas as *something that is invisible* because some gases can be seen (smog, chlorine) and solids become invisible if divided finely enough. The following operational definitions will help us recognize each state of matter:

- **Solid** Has a definite shape and volume. All true solids retain their shape and take up a certain amount of space (in three dimensions) for a given mass, unless something causes their shapes or volumes to change (such as a change in temperature or a force).

- **Liquid** Has an indefinite shape, but has a definite volume for a given mass. A liquid flows to take the shape of its container, or seeks its own level under the attraction of gravity. Most liquids are not easily compressible.

3.1 Which of the following are matter and which are energy?

a. radio waves
b. the glow from a computer screen
c. an aluminum can
d. the motion of flowing water
e. gasoline
f. a spark from flint and steel
g. baking soda

- **Gas** Expands to completely fill its container. A gas has no definite shape or volume of its own. If its container changes shape or volume, so does the gas. Gases are easily compressible.

CHANGES OF STATE

Changes of state are caused by changes in temperature or pressure (▶ Figure 3.4). The solid state of a given substance exists at lower temperatures, the liquid state at intermediate temperatures, and the gaseous state at higher temperatures, although not all substances exist in the same state at the same temperature. Water is a liquid at room temperature whereas oxygen exists as a gas and iron as a solid.

In our everyday experience we usually do not recognize the effect that pressure has on changes of state, but one effect is common. A butane lighter with a transparent reservoir reveals a liquid inside, whose level decreases as the lighter is used. This liquid is butane, which is a gas at room temperature and pressure. To remain liquid in the reservoir, the butane must be slightly

(a)

(b)

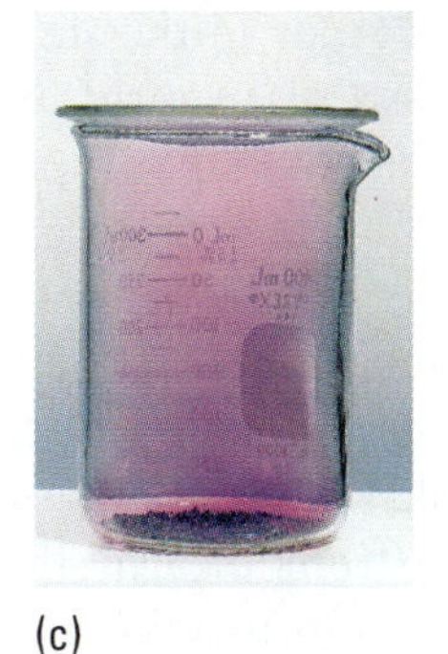
(c)

▶ Figure 3.4
States of Matter. *(a) At room temperature, benzene is a colorless liquid. (b) Benzene freezes to a solid when chilled in an ice/salt mixture. (c) Solid iodine crystals sublime to form a purple gas when heated.*

3.2 In what state of matter are the following substances usually found?

a. a glacier
b. bubbles in soda water
c. sand
d. a rubber tire
e. automobile exhaust
f. shaving lotion

pressurized. When the lighter is used, the butane gas above the liquid in the reservoir rushes out and is ignited by a spark. Propane gas, which is used as a fuel for gas grills, stoves, and heating and cooling appliances, also shares this property and is stored in the liquid state in tanks under pressure.

The **melting point** and the **freezing point** of pure substances have the same value.

Transitions between states of matter have special names that tell us what the process is. The term **melting** is used to describe the change from the solid to the liquid state as the temperature of a substance is increased. **Freezing** is the reverse of this process; that is, a liquid becomes a solid as the temperature is lowered. For pure substances, the temperature at which these transitions occur is the same; in other words, the **melting point** (a temperature) and the **freezing point** (a temperature) have the same value. When a liquid is heated until bubbles rise rapidly from beneath its surface to escape as vapor, it is said to be **boiling.** The temperature at which this occurs is called the **boiling point.** The reverse of this process, in which a gas is cooled and becomes liquid, is **condensation,** and the corresponding temperature is the **condensation point.** For pure liquids, the boiling point and the condensation point are the same as long as the pressure on the liquid stays the same.

It is not necessary that liquids boil for them to exist in the gaseous state. There is always some gas above the surface of a liquid, even when it is well below its boiling point. Water, for example, rapidly evaporates from ponds and clothes on a warm summer day (and more slowly on a cold day), forming water vapor in the air. This gaseous water often condenses into dew in the coolness of an early morning. Gasoline is an explosion hazard because its combustible vapor is present above the liquid. And liquid mercury, though it has a high boiling point, poses some risk to health, even if it does not come in contact with our skin, because even at room temperature a measurable amount of mercury vapor is associated with it. This vapor, which is toxic on long-term inhalation, can pass through water and even plastic; so mercury should not be stored under water or in plastic bottles. Also, any spilled mercury (from broken thermometers, thermostats, etc.) should be properly cleaned up to minimize the presence of mercury vapor in laboratory and household environments.

Some solid substances do not melt when heated but convert directly into the gaseous state. When their vapors are cooled, the solid phase reforms. In neither heating nor cooling is the liquid phase observed. Substances that proceed from solid to gas, skipping the liquid phase, are said to sublime, and the process is called **sublimation.** Familiar examples of substances that sublime are solid carbon dioxide (dry ice), moth flakes, and iodine crystals; the freeze drying of foods also takes advantage of this process. The formation of frost and snowflakes from water vapor in the air are examples of **deposition,** the reverse of sublimation.

PURE SUBSTANCES AND MIXTURES

Matter usually is divided into two categories, pure substances and mixtures, or impure substances. Impure substances are not as easy to study as pure substances, so chemists prefer to work with pure substances.

A **pure substance** is a form of matter that has a definite, constant composition. Moreover, it has a set of properties that is invariant and unique. **Properties** are attributes of a substance such as its color, crystalline form (salt crystals are cubes), its usual state of matter, melting point, boiling point, density, taste, odor, how well it conducts heat and electricity. The properties mentioned here are called **intensive properties.** That is, they do not depend on the amount of the substance present. Examples of pure substances are copper, baking soda, aluminum, water, table sugar, and diamond (▶ Figure 3.5a). Pure substances usually are homogeneous; but a mixture of two states of a pure substance, such as pure water and ice, is heterogeneous.

Each **pure substance** has its own set of properties.

Mixtures consist of two or more pure substances in various proportions. Unlike pure substances, mixtures have no fixed composition; their various components maintain their identities and usually can be separated. Likewise, mixtures have no unique set of properties. A salt solution that contains only a little salt freezes at a temperature only slightly below that for pure water; but one that contains a large amount of salt has a much lower freezing temperature. This is the reason a mixture of salt and ice (which makes a very concentrated salt solution mixed with ice and solid salt) becomes very cold and can be used to freeze homemade ice cream.

Mixtures have no fixed composition.

Mixtures can be either homogeneous or heterogeneous (▶ Figure 3.5b). The term, **homogeneous,** means "of one kind," or "uniform throughout." Samples of a substance that is homogeneous have the same composition no matter what part of the substance they come from. For example, pure sugar (a pure substance) is the same regardless of which part of the bag we sample. A solution of salt in water (a homogeneous mixture) does not consist of crystals of salt spread like little cubes throughout the water. Instead, it appears to be of one kind and has the same proportion of salt to water in any part of the solution we examine. By contrast, **heterogeneous** substances are not uniform

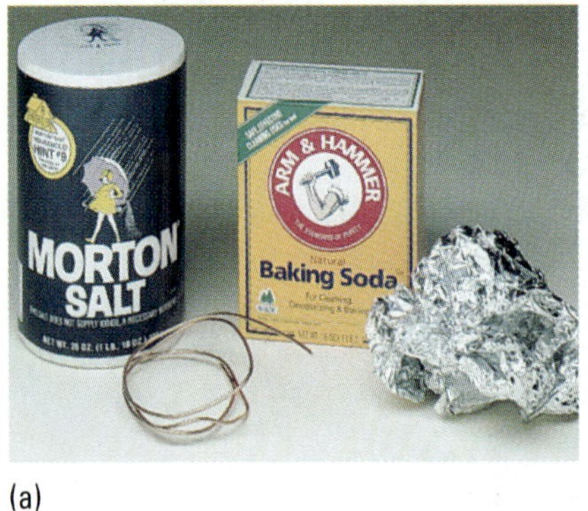

(a)

(b)

► Figure 3.5
Pure Substances and Mixtures. *(a) Pure substances found around the house include compounds, such as salt and baking soda, and elements, such as copper (wire) and aluminum (foil). (b) Mixtures that we eat include gelatin, which is homogeneous, and pizza and mixed fruit, which are heterogeneous. Milk appears to be homogeneous but is actually heterogeneous, as can be seen with a microscope.*

throughout and contain distinctly different particles distributed in one another. A mixture of salt and sand is heterogeneous, and we easily can identify the individual particles of salt as distinct from those of sand. If it is not thoroughly mixed, one part of the mixture may have a different proportion of salt to sand than another. Probably everyone has opened a box of cereal with fruit to find that the fruit has settled to the bottom. In practice, whether or not we call a substance homogenous sometimes depends on how closely we look at it. Homogenized milk appears to be uniform until we examine it under a microscope. Then we see the tiny droplets of butterfat suspended in the water solution that comprises most of the rest of the milk.

A solution of sugar or salt in water is a homogeneous mixture; so is air, gasoline, stainless steel, well water, dental amalgam, and gelatin. Examples of heterogeneous, or nonuniform, mixtures are beef stew, paint, blueberry muffins, diesel smoke, sewage, and carbonated beverages. ► Figure 3.6 illustrates how the different classes of matter are related.

► Figure 3.6
Classifications of Matter.

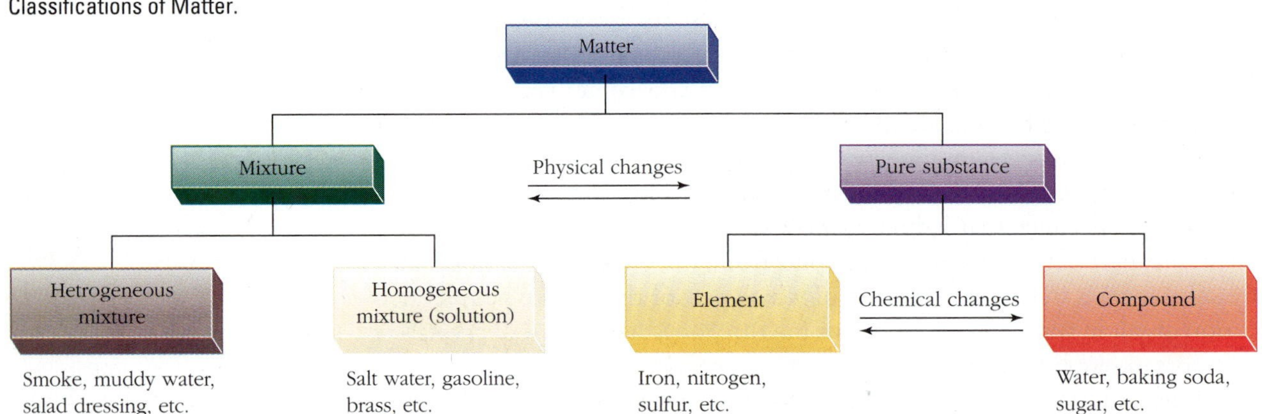

3.3 Decide whether each of the following is a pure substance or a mixture.

a. salt water
b. sugar
c. motor oil
d. silver
e. dry ice

3.4 Decide whether each of the following is homogeneous or heterogeneous.

a. strawberry jam
b. soda water
c. vegetable oil
d. paint thinner
e. concrete

ELEMENTS, COMPOUNDS, AND CHEMICAL CHANGE

Chemists further classify matter by identifying certain pure substances as those from which all other substances can be formed. The ancient Greeks, Empedocles and Aristotle, thought that these elemental substances were air, earth, fire, and water. This idea persisted in one form or another until 1789 when Antoine Lavoisier, a French chemist, published, in his *Elementary Treatise on Chemistry,* a list of 33 substances he thought to be elements in the sense we know them today. **Elements** are pure substances that cannot be broken down into other pure substances by chemical means. Nitrogen (gas) and silver (metal) are elements; we cannot use chemical change to simplify them. If by some nonchemical means we are able to break them down, they lose their identity as nitrogen and silver. Substances that arise from the chemical combination of two or more elements are called **compounds.** When the element magnesium burns in air, producing a very bright, hot flame, it combines chemically with the element oxygen to form a white powdery solid called magnesium oxide, a compound (Figure 3.7). At this writing, over 11 million chemical compounds are known.

What is a chemical change? A **chemical change** is a process by which the composition of matter is changed, and new pure substances with different properties are formed. In the preceding example, metallic magnesium reacted with the invisible gas oxygen to produce heat, light, and the white magnesium

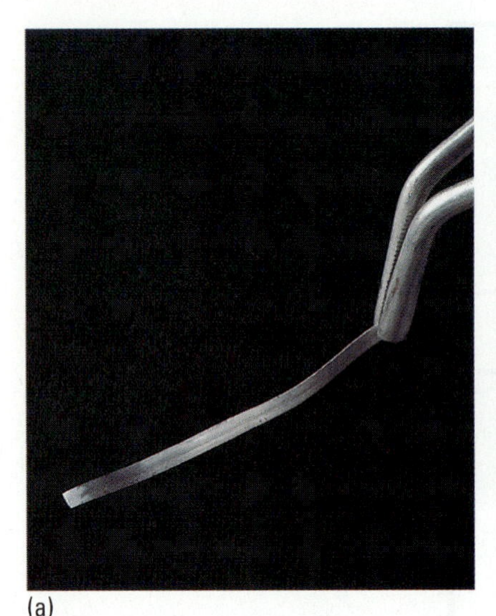
(a)

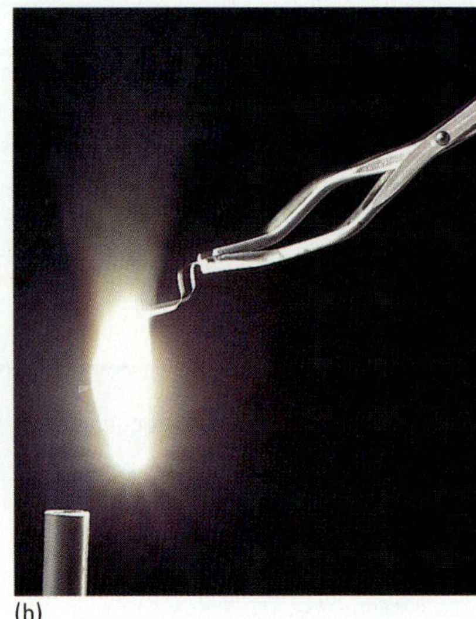
(b)

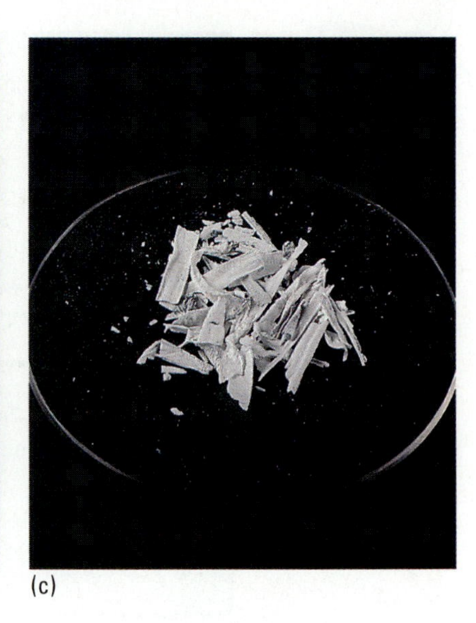
(c)

▶ Figure 3.7
Chemical Combination: Burning Magnesium in Air (Oxygen). *(a) A strip of magnesium metal is held with a set of tongs. (b) Ignited, the magnesium burns with a blinding white light. (c) The white magnesium oxide remains after the burning of several strips of magnesium.*

A **chemical change** alters the composition of matter. A **physical change** does not.

oxide. Although the oxide contains magnesium and oxygen, its composition and properties are different from those of either magnesium or oxygen: It resembles neither the magnesium nor the oxygen from which it formed.

PHYSICAL AND CHEMICAL CHANGES COMPARED

Physical changes do not alter the composition of a substance. Examples include grinding, breaking, melting, boiling, or passing an electric current through a wire. Even salt or sugar dissolving in water is a physical change, although it may not appear to be. By simply evaporating the water (and collecting it if desired), the salt or sugar is left behind, unchanged.

The difference between a chemical change and a physical change can be illustrated by an intimate mixture of finely divided iron (an element) and powdered sulfur (another element) (▶ Figure 3.8). The metallic, gray iron filings are distinct from the bright-yellow sulfur particles, and the iron and sulfur pieces easily can be separated mechanically because the iron is attracted to a magnet and the sulfur is not. This separation is a simple example of a physical change, because it does not involve a change in the composition of matter. The iron is still iron and the sulfur is still sulfur.

If, however, this same mixture of iron and sulfur is heated, it begins to glow and give off heat, leaving behind a gray substance that is not attracted to a magnet. This compound, iron sulfide, has different properties from either iron or sulfur; and although it contains both of these elements, they are not physically separable. Like the burning of magnesium, this process has caused a change in the composition of matter. A chemical change has occurred.

(a)

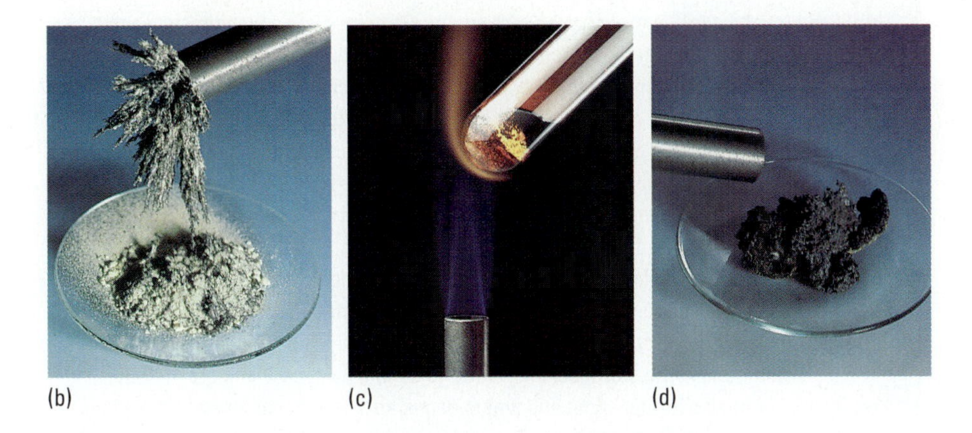

(b) (c) (d)

▶ Figure 3.8
Comparison of Physical and Chemical Change. *(a) Iron (gray filings), sulfur (yellow powder), and a mixture of iron filings and sulfur. (b) Simple physical change. Both components of the mixture retain their own properties. Because the iron is attracted to a magnet, it can be separated from the sulfur. (c) Chemical reaction. When heated, the iron-sulfur mixture glows and gives off heat, indicating that a chemical reaction is taking place. (d) Chemical change. The product is iron sulfide, a gray compound that is not attracted to a magnet. Its properties are different from those of either iron or sulfur.*

THE ENERGY HILL

In nature, almost all elements generally exist in combined form (as compounds); the **noble gases**—helium, neon, argon, krypton, and xenon—are exceptions. The nonmetals nitrogen, oxygen, carbon, and sulfur, and the **noble metals** copper, silver, gold, and platinum are some of the elements that occur

3.5 Decide whether each of the following is a physical change or a chemical change.

a. Rubbing alcohol evaporates.
b. Wine turns sour.
c. A candle burns.
d. A piece of paper is torn.
e. Pancake batter is baked.

CHEMICAL CHANGES

Breakfast. You're the first one up. Why does chemistry class have to be so early? You fill the kettle and light the stove with a wooden match (you and your housemates have agreed not to use the pilot light to save gas).

Since you were a child, you've struck matches without a thought, but now, after a week or so of chemistry, you realize that striking a match initiates a chemical reaction, and you're curious about it. You take another match out of the box: white on the tip and red below. You know from experience with flawed matches that the red part doesn't strike. So what's it for? A thought about the energy hill gives you a hypothesis. To test it, you strip the red part off the matchhead with a paring knife, carefully leaving the white tip and keeping your fingers away from the head in case it lights. When you strike the tip, it flares and goes out. That's just what you expected. You strike an unaltered match and watch it carefully. The white tip flares, flammable easily enough to be started down its energy hill by the heat generated by mere friction (it's a mixture of P_2S_3, which is easily flammable, $KClO_3$, which supplies extra oxygen, paraffin or sulfur for fuel, ground glass to add friction, and a binding material to hold it together). Then, boosted down its energy hill by the heat of the white tip, the red part flares ($KClO_3$ with paraffin or sulfur to support the reaction), which, in combination with the white tip, provides enough heat to start the wood down *its* energy hill.

The water in the kettle starts boiling. Water to steam, a vapor: a change of state. You hold your hand over the spout and it gets damp: another change of state. You pour hot water over coffee grounds in the filter, making a mixture of water and grounds, and separate the two components of the mixture with the filter–just as it says in the chemistry text. But what makes coffee smell so good? You put bread in the toaster, put the skillet on the stove to heat with a bit of oil, and get eggs from the refrigerator. The science major who talked you into taking this course comes sleepily into the kitchen as you crack eggs into the skillet.

"Rhonda," you ask, "what is it that smells?"

"The coffee."

"I know *that!* I mean, what makes it smell?"

"Volatiles." You shrug at that. "Vapors of about a hundred different chemicals in the steam."

"A hundred?"

"Organic chemistry is complex."

"Hm. I guess my next question is organic too: when the egg white turns solid, is that a change of state?"

"Nope. Egg white is denatured by heat."

"That's beyond me."

"Well, let's see . . . you know what a compound is? OK. Well, egg white is an organic compound called protein and heat changes the chemical nature of some proteins: egg white becomes opaque white and solid, among other less obvious changes, and we say it's denatured. But denaturing's not like boiling water. You can't turn egg white back into liquid, right?"

"So it's a chemical reaction?"

"Yes, it's the same basic substance, but heat has changed the arrangement of the atoms and the ways they're held together. The chemical behavior of a lot of proteins is like that. Your toast smells done." You snap the lever on the broken toaster and retrieve your toast.

"So, does toast get denatured and turn brown?"

"No, that's carmelization." You shrug again. "It's a complicated series of chemical reactions of starch and sugar. They contain polyhydroxycarbonyl compounds . . . ."

"Poly . . . ."

"Never mind; you'll learn about the names later. Some of these compounds are converted into brown substances by heat. It's called the browning reaction, and we like the taste of the result. Any more questions?"

"One more." She sighs with exaggerated patience. "Hey, you got me into this . . . be happy! I just want to know if it's the browning reaction when people get darker in the sun?"

"No!" She laughs. "The browning reaction is caused by cooking . . . it takes high temperatures. But I have to admit the result is similar. The sun's ultraviolet rays cause special cells in our skin to convert a chemical called tyrosine into melanin to try to protect skin from the sun's rays. Now eat!"

"I didn't realize there's so much chemistry in ordinary cooking."

"Actually, cooking is chemistry. We don't think of it that way because the reactions are so complex that we can hardly ever repeat a process exactly. It seems to me that's one definition of art."

naturally in both elemental and combined forms. Is there some kind of driving force that causes most elements to form compounds? Is there some underlying reason why they do this?

To get an idea, we look at what we must do to separate many of the elements from their ores. Separating pure elements from their ores is the reverse of combination, and it is usually accomplished by *smelting* or *refining*, processes that require the use of heat or some other form of energy.

The smelting of iron ore in a blast furnace can be represented by the general reaction,

Coke is a form of carbon.

$$\text{Iron ore} + \underset{\text{(a form of carbon)}}{\text{Coke}} \longrightarrow \text{Iron metal} + \text{Carbon dioxide}$$

Here the iron ore and coke are **reactants,** and the iron metal and gaseous carbon dioxide are **products.** The + sign between the reactants means *reacts with,* the + sign between the products means *and,* and the arrow (⟶) means *yields* or *produces.*

The reactants must be strongly heated to make this reaction proceed. Extra coke is added to the reaction mixture, and oxygen (in the form of compressed air) is forced through it to make a very hot fire, much like burning charcoal in a blacksmith's forge.

Energy is required to separate elements from their compounds.

This procedure is typical of many chemical processes used to liberate elements from their ores (▶ Figure 3.9). A considerable amount of energy, usually in the form of heat or electricity, must be applied to the reactants.

▶ Figure 3.9
Refining Iron. *Energy is used to purify iron metal. A typical furnace is charged with 200 tons of impure (pig) iron, 100 tons of scrap iron, and 20 tons of limestone.*

By contrast, given the proper conditions, heat is given off when most elements combine with other elements or react with suitable compounds to form new compounds. We have seen this in the burning of magnesium in air: a ribbon of magnesium burns with an intensely hot, white flame.

$$\text{Magnesium} + \text{Oxygen} \longrightarrow \text{Magnesium oxide} + \text{Heat}$$

Because heat is given off in this kind of process, it can be considered to be a *product* of the reaction. The chemical combination of an element with another element or substance, then, can be represented by a general equation showing the productlike nature of heat,

$$\text{Element} + \text{Other substance} \longrightarrow \text{"Ore"} + \underset{\text{(Product)}}{\text{Heat}}$$

Such a reaction is said to be **exothermic.**

METALLURGY AND ENERGY

Historically, copper was the first metal widely used for tools, because little heat, relative to other metals, is needed to obtain the pure element from some of its ores (Figure 1). Groups of primitive metallurgists probably smelted copper in underground hearths by blowing rhythmically through hollow sticks to make a very hot fire (about 1100 °C). Occasionally, copper is found in its elemental or native state as are silver and gold.

Bronze, an alloy of copper and tin, eventually replaced copper for tools because it is much harder than copper. Tin also is easily smelted from its ores. The widespread use of iron did not occur until later because much more heat and much higher temperatures (1600 °C) are necessary to release free iron from its combined state. The technology for smelting iron, like that for copper, involved blowing air through a mixture of ore and burning charcoal. The use of aluminum developed from the increasing availability and use of high-intensity electrical energy during the twentieth century. Refining aluminum ore is a particularly energy-intensive process.

HISTORICAL USE OF METALS FOR TOOLS AND IMPLEMENTS

Metal	Date of Common Use	Comments
Cooper	5000 B.C.	Easily obtained from its ores by heating.
Bronze	3800 B.C.	Alloy of copper and tin. Tin also easily obtained from its ores by heating.
Iron	1500 B.C.	Higher temperatures, more heat required than for smelting of tin. Impure. High carbon content. Weak and brittle.
Steel	1000 B.C.	Pure iron containing small amounts of carbon. Alloying elements may be present. Much stronger. More heat, better technology required.
Aluminum	ca. A.D. 1900	Special technology, large amounts of energy required.

Figure 1
A Nineteenth-Century Bakota Mask Made of Copper. *The Bakota people in the Central Congo developed a highly unique style of African art. Their striking masks are made of flattened wooden faces covered with thick metal sheets, usually of copper or brass. The technology for working copper, much of which came from Niger, was practiced by African people for at least 3000 years.*

An **exothermic** process liberates heat. An **endothermic** process absorbs heat.

Similarly, because heat is necessary to free an element from its ore, heat can be considered to be a *reactant* in this process. This kind of reaction is said to be **endothermic.** The reactant-like role of heat can be represented by the general equation,

$$\text{"Ore"} + \text{(Other reactants)} + \underset{\text{(Reactant)}}{\text{Heat}} \longrightarrow \text{Element} + \text{By-products.}$$

For our purpose here, we shall say that chemical processes in which heat is liberated (exothermic) tend to be spontaneous, or proceed on their own. Conversely, processes in which heat is absorbed (endothermic) tend not to be spontaneous. This is the reason that ores do not spontaneously revert to their respective elements in nature and why few elements are found chemically uncombined. They need the help of the refiner's fire to be converted to their elemental state. We can model these processes as taking place on an energy hill (▶ Figure 3.10).

Stability is relative.

Figure 3.10 introduces the concept of stability. **Stability** refers to relative levels of energy. A lower energy level (or state) is generally more stable than a higher energy level. Conversely, higher energy states are less stable than lower energy states. Because stability is not absolute, it is necessary to have a point of reference, so that we can discuss stability with respect to it. In diagrams depicting energy changes it is customary to place higher energy levels

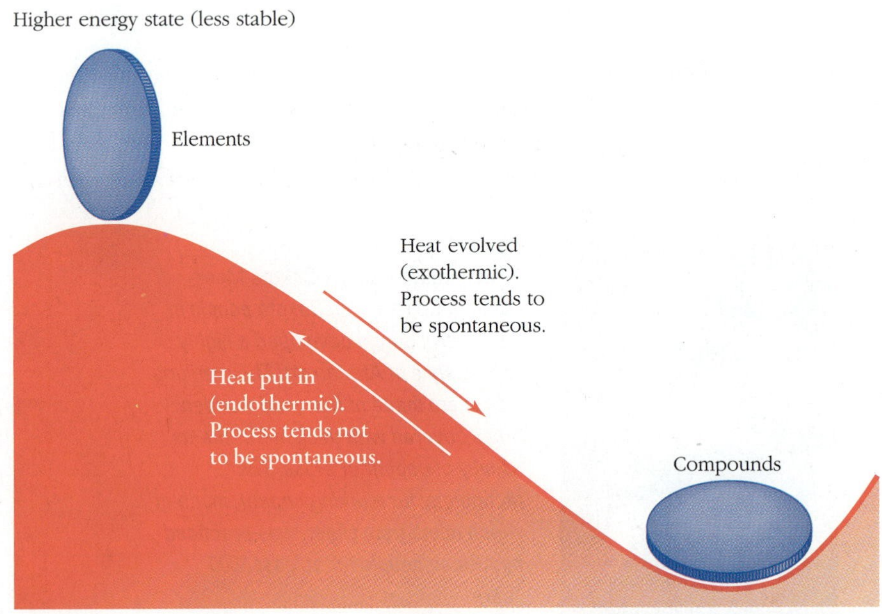
Higher energy state (less stable)
Elements
Heat evolved (exothermic). Process tends to be spontaneous.
Heat put in (endothermic). Process tends not to be spontaneous.
Compounds
Lower energy state (more stable)

▶ Figure 3.10
An Energy Hill. *Compounds are usually more stable than the elements from which they are formed.*

(less stable states) above lower energy levels (greater stability). Thus, moving upward on the diagram in Figure 3.10 means going to higher energy (lower stability) and moving downward means going to lower energy (greater stability). An analogy is a boulder resting on a mountain as illustrated in ▶ Figure 3.11. Its stability is less at the top of the mountain (higher energy state with respect to the valley floor) than it would be if it were to roll down the mountainside and come to rest on the valley floor below, where its stability would be greater (lower energy state with respect to the mountain top).

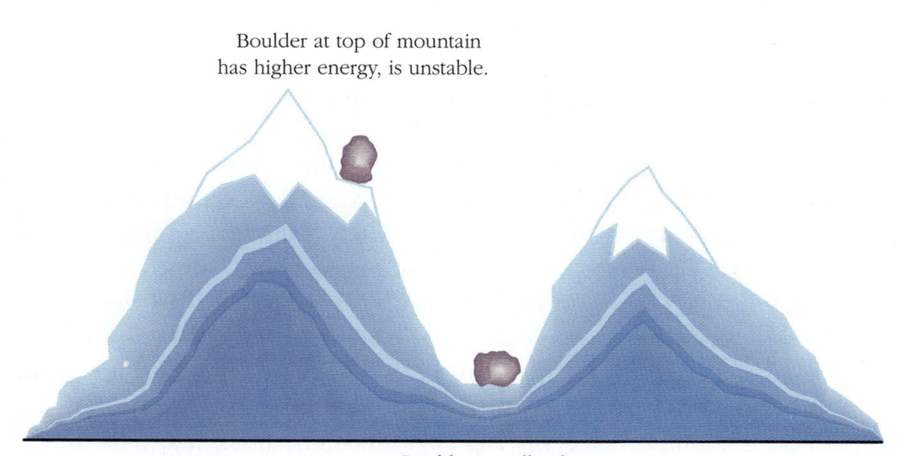
Boulder at top of mountain
has higher energy, is unstable.
Boulder in valley has
lower energy, is more stable.

▶ Figure 3.11
Stability. *A boulder on a mountain top is less stable than one on the valley floor.*

Note that in talking about whether a reaction will be spontaneous, we have used the term *tendency.* A process that is exothermic tends to be spontaneous; but this does not necessarily mean that it actually will proceed. For example, the boulder on the mountain could be resting in a crevice, and until some force pushes it upward out of the crevice it will not roll down the mountain. If the crevice were very deep, the boulder might remain in it for a long time. Chemical reactions that liberate heat behave similarly. Most of them need some kind of push, an input of energy to get them started. (A wooden match is a good example.) It is fortunate that this is so, because otherwise all chemical reactions would long ago have run their course.

Similarly, a process that is endothermic can be *made* to proceed. A boulder on the valley floor can be pushed up the mountain to a higher level. Although this requires a considerable amount of energy, it can be done. Likewise, chemical reactions that require energy input, such as the production of iron from iron ore, can be made to proceed if sufficient energy is available.

3.6 Which of the following is in the more stable energy state?

a. a book on a top shelf or a book on a bottom shelf
b. water behind a dam or in the river below
c. a skier at the bottom of a run or at the top of the lift
d. a piece of wood or its ashes
e. a fresh battery or a "dead" one

ABOUT THE ELEMENTS: THEIR NAMES AND SYMBOLS

The universe is composed mostly of hydrogen.

Although there are 89 naturally occurring elements, most of the universe is made up of hydrogen. Hydrogen is the most abundant element in the universe, comprising 75% of the mass of all matter, followed by helium with 23% (▶ Figure 3.12a). Thus, all of the other elements combined constitute only 2% of the mass of the universe. At this writing a total of 109 elements is known, 20 of which have been made in the laboratory.

The abundances of the elements in the earth's crust, its atmosphere, and the human body are very different (▶ Figure 3.12 b, c, d). The crust (10–25 miles thick) is about 47% oxygen and 28% silicon by mass; it also has much larger amounts of other heavier elements, such as aluminum, iron, calcium, sodium, potassium, and magnesium. Less than 1% of the crust is hydrogen. The earth's atmosphere is about 75% nitrogen and 23% oxygen by mass. This mix of elements provides an environment uniquely suited to supporting life on earth. The human body is made up mostly of oxygen (65%), carbon (18%), hydrogen (10%), and nitrogen (3%).

Every element has been assigned a unique name and a unique **chemical symbol** that is a shorthand for its name. The origins of the names and symbols for many of the elements are quite interesting. For example, lead (from the Anglo-Saxon *lead*) has the symbol Pb, which derived from the Latin *plumbum.* Lead was used for piping water in ancient Rome. The symbol for mercury (named after the planet Mercury), Hg, also comes from the Latin *hydrargyrum,* or liquid silver. A number of interesting derivations representative of how the names and symbols of the elements were obtained are listed in ■ Table 3.1 The photographs accompanying this table (▶ Figure 3.13) illustrate how several of the elements were named.

Note in Table 3.1 that some of the chemical symbols for the elements have only one letter, which is capitalized, such as N for nitrogen, P for phospho-

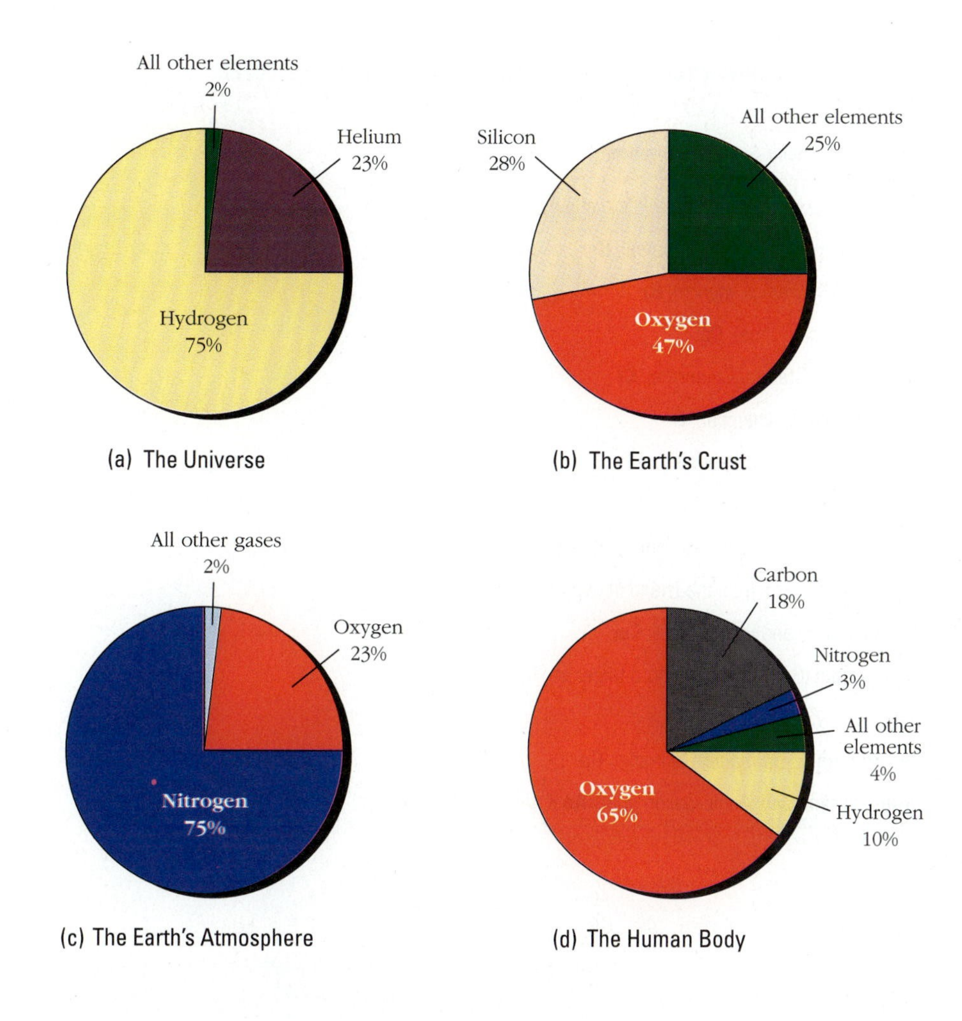

▶ Figure 3.12
Distribution of the Elements in Various Realms (Percent by Mass). *(a) The universe. (b) The Earth's crust. (c) The Earth's atmosphere. (d) The human body.*

rus, and B for boron. There are fourteen of these one-letter symbols. Eighty-nine elements have two-letter symbols. (The symbols for the last six laboratory-made elements have three letters). The first letter is always capitalized and the second letter never is. Examples are Br for bromine, Ne for neon, and Fe for iron. It is important that the second letter be written accurately because many abbreviations start with the same letter. Note that **mag**nesium is Mg, and **man**ganese is Mn. **Co**balt is Co, **chr**omium is Cr, **chl**orine is Cl, **cal**cium is Ca, and **curium** is Cm. It also is important to remember that some elements have symbols that do not resemble their English names at all. A few of these are sodium, Na; potassium, K; iron, Fe; gold, Au; and tin, Sn.

TABLE 3.1 DERIVATIONS OF THE NAMES AND SYMBOLS OF SELECTED ELEMENTS

Element (Derivation), **Symbol** (Derivation)
Arsenic (L. *arsenicum*, Gr. *arsenikon*, yellow ointment), **As**
Boron (Arabic *buraq*, Persian *burah*), **B**
Bromine (Gr. *bromos*, stench), **Br**
Calcium (L. *calc*, lime), **Ca**
Californium (State of California), **Cf**
Carbon (L. *carbo*, charcoal), **C**
Chlorine (Gr. *chloros*, greenish-yellow), **Cl**
Chromium (Gr. *chroma*, color), **Cr**
Cobalt (Ger. *kobold*, goblin or evil spirit), **Co**
Copper (L. *cuprum*, from the island of Cyprus), **Cu**
Curium (Marie and Pierre Curie), **Cm**
Dysprosium (Gr. *dysprositos*, hard to get at), **Dy**
Gallium (L. *Gallia*, France), **Ga**
Germanium (L. *Germania*, Germany), **Ge**
Gold (Sanskrit *Jval;* Anglo-Saxon *gold*), **Au** (L. *aurum*, shining dawn)

▶ Figure 3.13
Some Elements Were Named for Their Properties.

(a) Orpiment (As_2S_3), a yellow ore of arsenic. (L. arsenicum, *yellow ointment)*

(b) Some colorful compounds of chromium. (Gr. chroma, *color) Clockwise from upper left: potassium chromate (K_2CrO_4), chromium(VI) oxide (CrO_3), hydrated chromium(III) chloride ($CrCl_3 \cdot 6H_2O$), ammonium dichromate ($(NH_4)_2Cr_2O_7$), and chromium(III) oxide (Cr_2O_3).*

Element (Derivation), **Symbol** (Derivation)

Helium (Gr. *Helios,* the sun), **He**

Hydrogen (Gr. *hydro,* water, and *genes,* forming), **H**

Iodine (Gr. *iodes,* violet), **I**

Iridium (L. *iris,* rainbow), **Ir**

Iron (Anglo-Saxon *iron*), **Fe** (L. *ferrum*)

Krypton (Gr. *kryptos,* hidden), **Kr**

Lithium (Gr. *lithios,* stone), **Li**

Magnesium (Magnesia, district in Thessaly), **Mg**

Manganese (L. *magnes,* magnet), **Mn**

Neon (Gr. *neos,* new), **Ne**

Nitrogen (L. niter forming), **N**

Osmium (Gr. *osme,* odor), **Os**

Oxygen (Gr. *oxys,* acid, and *genes,* forming, acid former), **O**

Phosphorus (Gr. *phosphoros,* light bearing), **P**

Platinum (Sp. *platina,* little silver), **Pt**

Potassium (English *potash*), **K** (L. *kalium*)

Radium (L. *radius,* ray), **Ra**

Rhodium (Gr. *rhodon,* rose), **Rh**

Ruthenium (Ruthenia, Ukraine), **Ru**

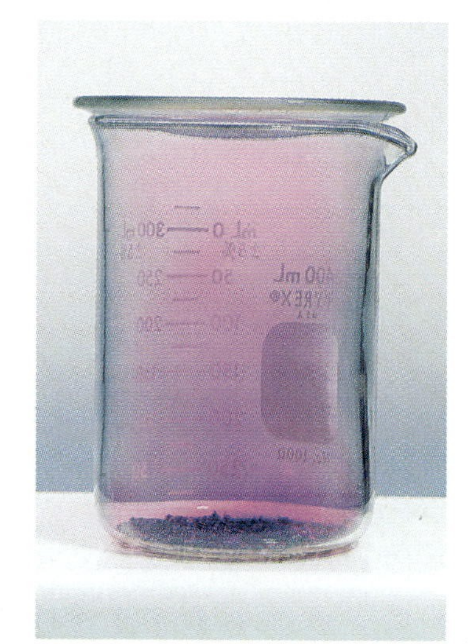

(c) Violet crystals and vapor of iodine. (Gr. iodes, *violet)*

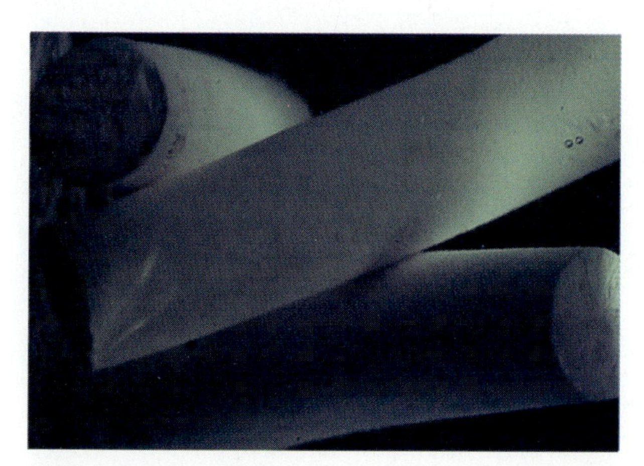

(d) White phosphorus, glowing when exposed to air at room temperature. (Gr. phosphoros, *light bearing)*

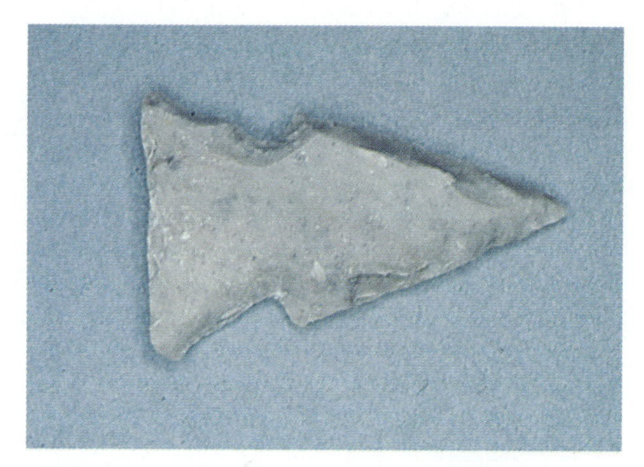

(e) Flint, a compound of silicon and oxygen. (L. silex, *flint)*

TABLE 3.1 DERIVATIONS OF THE NAMES AND SYMBOLS OF SELECTED ELEMENTS (Continued)

Element (Derivation), **Symbol** (Derivation)

Samarium (Samarski, a Russian), **Sm**

Scandium (Scandinavia), **Sc**

Selenium (Gr. *Selene,* moon), **Se**

Silicon (L. *silex,* flint), **Si**

Silver (Anglo-Saxon *soelfor*), **Ag** (L. *argentum*)

Sodium (English *soda*), **Na** (L. *natrium*)

Strontium (Strontian, a town in Scotland), **Sr**

Thorium (*Thor,* Scandinavian god of war), **Th**

Tin (Anglo-Saxon *tin*), **Sn** (L. *stannum*)

Titanium (L. *Titans,* the first sons of the Earth, myth.), **Ti**

Tungsten (Sw. *tungsten,* heavy stone), **W** (Ger. *Wolfram*)

Uranium (Planet Uranus), **U**

Vanadium (Scandinavian goddess, *Vanadis*), **V**

Xenon (Gr. *xenon,* stranger), **Xe**

Zinc (Ger. *zink*), **Zn**

Zirconium (Arabic *zargun,* gold color), **Zr**

Source: *Handbook of Chemistry and Physics* 39th Edition, Chemical Rubber Publishing Co., Cleveland, OH, 1957.

DID YOU LEARN THIS?

For each statement, fill in the blank with the most appropriate word or phrase from the list. Do not use any word or phrase more than once.

elements	liquid	mixture
compounds	gas	homogeneous
energy	chemical change	heterogeneous
matter	physical change	boiling point
mass	exothermic	melting point
weight	endothermic	sublimation
solid	pure substance	refining

a. _______ Materials that do not have a discrete set of properties
b. _______ The part of the makeup of the world that makes things move
c. _______ Substances formed from the chemical combination of two or more elements
d. _______ The kind of process that occurs when salt dissolves in water
e. _______ The resistance to being moved by a force or push
f. _______ Describes a process in which heat is absorbed
g. _______ A property of matter that depends on the attraction of gravity for an object
h. _______ Expands completely to fill its container
i. _______ The temperature at which a solid, when heated, begins to form the liquid phase
j. _______ Has a definite volume but no definite shape
k. _______ The process by which a solid passes into the gaseous state, bypassing the liquid state
l. _______ A descriptive term that applies to new motor oil but not to used motor oil

EXERCISES

1. What differentiates chemistry operationally from the other natural science disciplines?
2. Distinguish between matter and energy. The definition of matter is easy to visualize and understand. What do you think it is about the nature of energy that does not lend itself to an easily visualized definition?

3. Classify the following as matter or energy.

   a. lightning
   b. salt
   c. iron metal
   d. microwaves
   e. tidal motion
   f. an egg
   g. sunlight
   h. polyethylene plastic
   i. electricity
   j. seawater

4. Distinguish between mass and weight. Explain how an object can be "weightless" in outer space and yet have mass.
5. Define the three common states of matter. What is the usual state of matter of each of the following substances?

   a. gasoline
   b. a bubble in boiling water
   c. a human hair
   d. butter
   e. copper wire
   f. helium in a balloon
   g. a snowflake
   h. soot
   i. diamond
   j. the ozone layer
   k. corn syrup

6. What are the common changes of state? What names are given to the temperatures at which these processes occur? What is the relationship between melting and freezing points for pure substances? Of boiling and condensation points? What is sublimation? Give examples from your everyday experience.
7. Freezing compartments in refrigerators and freezers often need to be defrosted because of a buildup of ice and frost. Where does this frost come from and by what process is it formed?
8. Which of the following classes of matter—pure substances, homogeneous mixtures, and heterogeneous mixtures—fit the following criteria?

   a. fixed composition
   b. separable into two or more simpler substances by physical means
   c. unique, nonvarying properties
   d. two or more substances distinguishable from one another
   e. variable properties

9. Which of the following can be classified as homogeneous and which as heterogeneous? Which might be either, depending on magnification? Which are pure substances?

   a. vinegar
   b. mayonnaise
   c. soda water
   d. varnish
   e. potato chips
   f. ice

g. hot chocolate
h. dry ice
i. tea with sugar
j. aluminum foil
k. blood
l. hamburger
m. cold tablet
n. ice cream
o. air
p. wood ashes

10. Tell whether each of the following is an example of a chemical property or a physical property of matter.
    a. Graphite is a good conductor of electricity.
    b. Mercury is a liquid at room temperature.
    c. Silver metal turns black when exposed to polluted air.
    d. Ammonium carbonate is soluble in water.
    e. Bronze statues turn green when allowed to stand outdoors for a period of time.
    f. Chlorine is a pale green gas.
    g. Tungsten carbide is one of the hardest substances known.
    h. Sodium chloride crystals are brittle and easily broken when pressure is applied to them.
    i. When exposed to the air, aluminum forms a coating that protects the metal underneath.
    j. Copper metal is easily drawn into wire and pounded into various shapes.

11. Indicate which of the following is a chemical change and which is a physical change.
    a. Spray paint is expelled from its container.
    b. A slice of bread is toasted.
    c. The aroma of spices comes from the spice cabinet when the door is opened.
    d. A turkey is carved.
    e. Varnish hardens.
    f. Sugar dissolves in iced tea.
    g. Gasoline is burned in an automobile engine.
    h. Frost is formed.
    i. A child grows into an adult.
    j. Potassium metal violently produces hydrogen gas when placed in water.

12. Iso-octane, a component of gasoline, combines with oxygen in air to form two compounds, carbon dioxide and water. Is iso-octane an element or a compound? Iso-octane is a colorless liquid that has a boiling point of 99.3 °C, a freezing point of –107.4 °C, and a density of 0.6918 g/mL. Can you tell whether it is a pure substance from the information given?

13. Which of the following are pure substances and which are probably mixtures? (Look at all of the examples before deciding.)
    a. Halite [density 2.16 g/mL; hardness 2.5 (Moh's scale, talc = 1); white crystals; index of refraction (how much the substance bends light) 1.544]
    b. Camphor [melting point 179.8 °C; sublimes at 204 °C (sea level pressure); density 0.990 g/mL; colorless crystals; index of refraction (how much the substance bends light) 1.5462]
    c. Lepidolite [density 2.80–2.90 g/mL; hardness 2.5–4 (Moh's scale, talc = 1); pale pink to pale purple solid; index of refraction (how much the substance bends light) 1.525–1.548]
14. Which are the noble metals? Which are the noble gases? What do you suppose is the basis for classifying them in this matter?
15. What is the reason that few of the elements are found uncombined in nature? What must we do to obtain most of the elements from their ores?
16. Although heat is a form of energy and not a substance (matter), it can be considered to be a reactant or a product in chemical reactions. What is the reason for doing this?
17. How is a particular energy state determined to be more stable than another? Which is more stable, a man on the roof of a house or a man on the ground? Can you tell for sure? Why or why not? What must be known before stability of a state can be determined?
18. Under what conditions does a process tend to be spontaneous? Does this mean that the process actually must proceed? Give some examples of processes that tend to be spontaneous but require "assistance" to get underway? Under what conditions does a process tend to be nonspontaneous? Does this mean that a process of this kind never occurs? If not, what is required to make it proceed? From your own experience, give some examples of processes that tend not to be spontaneous.
19. Supply the correctly spelled name of each of the elements whose symbols are given.
    a. Ne ________
    b. K ________
    c. P ________
    d. Mg ________
    e. F ________
    f. Ag ________
    g. Be ________
20. Supply the correct chemical symbol for each of the elements that follow.
    a. aluminum ____
    b. sodium ____
    c. silicon ____
    d. mercury ____

e. iron ____
f. lead ____
g. bromine ____
h. sulfur ____
i. gold ____
j. boron ____
k. tin ____

THINKING IT THROUGH

Comparing the Sciences

Compare the science requirements for the baccalaureate degrees in the various natural and physical sciences—physics, biochemistry, microbiology, ecology, geology, botany, chemistry, etc.—in your college/university catalog. What kinds of science courses outside the specific discipline does each science program require? For example, what kinds of physics, chemistry, and biology courses are geology majors required to take? Compare these "outside" courses to those required, say, of ecology majors. Continue this comparison for a number of sciences. Do you see any trends or patterns in the kinds and numbers of these required courses in relationship to the spectrum-of-science disciplines presented in the first part of this chapter? What are they? In a similar manner, examine the mathematics requirements for each of the physical and natural science programs. What kind of trend do you see that reveals the role of mathematics in the various science disciplines?

ANSWERS TO "DID YOU LEARN THIS?"

a. mixture
b. energy
c. compounds
d. physical change
e. mass
f. endothermic
g. weight
h. gas
i. melting point
j. liquid
k. sublimation
l. homogeneous

A JOURNALIST FINDS CHEMISTRY ESSENTIAL

Mary Garvey Verrill

Journalist and Writer

B. A. English, Viterbo College

M. A. English, University of Wisconsin—Madison

As a journalist, I owe a lot to my high school and college chemistry teachers. They not only shared my strong belief in the power of words but also gave me an awareness of the power of physical matter. I learned many lessons in my chemistry classes, all of which still have real-world applications.

A full chemistry course in high school under the stern guidance of Sr. Maxine Zimmer (airplane pilot, violinist, and former wrestler) taught me a lesson in personal responsibility. Despite her warnings, I was one of those students who, in the final exam, decided to "test" for a salt by tasting the unknown substance on my lab table. Luckily, I suffered only a mild asthma attack, brief hallucinations, and a wild jump in temperature and heart rate. Never try this—it can be dangerous even if your uneducated guess proves correct. Thus, I learned my first lesson: You cannot cheat when using the primary materials of the universe, for this is serious stuff.

Although I was an unlikely candidate for a college chemistry course, it was required for a bachelor of arts degree. My professor, Ron Amel, began by explaining that *potential* has nothing to do with the ability to learn. In fact, my second lesson was that words take on a whole new meaning in chemistry: all *free radicals* did not go to the University of California at Berkeley; *models* did more than appear in magazines; *formula* was not just for babies to drink; *pi* was not apple or cherry; *state* did not mean Wisconsin; *nickel* was not a coin; *mole* was not an animal; *concentration* did not mean hard thinking; an *agent* was not always secret; an *orbital* did not reside in outer space; *metal* was not our favorite music; *mass* had nothing to do with Sunday; and *absolute zero* was not my blind date from last Friday night. A whole new world was out there, full of protons, neutrons, and electrons bounding around in frantic chaos, and I was expected to put them all in order. At first, I was so overwhelmed that I wanted to quit.

At that point, Professor Amel stepped in. He realized that a quiz at 4 P.M. every Friday afternoon did not give us the chance to show off our knowledge (what little we had on Friday afternoons). He also realized that arts majors in particular needed help. So once a week, he began to hold evening seminars that we could attend on a voluntary basis. The whole class, science and arts majors alike, turned out for every seminar. He did not let us out the door until we all understood every bit of that mathematics. I had received my third lesson: there is no subjective favoritism in chemistry, for we are all equally vulnerable; as humans, we are all in this together.

Laboratory work drove home the reality of physical matter. I held compounds, then altered, froze, and burned them as master of my test-tube universe. The test tube was fragile, and sometimes the experiments bore no resemblance to what should be happening. But as I measured, weighed, and controlled the powders and liquids, chemicals were no longer abstractions that only lived on Three Mile Island. They were everywhere—in every food item and consumer product that I used every day. My fourth lesson: humans are not helpless victims of their environment; they can learn about it, change it, clean it up, and refuse to comply with the short-term schemes of those who would ruin it for them.

When I graduated from college, I was surprised that I used chemistry on a daily basis and even more surprised that it actually paid the rent. With my knowledge of dry cells and simple electronic circuits, I took a short course in electronics and passed the examination for my Federal Communications Commission Third Class Radio Operators License (no longer required) to become an FM radio rock music disk jockey/air personality. I had fulfilled a lifelong dream, but I would not have accomplished it without that college chemistry course.

When I set out to work toward my next goal of obtaining a master's degree in English, I found that my undergraduate work satisfied all my science requirements. Within two semesters, I had obtained my degree, and once again chemistry followed me into the job market. Shortly after graduation, I became a technical writer for a corporation that manufactures liquid handlers for use in laboratories and chromatography—and I knew exactly what I was writing about. Here I began to use a foolproof journalistic method for explaining complex information to those who are not familiar with the subject: compile your data; choose what is important in the data and analyze the past, present, and future implications; interview all parties involved in the subject at hand; define all terms; and present all information clearly and concisely so that anyone can understand it. In other words, I had developed a scientific method in my reporting or, at least, a method based on objectivity. My last lesson: if you know the subject, you can pass it on, and if you don't know something, you can find out about it.

Since that time I have worked at several jobs and written many articles ranging from studies of digital recording to theater reviews. But ongoing environmental problems are my special interest, and I have begun to incorporate these issues into my fiction writing. I have won local awards for a series of articles on a new garbage incinerator located in downtown Minneapolis; the contamination of the Mississippi River; and lead contamination in children in the metropolitan area. One topic leads to another, and anyone who is similarly motivated can find seemingly endless problems waiting for investigative reporting. I am not writing for the awards, the money, or the fun, nor do I see great changes as a result of my work. Yet, I believe in the written word and that (1) you cannot cheat with the primal materials of the universe; (2) you must command the language of chemistry when dealing with chemical subjects; (3) we are all in this together; (4) we are not helpless victims of our environment; and (5) if you know something, you can pass it on.

My next concern is in my own neighborhood: for many years, an architectural metals manufacturing has been emitting 1,1,1-trichloroethane and methyl isobutyl ketone (MIBK) from a low stack within two blocks of an elementary school and community playground, without a permit from the Minnesota Pollution Control Agency. I can publicize and inform, but the chemists will have to do the rest to rectify the situation. As they search for alternatives and solutions, no doubt they will find their chemistry courses as essential as mine have proved to be.

4

Scientific Models: The Finer Structure of Matter

What distinguished Mendeleev was not only genius, but a passion for the elements. They became his friends; he knew every quirk and detail of their behaviour.

Jacob Bronowski, "The Ascent of Man"

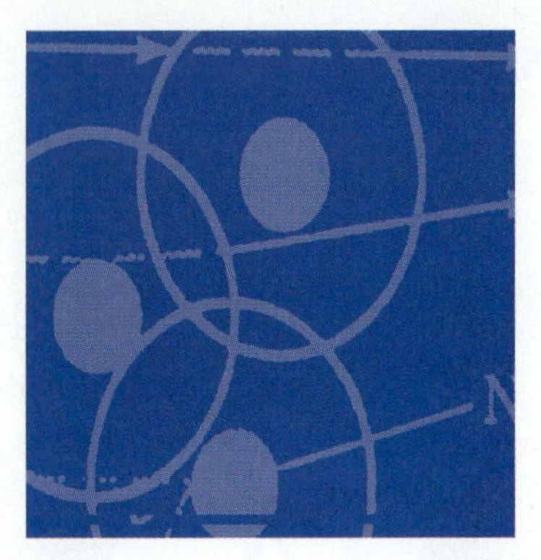

To understand the properties of the elements we need to have a theory of what their finer structure is like. Almost all scientific theories are based on models of some kind, and constructing models (or *modeling,* as it is sometimes called) is an important part of doing science.

KINDS OF SCIENTIFIC MODELS

Several kinds of models are used in formulating scientific theories. Everyone is familiar with model trains and, better for our purposes, models of bridges and airplanes. We make these **mechanical,** or *physical,* **models** to test the properties of a design (say, for the ability to withstand load or fly in a wind tunnel) without the expense or risk of building the real thing. Chemical pilot plants are built to test scale-up of chemical reactions that have been discovered in the laboratory before constructing large-scale production facilities. The pilot plants serve as mechanical models of intermediate size to work out the many technical problems associated with converting laboratory reactions in gram-sized quantities to commercial production in metric tons.

Mathematical models, which use equations or other mathematical concepts to describe processes or relationships, are also used extensively in science. The behavior of electricity, the pressure in a balloon, and the path of a rocket in space can all be described (and predicted) by mathematical models. The shapes of crystals, such as snowflakes and sugar cubes, are geometric forms (► Figure 4.1).

Conceptual models describe things that cannot be observed directly.

Conceptual, or *mental,* **models** are used to describe things that are too large, like the universe, or too small, like a bacterium, to be observed directly or entirely. Because we have no idea what the finer structure of matter actually looks like, we must envision things in our everyday experience that have attributes similar to those we observe in the laboratory and then, by analogy and inference, construct a model of what this structure might be like. Our model may not represent reality at all. It is unlikely that we shall ever know with certainty what this tiny world looks like, so we shall never be able to compare it directly with our model, side by side.

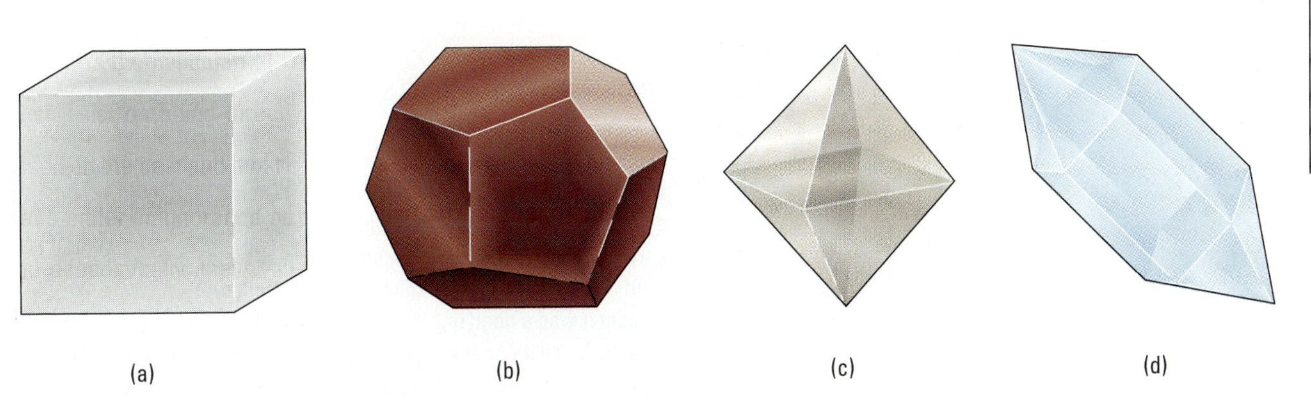

▶ Figure 4.1
Geometric Form. *Crystals of minerals occur in a variety of geometric forms. (a) Ordinary salt (halite) and pyrite (fool's gold) have cubic crystals. (b) Garnet, used in sandpaper and as a gemstone, forms crystals having the shape of a dodecahedron, a figure with twelve sides. (c) Diamond has octahedral, or 8-sided, crystals. (d) Quartz forms prisms that have ends that are pyramids.*

Often it is useful to make **pictorial models** of very large, very small, or complex things so that we can visualize them better. Today, computers enable us to draw pictorial models in three dimensions, change their perspective, rotate them in space, modify their structures, and even put them in motion. For example, the behavior of enzymes in biochemical systems and the dynamics of the wetting of silicone surfaces by water are modeled in this manner. Computer modeling also is applied to fields such as architecture, landscaping, municipal planning, criminal investigation, aircraft pilot training, and hair styling. With computer-aided design (CAD) you can visualize how your new house will look on your chosen site before you build it or where a maple tree will look best in the front yard. If you want to know how you would look with a new hair style, you can try it on before you cut!

THE DEVELOPMENT OF THE IDEA OF ATOMS

The idea of atoms is a model, a conceptual model of things we cannot see. It is also an old one, dating back to Leucippus and Democritus of Greece in the fifth century B.C. According to Democritus, all atoms were composed of the same substance but drew their properties from having many differences in size and shape. They were also thought to be indivisible and everlasting. Later, the Roman Lucretius also argued that matter was atomic, but a more widely held view at the time was that matter was continuous (infinitely divisible). The Greeks and Romans seldom carried out experiments to test their ideas, however, and probably could not have done so in this instance. In fact, it was not until the late eighteenth century that experimental evidence for the model of atoms was obtained.

Once it was realized that elements combine chemically to form compounds, chemical techniques could be used to determine which elements a substance contained. Water is made up of the elements hydrogen and oxygen; baking soda of sodium, hydrogen, carbon, and oxygen; and pyrite (fool's gold) of iron and sulfur. This important chemical evidence results from a kind

THREE EVERYDAY MODELS

His mother picked up a cantelope from the kitchen counter, smelled it intently for a few seconds, and remarked, "Ahhh! That's going to be a good one!" The author had purchased the melon earlier—after smelling it, thumping it, squeezing it, looking at the stem end, shaking it, and checking its skin for color and the minor soft spots indicative of ripeness. He had thought it was a good one, too. It turned out to be so-so.

Geologists are exploring for oil along the Wyoming–Idaho border in a geological region called the Overthrust Belt (Figure 1). The eastern part of this area, in western Wyoming and northeastern Utah, has a number of oil fields and coal deposits. Here, geologists have a good model of the geological structures in which these deposits are found, and have been able to exploit this knowledge to locate new sources of coal and oil. The western part of the Overthrust Belt, along the eastern Idaho border, is different. It is a region in which disruptive geological events, such as recent volcanic activity, have taken place. Although this area has been studied extensively, the geological relationships are so complex that it has been hard to piece together a model that predicts where usable amounts of petroleum might be found. A number of exploratory wells have been drilled to find oil, but so far no large oil reservoirs have been discovered.

An outfitter and his guides were camped in preparation for a float trip on the Owyhee River in a remote area of southeastern Oregon; it had been a strenuous day. After supper the outfitter, a man in his early 40s, complained of chest pains and difficulty with breathing. The symptoms persisted, so his guides decided to drive him that night to the nearest hospital 135 miles away. There, doctors examined him, took blood samples, gave him an electrocardiogram, and placed him under observation. The tests were completed in a short time, and the doctors made their diagnosis (a model of his condition): his heart was fine, but he had a hiatal hernia that was acting up. (A hiatal hernia is an opening in the muscle of the diaphragm that allows the intestines to enter the lung cavity.) It was not serious and did not need further attention, but he was told not to eat hot chili in camp before bedtime.

How does the use of models affect you?

Figure 1
Overthrust Belt.
Series of anticlines and synclines in the Absaroka Thrust Plate, Idaho-Wyoming Thrust Belt, part of the Overthrust Belt southwest of Geneva, Idaho. Geological relationships are so complex in this part of the Overthrust Belt that making a model to help locate petroleum reservoirs has been difficult.

of analysis called **qualitative analysis,** which answers the question, "What is it made of?"

It also was necessary to accurately measure the amount of each element in a compound. Pyrite contains 46.5% iron and 53.5% sulfur by mass. Triolite, another compound of iron and sulfur, contains 63.5% iron and 36.5% sulfur. These kinds of data are obtained by **quantitative analysis,** which answers, "How much does it contain?" The analytical balance is probably the most important instrument for making these measurements in the chemical laboratory (▶ Figure 4.2). Until the late 1700s, however, balances lacked the precision to measure the small differences in mass required to do quantitative chemistry. Using sensitive analytical balances of his own design, which were capable of a precision of 0.5 milligrams (mg), Antione Lavoisier (1743–1794) was the first to make extensive quantitative measurements of chemical reactions.

The analytical balance is the most important instrument for doing **quantitative analysis.**

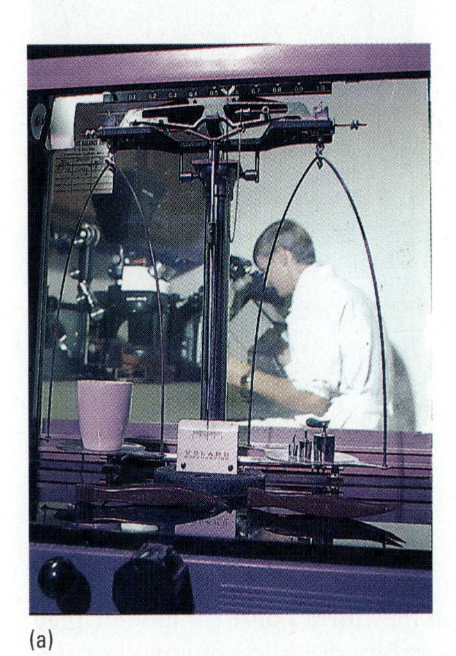

(a)

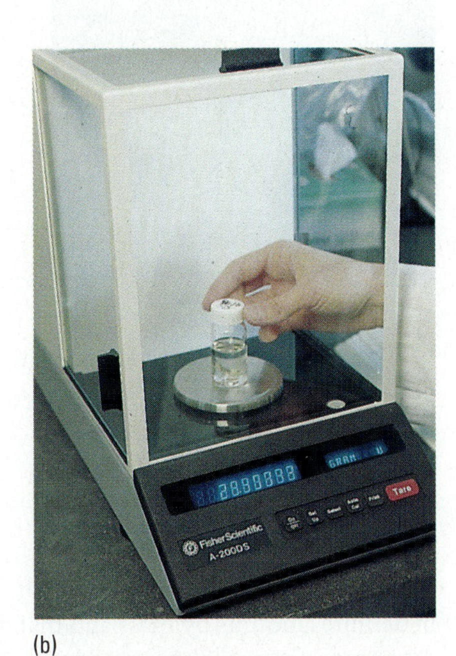

(b)

▶ Figure 4.2
The Analytical Balance. *(a) Two-pan balances of this kind were widely used in chemical laboratories before the development of electronic balances. With this two-pan balance, the mass of the sample is exactly balanced with a known standard mass to cancel out the effect of gravity. (b) With a modern analytical balance, the mass of the sample is balanced electronically.*

THE LAW OF CONSERVATION OF MASS

Lavoisier found that when a chemical reaction was carried out in a closed system, such as a sealed jar or flask, the mass of the materials inside did not change. That is, the total mass of the products of a reaction was equal to the total mass of the starting materials (reactants) before the reaction occurred. Lavoisier repeated such experiments many times, with the same results, and confirmed that matter is neither created nor destroyed in a chemical reaction. This statement is now called the **law of conservation of mass.**

Matter is neither created nor destroyed in a chemical reaction. This is the **law of conservation of mass.**

Lavoisier was not the first to set forth a principle of conservation of matter. According to science historian Frederic L. Holmes, Yale University School of Medicine, the idea of a "Conservation Law" had been used in chemical practice since the early seventeenth century.

For example, Lavoisier would weigh a small amount of the red oxide of mercury, a solid that decomposes (comes apart) when it is heated, becoming liquid mercury metal and oxygen gas. When heating the red oxide, he would carefully recover all of the mercury, trap the oxygen gas in a suitable container, and weigh them both. If he started with, say, 1.083 grams (g) of red oxide, he would obtain 1.003 g of mercury and 0.080 g of oxygen gas, or a total of 1.083 g of products. Likewise, 4.332 g of oxide produced 4.012 g of mercury and 0.320 g of oxygen, within experimental error (▸ Figure 4.3).

(a)

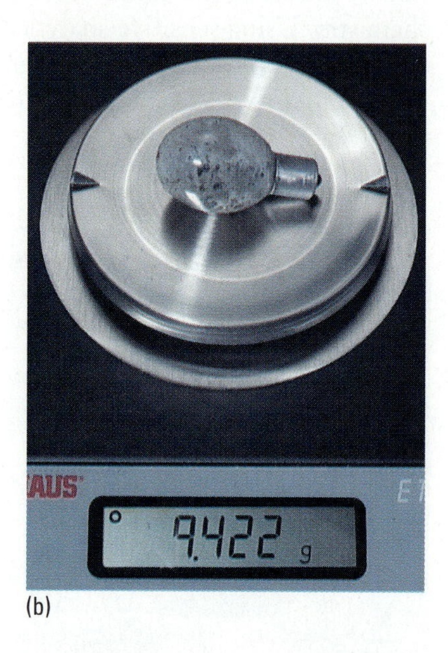

(b)

▸ Figure 4.3
Demonstration of the Law of Conservation of Mass. *Setting off a flash bulb demonstrates this law. (a) The unused bulb contains magnesium and oxygen. (b) The magnesium and oxygen have reacted and the used bulb contains magnesium oxide, but the mass of the flash bulb has not changed.*

THE LAW OF DEFINITE PROPORTIONS

In 1799 Joseph Proust, also a French chemist, showed that elements combine in definite amounts by mass and that when they form a pure compound, they are always present in the same fixed proportions, regardless of how the compound is made. For example, 12 g of carbon always combines with 32 g of oxygen to produce 44 g of carbon dioxide (▸ Figure 4.4a). If 12 g of carbon reacts with 40 g of oxygen, 44 g of carbon dioxide is formed and 8 g of oxygen is left over (▸ Figure 4.4b). When carbon dioxide is chemically taken apart, or decomposed into its elements, each 44 g of carbon dioxide always yields 12 g of carbon and 32 g of oxygen (▸ Figure 4.4c).

Note that in the three reactions shown (see figure), the total of the masses of the reactants exactly equals the total of the masses of the products: (a) 12 g of carbon plus 32 g of oxygen makes 44 g of reactants, exactly the amount of carbon dioxide produced; (b) 12 g of carbon plus 40 g of oxygen makes 52 g

(a) Carbon + Oxygen ⟶ Carbon Dioxide
12 g 32 g → 44 g
44 g Total Mass

(b) Carbon + Oxygen ⟶ Carbon Dioxide + Oxygen (unreacted)
12 g 40 g → 44 g 8 g
52 g Total Mass → 52 g Total Mass

(c) Carbon Dioxide ⟶ Carbon + Oxygen
44 g → 12 g 32 g
44 g Total Mass

▶ Figure 4.4
Combining Masses of Carbon and Oxygen Illustrate the Law of Definite Proportions. *(a, b) Carbon and oxygen combine in definite proportions by mass. (c) Carbon dioxide contains fixed amounts of carbon and oxygen. Mass is conserved in every reaction.*

of reactants, exactly the mass of the products (44 g of carbon dioxide plus 8 g of oxygen); and (c) the decomposition of 44 g of carbon dioxide gives a total of 44 g of products (12 g of carbon and 32 g of oxygen)—all in accordance with the law of conservation of mass. The total of the masses of all of the products is equal to the total of the masses of all of the reactants.

In a similar manner, 2 g of hydrogen combines with 16 g of oxygen to form 18 g of water (▶ Figure 4.5a). Combining 2 g of hydrogen with 20 g of oxygen also produces 18 g of water, but leaves 4 g of oxygen unreacted (▶ Figure 4.5b). When 2 g of hydrogen combines with 8 g of oxygen, only 1 g of the hydrogen reacts, producing 9 g of water and leaving 1 g of unreacted hydrogen (▶ Figure 4.5c). When water is decomposed to hydrogen and oxygen, 18 g of water gives 2 g of hydrogen and 16 g of oxygen; 1.8 g of water produces 0.2 g of hydrogen and 1.6 g of oxygen; and so forth (▶ Figures 4.5d, 4.6).

(a) Hydrogen + Oxygen ⟶ Water
2 g 16 g → 18 g
18 g Total Mass

(b) Hydrogen + Oxygen ⟶ Water + Oxygen (unreacted)
2 g 20 g → 18 g 4 g
22 g Total Mass → 22 g Total Mass

(c) Hydrogen + Oxygen ⟶ Water + Hydrogen (unreacted)
2 g 8 g → 9 g 1 g
10 g Total Mass → 10 g Total Mass

(d) Water ⟶ Hydrogen + Oxygen
18 g → 2 g 16 g
1.8 g Total Mass
1.8 g → 0.2 g 1.6 g
18 g Total Mass

▶ Figure 4.5
Combining Masses of Hydrogen and Oxygen also Illustrate the Law of Definite Proportions. *(a, b, c) Hydrogen and oxygen also combine in fixed proportions by mass. (d) Water contains a definite ratio of hydrogen and oxygen by mass. Mass is again conserved.*

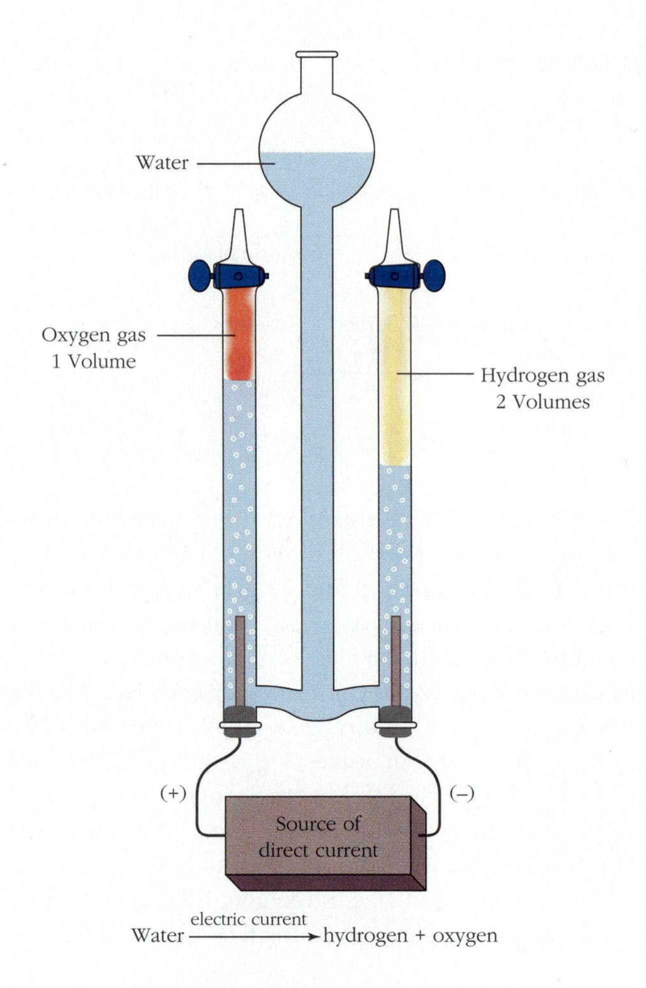

▶ Figure 4.6
The Decomposition of Water Produces Hydrogen and Oxygen in a Fixed Ratio. *Decomposition of water with an electric current yields two volumes of hydrogen gas for every volume of oxygen gas. The mass ratio of hydrogen to oxygen produced is 1:8.*

The combining masses of carbon with oxygen and of hydrogen with oxygen, and the decomposition of carbon dioxide and water into their respective elements, provided the kind of evidence that led Proust to formulate his **law of definite proportions,** often called the *law of constant composition.* In every pure compound the constitutent elements are present in definite (constant) proportions by mass; and as a result, compounds have fixed formulas.

The elements in a pure compound are present in fixed proportions by mass. This is the **law of definite proportions.**

THE LAW OF MULTIPLE PROPORTIONS

While attempting to develop a theory that could account for all these quantitative data, John Dalton, an English schoolteacher, discovered another law. Some elements such as carbon and oxygen combine in more than one ratio to form two or more compounds. For example, when carbon reacts with a limited supply of oxygen, a compound having a carbon-to-oxygen mass ratio of about 3:4 is formed (▶ Figure 4.7a). But when carbon reacts with an abundance of oxygen, a different compound, having a carbon-to-oxygen mass ratio of about 3:8, is formed instead (▶ Figure 4.7b). Thus, there is twice as much oxygen per gram of carbon in the second compound as in the first, or a whole-number ratio of 2:1.

Similarly, sulfur can form two compounds with oxygen, one having a sulfur-to-oxygen mass ratio of about 1:1 (▶ Figure 4.7c) and another having a mass ratio of about 2:3 (▶ Figure 4.7d). There is 1.5 times more oxygen per gram of sulfur in the second compound than in the first, a whole-number ratio of 3:2. In 1803 Dalton proposed the **law of multiple proportions,** which states that if two elements combine to produce two or more compounds, the masses of one element that can combine with a fixed mass of the other can be reduced to a ratio of small whole numbers, such as 2:1, 3:1, and 2:3.

(a) Carbon + Oxygen ⟶ Carbon Monoxide
12 g Limited amount 28 g
(16 g of oxygen in compound)

Mass ratio of C to O is 12 g:16 g, or 3:4.

(b) Carbon + Oxygen ⟶ Carbon Dioxide
12 g Excess amount 44 g
(32 g of oxygen in compound)

Mass ratio of C to O is 12 g:32 g, or 3:8.

(c) Sulfur + Oxygen ⟶ Sulfur Dioxide
32 g Limited amount 64 g
(32 g of oxygen in compound)

Mass ratio of S to O is 32 g:32 g, or 1:1.

(d) Sulfur + Oxygen ⟶ Sulfur Trioxide
32 g Excess amount 80 g
(48 g of oxygen in compound)

Mass ratio of S to O is 32 g:48 g, or 2:3.

▶ Figure 4.7
Dalton's Law of Multiple Proportions. *Two elements sometimes can combine in different ratios with one another.*

DALTON'S ATOMIC THEORY OF MATTER

The laws of definite proportions and of multiple proportions led Dalton to propose his model for the finer structure of matter—namely, that matter must come in small pieces called **atoms,** and that they must combine in certain ratios to form compounds. Dalton published his **atomic theory** of matter in 1808. In light of our considerable knowledge about atoms today, his theory can be summarized as follows:

Matter is composed of **atoms.**

1. All matter is made up of very small building blocks called atoms (Gr. *atomos,* indivisible).

2. All atoms of an element are very similar to each other and are quite different from the atoms of all other elements.
3. Atoms of different elements combine in simple whole-number ratios to form compounds.
4. A chemical change produces different substances by joining, separating, or rearranging atoms.
5. Atoms cannot be created, destroyed, broken apart, or changed to another kind of atom in a chemical reaction.

Initially, acceptance of Dalton's atomic model of the structure of matter was slow, but no one could disprove it and no one could find compelling evidence that matter is continuous. By the mid-1800s his atomic theory was in general use, primarily because it worked as a unified explanation for the various laws: conservation of mass, definite proportions, and multiple proportions.

With Dalton's theory, it could be understood that atoms are simply rearranged by a chemical reaction, not created or destroyed, so the overall mass does not change in the process; that is, mass is conserved. The fact that elements combine with one another in definite ratios by mass is explained by the idea that all atoms of an element have the same mass (or nearly so) and that the masses of atoms of other elements are different. Each atom of that particular element, when present in a compound, contributes a specific unit of mass to the compound. Finally, the fact that some elements combine to form two or more compounds also is easily explained: for example, a compound that contains twice as much oxygen for a fixed mass of carbon than another compound simply contains twice as many oxygen atoms per carbon atom. ▶ Figure 4.8 illustrates how the atomic theory works.

RELATIVE ATOMIC MASSES

Near the beginning of the twentieth century, it became evident that different elements also had different *relative* combining masses. In Dalton's time hydrogen, as the lightest element, was assigned a mass of about 1, then sulfur was thought to be about 16 times heavier than hydrogen, oxygen about 8 times heavier, mercury about 200 times, chlorine about 35.5 times, and so forth. With the acceptance of Dalton's atomic theory, it became apparent that these combining masses were related to the relative masses of elements' atoms (that is, their atomic masses), and that combining masses depended on two factors:

- the relative masses of the atoms of the elements in combination and
- the relative numbers of their respective atoms.

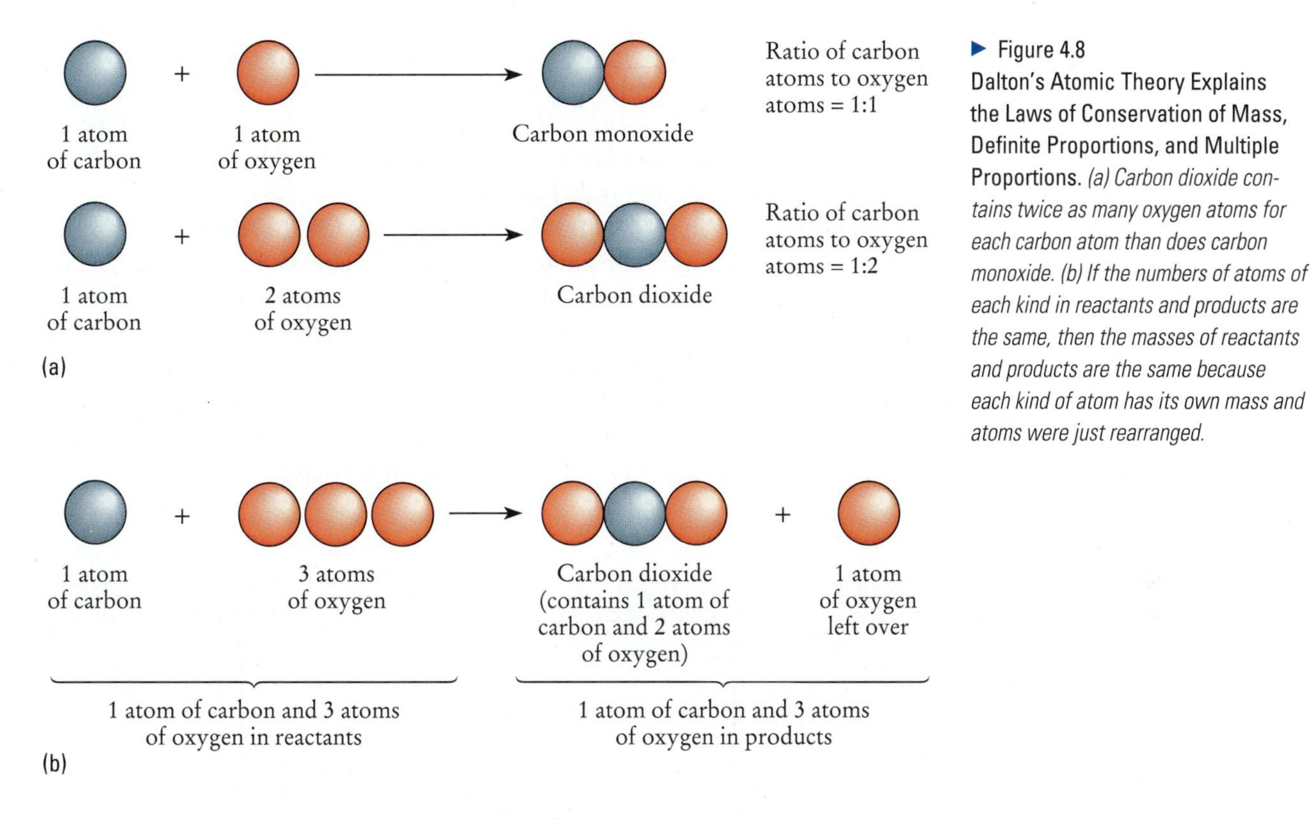

▶ **Figure 4.8**
Dalton's Atomic Theory Explains the Laws of Conservation of Mass, Definite Proportions, and Multiple Proportions. *(a) Carbon dioxide contains twice as many oxygen atoms for each carbon atom than does carbon monoxide. (b) If the numbers of atoms of each kind in reactants and products are the same, then the masses of reactants and products are the same because each kind of atom has its own mass and atoms were just rearranged.*

In water, for example, 2 g of hydrogen combines with 16 g of oxygen; but on a scale of relative masses, one oxygen atom is not 8 times heavier than a hydrogen atom (16/2). Instead, it is 16 times heavier, meaning that *two* atoms of hydrogen must have combined with one atom of oxygen to make the 16/2 combining-mass ratio. (Dalton thought that water contained one atom of hydrogen for each atom of oxygen.) Likewise, an oxygen atom is not 8/3 (32/12) times heavier than a carbon atom, but only 4/3 (16/12), because two atoms of oxygen combine with one atom of carbon to make carbon dioxide. Also, because two atoms of hydrogen combine with one atom of sulfur, the latter is actually 32 times heavier than hydrogen, not just 16 times.

FAMILY RESEMBLANCE

Certain elements resemble one another in their properties, much like persons in a family do.

Certain elements seemed to have considerable similarity to one another in their physical and chemical properties. Elemental lithium (Li), sodium (Na), and potassium (K), for example, are physically similar in that they are soft, silvery metals that are easily cut with a knife. All are chemically very reactive and react vigorously with water to produce alkaline (basic) solutions (Chapter 8) and hydrogen gas. When exposed by cutting, the shiny surfaces of these metals quickly corrode in the air; and when these elements combine with others, they form compounds that are also strikingly alike. Their compounds with chlorine (lithium chloride, sodium chloride, and potassium chloride) all have the same crystalline shape (cubes) and have a similar taste. In fact, potassium chloride is often used as a substitute for sodium chloride (table salt) for persons on a low-sodium (salt-free) diet.

Likewise, oxygen (O), sulfur (S), and selenium (Se) have common properties, as do chlorine (Cl), bromine (Br), and iodine (I) (▶ Figure 4.9). The last three elements, for example, make similar disinfectant solutions when dissolved in water. Chlorine is familiar to us as a disinfectant of municipal water supplies, swimming pools, and hot tubs; but bromine is often substituted in the last two applications. Iodine is used to sterilize water for drinking when hiking, camping, or working away from potable water sources; and an alcohol solution of iodine (a tincture) is used as an antiseptic.

▶ Figure 4.9
The Halogens: A Family of Elements. *Chlorine, bromine, and iodine belong to the family of elements called the halogens (Group 7A). Under ordinary laboratory conditions, chlorine is a pale yellow-green gas, bromine is a reddish-brown liquid that vaporizes easily, and iodine is a violet-black crystalline solid that sublimes when warmed. The halogens are typical nonmetals.*

THE PERIODIC LAW

By the middle nineteenth century, it was clear that a number of groups of elements show family resemblances. Though these relationships had been recognized by several chemists, Dmitri Mendeleev of Russia usually is credited with the first formulation of the **periodic law,** in 1869. His original law stated that the properties of the elements were periodic functions of their atomic masses. When Mendeleev arranged the elements in order of increasing atomic mass, he noticed a periodic repetition of their properties. When he arranged the elements in horizontal rows of eight elements, those with family resemblance appeared together in vertical columns. Lithium, sodium, and potassium thus made up a family, or vertical column, as did oxygen, sulfur, and selenium, and chlorine, bromine, and iodine. Mendeleev published his original arrangement of the elements, called the **Periodic Table,** in 1872 (▶ Figure 4.10).

Reihen	Gruppe I.	Gruppe II.	Gruppe III.	Gruppe IV.	Gruppe V.	Gruppe VI.	Gruppe VII.	Gruppe VIII.
	—	—	—	RH^4	RH^3	RH^2	RH	—
	R^2O	RO	R^2O^3	RO^2	R^2O^5	RO^3	R^2O^7	RO^4
1	H=1							
2	Li=7	Be=9,4	B=11	C=12	N=14	O=16	F=19	
3	Na=23	Mg=24	Al=27,3	Si=28	P=31	S=32	Cl=35,5	
4	K=39	Ca=40	—=44	Ti=48	V=51	Cr=52	Mn=55	Fe=56, Co=59, Ni=59, Cu=63.
5	(Cu=63)	Zn=65	—=68	—=72	As=75	Se=78	Br=80	
6	Rb=85	Sr=87	?Yt=88	Zr=90	Nb=94	Mo=96	—=100	Ru=104, Rh=104, Pd=106, Ag=108.
7	(Ag=108)	Cd=112	In=113	Sn=118	Sb=122	Te=125	J=127	
8	Cs=133	Ba=137	?Di=138	?Ce=140	—	—	—	— — — —
9	(—)	—	—	—	—	—	—	
10	—	—	?Er=178	?La=180	Ta=182	W=184	—	Os=195, Ir=197, Pt=198, Au=199.
11	(Au=199)	Hg=200	Tl=204	Pb=207	Bi=208	—	—	
12	—	—	—	Th=231	—	U=240	—	— — — —

MENDELEEF'S PERIODIC TABLE FROM HIS PAPER IN LIEBIG'S ANNALEN SUPP. 8, 133.

▶ Figure 4.10

Mendeleev's Periodic Table. *Mendeleev published his Periodic Table in* Annalen, *a German journal, in 1872. In German,* Gruppe *means group,* Reihen *means row, and* J *is the symbol for iodine. The letter* R *in the formulas at the head of each group represents any of the elements in that column, and the superscripts in the formulas are subscripts today. The elements are arranged in order of increasing atomic mass, which is shown after the equal signs following the symbols for the elements. In 1872 the elements with atomic masses 44, 68, 72, and 100 were unknown, and Mendeleev left blanks in the table for them.*

Lothar Meyer of Germany published a similar law in 1870, but Mendeleev is given the credit because he used his formulation of the law to predict the existence and properties of elements that were unknown at the time. He courageously left some gaps in his table where he thought there might be undiscovered elements. The missing elements should have atomic masses 68, 44, and 72. He named these *ekaaluminum, ekaboron,* and *ekasilicon,* respectively; we know them as gallium (Ga), scandium (Sc), and germanium (Ge). Already a well-known chemist outside his native Russia, Mendeleev became internationally famous when these elements were later discovered, for he also had predicted their properties with astounding accuracy! His most spectacular success was in predicting the properties of germanium, discovered in 1885 (■ Table 4.1).

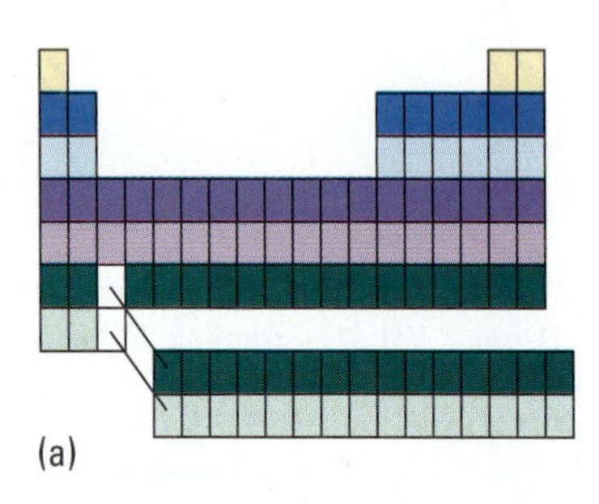

(a)

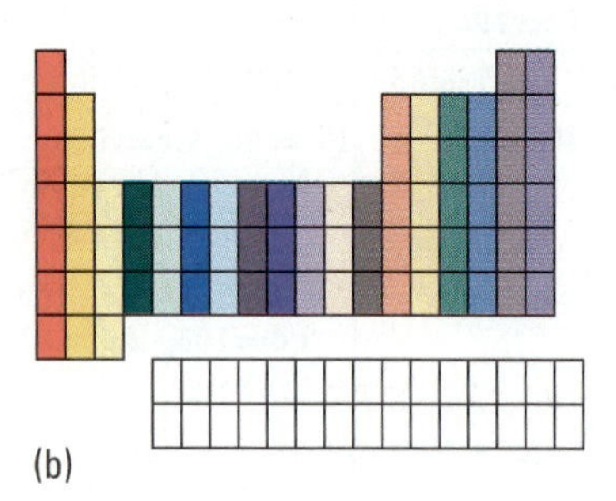

(b)

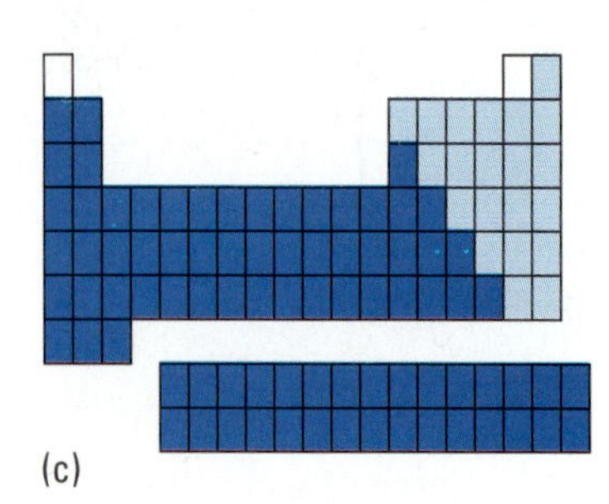

(c)

▶ Figure 4.11
Periods and Groups of Elements; Location of Metals and Nonmetals in the Periodic Table. *(a) Periods on elements. (b) Groups (families) of elements. (c) Metals (dark blue) and nonmetals (light blue).*

■ TABLE 4.1 MENDELEEV'S PREDICTIONS FOR EKASILICON AND THE MEASURED PROPERTIES OF GERMANIUM

Property of Element	Predicted for Ekasilicon (Es)	Measured for Germanium
Atomic mass	about 72	72.6
Appearance	dark gray metal	gray metal
Density	5.5 g/mL	5.36 g/mL
Melting point	"high"	958 °C
Formula of chloride	$EsCl_4$	$GeCl_4$
Boiling point of chloride	under 100 °C	86 °C
Density of chloride	1.9 g/mL	1.887 g/mL
Formula of oxide	EsO_2	GeO_2
Density of oxide	4.7 g/mL	4.703 g/mL

THE MODERN PERIODIC TABLE

This periodic repetition of the properties of the elements forms the basis for today's Periodic Table. In the modern Periodic Table (▶ Figures 4.11, 4.12) the elements are arranged in rows, called **periods,** which run from left to right across the table. Two elements, hydrogen and helium, comprise the first period. The second period contains the eight elements lithium (Li) through neon (Ne), the third starts with sodium (Na) and ends with argon (Ar), and so forth. The **groups** (or families) are arranged vertically, and the elements in each group bear striking similarities to one another in their physical and

PERIODIC TABLE OF THE ELEMENTS

GROUP NO.	1 1A	2 2A	3	4	5	6	7	8	9	10	11	12	13 3A	14 4A	15 5A	16 6A	17 7A	18 8A
PERIOD NO. 1	1.01 H 1																1.01 H 1	4.00 He 2
2	6.94 Li 3	9.01 Be 4											10.8 B 5	12.0 C 6	14.0 N 7	16.0 O 8	19.0 F 9	20.2 Ne 10
3	23.0 Na 11	24.3 Mg 12				Transition Elements (B-Group Elements)							27.0 Al 13	28.1 Si 14	31.0 P 15	32.1 S 16	35.5 Cl 17	40.0 Ar 18
4	39.1 K 19	40.1 Ca 20	45.0 Sc 21	47.9 Ti 22	50.9 V 23	52.0 Cr 24	54.9 Mn 25	55.8 Fe 26	58.9 Co 27	58.7 Ni 28	63.5 Cu 29	65.4 Zn 30	69.7 Ga 31	72.6 Ge 32	74.9 As 33	79.0 Se 34	79.9 Br 35	83.8 Kr 36
5	85.5 Rb 37	87.6 Sr 38	88.9 Y 39	91.2 Zr 40	92.9 Nb 41	95.9 Mo 42	Tc 43	101 Ru 44	103 Rh 45	106 Pd 46	108 Ag 47	112 Cd 48	115 In 49	119 Sn 50	122 Sb 51	128 Te 52	127 I 53	131 Xe 54
6	133 Cs 55	137 Ba 56	57- 71	178 Hf 72	181 Ta 73	184 W 74	186 Re 75	190 Os 76	192 Ir 77	195 Pt 78	197 Au 79	201 Hg 80	204 Tl 81	207 Pb 82	209 Bi 83	Po 84	At 85	Rn 86
7	Fr 87	Ra 88	89- 103															

Lanthanide Series	139 La 57	140 Ce 58	141 Pr 59	144 Nd 60	Pm 61	150 Sm 62	152 Eu 63	157 Gd 64	159 Tb 65	162 Dy 66	165 Ho 67	167 Er 68	169 Tm 69	173 Yb 70	175 Lu 71
Actinide Series	Ac 89	232 Th 90	Pa 91	238 U 92	Np 93	Pu 94	Am 95	Cm 96	Bk 97	Cf 98	Es 99	Fm 100	Md 101	No 102	Lr 103

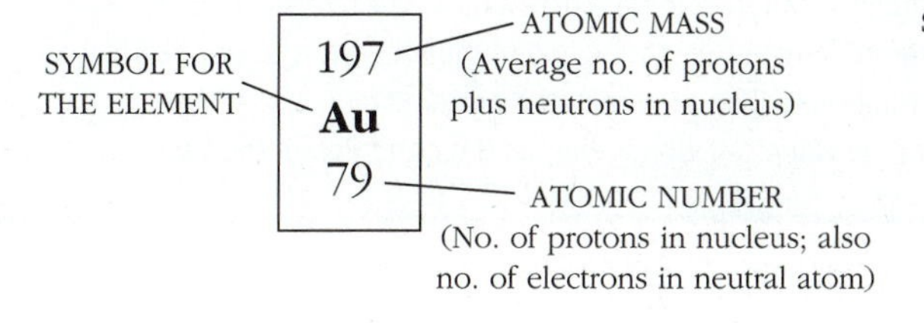

Some Family Names: Group 1A; The Alkali Metals
Group 2A; The Alkaline Earth Metals
Group 5A; The Pnicogens
Group 6A; The Chalcogens
Group 7A; The Halogens
Group 8A; The Noble Gases

Note: Elements for which no atomic mass number is given have no known stable isotopes.

▶ Figure 4.12
The Modern Periodic Table.

COINAGE

You drop coins in a vending machine for a soda. When you were little, you thought coins had been made to use in vending machines. You've always known–it seems intuitive–what metals are: they're shiny and heavy, and feel colder and seem to get hotter than other kinds of matter. Now you know why: metals are good conductors of heat. And electricity. They're also ductile, which means they don't break, which is why they're used for making coins.

The least reactive metals–gold, lead, silver, and copper–were discovered first because they're found in "native" form, as fairly pure metals. All metals are shiny when freshly cut or polished (as coins are by constant handling), and humans like shiny things and pretty rocks such as rubies and emeralds. Our first use of metals was probably as body decoration. The first gold nugget ever found was probably as valuable, relatively, to its finder as to any modern prospector: she or he wore it and fascinated the whole band, gaining instant status. And the first "money" was probably chunks of native metal, other minerals, and pearlescent shells–so prized they could be traded for salt, a valuable commodity to primal humans.

Gold, too soft for earlier tools or weapons, was useless for anything but art and coins until the modern age, when its low reactivity and good conductivity make it useful for some modern tools of our high-tech civilization, such as contact points in electronics and heat shields for instruments and people in space. And yet it has always been the most valued metal.

Gold, silver, and copper were useful as money because they were valuable and because they could be stamped with a value, date, and the likeness of a king to identify a piece of it as proper exchange in a country. Identifiable and dated coins can acquire a reputation for being trustworthy, like a truthful person, but an unidentified lump of mineral or a jewel, however valuable, doesn't make good money because there's no easy way for everyone to agree on its worth.

In ancient Greece, a drachma had to contain a drachma's worth of silver, 67.5 grains (4.38 grams). So, later in Rome, for instance, when the Emperor Hadrian put his face on a coin, he was guaranteeing that it contained a certain amount of silver or gold. The first coins were of full value but, depending on the condition of the treasury,

chemical properties. As we proceed through this course, we shall add to our knowledge of the Periodic Table. It contains a great deal of information because of the way it is structured, and once we learn to read it, we shall be able to extract its information at a glance.

Hadrian was quite likely to cheat by slightly decreasing the amount of precious metal in the coins. Coiners of monies, legitimate and otherwise, have often "debased" a country's coinage. Coins used to get a bit ragged around the edges, too, because people "made change" by clipping bits off them, two bits or four bits. You've heard the expression.

But there's a way of finding out if a metal is debased. A Greek scientist, Archimedes, was asked to determine if a crown given to the king was of pure gold, as represented, or gold debased with silver. Archimedes was stymied by the problem until he hopped into his bath one day and overflowed it. He was so excited by the insight this sloshing gave him that he forgot his clothes and ran naked to the king shouting "Eureka," which meant "I have found it." The point of the story is that the density–the ratio of weight to volume of a material–differs among metals: gold is more dense than silver. So, put a crown you want to test in a tub of water and mark the water level, then put a lump of pure gold of the same weight in the tub–and you can probably figure out the rest.

Gold is so soft that it wears away and loses value. People used to check the legitimacy of a coin by biting it to see if it had gold's softness. But the main problem with putting a dollar's worth of any metal into a dollar is that the values of metals change. If the metal in a dollar gets to be worth more than a dollar, then dollars are melted down and sold for the value of the metal. The coin of the realm disappears, at great expense to the government. (It's illegal to deface American coins, but, as you might expect, that law didn't stop the melting of coins when the price of silver or gold went up.)

As more was learned about economics and as business became more global during the nineteenth century, governments gradually had to adopt the radical idea of being honest. No king or government could benefit by debasing a currency, and the value of a country's monies was based on the strength of its economy. If a government is dependably willing to buy its own money back at face value, then coins don't have to contain their value in precious metal, and even a plug nickel is worth whatever the government says it's worth.

There's been no gold in American coins since 1934 or silver since 1971, and our pennies are minted from zinc discs with a thin plating of copper. Our other coins are made of an alloy of 75% copper and 25% nickel. They still shine. They *have* to shine!

Though harder and more durable than gold, silver, or copper, nickel itself isn't as hard or durable as steel. So, why don't they make steel coins?

As an example, one prominent feature of the Periodic Table is that the metallic elements, which comprise about 80% of all of the elements, are grouped roughly in the left three-fourths of the chart. The nonmetallic elements are grouped to the upper right of an imaginary line roughly drawn on

Metals make up about 80% of all the elements.

a diagonal from boron (B) through astatine (At). On some printings of the Periodic Table a dark, zig-zag line, called the **Zintl border,** appears here (Figure 4.12). Note that the form of the Periodic Table seems to reflect some inner structure in nature. Indeed, we shall discover later that the architecture of the atom itself is mirrored here.

METALS AND NONMETALS

Because the Periodic Table has two rough categories of elements, metals and nonmetals, we need to look further into the properties that characterize each category. We can recognize a **metal** almost immediately when we see it. The property of *metallic luster,* or metallic reflectivity, is something we simply seem to "know" without having to define it (▶ Figure 4.13). Other properties of metals are *malleability* (they can be hammered into various shapes), *ductility* (they can be drawn into rods or wire), good *thermal and electrical conductivity* (heat and electricity pass through them easily), and high *densities* (they have a high mass per unit volume) compared with nonmetals. With the exception of mercury (Hg), all of the metals are solids at room temperature. Some metals tend to react with other elements or compounds in their environment to form a surface coating, or film, which masks their metallic luster. When we try to identify an element, we should examine a freshly exposed surface of it in an inert (chemically unreactive) environment, and we should look for the other metallic properties as well.

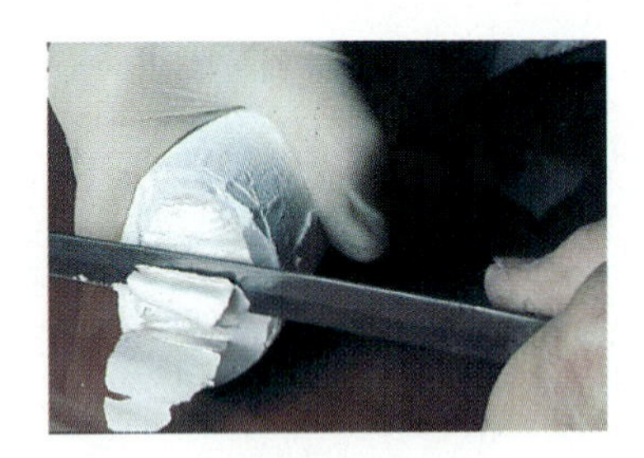

▶ Figure 4.13
Metallic Luster. *Sodium metal is sufficiently soft to be cut with a knife. When freshly cut, it has a soft silvery luster, but it quickly dulls in the air.*

Properties of the **nonmetals** (those substances that are not metals and comprise most of the remaining 20% of the elements) include lack of metallic luster (although some are shiny), *brittleness* (they shatter when hammered or drawn), poor thermal and electrical conductivity (they are insulators), and low densities compared with the metals. Many nonmetals are gases at room temperature. Bromine is the only liquid nonmetal (see Figure 4.9).

A few elements have some of the properties of both metals and nonmetals. For example, arsenic (As) appears to be nonmetallic, but certain crystals of it conduct electricity in one direction of the crystal but not very well in another. Antimony (Sb), usually considered a metal, is extremely brittle and is a poor conductor of heat and electricity. The elements that have properties of both metals and nonmetals are found along the Zintl border and are called **metalloids.** Many of them are semiconductors and thus are used as raw materials in making microchips.

Metalloids have properties of both metals and nonmetals.

Hydrogen is an element that has chemical properties resembling those of metals and nonmetals to about an equal degree. Because of its special properties and because it is the first element, it sometimes is placed top center of all of the elements to emphasize this uniqueness. In Figure 4.12 it is placed over *both* the metals (lithium) and the nonmetals (fluorine) to show its dual properties.

DEVELOPING A CHEMICAL VOCABULARY

To understand the language of chemistry you need to develop a chemical vocabulary and become comfortable with the symbols and conventions used to express it. This vocabulary begins with the names of the elements. It is important for you to learn the chemical symbol and the properly spelled name for each of the elements you will encounter on a regular basis. Because you will learn them better by using them repeatedly rather than by rote memorization, you will be asked to learn only a few for now. As you encounter new ones, you should make it a habit to learn them also and build on what you already know.

4.1 Using the table inside the back cover of this text, practice writing the correct symbol for the following elements:

a. hydrogen
b. lithium
c. boron
d. nitrogen
e. fluorine
f. sodium
g. aluminum
h. phosphorus
i. chlorine
j. potassium
k. iron
l. zinc
m. silver
n. iodine
o. mercury
p. helium
q. beryllium
r. carbon
s. oxygen
t. neon
u. magnesium
v. silicon
w. sulfur
x. argon
y. calcium
z. copper
aa. bromine
bb. tin
cc. gold
dd. lead

Once you have mastered chemical symbols, your next step is to learn the shorthand used for writing chemical formulas and expressing chemical reactions (chemical changes). Writing out in words the kinds of changes that occur in a chemical reaction is laborious and time-consuming, so symbols and formulas are used in place of words to describe the process. Such a written statement is called a **chemical equation.** For example, the chemical equation for the reaction of zinc with sulfur can be written,

A **chemical equation** is a shorthand used to describe a chemical process.

$$Zn + S \longrightarrow ZnS$$

In words, this chemical equation reads, "Zinc reacts with sulfur to yield zinc sulfide." Recall that the + means "reacts with" and the arrow means "to yield" or "to produce." The reactants (Zn and S in this case) always lie to the left of the arrow, and the products (ZnS in this case) lie to the right.

Note that we have introduced a new notation, ZnS, which is called a chemical formula. A **chemical formula** of a compound tells which elements are present and in what proportions. Numerical *subscripts* following the symbol of each element in a formula tell the number of atoms of that element present in the compound (the lack of a subscript means that *one* atom is present). Thus, ZnS contains one atom of zinc and one atom of sulfur in a **formula unit,** ZnS. The chemical formula for water, H_2O, means that the compound consists of two atoms of hydrogen and one atom of oxygen in one formula unit, H_2O. Likewise a formula unit of CCl_4 contains one atom of carbon and four atoms of chlorine, and one of $C_{12}H_{22}O_{11}$ contains 12 atoms of carbon, 22 atoms of hydrogen, and 11 atoms of oxygen. Sometimes parts of formulas are grouped together in parentheses: for example, $(NH_4)_3PO_4$. In this case the grouping of atoms within the parentheses is repeated three times, or the whole formula contains 3 atoms of nitrogen (3 times 1 atom of nitrogen), 12 atoms of hydrogen (3 times 4 atoms of hydrogen), 1 atom of phosphorus, and 4 atoms of oxygen.

4.2 Tell how many atoms of each element are present in one formula unit of each of the following compounds:

a. $CaCl_2$
b. N_2H_4
c. OF_2
d. CH_2Br_2
e. SiH_4
f. BF_3
g. $(NH_4)_2SO_4$
h. $Al_2(SO_4)_3$
i. NH_3
j. Na_2S
k. C_2H_6O
l. Al_2O_3
m. K_2SO_4
n. HNO_3
o. $Ca(NO_3)_2$
p. $Cu_2(CN)_2$

BALANCING CHEMICAL EQUATIONS

In a chemical reaction, the masses of the products equal the masses of the reactants.

Another thing to remember when writing a chemical equation is that you must *balance* it, because in a chemical reaction the total of the masses of the reactants must be equal to the total of the masses of the products. (This is the law of conservation of mass.) That is, the number of atoms of every element that comprises the reactants (on the left) must be exactly equal to the number of atoms of that same element in the products (on the right). In the reaction of zinc with sulfur, there is one atom of zinc on the reactant side of the arrow and also one atom of zinc in the product, combined with the sulfur. Likewise, the sulfur is balanced, one atom being in the reactants and one atom in the products.

$$\underset{\text{one atom Zn}}{Zn} + \underset{\text{one atom S}}{S} \longrightarrow \underset{\text{one atom Zn, one atom S}}{ZnS}$$

Often, simply writing down the symbols of the reactants and products of a reaction gives us a chemical equation that is not balanced. A good example of this is the combination of hydrogen gas with oxygen gas to form water,

$$H_2 + O_2 \longrightarrow H_2O$$

It may seem strange that the elements hydrogen and oxygen are represented by chemical formulas indicating two atoms of hydrogen and two atoms of oxygen, respectively, whereas most of the other elements, such as zinc, iron, and sodium, are not. For reasons we shall discuss later, these elements exist as *diatomics*—that is, two atoms together. Not only hydrogen and oxygen exist in nature this way, but nitrogen, fluorine, chlorine, bromine, and iodine are also diatomics. Thus, we write N_2, F_2, Cl_2, Br_2, and I_2, as well as H_2 and O_2, when we represent these elements in chemical equations.

In the reaction of hydrogen with oxygen as just written, there are two atoms of hydrogen on the left-hand side of the equation and two atoms on the right, so hydrogen is balanced. However, oxygen is not balanced because there are two atoms of oxygen on the left and only one atom on the right.

To balance the equation, we must place an equal number of atoms of oxygen on both sides of the equation while retaining the balance of hydrogen. At the same time we must remember to keep the chemical *formulas* intact (to obey the law of constant composition), and to *not* change the ratios of the elements in either formula. This can be done by placing numbers called *coefficients* in front (to the left) of the appropriate chemical formulas. These coefficients multiply the amount of the substance but do not change its identity. (That is, they do not change the subscripts in a chemical formula.) For example, writing the coefficient 2 in front of the formula H_2O gives $2H_2O$. This representation means that there are two formula units of H_2O, and that these two formula units of H_2O together contain a total of four hydrogen atoms (coefficient 2 times hydrogen subscript 2) and two atoms of oxygen (coefficient 2 times oxygen subscript 1) (► Figure 4.14).

Coefficients multiply chemical *formulas.*

(a)

Li_2CO_3

2 atoms Li ↗ ↑ ↖3 atoms O

1 atom C

(b)

$\rightarrow 3\ Li_2CO_3$

3 formula units Li2CO3

► Figure 4.14
Meaning of (a) subscripts and (b) coefficients.

Thus, by placing the coefficient 2 in front of the H_2O, we have a way of getting two atoms of oxygen on the right side of the preceding equation, and

$$H_2 + O_2 \longrightarrow 2H_2O$$

But doing this makes hydrogen unbalanced, there now being four atoms of hydrogen on the product side and only two atoms on the reactant side. This damage is easily repaired by placing a coefficient 2 in front of the H_2 on the left to restore hydrogen balance, and

$$2H_2 + O_2 \longrightarrow 2H_2O$$

The equation is now balanced, but it is always a good idea to check to make sure. Four atoms of hydrogen on the left means there should be four on the right. That checks out. Two atoms of oxygen on the left and two on the right? That checks out, too.

4.3 Some of the following equations are balanced and some are not. First, check to find those that are balanced, and then balance those that are not.

a. $H_2 + Cl_2 \longrightarrow HCl$
b. $S + O_2 \longrightarrow SO_2$
c. $Ca + O_2 \longrightarrow CaO$
d. $SO_3 + H_2O \longrightarrow H_2SO_4$
e. $I_2 + Cl_2 \longrightarrow ICl$
f. $Na + Br_2 \longrightarrow NaBr$
g. $K + O_2 \longrightarrow K_2O$
h. $Ca + S \longrightarrow CaS$
i. $H_2 + S \longrightarrow H_2S$

When iron rusts in air, the red-brown iron oxide, Fe_2O_3, is formed,

$$Fe + O_2 \longrightarrow Fe_2O_3$$

In this instance neither the iron nor the oxygen is balanced. It is a good idea to balance one element at a time. There are two atoms of oxygen on the left and three on the right. Because the element oxygen comes two atoms at a time, we cannot divide an O_2 in half to get another O for the left-hand side; but we can make the total number of atoms of oxygen on the right an *even* number. This makes the oxygen on the left side easy to balance because it also has an even number of atoms. Placing the coefficient 2 in front of Fe_2O_3 makes six atoms of oxygen on the right, an even number,

$$Fe + O_2 \longrightarrow 2Fe_2O_3$$

From here on, it is easy to see that we need six atoms of oxygen on the left, and that $3O_2$ gives this number. To balance iron we note that there are four atoms of iron on the right, so we place a coefficient of 4 in front of the Fe on the left, and

$$4Fe + 3O_2 \longrightarrow 2Fe_2O_3$$

4.4 Some of the following equations are balanced and some are not. First, check to find those that are balanced, and then balance those that are not.

a. $Al + O_2 \longrightarrow Al_2O_3$
b. $Al + Cl_2 \longrightarrow AlCl_3$
c. $S + O_2 \longrightarrow SO_3$
d. $2P + 3Cl_2 \longrightarrow 2PCl_3$
e. $Mg + N_2 \longrightarrow Mg_3N_2$
f. $2P + O_2 \longrightarrow P_2O_5$

DID YOU LEARN THIS?

For each of the given statements, fill in the blank with the most appropriate word or phrase from the list. Do not use any word or phrase more than once.

metals	quantitative analysis	nitrogen
nonmetals	hydrogen	chemical change
metalloids	silicon	chemical formula
Zintl border	intrinsic properties	chemical balance
group	physical change	constant composition
period	oxygen	multiple proportions
qualitative analysis	helium	chemical equation

a. ______ Tells what elements are present in a compound and in what proportions
b. ______ Attributes such as color, conductivity, and melting point that characterize a pure substance
c. ______ The most abundant element in the universe
d. ______ An abbreviated statement of what happens in a chemical reaction
e. ______ Elements such as sodium, calcium, iron, and lead
f. ______ States that a pure compound is always made up of the same ratio of elements
g. ______ The law of conservation of mass
h. ______ An imaginary dividing line between the metals and the nonmetals in the Periodic Table
i. ______ Makes up most of the atmosphere
j. ______ Are found in the upper right-hand part of the Periodic Table
k. ______ Substances that may conduct electricity but are hard and brittle
l. ______ A family of elements
m. ______ Finding out that a sample of pyrite ore contains iron and sulfur

EXERCISES

1. List the kinds of models used in science. Briefly describe each one. What problems do we face when we make models of things that are too small or too large to see? How does this affect the utility of these models? Their relationship to reality?
2. List several examples of the use of models from your everyday experience. What purpose does each serve? How do you use models?

3. Distinguish between qualitative and quantitative analysis. What kind of information does each give about a substance?
4. If 12.2 g of magnesium reacts completely with 8.0 g of oxygen to give one product, magnesium oxide, what mass of it is formed?
5. A sample of copper sulfide, which had a mass of 9.56 g, was decomposed by heating to give 6.35 g of elemental copper and sulfur as the only products. What mass of sulfur was produced?
6. When 9.0 of water was decomposed into its elements it yielded 1.0 g of hydrogen and 8.0 g of oxygen. Hydrogen and oxygen also form another compound, hydrogen peroxide. When 17.0 g of hydrogen peroxide was decomposed, it gave 1.0 g of hydrogen and 16.0 g of oxygen. Show how these data support the law of multiple proportions. The chemical formula for water is H_2O. What could be some possible chemical formulas for hydrogen peroxide?
7. Give the correctly spelled name for the elements whose symbols are listed here.

   a. He
   b. Be
   c. C
   d. O
   e. Ne
   f. Mg
   g. Si
   h. S
   i. Ar
   j. Ca
   k. Cu
   l. Br
   m. Sn
   n. Au
   o. Pb
   p. H
   q. Li
   r. B
   s. N
   t. F
   u. Na
   v. Al
   w. P
   x. Cl
   y. K
   z. Fe
   aa. Zn
   bb. Ag
   cc. I
   dd. Hg

8. Using Figure 4.11 and the Periodic Table (Figure 4.12), determine what the following statements refer to: group (family), period, metal, nonmetal, or metalloid.

   a. The elements Na and Cs belong to the same _____. They are classified as _____.
   b. The elements O and Se belong to the same _____. They are classified as _____.
   c. The elements Ge and As belong to the same _____. They are classified as _____.
   d. The element K starts a _____. It is classified as a _____.
   e. The elements Ba and Pb belong to the same _____. They are classified as _____.
   f. S, P, and I are classified as _____.
   g. _____ belongs to the same family as Mg and Ba. (Several answers possible.) They are classified as _____.
   h. _____ belongs to the same family as Ar and Rn. (Several answers possible.) They are classified as _____.
   i. Although Na is a _____, and Cl is a _____, they both belong to the same _____.

9. List the properties common to metals. Compare them with those of the nonmetals. What are metalloids? What kinds of properties do they have?
10. How is the uniqueness of hydrogen represented on many Periodic Tables?
11. Based on the information given for each compound that follows, write its chemical formula.
    a. Pentane contains 5 atoms of carbon and 12 atoms of hydrogen.
    b. Phenol contains 6 atoms of carbon, 6 atoms of hydrogen, and 1 atom of oxygen.
    c. Maltose contains 12 atoms of carbon, 22 atoms of hydrogen, and 11 atoms of oxygen.
    d. Magnesium sulfate (Epsom salts) contains 1 atom of magnesium, 1 atom of sulfur, and 4 atoms of oxygen.
    e. Aluminum sulfate contains 2 atoms of aluminum, 3 atoms of sulfur, and 12 atoms of oxygen. A sulfate unit contains 1 atom of sulfur and 4 atoms of oxygen. Write this formula using parentheses.
12. Indicate how many atoms of each element are contained in the following chemical formulas.
    a. $HClO_4$
    b. B_2H_6
    c. $N(CH_3)_3$
    d. $(NH_4)_3PO_3$
    e. $Ca(C_2H_3O_2)_2$
13. How many atoms of each element are there in the number of formula units represented for each compound?
    a. 2 $NaOH$
    b. 4 $C_{12}H_{26}O$
    c. 3 Na_2SO_4
    d. 100 O_2
    e. 5 $Al_2(CO_3)_3$
14. Answer the following questions about this reaction, which occurs in an ordinary dry cell:

$$Zn + 2MnO_2 + 2NH_4Cl \longrightarrow ZnCl_2 + Mn_2O_3 + 2NH_3 + H_2O$$

    a. Is the equation balanced?
    b. How many different products are formed in this reaction?
    c. How many of the substances in this reaction are elements? Which one(s)?
    d. Which substances in this reaction are compounds?
    e. Which substances in this reaction are the reactants?
    f. How many formula units of zinc would it take to react with eight formula units of MnO_2?

15. Balance the following equations.

    a. $C_3H_8 + O_2 \longrightarrow CO_2 + H_2O$
    b. $C_6H_{14} + O_2 \longrightarrow CO_2 + H_2O$
    c. $P_4 + Cl_2 \longrightarrow PCl_3$
    d. $SO_3 + H_2O \longrightarrow H_2SO_4$
    e. $P_2O_5 + H_2O \longrightarrow H_3PO_4$

16. List the five propositions of Dalton's atomic theory. How does the theory account for the law of conservation of mass? The law of definite proportions (constant composition)? The law of multiple proportions?
17. How do the atomic masses of the elements differ from their relative combining masses? What two factors do the relative combining masses of the elements depend on?

THINKING IT THROUGH

Finding a Substitute for Lead

The use of lead sinkers in fishing may be prohibited in the future because metallic lead is harmful to waterfowl, particularly bottom-feeding ducks. These birds ingest solid objects, including sinkers, from the bottoms of lakes and ponds to help their gizzards grind the food they have eaten. Under "gizzard" conditions, lead forms compounds that are toxic and can cause ducks to die.

What metal (or metals) could be used to make fishing weights that would be as effective, inexpensive, and easily fabricated as those made of lead but would also be nontoxic to waterfowl and other forms of aquatic life? Several references that may be of use are the *CRC Handbook of Physics and Chemistry* (new edition each year), various editions of *Lange's Handbook of Chemistry,* and various editions of the *Merck Index.* Your instructor will be able to suggest others that are available in your library.

ANSWERS TO "DID YOU LEARN THIS?"

a. chemical formula
b. intrinsic properties
c. hydrogen
d. chemical equation
e. metals
f. constant composition
g. chemical balance
h. Zintl border
i. nitrogen
j. nonmetals
k. metalloids
l. group
m. qualitative analysis

APPLYING CHEMISTRY AS A TECHNICAL ILLUSTRATOR

Brian C. Betsill

Graphic Designer and Technical Illustrator

B. A. Computer Science, DePauw University

The second semester of my junior year in college, I changed my major from chemistry to computer science. The previous semester I had tried to muddle through a course in inorganic chemistry. During the course we had conducted analytical experiments to find out how much iron there was in spinach and how much calcium there actually was in powdered milk. I could not get excited about these or any other such experiments and could not fathom another year and a half as a chemistry major. So I made what seemed the logical choice and changed majors.

After changing majors, I thought I would never have to apply chemistry in a practical sense again. However, as a technical illustrator of math and science textbooks, I have frequently found myself trying to recall the molecular structures of organic compounds and the various laboratory apparatuses. Although I did not pursue chemistry as a major, let alone career, I have found it useful in my chosen profession.

As a technical illustrator, I am responsible for supplying art for math and science textbooks. I have recently illustrated general chemistry books and chemistry laboratory manuals. In preparing the illustrations, I must work very closely with the authors who provide the general guidelines for the art. My background in chemistry has helped me communicate with the authors and understand the subjects of my illustrations.

In my line of work, it is important to establish a good rapport with the authors. We need to be able to communicate comfortably with one another in order to complete the complex process of publishing a book. I find that when I am working on these projects, I have more credibility with the authors once they discover that I have a background in chemistry. We are speaking the same language, as it were, and they are relieved to find they can discuss problems with me directly without constantly having to search for a lay expression to make a point.

Just as a medical illustrator would find it difficult to do his or her job without understanding how the human body works, I would have problems doing chemistry drawings without a basic understanding of chemistry. When I am asked to draw a given molecule or chemical compound, I often begin with only a rough sketch or a brief description. I must then transform this sketch into a detailed illustration. Sometimes I find it helpful to construct visual aids such as Styrofoam models to assist in the process. Since I do the vast majority of my drawings on the computer, I am also able to take advantage of software programs that are designed specifically for molecular modeling. They require the user to input formulas, atoms, or the various substructures needed to create the desired compounds. Without these visual aides or actual computer-generated models, it would be difficult to arrive at a final product, and without a general understanding of chemistry, it would be virtually impossible for me to make use of these tools.

Although I thought I would not be using chemistry once I finished school, I have found it useful time and again in my career as a technical illustrator. I suppose the lesson to be learned is that academic subjects such as chemistry can be helpful in the "real world." Biochemists and pharmacists certainly find chemistry directly relevant to their work, but chemistry can also be applied indirectly in a number of other careers.

5

A Model for the Atom

[Those atoms] which are able to affect the senses pleasantly consist of smooth round elements; while all those on the other hand which are found to be bitter and harsh are particles more hooked and for this reason want to tear open passages into our senses. . . .

Lucretius "Concerning the Nature of Things"

. . . . Now it develops that our replacement picture [of the atom] is not a picture at all, but an unvisualizable abstraction. This is uncomfortable because it reminds us that atoms were never "real" things anyway. Atoms are hypothetical entities constructed to make experimental observations intelligible. No one, not one person, has ever seen an atom. Yet we are so used to the idea that an atom is a thing that we forget that it is an idea. Now we are told that not only is an atom an idea, it is an idea we cannot even picture.

Gary Zukav "The Dancing Wu Li Masters"

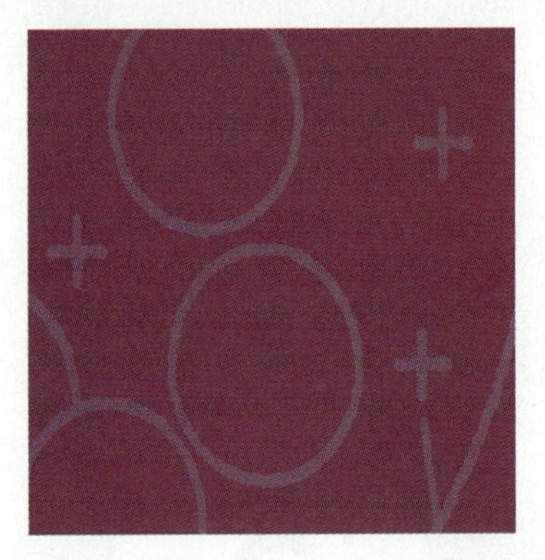

By the late 1800s many scientists thought that science had discovered just about everything fundamental that could be known, and that new discoveries in science would be scarce or nonexistent. Not much was left to do but to refine measurements and tie up loose ends. Even Albert Michelson, who later was to win the Nobel Prize for Physics, said in 1894 that ". . . the future truths of physics are to be looked for in the sixth place of decimals."

Atoms had become accepted as part of the model of structural chemistry by this time, but Dalton's theory that they were indivisible persisted. In the mid-1800s Michael Faraday, a pioneer in the study of *electrochemistry* (the interaction of electricity with matter), first showed that matter was electrical in nature and proposed that atoms might have smaller parts. Although he was a well-known and respected scientist, his ideas met with considerable resistance.

THE ATOM COMES APART

Atoms contain **electrons** as part of their structure.

Science was becoming anything but boring, despite those who thought it was complete. In 1879 William Crookes discovered *cathode rays,* later to be called **electrons.** He thought these probably were small particles, each carrying a negative charge of electricity. Apparently they could move through some substances, such as metals, and carry an electric current. If this were the case, atoms might contain electrons as part of their structure. Because atoms are electrically neutral (have no electrical charge), the rest of an atom would have to be "something positive" to balance the negative charge of the electrons.

Wilhelm Röntgen discovered X-rays in December 1895, and within several weeks these rays were causing a sensation all over the world. A few months later, March 1896, Henri Becquerel discovered **radioactivity,** a process in which large atoms, like those of uranium (U), spontaneously fall apart into smaller atoms and emit several kinds of energetic rays, which can fog a photographic plate.

On the heels of this discovery came that of J. J. Thomson (Cambridge University, England), who in 1897 showed that the electron has a charge equal in magnitude to that on a hydrogen ion, but of opposite sign, probably 1−. (**Ions** are atoms, or small groups of atoms bound together, that carry a positive or negative charge.) More backing for the idea that electrons are part of an atom came in 1900, when Becquerel showed that electrons also came from radioactive decay. In 1909, Robert A. Millikan at the University of Chicago actually measured the charge and mass of the electron (9.1×10^{-34} g). It is a very tiny particle, having only about 1/1837th the mass of the hydrogen atom. A hydrogen atom contains only one electron. The remainder of the hydrogen atom (containing almost all of its mass), formed when the atom loses its electron, is positively charged (1+) and is called a hydrogen ion, or **proton** (▶ Figure 5.1).

$$\underset{\text{Hydrogen atom}}{H} \longrightarrow \underset{\text{Proton (hydrogen ion)}}{H^+} + \underset{\text{Electron}}{e^-}$$

▶ Figure 5.1
The Hydrogen Atom. *The hydrogen atom is made up of a proton and an electron.*

THE PLUM PUDDING MODEL

If an atom contains a certain number of electrons, it also must contain the same number of positive charges. This is necessary to keep the atom electrically neutral. The first model of the atom to incorporate these ideas was proposed by Thomson and was called the *plum pudding model* of the atom (▶ Figure 5.2). Thomson thought that the atom's mass was distributed evenly throughout its volume and that the atom was a spherical ball consisting of a homogeneous mixture (a pudding) of positive charges (containing most of the mass of the atom) and an equal number of electrons. The electrons (plums) were somehow imbedded in and among the positive charges (dough) to prevent the latter from flying apart. (Like charges repel each other; opposites attract, ▶ Figure 5.3.)

▶ Figure 5.2
Thomson's Plum Pudding Model of the Atom. *A number of negative electrons (plums) are imbedded in a homogeneous mass containing an equal number of positive charges (dough).*

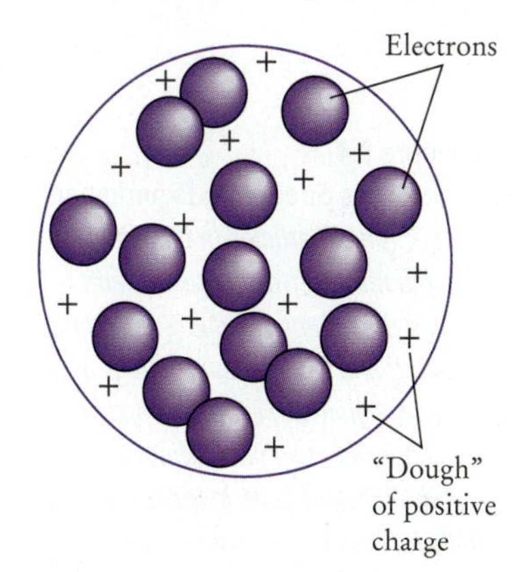

RUTHERFORD'S SCATTERING EXPERIMENT

Under Thomson's direction, the Cavendish Laboratories at Cambridge had by the turn of the century become the focal point for the study of atomic structure. Ernest Rutherford, a New Zealander, joined Thomson in 1895 to study the rays produced from radioactive decay. He found that the radiation discovered by Becquerel came in two forms, α-rays and β-rays (These are also called α-particles and β-particles. γ-Rays were discovered later.) Moving to McGill University in Canada in 1897, Rutherford continued his

β-Particles are electrons produced when radioactive atoms decay.

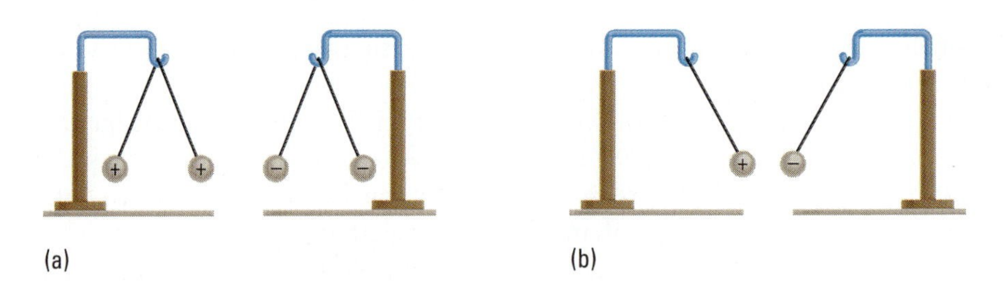

► Figure 5.3
Repulsive and Attractive Forces Between Charged Objects. *Balls having the same charge (a) repel each other, whereas those with opposite charges (b) are attracted to one another.*

investigation of **α-particles.** By 1902 he had confirmed that these were positively charged helium ions, He^{2+}, and that their mass was four times that of the hydrogen atom.

Returning to England and taking a position at the University of Manchester in 1907, Rutherford began experiments that would soon dramatically change the model of the atom. What, wondered Rutherford, would happen if a beam of α-particles were shot at atoms? When they approach an atom, the positively charged α-particles should be repelled by the positive charges (dough) in the atom and should be deflected slightly from their paths—that is, scattered. In fact, Rutherford already had reason to suspect that this was happening when α-particles struck atoms in the air as they passed through it.

α-Particles were known to cause little flashes of light, called **scintillations,** one flash each time one of them struck a screen coated with a substance, such as zinc sulfide, called a **phosphor.** Ernest Marsden, one of Rutherford's students, directed a beam of α-particles at a piece of very thin platinum foil (platinum has large, heavy atoms close together). Behind the foil he had placed a screen coated with zinc sulfide to see the flashes the α-particles would produce (► Figure 5.4). A few of the α-particles were scattered some-

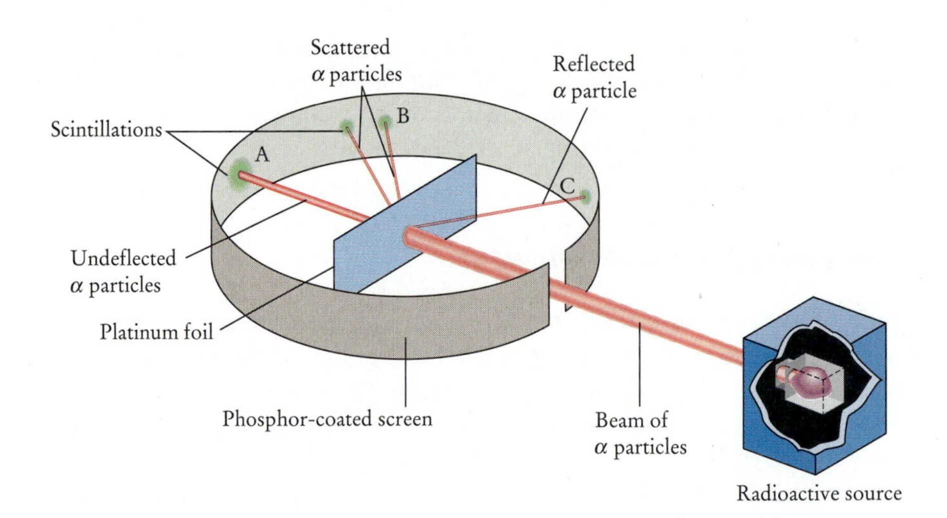

► Figure 5.4
Rutherford's Scattering Experiment. *Most of the α-particles in the beam from the radioactive source passed through the platinum foil without being deflected and caused scintillations at* A. *Some of them were scattered and produced flashes of light at various places near* B. *A few (about 1 in 8000) were actually reflected back toward the source* C. *Although most texts state that gold foil was used in this experiment, the original paper reports that the target was made of platinum.*

what, but most of them were not, and instead passed straight through the platinum foil without being deflected at all. Much more surprising was the observation that some of the α-particles were scattered in all directions. A few (about 1 in 8000) were even scattered backward toward the source of the beam. Rutherford was totally astounded. "It was quite the most incredible event that has ever happened to me in my life," he said. "It was almost as incredible as if you fired a fifteen-inch shell at a piece of tissue paper and it came back and hit you."

THE NUCLEAR ATOM

Rutherford repeated the experiment a number of times and then mulled it over for several years. The only explanation he could think of was that most of an atom was empty space! The positive charges, even though they all repelled each other, must be stuck together somehow, with their total mass in one small, dense, positive core, which he called the **nucleus.** The electrons in the atom, then, must surround the nucleus (▶ Figure 5.5).

Most of the atom is empty space.

From his experiments Rutherford could calculate that the diameter of the nucleus was roughly 1/100,000 that of the atom itself; that is, the nucleus in the middle of an atom is roughly comparable to a fly in the middle of a domed athletic stadium (▶ Figure 5.6). The nucleus is so incredibly dense that if all of the matter in the earth were as dense as that in the nucleus, it would occupy a volume only the size of a softball! A piece of nuclear matter the size of the end of your thumb would weigh 60,000,000 tons.

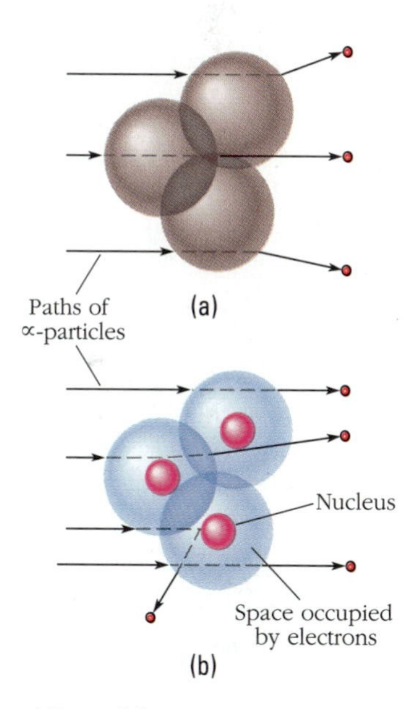

▶ Figure 5.5
α-Particles Interacting with Atoms.
(a) If the atom were a pudding of diffuse positive charge with electrons imbedded in it, its mass and charge would not be dense enough to strongly deflect positive α-particles. (b) The nuclear atom's mass and positive charge are concentrated in the small nucleus. Most of the α-particles in the beam do not strike it, but those that do are strongly deflected.

A NEW MODEL OF THE ATOM: BOHR'S EXPLANATION

Rutherford proposed his model of the nuclear atom in March 1912, but it was largely discounted, primarily because it was not consistent with other accepted theories at that time. What kept the positive matter in the nucleus from flying apart? What kept the negative electrons from falling into the positive nucleus or kicking each other out of the atom? Besides, everyone knows that matter is hard. An atom made of empty space would be expected to be mushy, and so would matter made from it.

Niels Bohr, a young Danish physicist who had begun work with Rutherford at Manchester early in 1912, found these questions intriguing. Were the old models right, or could he find a new model that could help explain how the nuclear atom holds together?

A major question Bohr had to address was how the electrons could be arranged in the empty space around the dense nucleus and not spiral into it because of the attraction of opposite charges. Like many creative scientists,

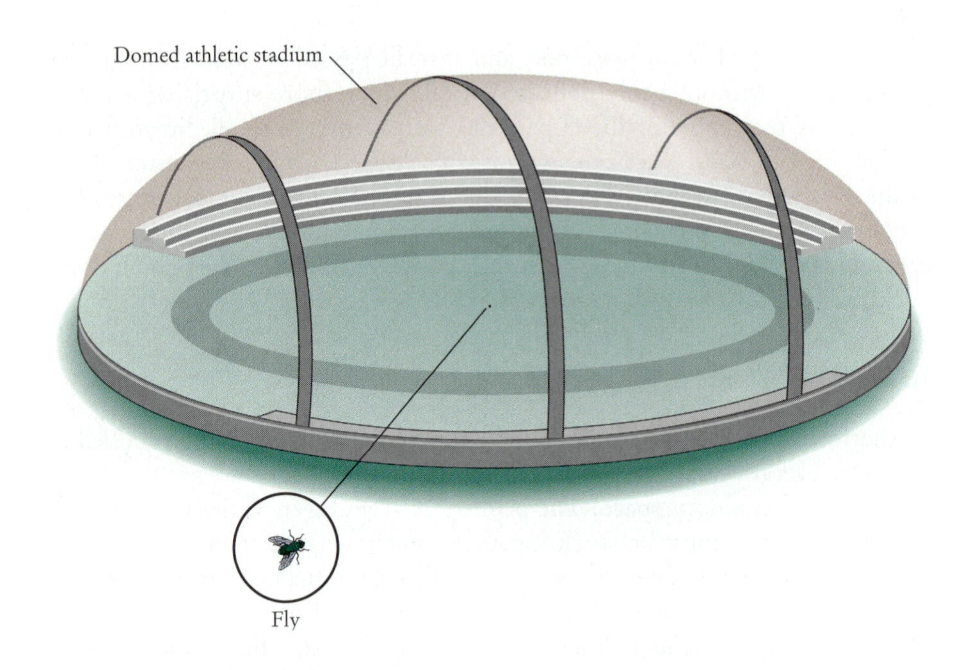

▶ Figure 5.6
Size of the Nucleus. *The diameter of the nucleus is approximately 1/100,000 that of the atom itself. This relationship is about the same as the size of a fly compared with the size of a domed athletic stadium.*

Bohr solved this problem by putting together, or synthesizing, various scientific principles that at first seemed to be unrelated. This required that he look beyond the immediate question and take an *outside viewpoint.* That is, he looked at other scientific phenomena to gain an understanding of the problem he was working on. As it turned out, Bohr had to try several of these perspectives before he could make a model of the nuclear atom that worked.

EMISSION SPECTRA

Neon signs are discharge tubes.

One of Bohr's outside viewpoints came from the study of light from discharge tubes, such as is given off by neon signs. Applying a high voltage to neon gas inside a glass tube causes the neon to give off orange-red light. Under similar conditions, hydrogen gas glows with pale blue light, and other elements with other colors. Sodium vapor produces the familiar yellow of some street lamps and mercury vapor the eerie blue of others (▶ Figure 5.7). Strontium imparts its red color to highway flares.

It turned out that the colors emitted by these gases are related to the colors of the rainbow. We see the familiar rainbow whenever ordinary white light from a lamp filament or the sun (or any glowing solid or liquid) passes through a prism (or a raindrop); it is called a **continuous spectrum.**

But when light from a discharge tube containing a particular element passes through the slit and prism of a spectroscope (▶ Figure 5.8), we see nar-

(a)

(b)

▶ Figure 5.7
Uses of Discharge Tubes. *(a) Mercury vapor lamps and (b) sodium vapor lamps are discharge tubes. When an electric current is passed through a gaseous element, its atoms gain extra energy and become excited. Mercury and sodium produce their characteristic colors because excited mercury and sodium atoms emit different colors of light.*

row bright lines of that element's characteristic colors, separated by dark spaces. This is called a **line spectrum** (or **emission spectrum.**) It looks rather like a bar code in rainbow colors and can be used to identify an element. Line spectra are produced only by glowing gases.

The line spectrum and the continuous spectrum are related as shown in ▶ Figure 5.9. Each colored line is in the place it would occupy if it were part of a continuous spectrum. That is, a red line of the line spectrum appears in the red region of the continuous spectrum, a yellow line in the yellow region, and so on.

▶ Figure 5.8
A Simple Spectroscope in Operation. *The spectroscope is an instrument in which the light from a discharge tube (or any glowing source) is passed first through a narrow slit (to give a very narrow beam of light) and then through a prism (which separates, or refracts, the light into its colors); the colors can be projected onto a screen for viewing. A continuous spectrum is shown.*

THE NATURE OF LIGHT

Before Bohr could build his model of the atom, he had to consider another outside view, namely, the nature of light itself.

Light as Waves

Light is a form of energy and usually is described as being wavelike. Waves in our common experience are those in water and in air. These waves transmit their energy through their respective media to produce, for example, tides and sound. Light is not so obviously wavelike, but it radiates from a source, like waves in air and water; and it undergoes refraction (bending through a medium), diffraction (bending around a corner), and interference (canceling itself and reinforcing itself) in an analogous manner (▶ Figure 5.10). Most importantly, however, the mathematical equations that describe the behavior of water and sound waves also work to describe the behavior of light. It is

Light is a form of energy and behaves like the waves in water and air.

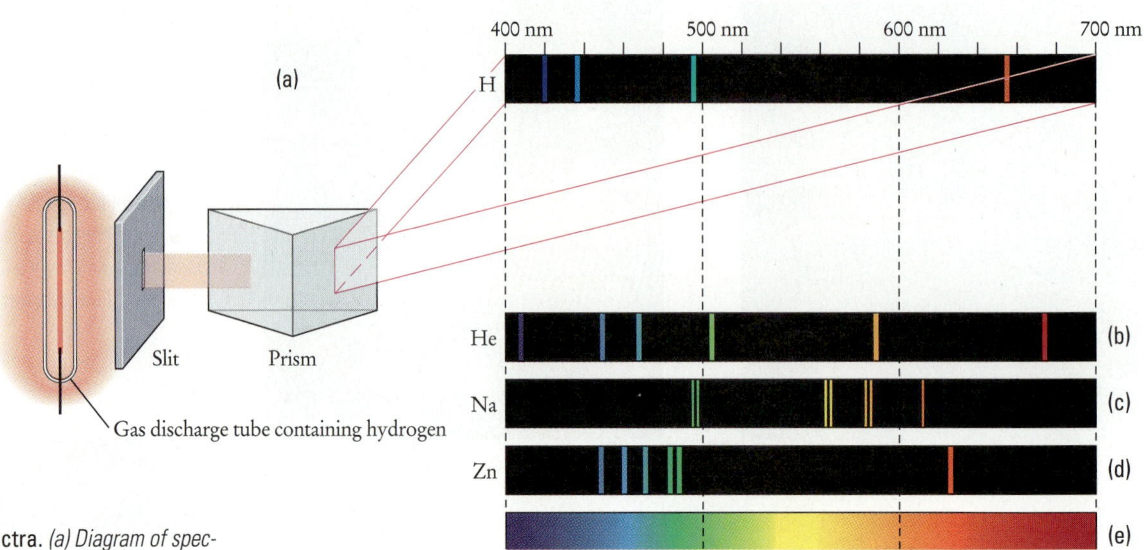

▶ Figure 5.9
Emission Spectra. *(a) Diagram of spectroscope producing an emission (line) spectrum of hydrogen. Below the hydrogen spectrum are emission spectra of helium (b), sodium (c), and zinc (d). Each element produces its own characteristic emission spectrum. Note that the colored lines remain in register with the continuous spectrum (e).*

very comfortable scientifically when the behavior of a new phenomenon "fits the equations" of other phenomena more familiar to us.

Waves of light are different, however, because they do not need a medium to travel through (though they are able to pass through many substances). Light travels easily through a **vacuum,** the absence of any matter at all.

We need to look briefly, but carefully, at some simple equations that light obeys and define some terms that describe it. Let's visualize a light wave being propagated from left to right across this page, staying in the plane of the paper on its path (▶ Figure 5.11).

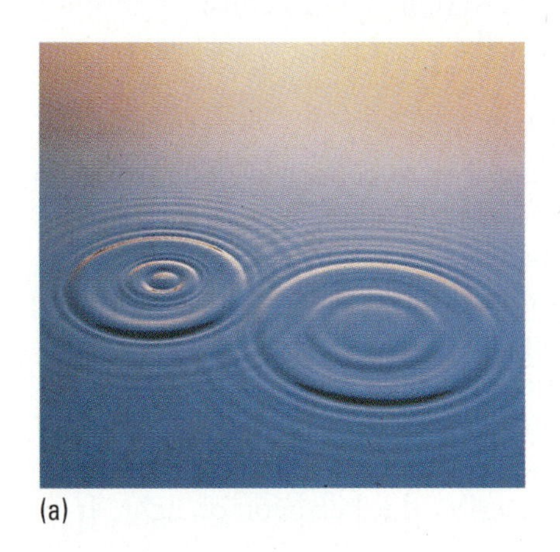
(a)

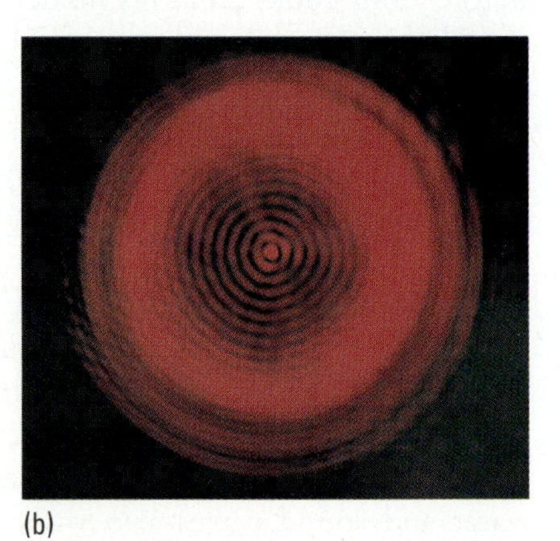
(b)

▶ Figure 5.10
Interference Patterns. *(a) Dropping two stones in water produces waves, which meet and form interference patterns. (b) An interference pattern (or diffraction pattern) made by light coming through a pinhole is evidence for the wave nature of light.*

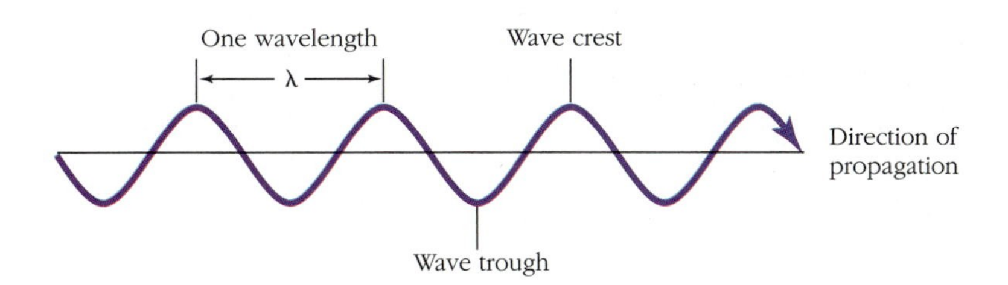

► Figure 5.11
A Light Wave Moving From Left to Right.

As the light moves on its way, it undulates up and down, crest-to-trough-to-crest. The spacing of the crests (or troughs) can be close together or far apart. The distance between the crests (or troughs) of the wave is called the **wavelength** and is represented by the symbol λ (pronounced "lambda"). Wavelength has units of length. The timing of the passage of the crests or troughs is called the **frequency,** represented by the symbol ν (pronounced "nu") and is the number of waves passing a given point in a second. Frequency is measured in cycles (repetitive events) per second. One cycle per second is called a hertz, abbreviated Hz. Thus, a hertz has the dimensional units, 1/s.

All light travels at the same velocity in a given medium. The number of waves that pass a point in one second (frequency) depends on the spacing of the crests (wavelength) and the speed of propagation of the wave, or its velocity. Wavelength and frequency are inversely proportional to each other. That is, as the wavelength increases, the frequency decreases, and vice versa (► Figure 5.12).

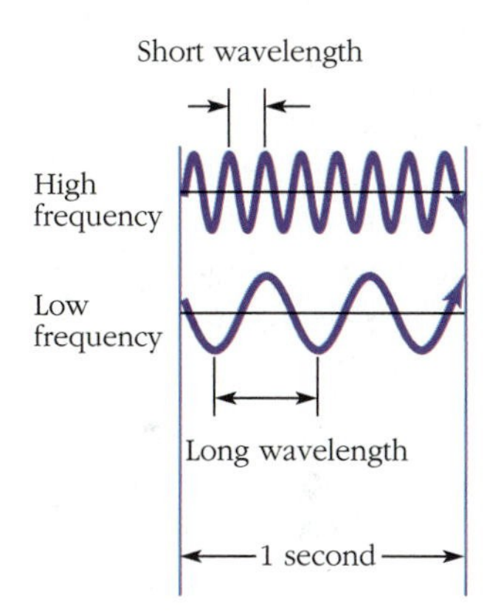

► Figure 5.12
The Relationship of Wavelength and Frequency. *For two light waves traveling at the same velocity, the one with the higher frequency will have the shorter wavelength.*

5.1 Complete the following statements regarding the nature of light.

a. If the frequency of light is high, its wavelength will be relatively ________.
b. If λ is small, ν will be relatively ________.
c. If the frequency is low, the wavelength of light will be relatively ________.
d. If the wavelength of light is large, its frequency will be relatively ________.
e. If ν is low, λ will be relatively ________.
f. The velocity of light is represented by the symbol, ______ and its value in a vacuum is a ________.

Wavelength and frequency are inversely proportional to one another.

The simple mathematical relationships among the wavelength of a wave, its frequency, and its velocity are as follows:

$$\lambda \times \nu = c$$

$$\nu = \frac{c}{\lambda}$$

$$\lambda = \frac{c}{\nu}$$

The velocity of light in a vacuum, c, is a constant (3.00×10^8 m/s, or 186,000 mi/s). As the wavelength (λ) gets larger, the frequency (ν) gets smaller, and vice versa. Likewise, if the frequency (ν) increases, the wavelength (λ) decreases (gets shorter).

Light As Particles

When a piece of metal, such as sodium, potassium, or zinc, is placed in a vacuum and irradiated with light, it loses electrons and becomes positively charged. This effect, called the **photoelectric effect,** is used today in electric-eye door-openers and other photocell devices. Another familiar device that employs this phenomenon is the solar cell, in which a piece of silicon, or other semiconductor, produces an electric current when light strikes it. In both cases, moving electrons are produced by light.

light in ⟶ metal ⟶ electrons out

The energy of the electrons produced depends only on the energy (color) of the light striking the metal. Brighter light produces more electrons but does not give them greater energy. That is, blue light (which has a higher energy than red light) releases electrons that are more energetic than those produced by the same intensity of red light, but it does not produce any more electrons. Even very bright red light does not eject any electrons from potassium, but very dim blue light does. Bright blue light and dim blue light produce electrons of the same energy. Thus, when the energy of the light striking the metal is below a certain threshold, no electrons are produced, regardless of the brightness of the light.

blue light (higher energy) ⟶ higher energy electrons
red light (lower energy) ⟶ lower energy electrons (or none)

Energy comes in bundles, called **quanta.**

In 1900, Max Planck, a German physicist, proposed that energy is not continuous, but rather comes in bundles, or packets, of energy called **quanta.** Taking Planck's idea that energy came in little bundles another step, Albert Einstein proposed in 1907 that if light were actually composed of these small packets of energy, in the form of massless particles of light, the photoelectric effect could be explained. These particles of light were later named **photons.** In Einstein's model, a high-energy photon interacts in some manner with an atom in the metal it strikes to generate a high-energy electron. A lower energy photon generates a lower energy electron. When the energy of the photon is too low, below the amount required to eject an electron from the atom, nothing happens. In this model we have the idea of particle in, particle out (▶ Figure 5.13). All the energy of the incoming particle (photon), minus the amount of energy needed to eject the electron from the atom, is imparted to the outgoing particle (electron). For his idea that light could be particles (not for his theory of relativity) Einstein was awarded the Nobel Prize for Physics in 1921.

A **photon** is a massless particle of light.

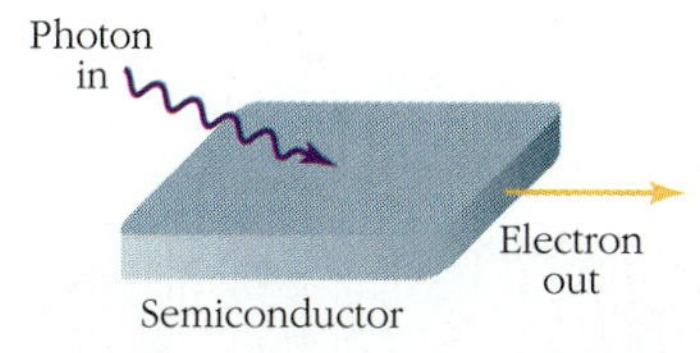

▶ Figure 5.13
A photon of light causes an electron to be ejected from an atom. In the photoelectric effect, light energy is converted into electrical energy.

The energy of a quantum of light is directly proportional to its frequency, ν, and is expressed in *Planck's relationship:*

$$E = h\nu$$

where h is Planck's constant (a very small number) and E is the energy of the particular color of light.

As the frequency of light goes up, so does its energy, and vice versa. If the frequency is doubled, for example, the energy is doubled also. Because frequency and wavelength are inversely proportional, the energy of light also is inversely proportional to its wavelength. That is, when we triple the wavelength of the light, we find its energy to be one-third of the previous value.

$$E = \frac{hc}{\lambda}$$

This concept of the energy of light being related to its frequency and wavelength is very important, not only in understanding atomic structure but also in practical application to our everyday lives (▶ Figure 5.14). We hear almost daily that too much exposure to the ultraviolet radiation from the sun causes skin cancer or damage to our eyes, or that X-rays are harmful to us if the exposure is too great. These kinds of light have very short wavelengths (billionths of a meter or smaller) and very high frequencies, which means that they are highly energetic. Most often, X-rays have so much energy that they pass right through our tissues, which makes them useful, for example, in medicine (▶ Figure 5.15). However, those few rays that do strike something in our bodies and transmit their energy to our tissues can do a lot of damage, causing tissue destruction and sometimes cancer. Radio waves, on the other hand, are not considered to be particularly dangerous. These have very long wavelengths (centimeters or more) and relatively low frequencies. Thus, their energy is low. (There is now concern, however, that exposure to certain kinds

X-Rays have very short wavelengths. This makes them very energetic.

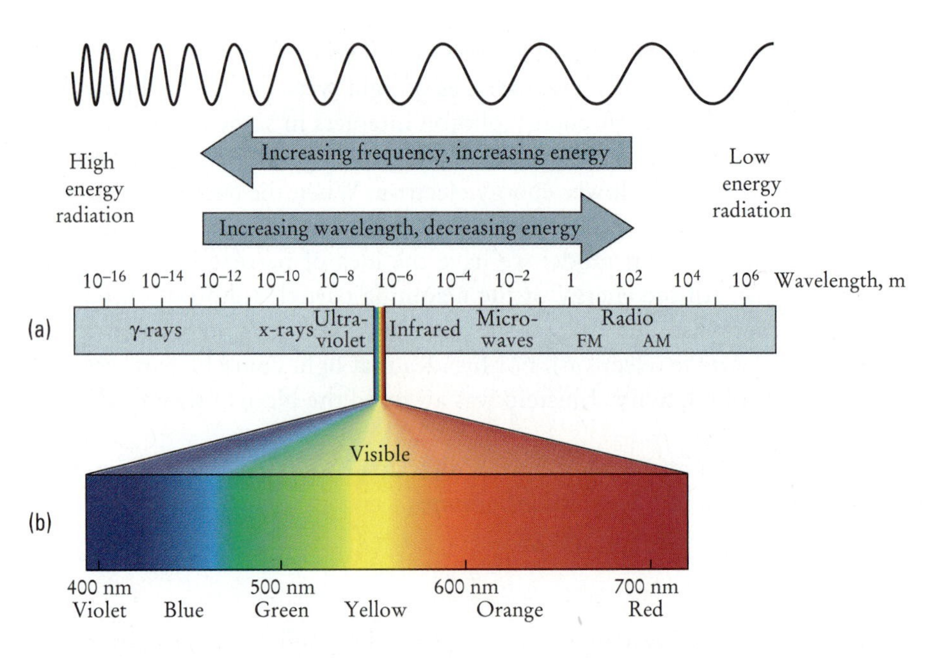

▶ Figure 5.14
The Electromagnetic Spectrum. *(a) The relationship of wavelength, frequency, and energy of electromagnetic radiation. (b) The visible spectrum, which corresponds to only about one "octave" of frequencies in the entire spectrum, is expanded.*

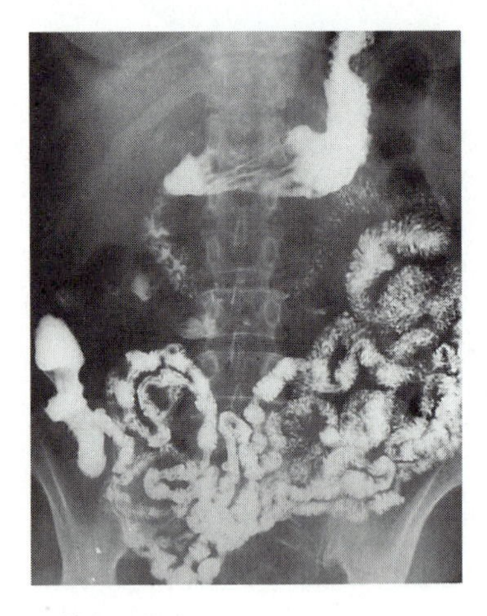

▶ Figure 5.15
Highly energetic X-rays pass through most tissues in the human body. However, the stomach and intestines appear white in this X-ray because the patient had drunk a suspension of barium sulfate, which is opaque to X-rays.

5.2 Complete the following statements, which deal with the relationship of the frequency and wavelength of light to its energy. (Use Figure 5.14 if you need to.)

a. Blue light has __________ energy than red light.
b. Light having a relatively low frequency has a relatively __________ energy.
c. Light of short wavelengths has an energy that is relatively __________.
d. Light having a high energy has relatively __________ frequencies.
e. Light with long wavelengths has energies that are relatively __________.
f. Radio waves have __________ wavelengths compared with those of visible light. Therefore, their energies are relatively __________.
g. γ-Rays and cosmic rays have very __________ wavelengths. Thus, their frequencies are very __________ and their energies are very __________.
h. Red light with a wavelength of 750 nanometers (nm) has a frequency that is __________ than that of green light with a wavelength of 500 nm. The energy of the red light is __________ than that of the green light.

of low-frequency electromagnetic radiation, called EMFs, may have some adverse health effects in humans. Also, microwaves can be dangerous to persons wearing pacemakers.)

PUTTING THE MODEL TOGETHER: BOHR'S ATOM

To explain emission spectra, Bohr reasoned that the high-energy electrons in a discharge tube are jumping, or leap-frogging, from one gaseous atom to another through space. (Gaseous atoms are far apart, not touching as metal atoms are.) As the electrons jump, they give up some of their energy to the atoms, putting these atoms into an **excited state** (a state of extra energy). The excited atoms, being unstable (higher on the energy hill), emit (give off) this extra energy as light and then return to their **ground state** (their most stable, or lowest, energy state) (Figure ▶ 5.16).

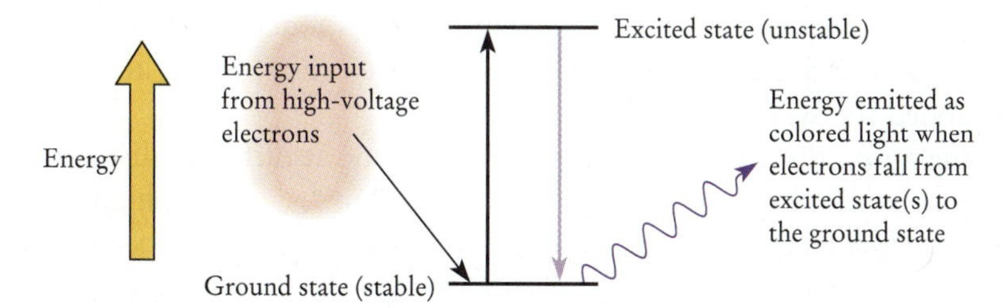

▶ Figure 5.16
High-voltage electrons in a discharge tube impart extra energy to electrons in atoms, raising them to an excited state. Electrons fall from this higher-energy, unstable state to the ground state, which is more stable. The energy released is emitted as colored light.

Light emitted from discharge tubes has discrete lines of color (definite and separate colors); and because the color of light is related to its energy, this light must also have discrete energies (definite and separate energies). Bohr inferred from this observation that instead of having just one excited state, atoms must have many of them, and that as an excited atom decays from these excited states to the ground state, it loses energy in discrete energy steps. The model then looked like this: Electrons in atoms are given extra energy by the high-voltage electrons in the discharge tube and are kicked from the ground state up to some excited state, which is unstable. These electrons then cascade downward in discrete energy steps, emitting as photons of light the energy that corresponds to each step, until they again reach the stable ground state.

The **energy levels** of electrons in atoms are **quantized**.

The idea that the electrons in an atom could occupy only certain energy levels, and not a continuous range of energies, was necessary to explain the emission spectra. Drawing on the ideas of Planck and Einstein, Bohr applied the concept of **quantized energy levels** to the atom. For an electron in an excited atom to go from one energy level to another required that a specific quantity of energy be given off in that transition (▶ Figure 5.17). Furthermore,

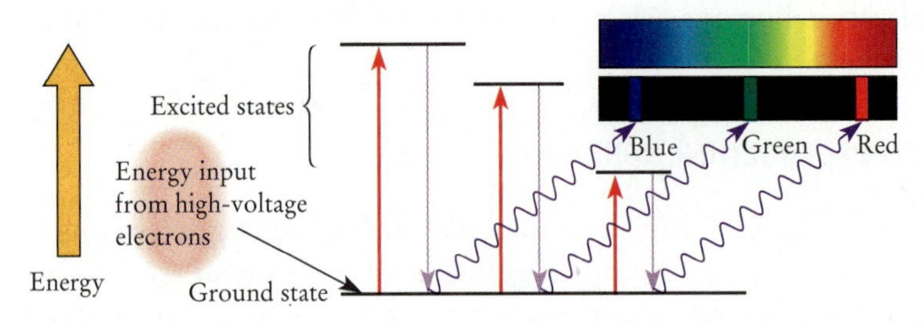

▶ Figure 5.17
Energy levels of excited states in atoms exist in discrete steps. Photons of specific energies—and specific colors—are emitted when electrons fall from one level to another level, or to the ground state.

Bohr was able to show a regular mathematical relationship between the energy of the steps and the energy of the light.

Another problem with the nuclear atom was that the electrons, being negatively charged, should be attracted to the positive nucleus and fall into it. To explain why they do not fall in, Bohr proposed his planetary model of the atom, an analogy with the solar system. In this model the small, dense nucleus is surrounded by the electrons in discrete orbits, much like the sun is surrounded by the orbiting planets.

An electron far from the nucleus has a greater energy than one close by.

In the atom, the electrons in their orbits reside at different distances from the nucleus, those being farther from the nucleus having the higher energy. The closer the electron's orbit is to the nucleus, the lower is its energy with respect to the nucleus. The ground state (most stable, lowest energy state) of any atom is one in which all of the electrons in it are as close to the nucleus as possible.

When a high-energy electron (from a discharge) hits an electron in its ground state orbit, that electron acquires extra energy and jumps to a new orbit of higher energy farther from the nucleus (an excited state). (It is possible by this process to impart enough energy to the electron to make it depart the atom completely, leaving a positive ion behind. This commonly occurs in discharge tubes.) Once the electron is in an unstable excited state, it will fall to an orbit closer to the nucleus (of lower energy) and emit a photon of light carrying a packet (quantum) of energy that is equal to the energy loss in the atom. The farther the electron falls in a given step, the more energy the photon will have (▶ Figure 5.18). This process is repeated as the electron cascades stepwise to its ground state orbit, closest to the nucleus.

Bohr knew there were deficiencies in his atomic model. It worked for the hydrogen atom and other one-electron situations, but even when modified somewhat the model could not account satisfactorily for the line spectra of larger atoms. Nonetheless, the Bohr model was useful. Even though it had its weaknesses and depended on arbitrary assumptions, it fit the behavior of atoms better than any previous model. The aim of a scientific theory is to

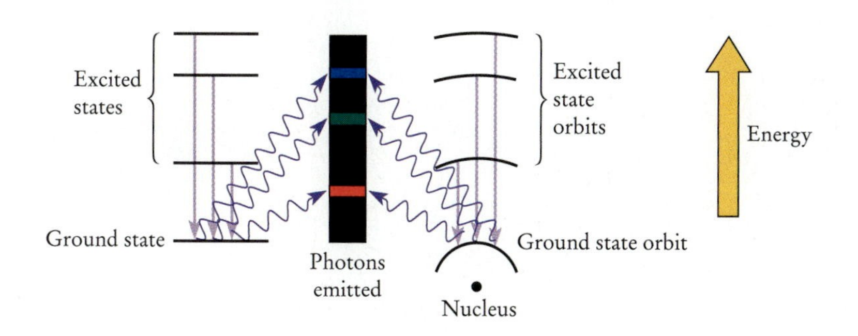

▶ Figure 5.18
In the Bohr model of the atom, electrons in excited state orbits have more energy the farther they are from the nucleus. The farther an electron falls toward the nucleus, the more energy the emitted photon will have.

make testable predictions, not to represent absolute truth, and the Bohr model did that in many instances. It also stimulated intense discussion and a vigorous search for a better model.

THE MODERN ATOM

Atoms, being very tiny, do not behave at all like the larger objects in the macro world of our everyday experience. Whatever reality is at the atomic level, it is very different from our own reality. So the atomic model we use today will at first seem a little strange. Rather than trace the development of atomic theory any further, we shall use hindsight to produce a picture of the structure of the atom.

ORBITALS

To begin, let's assume we could do a perfect experiment that looked at a single electron in a hydrogen atom and found where it was at any given instant in time. This would give us the location of the electron with respect to the nucleus at that instant. Then we would repeat the experiment at another instant in time. Then another, and another, and so on.

Every time we perform our experiment, we would record the location of the electron as a dot somewhere outside the nucleus (▶ Figure 5.19). Each successive time we asked the electron where it was, we would record a different dot, until our nucleus was surrounded by hundreds, then thousands, of dots (if we did the experiment enough times).

When we study the pattern that emerges, we see something puzzling. The dots we have drawn are all over the place in the space (in three dimensions) surrounding the nucleus. Some, surprisingly, are very close to the nucleus (very few, but some), and some are very far from the nucleus (again very few, but some), but most of the dots are clustered in a group around the nucleus in

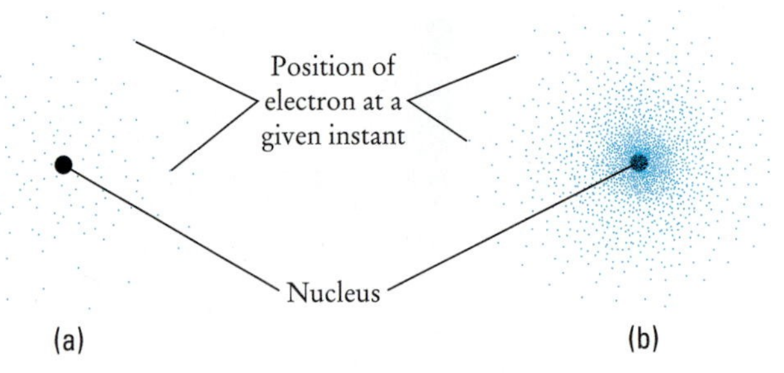

▶ Figure 5.19
Locating an Electron in the Atom. *(a) In a perfect experiment, the position of an electron about the nucleus would be indicated by a dot every time we make an observation of its location. (b) After many observations the dots would form a spherical pattern around the nucleus.*

what looks like a sphere. The sphere has fuzzy edges and does not have a sharp boundary, but most of the dots seem to be concentrated or grouped in this particular space.

How do we tell where the electron really is in the hydrogen atom? It turns out that we cannot. All we can really say is that on the average the electron will be in the fuzzy sphere most of the time. That is all. If we choose to draw a sharp boundary (in three dimensions) somewhere on the outside edge of the fuzzy sphere, we can *define* the volume that it encloses as the place we could expect to find the electron, say, 95% of the time. We cannot say where the electron actually will be at any instant because its position is unpredictable. All we can say is that there is a 95% probability of finding the electron inside the sphere. Neither can we say where the electron is *inside* the sphere at any instant (because 5% of the time it will be outside the sphere). And we cannot ask (do an experiment to tell) how the electron gets around inside the sphere, or outside of it. Our sphere around the nucleus is a volume in space in which we can reasonably expect to find the electron a high percentage of the time. In other words, there is a high probability of finding the electron inside this volume, which is called an **orbital** (▶ Figure 5.20).

An **orbital** is a volume in an atom in which there is a high probability of finding an electron.

MEASUREMENT OF VERY SMALL OBJECTS

The reason we must talk about the probability of finding an electron in a space or volume around the nucleus in our new model of the atom is that, in all of the experiments anyone has carried out, we find that we cannot simultaneously know the position and direction (actually, the momentum) of the electron in the atom with great precision. When, for example, we do an experiment that asks the electron where it is at a given instant, it will tell us; but it will not at that same time tell us where it is going. On the other hand, when we ask where it is going, we cannot simultaneously tell exactly where it is.

We cannot devise an experiment that will give us both kinds of information at once. This is because our measurement itself disturbs the electron, and interferes with what it is "doing" while we are trying to measure it. To "see"

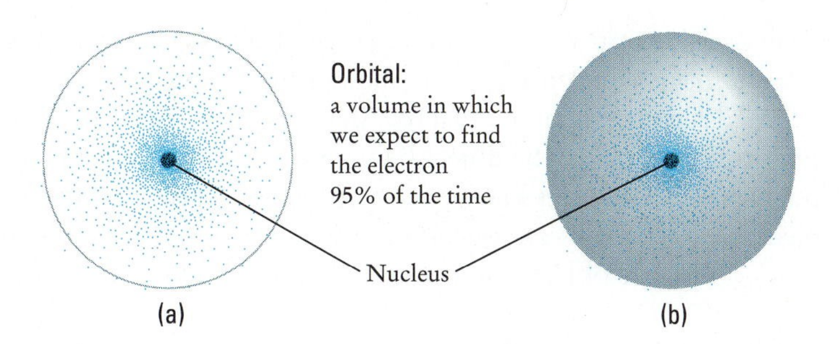

▶ Figure 5.20
Orbitals are Spaces. *(a) An orbital is defined as a volume about the nucleus in which the electron can be found 95% of the time. That is, 95% of the dots obtained from the perfect experiment are found inside the boundary that is drawn. (b) This volume often is shown as an enclosed space, in this case a sphere. Note that the electron is outside this sphere 5% of the time.*

any object we need to use light that has approximately the same wavelength as the size of that object (or preferably shorter). Because the electron is a very tiny object, very short wavelengths of light must be employed to observe it. As a consequence, this light has very high energy and disturbs the electron when it interacts with it. What we record is a very brief event, one of intense light energy interacting with an electron at that instant. We have no way of knowing what the electron was doing before or after the event (we weren't looking). During the event the electron is disturbed by our doing the measurement. When we ask it where it is, we knock it off course while we are determining its position; thus, we cannot tell where it was going. When we ask where it is going, we change its location while finding this out.

The act of measurement disturbs very small particles.

THE UNCERTAINTY PRINCIPLE

This phenomenon is observed in all the measurements we make of very small objects: the measurement always perturbs (disturbs) the object being measured. Larger objects also are perturbed by a measurement, but the perturbation (disturbance) is so small relative to the precision of the measurement that it is negligible. This is a statement, in words, of the **Heisenberg uncertainty principle.** This principle states that there are limits to our ability to make precise measurements, and that there is a built-in tolerance (error) associated with our measurements that is a part of the way nature works. This means that our measurements can never be exact and that there are limits to our ability to perceive reality. The late Jacob Bronowski, a mathematician and modern scholar of science, has called this the Principle of Tolerance. This conveys the idea that even though we can never measure anything exactly, we can know its properties *within certain limits of error.*

There is a built-in tolerance associated with *all* measurements.

In fact, the uncertainty principle enables us to know what the limits of this error will be and to actually calculate them using the equation:

$$\Delta x - \Delta p \geq h$$

The term Δx is the error in position or place of the object, and Δp is the error or uncertainty in its momentum (momentum = mass $\times$ velocity). The symbol $\geq$ means greater than or equal to. Again, Planck's constant, h, is a very small quantity. The smaller the object being measured, the greater the uncertainty in measurement. If x is known with a fair degree of certainty for a very small object, then p will be very uncertain. Likewise, if p is well defined, x will be uncertain. For large objects, both x and p can be known with a reasonable certainty.

Let's take two examples to illustrate this idea: a rifle bullet and an electron. Say that we know how fast each of these is going within one part per million, a high degree of precision. If the bullet weighs 10.0 g (about 0.35 oz) and is moving at 1.0×10^5 cm/s (1 km/s or about 2240 mph), we can know where it is within 6.6×10^{-27} cm.

The inherent uncertainty in position (Δx) for the bullet, and all other large objects like it, is so vanishingly small that it is negligible. Unless the bullet is traveling at an incredible velocity, it is possible for us to know where it is within very narrow limits. Although most rifle bullets travel much too fast to be seen, techniques such as high-speed photography make it commonplace for us to "stop" them at any instant in time.

We cannot know the position and direction of an electron at the same time.

By contrast, we can know the position of an electron having a mass of 9.1×10^{-28} g and moving at 3.0×10^8 cm/s (about 0.01 the speed of light, a fairly slow electron) with an error of 2.4×10^{-2} cm. This is very much larger (about 10^6, or a million times) than even the size of an atom (about 10^{-8} cm), and atoms *contain* electrons. This kind of uncertainty (tolerance) is inherent to all very small objects. If we know where one is going with any precision, we will have almost no idea where it may be, and conversely.

A NEW ANALOGY

As we build our model of the electronic structure of the atom, then, we find that we must describe an orbital in terms of the probability of finding an electron within it, and that this volume has rather fuzzy edges. We have presented this orbital as being spherical in shape. Larger atoms contain many electrons, however, and we want our orbital model to accommodate their numbers as well. (The Bohr model of the atom, we recall, did not work well for atoms larger than hydrogen.) How did the idea of a spherical orbital arise, and are there other conceivable shapes that orbitals can have? To answer these questions we must take yet another outside view—this one provided by the Austrian physicist Erwin Schrödinger, in 1925.

We know that emission spectra result when electrons cascade down from excited states in the atom, giving off photons of discrete energies as they fall from one energy level to another. Now, let's turn this process around and see what happens when an electron in an atom is bumped up from its ground state to a higher energy level in the atom.

The bumping up of electrons in atoms from a lower energy level to a higher one is called *excitation.* Input of energy in the form of a photon of light is needed to cause this. When this photon excites the electron, the photon itself is said to be *absorbed,* and it disappears. An atom in an excited state results. (Recall that this is in contrast to emission, in which an excited atom gives off energy in the form of a photon.) The energy necessary to excite an electron in the ground state to, say, the first excited state also is quantized. That is, only a photon having the amount of energy that corresponds exactly to the difference in energy between the ground state and the first excited state can cause the electron to jump. Too little energy will not make the electron jump part way, and too much energy will not make it jump to the first excited state, with a little energy left over. It takes a photon of light of *exactly the same energy* as the energy of the transition that is to occur.

Absorbtion

Photons of differnt energies absorbed by atom. Electrons moved to excited states.

Energy

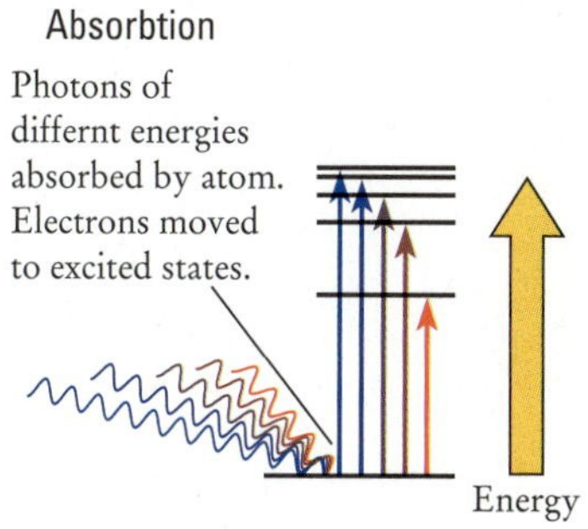

► Figure 5.21
Energy levels are shown for excited states in an atom due to its absorption of photons of the proper energies. Note that the higher excited states coalesce (lie ever closer together) in energy.

Photons can excite electrons to the second, third, or other excited states as well. ► Figure 5.21 shows the spacing that is typical of a series of excited states. Note that the higher the excited states are the closer they are in energy (they coalesce); they are not evenly spaced.

Chemical compounds can absorb photons of light in the same manner as atoms. Every compound has a unique set of excited states, just like atoms of different elements do. As a result, every substance has a characteristic **absorption spectrum,** which can serve as a fingerprint of that substance (► Figure 5.22). Theory also enables us to predict what an absorption spectrum will

Many substances can be identified from their **absorption spectra.**

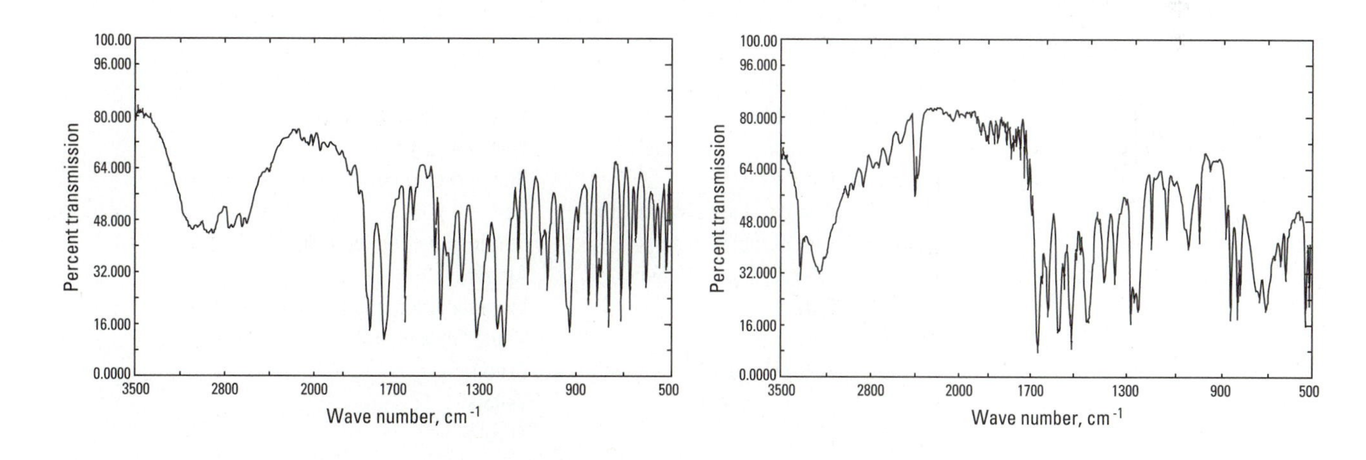

► Figure 5.22
Many organic substances can be identified by their infrared absorption spectra: (a) aspirin and (b) Tylenol. These spectra show the percent of transmitted infrared light (vertical axis) versus wave number in cm^{-1} (horizontal axis). The wave number is the reciprocal of the wavelength, 1/cm. Note that the peaks appear upside down. Each molecule has its own characteristic fingerprint.

COLORED GLASS

White light, such as that from the sun, a tungsten lamp, or a carbon arc, is actually a mixture of all colors of the visible spectrum. These colors can be seen if the white light is separated into a continuous spectrum by a prism. Sir Isaac Newton showed that once separated, this rainbow of colors could be put back together by using a second prism to make white light again. By contrast, the color black is the absence of all color; no visible light reaches the eye.

All wavelengths (and therefore all colors) of light can pass (be transmitted) through empty space, which contains no matter. When light passes through matter, however, some wavelengths are absorbed (taken up). For example, ordinary glass absorbs ultraviolet radiation but transmits all colors of visible light. Thus, the glass appears colorless. Because glass absorbs in the ultraviolet region, corrective lenses made of glass afford better eye protection from the sun's ultraviolet radiation than do many kinds of plastic lenses. Very dark sunglasses may or may not protect the eyes from ultraviolet light. Because these harmful wavelengths are invisible, it is impossible to tell whether they are being absorbed without using an instrument called a spectrophotometer.

Stained glass absorbs certain colors of visible light and transmits others. The color of the glass results from the combination of colors that are transmitted, not those that are absorbed (Figure 1). In other words, the color we perceive is the complement of the one that has been absorbed (Figure 2). For example, cobalt blue glass (containing CoO) absorbs red, orange, and yellow light, allowing violet, blue, and green light to pass through. Glass containing Cr_2O_3 absorbs red, orange, and violet wavelengths but transmits those of yellow, green, and blue, producing a green color. The following table gives other common ingredients used to produce various kinds of colored glass (Figure 3):

Ingredient	Color or Effect
Uranium compounds	Yellow and green
CuO	Blue, green, and red
Fine particles of selenium	Red
Fine particles of gold	Purple, blue, red, or pink
Fine particles of silver	Amber
PbO	Increases reflections: used in cut glass
CaF_2	White "milk glass"
Fe^{2+} compounds	Green
Fe^{3+} compounds	Yellow
MnO_2	Purple

Photochromic glass, which darkens on exposure to bright light, contains silver halides, usually AgCl or AgBr. When light strikes these compounds, it decomposes them into silver atoms and chlorine or bromine atoms, for example,

$$AgCl \xrightarrow{\text{bright light}} Ag + Cl$$

The tiny particles of metallic silver produced darken the glass. In the absence of bright light the atoms combine to reform the AgCl, and the glass becomes clear after a few minutes.

$$AgCl \xleftarrow{\text{dim light}} Ag + Cl$$

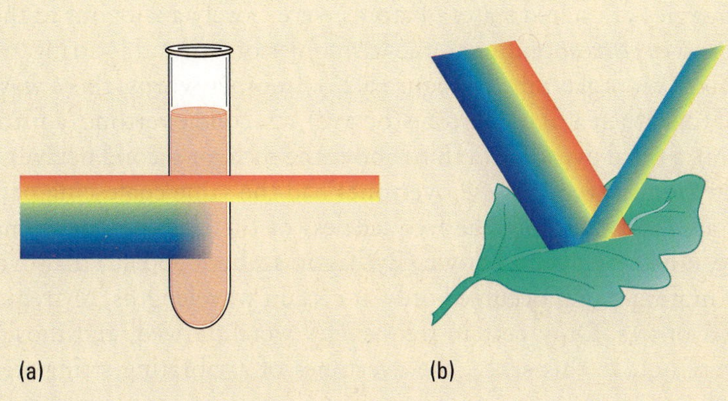

Figure 1
(a) An orange solution absorbs wavelengths of green, blue, and violet light. The remaining wavelengths of yellow, orange, and red light are transmitted *to the observer's eyes. (b) Light of red, orange, and violet wavelengths is absorbed when sunlight falls on a green leaf. The leaf appears green because light of yellow, green, and blue wavelengths is* reflected *to the observer's eyes.*

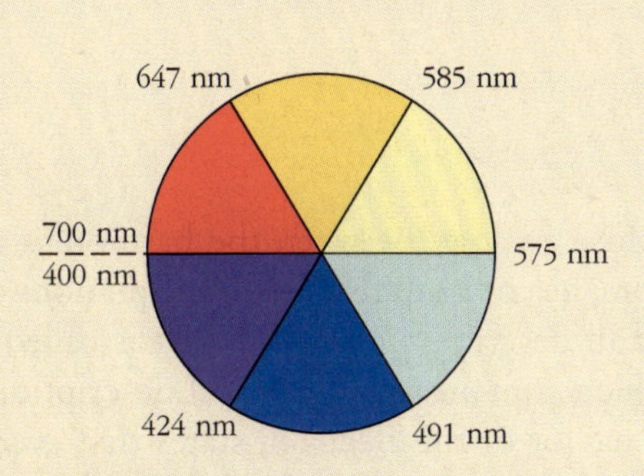

Figure 2
Complementary colors lie opposite each other on a color wheel.

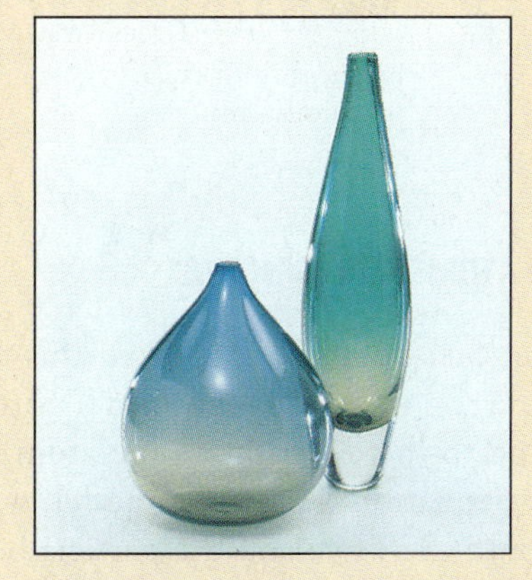

Figure 3
Colors in glass are produced by adding fine particles of metals or compounds of various metals to it.

look like for a compound that was heretofore unknown. Absorption spectroscopy is thus widely used to identify a host of chemical substances, including environmental pollutants, illicit drugs, natural products (compounds obtained from plants or animals, such as insulin), newly synthesized compounds of unknown structure, and even the composition of material in outer space and the distant stars.

The new view of atomic structure provided by Schrödinger grew out of his expert knowledge of the harmonics of sound. He observed that the patterns for the energies of excited states of atoms were closely analogous to those for the frequencies corresponding to the harmonics of a vibrating string, or a column of air in an organ pipe (► Figure 5.23). In other words, there was a fundamental tone, then a first overtone (octave), a second overtone (a fifth above the octave), a third overtone (a third above the fifth, or second octave), and so forth. The frequencies of these overtones had the same coalescing pattern as the energies (and therefore the frequencies) of the electrons in atoms. Furthermore, it had long been known (Pythagoras, about 350 BC) that overtones from a vibrating string occurred only at certain wavelengths, or frequencies, and not at others. Only certain frequencies were allowed, and those in between were not. In this sense, the overtones of a vibrating string were also quantized.

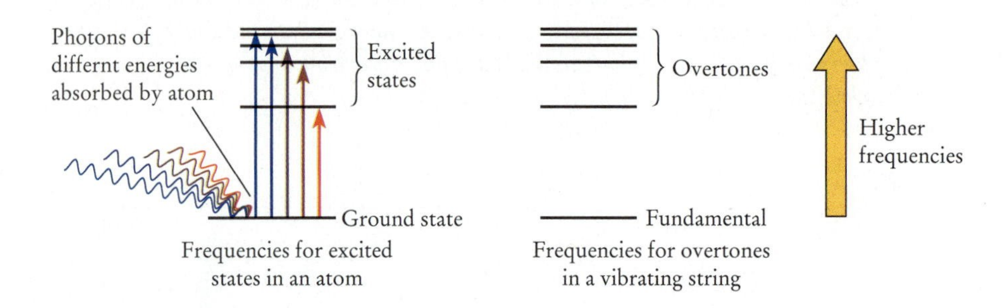

► Figure 5.23
An analogy exists between frequencies of excited states in an atom and those in a vibrating string. The higher frequencies in each have similar coalescing patterns.

THE ATOMS ARE "SINGING"

Once Schrödinger recognized this close analogy between the harmonics of sound and the energy levels in the atom, he could apply the wave equations of sound to the electrons in the atom and describe them with what is called a wave equation. His **wave model,** which contains mathematical descriptions of both the wave and the particle behavior of the electron, states that every electron in an orbital has a specific energy associated with it, and no other energy. It is an allowed value, much like the frequency associated with a particular overtone of a vibrating string. Every electron in an atom has a unique energy associated with it and with the orbital in which it resides.

Orbitals have the properties of standing waves.

The shapes of these orbitals (regions in space around the nucleus), each of which can contain two electrons, are analogous to the shapes produced by standing waves. How can a volume have the shape of a standing wave? Let's again consider the analogy of the vibrating string. The fundamental note is a vibration of the string that involves the whole string and encloses, or envelops, the space in which the string moves (► Figure 5.24). The orbital of lowest energy in an atom is analogous to this envelope, or enclosure of the string's vibrations, and it is shaped like a sphere. We must remember, though,

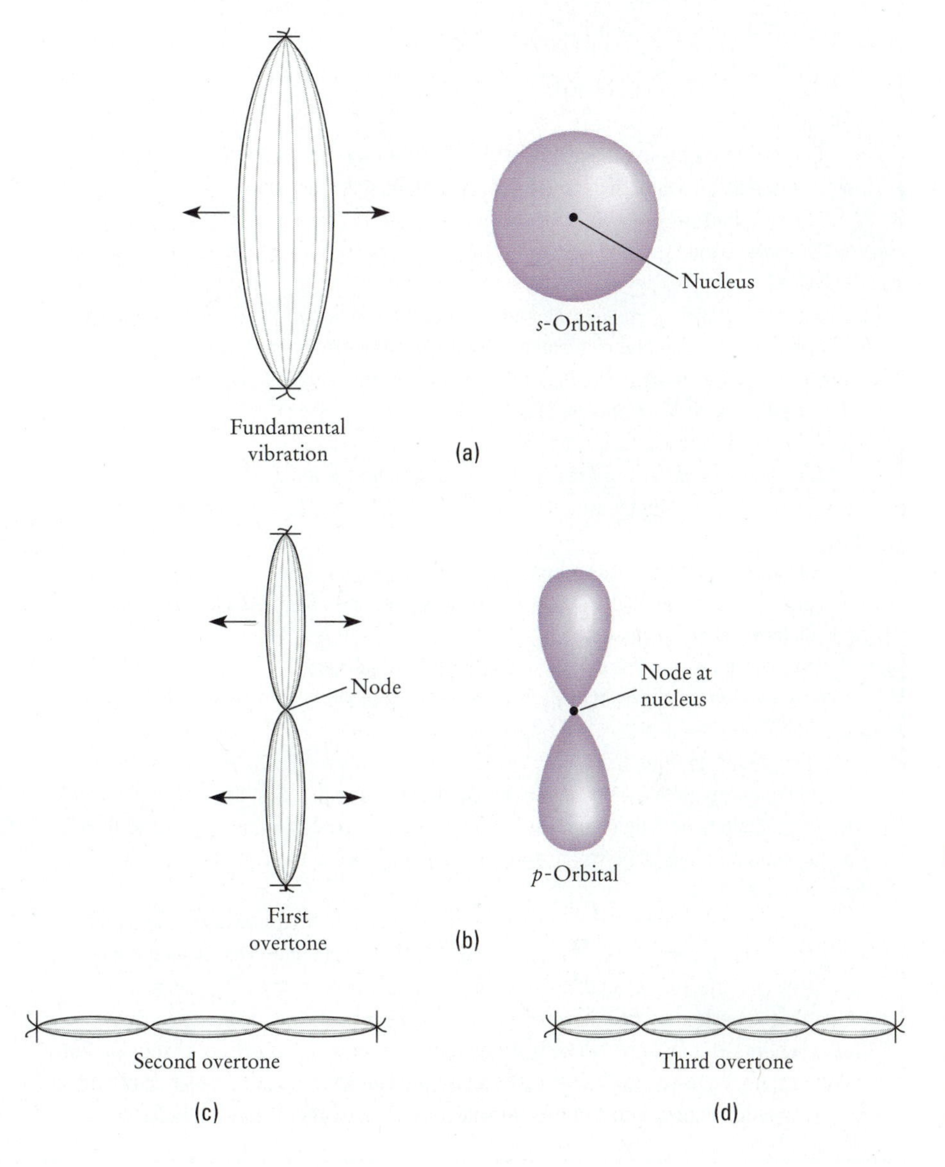

► Figure 5.24
Envelopes of standing waves in vibrating strings are shown along with their analogous orbitals. (a) An s *orbital corresponds to the fundamental vibration and (b) a* p *orbital to the first overtone. The second and third overtones, (c) and (d), have analogous* d *and* f *orbitals (not shown).*

PRETTY LIGHTS

"Hey everybody! Come out and look!" It's Carl, a transfer student from the University of Alaska. "Turn off the lights." Everyone hurries outside in various states of dress and alarm. "Look!" In the north you see shimmering lights, mostly greenish but with flickers of blue and purple, like a dim shifting curtain.

"What is it?"

"The aurora! Aurora Borealis!"

"This far south?"

"Oh, they're visible in Mexico when the sun flares up. Of course, this isn't like at home where they're right over your head." Everyone has a question.

"The sun does it?"

"The solar wind."

"Wind?"

"Not really wind . . . high-speed charged particles the sun throws out. When they hit Earth's magnetic field they make an electric current around the poles and that excites gases in the atmosphere. Ozone makes the greenish light, and nitrogen lower down makes the purplish and blue light."

"How do you know all this, Carl? You're a so-called artist."

"Ah, but the University of Alaska's the aurora capital of the world. You can't go near the place without learning something about it. Besides, I'm an aurora artist."

"He means he bends glass tubes into weird shapes, fills them with gas, and turns them on." That's Rhonda. The others don't know enough to be impressed, but you do, and you look at Carl surprised.

"The aurora is the same thing as neon lights?" He shrugs.

"Shoot an electric current through a gas . . . same thing."

Later you talk to Rhonda.

"Do I understand this right? If you run electricity through a gas, the electrons in the gas atoms jump to excited states and then drop to less excited states, and when they do, they emit particles of light . . . photons. And the color of the light depends on what kind of gas it is?"

"That's partly right, but come with me." You follow her into the kitchen and watch while she lights the stove and drops a pinch of salt into the flame. The flame turns yellow. "Any kind of energy, not just electricity, excites atoms. A little bit of the sodium chloride is vaporized by the heat, and gases at low pressure radiate the spectrum of their element. That color of yellow is the dominant color of the spectrum of sodium. But solids radiate too. If you take a mass of closely packed atoms, such as a piece of iron, and put enough energy into it to excite its atoms to their highest states, it radiates white light, and we say the iron's white hot." She waits while the scientific meaning of *white hot* soaks in. Your mind is racing:

"So the atoms in white-hot iron radiate at all the wavelengths of the spectrum, and all the colors together make white!"

"Bravo!"

"What about red hot . . . not as hot?"

"Right. Not as excited."

"Light bulbs: white hot?"

She nods: "Tungsten filament, a solid."

"What about the emission spectrum?"

"The atoms of a gas at low pressure can't be excited to white heat because they're too unconfined, so the electrons of a gas can be excited only to certain wavelengths. Light of particular wavelengths is the signature of an element in its gaseous state." She drops another pinch of salt into the flame. "If you shine sodium's light through a prism, the spectrum-rainbow is dark except for two pairs of bright lines where this yellow is, one orange line, and a double blue-green one. That's sodium's spectrum. All the elements show lines at particular wavelengths of the spectrum, and we can identify them that way. You understand?"

"I think so. Each kind of atom emits photons at certain energies. Are the colored flames from burning gift wrap because of that?"

"Yes. Mostly heavy metals, so don't burn it. They shouldn't even make it that way!"

"OK. Tell me about the absorption spectrum."

"It's the exact opposite of the emission spectrum. If you shine white light through a gas, the same electrons that make the emission spectrum when they're excited absorb photons of the same energies from the white light. So when you shine that light through a prism, you get the whole rainbow except for dark lines at the wavelengths of the gaseous element. In other words, you get a continuous spectrum with dark lines where the bright lines for that element would be in an emission spectrum. That's how we know what elements are in the sun."

"By the absorption spectrum?"

"The photosphere, the luminous surface of the sun, gives off white light . . . ."

"But isn't the sun made of gas?"

"Under extremely high pressure."

"Ah . . . closely packed atoms."

"Yes. But the chromosphere, it's outer atmosphere, is a low-pressure gas, which absorbs the wavelength of the elements in the light from the photosphere. That's how we've learned what elements are in the stars, and we've learned what elements and compounds are in the gases in space between us and the stars the same way."

that an orbital describes a volume in which there is a high probability of finding the electron, and it does not have a well-defined boundary like that of the envelope described by the vibrating string.

The first overtone of the string (octave) has two equal envelopes of space with a node in the middle, a place where the string is not "moving." The next orbital in an atom corresponds to this first overtone; it also has two envelopes with a node between them. This **node** occurs at the center of the nucleus and represents a place of zero probability of finding the electron. This orbital looks like a fat, squatty dumbbell. Spherical orbitals are called *s orbitals,* and the dumbbell-shaped orbitals are called *p orbitals.*

The second overtone of the vibrating string is divided into three equal envelopes with two nodes between them; another orbital, a *d orbital,* corresponds to this overtone. The third overtone of the string likewise has an analogous *f orbital.* The vibrating string is always divided into equal lengths, and in no case is there a fractional division of the string. To divide the string unevenly would result in vibrations that would give the note a strange sound and would cause the string to stop vibrating because of interference. This analogy can be carried over to the electrons in orbitals in the atom and the fact that their energies are quantized. In our new model of the atom, the electrons are "singing" in some kind of harmony, and their vibrations must be "allowed." Sour notes cannot be permitted because the standing wave will not continue to vibrate except at certain harmonic frequencies, and thus at specific energy levels.

ORBITALS IN LARGER ATOMS

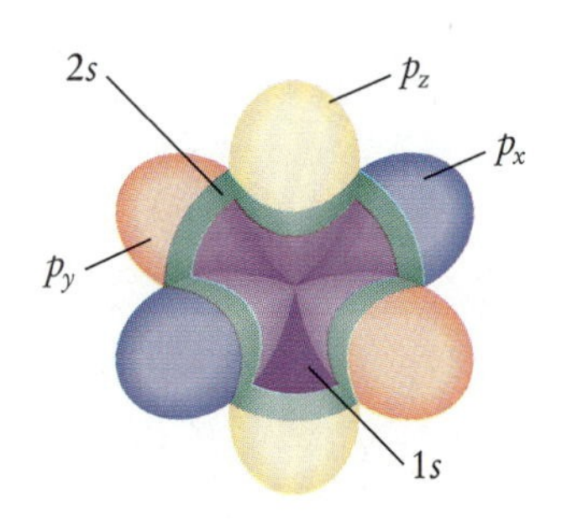

▶ Figure 5.25
Superimposition of Orbitals In A Large Atom. *Orbitals, being regions in space, can be placed upon one another and can be superimposed.*

Every orbital can hold two electrons, but that is all. For any atom larger than helium (which has two electrons), it is necessary to use more than one orbital to contain all of the atom's electrons. For example, an atom of silver (Ag) has 47 electrons, which means that we need a minimum of 24 orbitals to accommodate them.

Cramming 47 electrons in 24 orbitals around the nucleus means that we must somehow superimpose the orbitals. This at first seems impossible, but orbitals are not solid. Orbitals are regions of space, and it is very simple to place one volume, or many volumes on one another in space (space upon space). An orbital can be visualized as the territory of two electrons, and these territories can overlap one another and can even occupy the same space (▶ Figure 5.25).

WAVE PROPERTIES OF THE ORBITALS AND CHEMICAL PERIODICITY

Certain electrons in atoms, called **valence electrons,** are involved in forming chemical bonds. The orbital picture of how these fit into atoms accounts for many of the physical and chemical properties of the elements and explains the arrangement of the elements in the Periodic Table. Valence electrons occupy large orbitals on the periphery of atoms (outer orbitals) and usually are those farthest from the nucleus. Thus, they have the highest energy with respect to the nucleus and are the easiest to remove from the atom. (Or, they are least attracted to the nucleus because they are most distant from it.)

According to the wave model, valence electrons can occupy various kinds of outer orbitals in atoms, depending on the total number of electrons an atom contains. This is a periodic, mathematical relationship which repeats itself as the the number of electrons in atoms gets successively larger. In addition, as the number of electrons in atoms increases, more kinds of orbitals can be used to accommodate electrons. The kind of outermost orbitals that are occupied by valence electrons is what determines an element's place in the Periodic Table. (It is not necessary that all of an element's valence electrons reside in outermost orbitals; a chlorine atom has seven valence electrons, for example, but only five of them occupy outermost *p* orbitals.)

The properties of the elements depend on the kind of outermost orbitals their **valence electrons** occupy.

The Periodic Table in ▶ Figure 5.26 is divided into blocks of elements to show the kind of outermost orbitals that the valence electrons occupy. Note that valence electrons in the Group 1A and 2A elements occupy outermost *s* orbitals. For elements in Groups 3A through 8A, the outermost valence electrons are in *p* orbitals; for the transition elements (B-Group elements) they are in *d* orbitals; and for the rare earth elements, they are in *f* orbitals. Note also that as the period number increases, the number of orbitals available to hold electrons goes up dramatically.

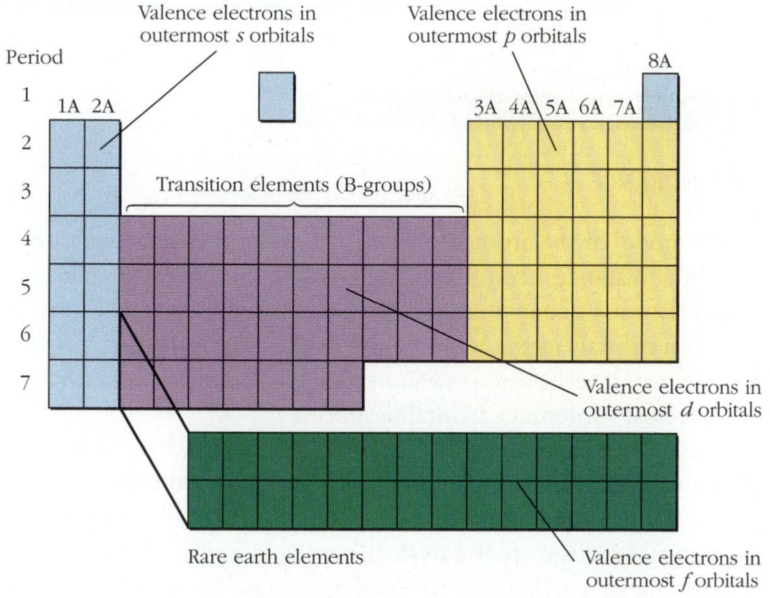

▶ Figure 5.26
Blocks indicate the elements in the Periodic Table for which the valence electrons are in the same kind of outermost orbitals.

5.3 What kinds of outermost orbitals do valence electrons occupy in each of the following elements?

a. cobalt
b. potassium
c. uranium
d. barium
e. chlorine
f. lead
g. sulfur
h. silver

The *Periodic Law* now can be restated in terms of modern theory: The properties of the elements are periodic functions of their atomic number. The **atomic number** of any element is simply the number of electrons contained in its uncombined neutral atom.

5.4 How many electrons are there in uncombined neutral atoms of the elements listed below?

a. magnesium
b. In
c. As
d. Te
e. silicon
f. bromine
g. iron
h. mercury
i. potassium
j. Mn

LIMITS TO SCIENTIFIC KNOWLEDGE

IMPOSED IGNORANCE

In the Bohr model of the atom, the energy levels of the electrons' orbits (and also the orbits' distances from the nucleus) are quantized, and energy-jumps between levels must occur in steps. The same is true for transitions of electrons between orbitals in the wave model of the atom: the electron can jump from one energy level (orbital) to another, but *it can never be in between!* (This is a necessary inference from line spectra.)

How the electron gets from one energy level to another, or what it does while going there, are questions that cannot be answered because we cannot devise an experiment to observe what happens. Somewhat differently, the uncertainty principle tells us that a built-in error is associated with all the measurements we make, particularly for small particles. Thus, there appears to be something inherent in nature that limits our ability to observe it. Our awareness of this "imposed ignorance," an attribute of nature that makes our knowledge about it incomplete, has had a profound effect on science.

This realization contrasts sharply with the view of Newtonian physics, which held that if we could obtain enough data and had the right equations for describing the laws of nature, we could know everything and even predict the future with certainty. In this view the universe is a machine and its entire story, including human history, is like a movie on a film strip set in motion at the beginning of time. The projector of events runs the film until its end, and everything that happens on the screen has been predicted beforehand but is not seen until its moment. (Parallel views common at this time were those of scientific determinism and religious predestination.)

WAVE/PARTICLE DUALITY: THE COPENHAGEN VIEW

The limits of scientific knowledge also are embodied in the dual nature of light: is it waves, or is it particles? The Newtonian view of physics said it could not be both. When experimenters all over the world looked for an answer to this question, their results were astounding. The nature of light

WHY MUST A PHOTON BE MASSLESS?

Energy is part of something that is moving and also manifests itself in motion: when energy is not in motion, we cannot observe it. When an object is made to go increasingly faster, more and more energy must be put into it (from an external source) to accelerate it. This means that the object contains more and more energy (in *it*) the faster it is made to go. Because energy is mass, the faster an object goes, the more mass it has (due to the extra energy it now contains). Thus, the object's mass increases continually as its velocity continues to increase, until at the velocity of light, c, its mass would become infinite. Einstein pointed out that this is the reason that nothing can travel faster than the velocity of light. The idea of a massless photon fits in nicely with this concept of energy/mass. A photon must be massless when it travels at the velocity of light (or it could not go that fast), but it also can behave as if it has mass by virtue of the fact that it has energy. When a photon strikes another object, it imparts its momentum (a particle property related to mass) to the object it hits and causes that object to move, much like a collision of one billiard ball with another. Because the photon has a very tiny momentum, the deflection of the object it strikes is very small and usually is not noticed. However, satellites traveling in outer space are sometimes observed to move *away* from the sun because of this effect.

(whether particles or waves) depends on the experiment. When an experiment was designed to "ask" light if it was wavelike, light said yes. When asked if it was particlelike, light also said yes. Light passing through a prism behaves as waves, but in the photoelectric effect it exhibits the behavior of particles. There was no experiment then, nor is there one today, that can be designed to ask light if it is both wavelike and particlelike at the same time. Here again, nature has imposed limits on what we can know.

It is our present model of light that gives light this dual nature—of being a wave and a particle at the same time—but we cannot say that duality is the actual nature of light. We are not dealing with absolute reality here, but rather with an idea that accounts for what we observe. Because they serve to explain and predict, our models of light cannot be called incorrect, but they are incomplete and imperfect because we cannot observe everything we need to complete the picture.

In this micro world the answer nature gives us depends on what we ask nature to tell us, and we the observers are not detached from the experiment. Being part of it, we influence the outcome. This realization about the nature of experimental knowledge makes it difficult to accept the notion of scientific objectivity and leads to the Copenhagen view of reality. It does not mean, however, that we cannot do science in any objective way. Rather, we can know what the limits of our scientific knowledge are.

Although this realization about scientific knowledge at first caused much concern, it forced scientists to break away from the cut-and-dried ideas that shackled Newtonian physics and to look at the world from a very different perspective. An atmosphere of excitement in science grew from a new-found freedom to explore ideas that would have been considered implausible, unfruitful, or even ridiculous before.

OTHER DUALITIES: THE TAO

If light has a dual nature, do other things in nature behave in an analogous way? The answer is yes. We are all familiar, for example, with the famous **Einstein equation,** $E = mc^2$. This equation relates mass *(m)* and energy *(E)* and is a mathematical expression that matter is energy, and energy is matter. This principle, that matter and energy are interconvertible, provides the basis for the conversion of part of the mass of uranium to energy in nuclear reactors.

Our new freedom to question traditional scientific models results in some other mind-boggling ideas. For example, DeBroglie (1924) took the matter of the wave/particle nature of light and the sameness of matter and energy (matter/energy) one step farther. Does matter have wavelike properties? This question was soon answered in the affirmative by Davisson and Germer (1933), who showed that electrons can display a diffraction pattern like light

(► Figure 5.27). Thus matter was shown to have wave/particle duality. Probably the most familiar application of this principle is the electron microscope, in which electrons (which are considered to be particles) do behave like light. **DeBroglie's equation,** $h\nu = mc^2$, shows the relationship between the mass (an attribute of particles) of any object and its frequency (an attribute of waves). As the mass of an object increases so does its frequency, in direct proportion.

The ideas of the wave/particle duality of light and the sameness of matter and energy, which were becoming accepted in science in the late 1920s, were incompatible with the common beliefs of the time. Western philosophical thought is reductionistic and was/is based on ideas of right/wrong, yes/no, black/white, male/female, either/or. In Western thought there is little room for a middle ground of values, and Western philosophers have spent many centuries trying to rationalize ethical, moral, and philosophical ideas that

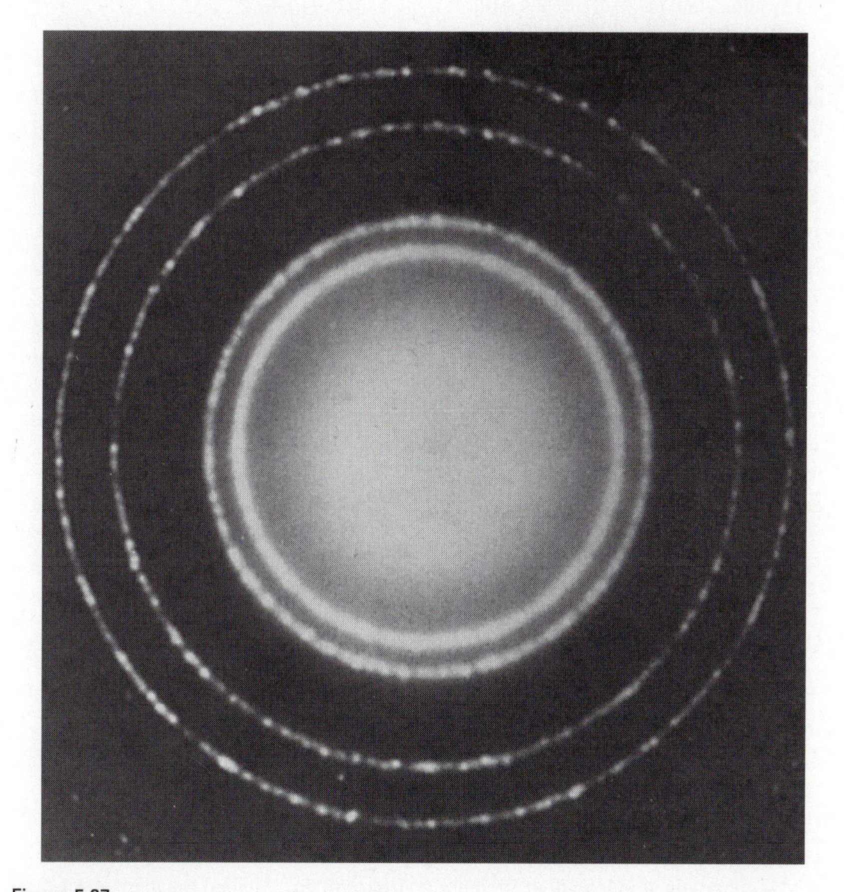

► Figure 5.27

This electron diffraction pattern is produced when aluminum foil is bombarded with a beam of electrons. It is an interference pattern, similar to the one shown for light in Figure 5.10 (b), and is evidence that electrons have wave properties.

ENERGY FROM THE NUCLEUS: MATTER/ENERGY CONVERSIONS

Nuclei of all atoms (except hydrogen) are made up of protons (p) and neutrons (n). Protons are very small atomic particles, called nucleons, which carry a 1+ charge. Neutrons are nucleons which have almost the same mass as protons and carry no electric charge. Combined in the nucleus, protons and neutrons constitute most of the mass of an atom.

The number of protons in the nucleus of an element is given by the atomic number of the element. (Recall that this also indicates the number of electrons the neutral atom contains.) Thus, a carbon atom has six protons in its nucleus, a barium atom has 56 and a mercury atom 80. The atomic number of an element (the number of protons and electrons its atoms contain) is often written as a subscript to the left of its chemical symbol, such as ${}_{6}C$, ${}_{56}Ba$, and ${}_{80}Hg$.

All nuclei of a given element contain the same number of protons. However, the number of neutrons in the nucleus can vary. For example, the nucleus of a carbon atom can contain six, seven, or eight neutrons, in addition to the six protons. Atoms of the same element that contain different numbers of neutrons in their nuclei are called isotopes. The element carbon, then, has three isotopes, one containing 12 nucleons (6p + 6n), another containing 13 (6p + 7n), and the third containing 14 (6p + 8n). This sum of nucleons (p + n) is called the mass number of an isotope and is indicated by a superscript placed before the element's chemical symbol, thus ${}^{12}_{6}C$, ${}^{13}_{6}C$, and ${}^{14}_{6}C$.

The nuclei of isotopes ${}^{12}_{6}C$ and ${}^{13}_{6}C$ are stable. However, the nucleus of ${}^{14}_{6}C$ (called *carbon-14*) is unstable and spontaneously undergoes a nuclear transformation (called radioactive decay) to produce ${}^{14}_{7}N$ and an electron, or,

$${}^{14}_{6}C \longrightarrow {}^{14}_{7}N + {}^{0}_{-1}e \text{ (or } \beta^{-})$$

In this particular process, called beta decay, a neutron in the nucleus of the carbon atom is converted to a proton, and an electron is expelled to maintain charge balance. (Recall that an electron and a β-particle are the same thing.) Adding a proton to a carbon nucleus increases the nuclear charge by 1+, to 7+, converting a carbon atom into a nitrogen atom. Nuclear transformations usually, but not always, result in the conversion of one element into another.

In chemical reactions the mass of the products equals the mass of the reactants (the law of conservation of mass). In spontaneous nuclear reactions, the total mass of the products is very slightly *less* than the total mass of the reactants. That is, a tiny amount of matter appears to be lost in the process. But it is neither lost nor destroyed. Instead, it has been converted into energy according to Einstein's equation, $E = mc^2$. Converting a small amount of matter produces an enormous amount of energy because the conversion factor, c^2, is a very large number (9.00×10^{16} m²/s²). As a rule, nuclear changes produce about 10^5 more energy than do chemical reactions involving the same amount of matter (in moles, Chapter 7).

Production of energy from the nucleus takes advantage of this conversion of matter to energy when nuclear reactions occur. Two kinds of nuclear processes, fission and fusion, can produce the large quantities of energy needed today.

In *fission,* the nucleus of a large atom, most commonly $^{235}_{92}U$ (uranium-235), breaks apart into nuclei of smaller atoms when bombarded with slow (low-energy) neutrons. When a slow neutron strikes a $^{235}_{92}U$ nucleus, it is absorbed and forms a $^{236}_{92}U^*$ (uranium-236*) nucleus, which is unstable. The uranium-236* nucleus undergoes fission and breaks apart (in one of several ways) to produce smaller atoms, neutrons, and energy, as shown in Figure 1.

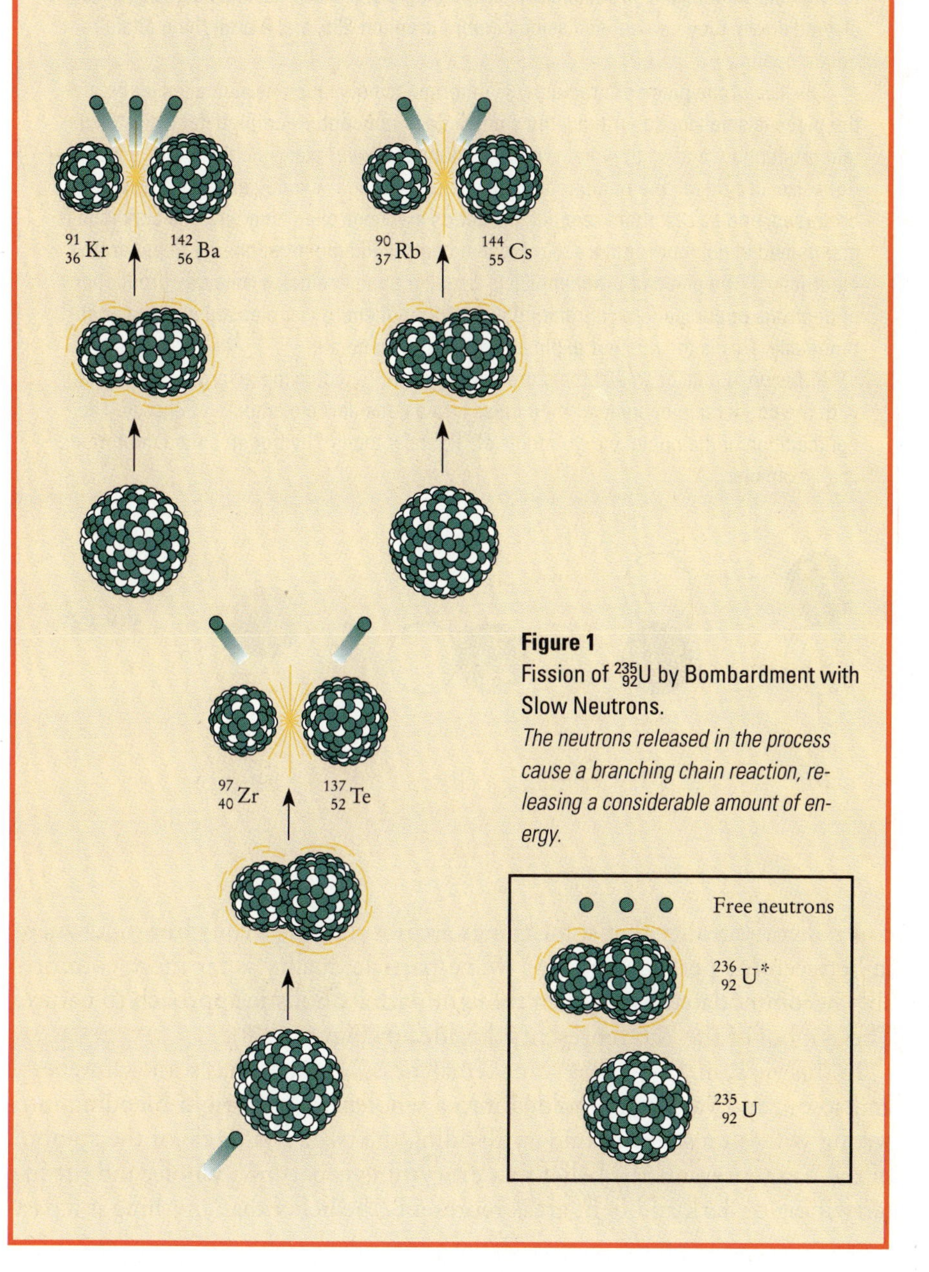

Figure 1
Fission of $^{235}_{92}U$ by Bombardment with Slow Neutrons.
The neutrons released in the process cause a branching chain reaction, releasing a considerable amount of energy.

ENERGY FROM THE NUCLEUS: MATTER/ENERGY CONVERSIONS (Continued)

Note that although there are various paths by which the uranium-236* nucleus can break apart, in every case two or more neutrons are formed from an investment of only one neutron. This causes the number of neutrons produced to increase rapidly. If these newborn neutrons are slowed down, they can cause fission of more uranium-235, and a branching chain reaction occurs.

The size of the piece of uranium-235 determines how much fission takes place. If the piece is small, or subcritical, the sample does not capture enough neutrons to sustain production, and so the chain reaction stops (or never starts). If the piece is just the right size, or critical, the number of neutrons consumed is exactly equal to the number produced, and so the chain reaction proceeds in a controlled manner. This condition is maintained in nuclear reactors to release a controlled amount of energy to generate electricity. If the piece of uranium-235 is large, or supercritical, a runaway production of neutrons occurs, and so a tremendous amount of energy is released almost instantaneously. This is the nuclear explosion of an atomic bomb.

In *fusion,* nuclei of small atoms are combined, or fused, to make larger ones, again with the conversion of some of their mass into a considerable amount of energy. The combination of the nuclei of two atoms of ${}^2_1\text{H}$, an isotope of hydrogen called deuterium, is an example:

+ + → + + **energy**

$${}^2_1\text{H} + {}^2_1\text{H} \longrightarrow {}^3_2\text{He} + {}^1_0\text{n}$$

contradict the reality that some things are not right or wrong but somewhere in between (or perhaps neither). Wave/particle duality is far more comfortably accommodated by Eastern thought with its holistic approach to nature. The symbol of the *Tao* represents this idea (► Figure 5.28).

► Figure 5.28
The *Tao.*

In this view, maleness, has some femaleness, blackness has some whiteness, and so on, and values are molded into a wholeness, with right blending into wrong without a sharp boundary dividing the two. The circle of the symbol of the Tao represents the belief that everything is part of a whole, and the intertwining of dark and light areas represents the belief that anything is a part

Although naturally occurring hydrogen contains only about 0.015% deuterium ($^{2}_{1}H$), enough of it is available on Earth (about 2.5×10^{13} tons) to provide an almost unlimited supply of energy. Although fusion reactions occur in hydrogen bombs, fine control of these reactions is difficult. For one thing, the repulsive forces between the two positively charged nuclei must be overcome before they will react with one another at all. This requires the very high kinetic energies and very high temperatures (about 10^8 K) of a plasma (a hot soup of electrons and nuclei). At the same time, the nuclei must be held together, or contained, long enough to react. Much work has been done trying to produce energy from controlled nuclear fusion, but without success. Whether this will eventually be possible is still uncertain.

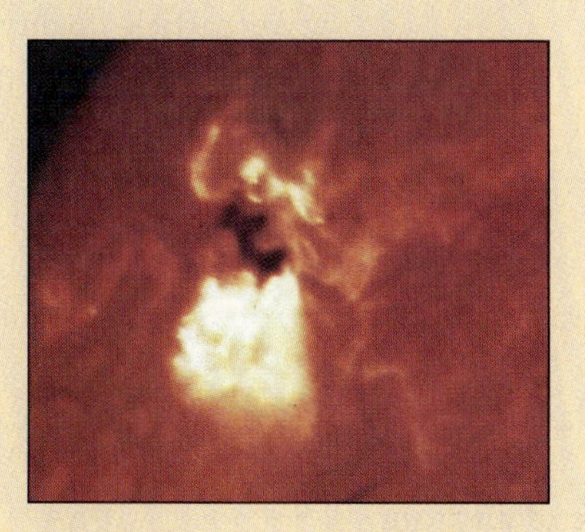

Figure 2
A Solar Flare.
The fusion reactions that take place in the sun are the ultimate source of most of our energy.

The sun is a huge nuclear fusion reactor (Figure 2). The sun's intense gravity holds everything together and its plasma core is intensely hot. Here, hydrogen is converted to helium and a tremendous amount of energy in a series of steps. The overall fusion reaction is,

$$4\,^{1}_{1}H \longrightarrow \,^{4}_{2}He + 2\,^{0}_{+1}e + \text{energy}$$

where $^{0}_{+1}e$ is a positive electron, or positron.

of everything else, and in fact *is* everything else. Someone once summarized this kind of thinking with the comment that "Everything is just a smaller or larger piece of something else."

Many ideas in science, from those of particle physics to those in environmental biology, seem to be pointing in the direction of Eastern philosophy, and a holistic view of nature is emerging in science. Science as we know it probably could not have originated in the East because it is necessarily based on a reductionistic view of nature. It is interesting that many modern scientific theories now look eastward for their philosophical reconciliation.

DID YOU LEARN THIS?

Part 1. For each of the statements given, fill in the blank with the most appropriate word or phrase from the following list. Do not use any word or phrase more than once.

frequency	emission spectrum	radioactive decay
wavelength	ground state	Bohr model of the atom
photon	excited state	photoelectric effect
velocity	continuous spectrum	color
prism	empty space	direct proportionality
scintillations	J. J. Thomson	inverse proportionality
nucleus	β-particle	wave/particle duality
quantized	E. Rutherford	matter/energy duality
M. Faraday	H. Becquerel	higher energy
α-particle	scattering	lower energy

a. _______ That which energy transitions of electrons within atoms are said to be
b. _______ An analogy to the solar system but on a much smaller scale
c. _______ $E = mc^2$
d. _______ The most stable state of an atom
e. _______ Part of a spectroscope
f. _______ Obtained when light from an incandescent lamp is passed through a spectroscope
g. _______ Proposed that matter is electrical in nature
h. _______ Flashes of light on a phosphor screen
i. _______ The relationship of an electron at some distance from the nucleus compared with one close by
j. _______ Has the symbol, λ
k. _______ The observed information that tells us electrons are falling from excited states of atoms in a discharge tube
l. _______ Proposed the plum pudding model of the atom
m. _______ The part of the atomic model required to explain why α-particles were scattered backward
n. _______ An indicator of the energy of light
o. _______ The condition of an atom having extra energy
p. _______ The relationship of the frequency of light to its energy
q. _______ A helium ion
r. _______ Has the units of 1/s
s. _______ A property of very tiny particles that prevents us from describing results of all measurements on them with one single model

t. _______ The process of larger atoms spontaneously falling apart into smaller ones
u. _______ The small, massless particle of light
v. _______ Makes up most of the volume of any atom
w. _______ The relationship of wavelength to frequency
x. _______ Conversion of light energy to electrical energy

Part 2. For each of the statements given, fill in the blank with the most appropriate word or phrase from the following list. Do not use any word or phrase more than once.

Niels Bohr	node	uncertainty principle
E. Schrödinger	orbital	Groups 1A–2A
W. Pauli	standing wave	B-Groups
M. Planck	absorption	Groups 3A–8A
W. Heisenberg	emission	position
probability	perturbation	momentum

a. _______ The part of the Periodic Table in which outermost *p* orbitals are occupied by valence electrons
b. _______ That which happens to tiny particles when we attempt to make measurements on them
c. _______ The likelihood of finding a particle at a particular place
d. _______ Used as the basis for the analogy on which the *shapes* of orbitals are based
e. _______ That which happens to a photon when it excites an electron in an atom from the ground state to an excited state
f. _______ The part of an orbital in which there is zero probability of finding an electron
g. _______ That which cannot be known for an electron if its momentum is known with certainty
h. _______ The part of the Periodic Table in which outermost *d* orbitals are occupied by valence electrons
i. _______ Noted the similarity between the harmonics of sound and the excited states of atoms
j. _______ An envelope in space in which there is a high probability of finding an electron in an atom

EXERCISES

1. When Michael Faraday proposed, in the mid-1880s, that matter is electrical in nature, his suggestion was resisted by the majority of the scientific community. Historically, what belief was held by

many scientists working at that time which would in part explain this dismissal of Faraday's idea?

2. Several discoveries in science that were made in the late 1800s provided evidence that atoms were not indivisible (that is, they had constituent parts). What were they, by whom were they made, and what was the significance of each?
3. Which part of the atom makes up most of its mass, the positive part or the negative part? Who first showed that this is the case?
4. What are α-particles? Where do they come from? What are phosphors? What are scintillations?
5. Why was the plum pudding model of the atom inconsistent with Rutherford's scattering experiments? What did Rutherford's experiment show about the nature of the atom that was completely unexpected? What observation led him to conclude that the atom must have a nucleus? Why was Rutherford's nuclear model of the atom greeted with skepticism when it was proposed—that is, considering the state of scientific knowledge at that time, what objections were brought against it?
6. What is a discharge tube? What is a spectroscope? (Make a rough drawing of them together.) Distinguish between a continuous spectrum and a line (or emission) spectrum. How are continuous spectra produced? What kind of spectrum is obtained from discharge tubes? What kinds of energies of light are present in a continuous spectrum? In a line (or emission) spectrum?
7. List some of the properties of light that lead to the conclusion that it behaves in a wavelike manner. What is the most conclusive demonstration that light is wavelike? What attribute of light is different from that of other wave phenomena, such as sound, earthquakes, and tides?
8. What is wavelength and what are its units? What is frequency and what are its units? What is the relationship between the two? Write the symbols for wavelength, frequency, and the velocity of light. How does the velocity of light fit into the relationship between wavelength and frequency? Write the equation that shows this relationship.
9. What is the photoelectric effect? What observation about the photoelectric effect led Einstein to propose that light existed as small bundles of energy called photons? How did the idea of photons help Bohr explain how line spectra were produced from excited states in atoms?
10. What is the relationship between the frequency of light and its energy? Between the wavelength of light and its energy? Why is it possible to get a sunburn from sunshine more easily than from an infrared heat lamp?

11. What is a ground state? An excited state? Which is more stable? Because the electrons in excited states of atoms, on returning to the ground state, produce line spectra instead of continuous spectra, Bohr concluded that there was something special about them. That is, the excited states had a particular relationship to one another. What term is used to describe this relationship?
12. The Bohr model of the atom is based on an analogy. What is it? In terms of relative energy, how do electrons in orbits far from the nucleus compare with those in orbits nearby? What is the most stable state of any atom? What is the distance of the electrons from the nucleus in this state? When an electron in an excited state of an atom falls to a lower level, what happens? What is the relative energy state of the atom afterward?
13. What deficiency did the Bohr model of the atom have?
14. What is it about the nature of measurement of very small objects that makes it impossible for us to determine both their position and momentum (direction) simultaneously? Why isn't this a problem in measuring objects large enough to see? What does the uncertainty principle tell us about the error in measurement of large objects as compared with very small ones?
15. What is an absorption spectrum? How is it different from an emission spectrum? How are absorption and emission spectra the same? How are absorption spectra used in practical application? What attribute of the excited states of substances makes this possible? Could emission spectra be used to obtain similar kinds of information? Explain.
16. Erwin Schrödinger noted an analogy, which led him to think that the electrons in atoms had wavelike properties. What was this analogy?
17. What is an orbital? Why must we use the idea of probability to describe an electron inside an orbital? What would an orbital look like if we chose to define it in such a way that we could find the electron there 99% of the time instead of 95% of the time? One hundred percent of the time? In the modern wave mechanical model, what property of electrons in atoms gives the orbitals their different shapes? How does this property relate to the model that the energy levels in atoms are quantized?
18. In an orbital, what is a node? How many electrons can occupy an individual orbital? An atom of Sn has 50 electrons. How many orbitals must a Sn atom have to accommodate all its electrons? How are all these orbitals arranged about the nucleus, and what feature of the model of orbitals permits them to be arranged like this?
19. How many kinds of orbitals are there? List them.
20. What are valence electrons? Where do they reside in an atom? What do they do?

21. What is atomic number? Restate the Periodic Law using the concept of atomic number. The structure of the Periodic Table is related to two attributes of the electrons in atoms. One is the number of electrons each atom contains. What is the other? Show the parts of the Periodic Table that correspond to the elements in which *s*, *p*, *d*, and *f* outermost orbitals are occupied by valence electrons.
22. The theory that energy levels in atoms are quantized means that an electron cannot be observed to be between them. That is, it cannot have an intermediate energy. How does this lead to the idea of imposed ignorance? What does this view tell us about the nature of scientific knowledge, and how does it differ from the Newtonian view of nature?
23. What does the Heisenberg uncertainty principle tell us about scientific knowledge? How certain can our scientific measurements be?
24. How does the concept of wave/particle duality support the idea of imposed ignorance? What is the relationship between dualities such as wave/particle, mass (matter)/energy, mass (matter)/frequency, and what actually exists in nature? (For example, what really *is* an electron?) How does the dualness of things in nature affect our belief in scientific objectivity? How does the Copenhagen view of reality accommodate these ideas?

EXERCISES FOR "DID YOU KNOW? ENERGY FROM THE NUCLEUS"

25. The following isotopes have been written without a chemical symbol for the element. What element does each represent? (xx = symbol for the element)

    a. $^{37}_{17}xx$ e. $^{7}_{3}xx$
    b. $^{8}_{5}xx$ f. $^{27}_{13}xx$
    c. $^{106}_{42}xx$ g. $^{115}_{46}xx$
    d. $^{82}_{35}xx$ h. $^{58}_{26}xx$

26. Which of the following pairs are isotopes?

    a. $^{15}_{7}xx$ and $^{15}_{8}xx$ c. $^{49}_{22}xx$ and $^{50}_{22}xx$
    b. $^{6}_{3}xx$ and $^{7}_{3}xx$ d. $^{238}_{92}xx$ and $^{238}_{93}xx$

27. How many neutrons does the nucleus of each of the following isotopes contain?

    a. $^{4}_{2}He$ d. $^{19}_{9}F$
    b. $^{35}_{16}S$ e. $^{128}_{53}I$
    c. $^{211}_{83}Bi$ f. $^{239}_{94}Pu$

THINKING IT THROUGH

A Particle Without Mass?

The idea that a photon is a massless particle appears to run counter to common sense because we usually think of particles as matter, which of course has mass. Show how you would accommodate (or explain) these seemingly contradictory attributes on the basis of either a western or an eastern philosophical viewpoint *(but not both simultaneously).* Then, take the philosophical viewpoint other than the one you chose and see if you can make a case for that side.

Do They Exist or Don't They?

Take a position (agree or don't agree) concerning the following statement made by David E. Henrie, and, using the knowledge you have gained in this course so far, defend your choice with a reasonable brief argument. "Electrons, atoms, photons, etc., all are nothing but thoughts used to interpret the reality of what we perceive using our five senses."

ANSWERS TO "DID YOU LEARN THIS?"

Part 1

a. quantized
b. Bohr model of the atom
c. matter/energy duality
d. ground state
e. prism
f. continuous spectrum
g. M. Faraday
h. scintillations
i. higher energy
j. wavelength
k. emission spectrum
l. J. J. Thomson
m. nucleus
n. color
o. excited state
p. direct proportionality
q. α-particle
r. frequency
s. wave/particle duality
t. radioactive decay
u. photon
v. empty space
w. inverse proportionality
x. photoelectric effect

Part 2

a. Groups 3A–8A
b. perturbation
c. probability
d. standing wave
e. absorption
f. node
g. position
h. B-Groups
i. E. Schrödinger
j. orbital

CHEMISTRY AND THE STARS

Sallie Baliunas

Astrophysicist, Harvard-Smithsonian Center for Astrophysics

B. A. Astronomy, Villanova University

A. M., Ph.D. Astronomy, Harvard University

A tabloid newspaper might have run the headline "Chemistry Stops Girl's Moon Journey." Here's how it happened. As a five year old, I caught "mission to the Moon" fever. Humanity was going to the moon. The next step would be Mars, then the stars! And I was angling to get in on the adventure.

I soon learned that only *boys* could become astronauts because they were pilots with many hours of flying time in advanced jet aircraft. I chose a backup career in aeronautical engineering, a field in which women did work. Then along came tenth-grade chemistry. I remember Chapter 7, "Spectroscopy," in our chemistry textbook. Imagine studying the spectrum of a gas and being able to deduce physical properties about the gas, such as the speed of its atoms and the relative abundances of different elements. The gas need not be present in the lab to infer its properties, only the spectrum. Thus, spectroscopy can be used as a remote-sensing device—distant stars and galaxies yield knowledge of their physical properties in our terrestrial laboratories where the spectra could be collected and analyzed. Here was a way to study space *without leaving earth.*

Chemistry also taught me about a fabulous scientist, Marie Curie. She was the first person to be awarded two Nobel prizes (one in chemistry, one in physics). Her example taught me that a woman could become a research scientist.

The preparation for my job as an astrophysicist at the Smithsonian Institution wasn't easy (many years of study were required), but the result is *fun.* The Smithsonian has several research labs (in addition to its museums), and I work at its Astrophysical Observatory in Cambridge, Massachusetts. I study the changes in the sun's magnetic activity (the 11-year sunspot cycle) by observing the magnetic phenomena of nearby stars and contrasting their behavior to the sun's. The sun's changing magnetism seems to be connected to its total energy output, which, in turn, affects the earth's climate. Although my work focuses on understanding the complicated engine that produces solar magnetism, it may also help us understand global change.

My research tools are telescopes, both on the ground and in space. I use satellite observations to do spectroscopy of the ultraviolet radiation from nearby stars. The ultraviolet spectrum is rich in emission lines that detail the behavior of magnetic activity and hot plasma on these stars. This information must be acquired above the earth's atmosphere, which blocks ultraviolet light.

From earth, I study the spectrum of singly ionized calcium in stars. The spectrum of starlight contains a pair of emission lines, the "H" and "K" doublet near 393 nanometers, which changes in intensity as the number of magnetic "starspots" waxes and wanes.

I have just started an experiment with the SETI (Search for Extra-Terrestrial Intelligence) program. We are seeking evidence of extraterrestrial civilizations by listening for their radio communications. A technically advanced civilization took over four billion years to develop in our solar system, so we should be targeting star systems at least as old as ours. I am determining the ages of over a thousand stars in the solar neighborhood to identify the stars old enough to have ETI.

My research has taken me from arid mountaintops (where most observatories are located) to satellite control stations at NASA labs. In the course of my research, I have run into hungry bears, rattlesnakes, and deranged computers!

I like the thrill of working on the edge of the unknown. Because the area of scientific exploration is full of surprises, an experiment is always *slightly* out of control. Not only the unexpected, but the *unexpectable* may happen—and sometimes does—whether or not I am prepared for it. Scientific experimentation is often depicted as a character in a lab coat shouting "Eureka" after the lab blows up. Although a bit exaggerated, this element of science is real and adds adventure to the quest.

Beyond research, my job has another enjoyable function that is very important—communication. If the results of an experiment are never explained to others, then the experiment might as well not have been conducted in the first place!

Now—what does your everyday astrophysicist do in her spare moments? I give older cars a new life as modern hot rods. I've rebuilt a 1933 Ford, a 1957 Chevy, and a 1970 Corvette. I'm always overhauling or upgrading at least one car. This is as close to speeding through the universe as I've come!

6

Chemical Bonding

If we then consider the . . . molecule as one in which the electrons belonging to the individual atoms are held by such constraints that they do not move far from their normal positions, while in the polar molecule [ionic substance] the electrons, being more mobile, so move as to separate the molecule into positive and negative parts, then all the distinguishing properties of the two types of compounds become necessary consequences of this assumption.

G. N. Lewis

Chemists, who are interested in the structure of matter, find that one of the most useful applications of modern atomic theory is to explain how the various kinds of chemical bonds join atoms to form compounds. Chemical bonds are strong attractions that hold atoms together, and it is these bonds that primarily determine the properties of substances: brittle crystalline salt, elastic rubber, fragrant wintergreen, or malleable copper. The elements' properties and the kinds of chemical bonds they form are determined largely by how their valence electrons behave.

IONIC AND MOLECULAR SUBSTANCES

Simple binary compounds of metals and nonmetals, such as Na_2O and $MgBr_2$, are **ionic substances** (they contain ions) and share common physical properties. (Recall that ions are atoms, or small groups of atoms bound together, that carry a positive or negative charge.) Most ionic substances are crystalline solids, which usually are brittle and have high melting points (► Figure 6.1).

Molecular compounds result from the combination of nonmetals with one another. They can exist as gases or liquids, as well as solids, and can ex-

(a)

(b)

► Figure 6.1
Ionic Substances are Crystalline and Brittle. *(a) Crystals of sodium chloride. (b) Sodium chloride crystals after being hit with a hammer.*

hibit a host of different properties (▶ Figure 6.2). As solids they often are neither crystalline nor brittle (although some are), and generally they have lower melting points than do ionic substances. Examples such as gasoline, sugar, water, epoxy glue, diamond, and carbon dioxide illustrate the variety of properties molecular substances can have.

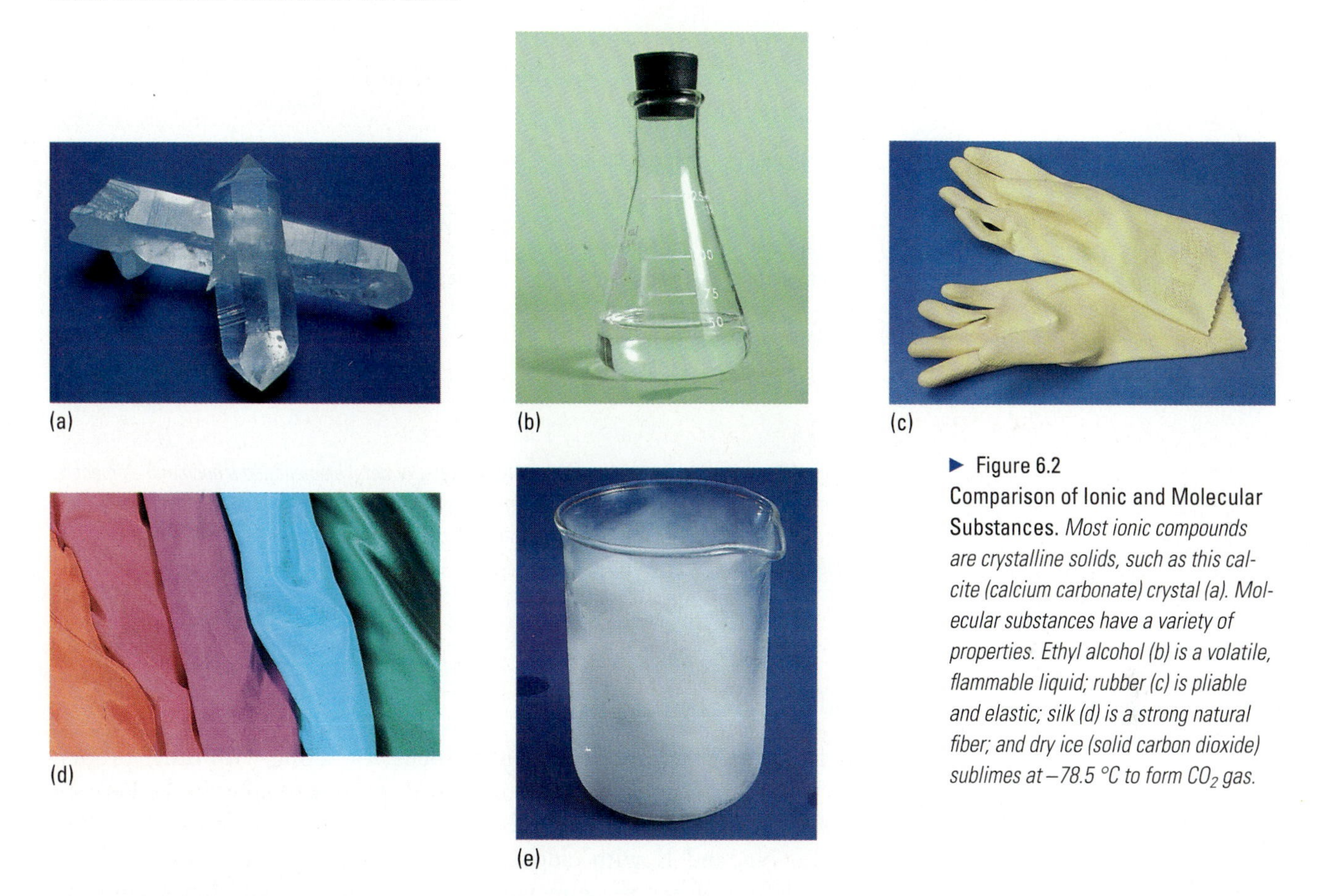

▶ Figure 6.2
Comparison of Ionic and Molecular Substances. *Most ionic compounds are crystalline solids, such as this calcite (calcium carbonate) crystal (a). Molecular substances have a variety of properties. Ethyl alcohol (b) is a volatile, flammable liquid; rubber (c) is pliable and elastic; silk (d) is a strong natural fiber; and dry ice (solid carbon dioxide) sublimes at −78.5 °C to form CO_2 gas.*

IONIC SUBSTANCES

Metals tend to give up their valence electrons in chemical reactions whereas nonmetals tend to accept them. So it is very common for metals to form compounds with nonmetals, and the two are often found combined this way in nature. On the other hand, the noble gases He, Ne, Ar, Kr, Xe, and Rn generally are chemically unreactive and neither lose nor gain electrons.

Compounds of some of the noble gases have been prepared, however.

Metal atoms are electrically neutral; that is, they contain an equal number of positive and negative charges. When they lose their negatively charged valence electrons, a positive metal ion remains. Positive ions are called **cations** (pronounced "cat́-ions").

Atoms are electrically neutral.

Nonmetal atoms are also electrically neutral. When they gain additional valence electrons, each having a negative charge, they become negatively charged ions. Negative ions are called **anions** (pronounced "anń-ions"). ▸ Figure 6.3 provides an overview of how metals and nonmetals, respectively, lose and gain electrons in chemical reactions.

	Elements				Compound		
(a)	metal element	+	nonmetal element	⟶	compound of a metal	and	nonmetal
(b)	metal atom *loses* electrons	+	nonmetal atom *gains* electrons	⟶	metal atom has *lost* electrons	and	nonmetal atom has *gained* electrons
(c)	metal atom (not charged)	+	nonmetal atom (not charged)	⟶	metal *ion* *positive ion* (cation)	+	nonmetal *ion* *negative ion* (anion)

▸ Figure 6.3
Formation of Compounds from Metals and Nonmetals. *A metal element and a nonmetal element react to form a compound (a). In the process, the metal atoms lose electrons and the nonmetal atoms gain electrons (b). The reaction of uncharged metal and nonmetal atoms forms positive metal ions (cations) and negative nonmetal ions (anions) (c).*

METAL IONS

Mendeleev used similarities in chemical behavior, along with other properties, to help him place the elements in their respective families in the Periodic Table. For example, all compounds formed from the reaction of the *alkali metals* Li, Na, and K with chlorine have the same general formula: LiCl, NaCl, and KCl. When these metals combine with oxygen, they form compounds having formulas Li_2O, Na_2O, and K_2O. Similarly, the *alkaline earth metals* Mg, Ca, and Sr form compounds of the formula $MgCl_2$, $CaCl_2$, and $SrCl_2$ with chlorine and MgO, CaO, and SrO with oxygen.

Metal atoms lose their valence electrons to form positive ions.

From such patterns, chemists deduced that atoms of certain metals always lose just one electron to form 1+ ions in chemical reactions, some lose two electrons to form 2+ ions, and some lose three electrons to form 3+ ions. For example, the lithium atom loses just one electron to form Li^+, the *lithium ion.* It never loses two or more electrons when it forms compounds. Likewise, the calcium atom loses two electrons to form Ca^{2+}, the *calcium ion.* It never loses one electron, nor does it lose three or more. Aluminum loses three electrons, no other number, to form Al^{3+}, the *aluminum ion.*

Note that when an atom such as Li loses one electron, a positive ion with 1+ charge is formed. The atom is electrically neutral (has zero charge) in the beginning. When an electron, which is negatively charged, leaves an atom (and goes, say, to a nonmetal), a metal cation remains, it must have an equal but opposite charge, or 1+ in this case (▶ Figure 6.4). Likewise, when an atom loses two electrons (2–), a metal cation of 2+ charge must remain. When three electrons (3–) are lost, a 3+ ion is formed.

Metal atom – electron(s) → **cation.**

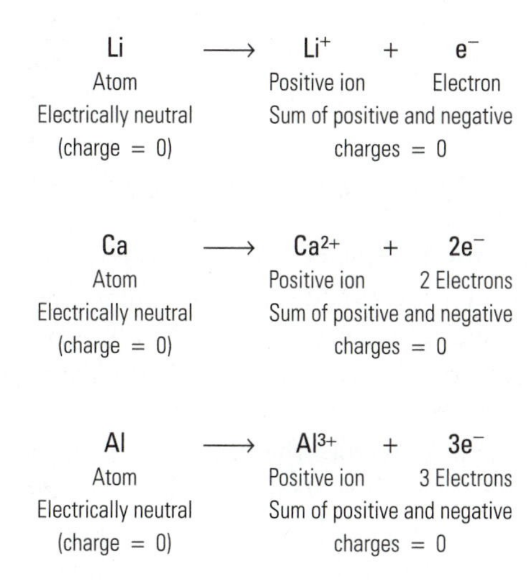

▶ Figure 6.4
Formation of Cations from Various Metal Atoms.

NONMETAL IONS

Nonmetal atoms are also electrically neutral in their uncombined state. When they react chemically, however, their tendency is to gain the negatively charged electrons rather than lose them. Again, group similarities appear. The *halogens* F, Cl, Br, and I, for example, combine with lithium to form only compounds with formulas LiF, LiCl, LiBr, and LiI; compounds with formulas like Li_2Cl or Li_3Br are unknown.

Nonmetal atoms gain valence electrons to form negative ions.

A nonmetal atom (having an electrical charge of zero) that gains one electron (negative one charge) will form a nonmetal ion with a 1– charge. A nonmetal atom that gains two electrons will form a nonmetal ion of 2– charge, and one that gains three electrons will yield an anion with a charge of 3–. Thus, Cl gains one electron to form Cl^-, the *chloride ion;* S gains two electrons to form S^{2-}, the *sulfide ion;* and N gains three electrons to form N^{3-}, the *nitride ion* (▶ Figure 6.5).

Nonmetal atom + electron(s) → **anion.**

► Figure 6.5
Formation of Anions from Various Nonmetal Atoms.

Cl	+	e^-	⟶	Cl^-
Atom (charge = 0)		Electron		Chlor*ide ion*
Sum of charges = 1–				Charge = 1–
S	+	$2e^-$	⟶	S^{2-}
Atom (charge = 0)		2 Electrons		Sulf*ide ion* Charge = 2–
Sum of charges = 2–				
N	+	$3e^-$	⟶	N^{3-}
Atom (charge = 0)		3 Electrons		Nitr*ide ion*
Sum of charges = 3–				Charge = 3–

PREDICTING CHARGE ON METAL AND NONMETAL IONS

The elements in Groups 1A through 8A (the A-Group elements) are often called the **representative elements** because their chemistry is relatively straightforward and predictable (► Figure 6.6). The B-Group elements, or **transition elements,** have a more complicated chemistry and will not be discussed in detail in this text.

There is an easy way to predict the usual charge on a simple (one-atom) ion formed from a representative element. For the metals in Groups 1A–3A, the charge on the metal ion usually is + its group number. The name of the metal ion is the same as that of the metal. Thus, cesium metal, in Group 1A, forms the cesium ion, Cs^+. (We do not write Cs^{1+} because the 1 is understood.) Magnesium metal in Group 2A forms the magnesium ion, Mg^{2+}, and aluminum metal in Group 3A forms the aluminum ion, Al^{3+}.

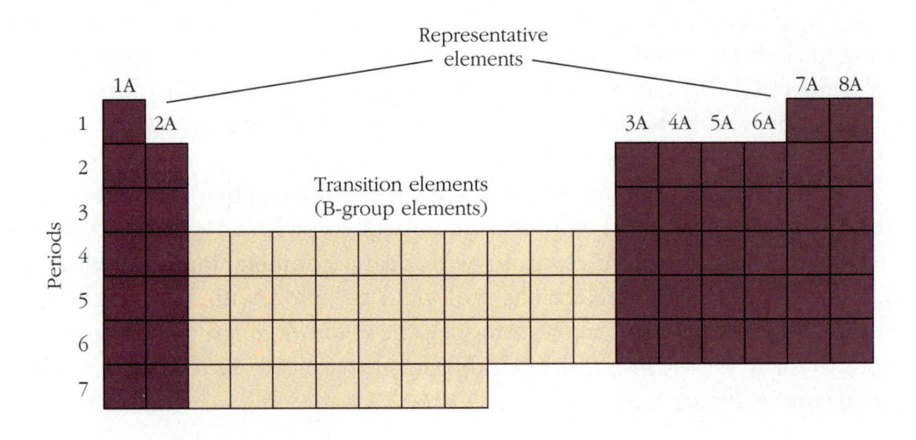

► Figure 6.6
The Representative Elements (in Burgandy) and the Transition Elements (in Beige).

6.1 Write the chemical symbol (with the proper charge) and give the correct name for the ions formed from each of the following metals.

a. Li
b. Al
c. Ba
d. K

For the nonmetals in Groups 5A–7A, the charge on the corresponding ion is 8– plus the group number. (Or you can count back from Group 8, adding 1– for each group as you go left in the Periodic Table.) The name of the ion contains the ending *-ide ion*—that is, chloride ion, sulfide ion, and so on. For example, selenium (Se) is in Group 6A. The charge on its ion is 8– + 6, or 2–. [Counting back from Group 8 we get 1– (at Group 7A), 2– (at Group 6A).] Selenium, then, forms Se^{2-}, the selenide ion. Nitrogen, in Group 5A, forms the nitride ion, N^{3-}.

What about the elements in Group 4A? Carbon, silicon, and germanium (Ge) do not form simple ions. Tin (Sn) and lead (Pb) usually form 2+ ions instead of 4+ ions. In other words, these elements do not behave as predictably. This is also true for the elements below aluminum in Group 3A and below arsenic in Group 5A. Gallium (Ga), indium (In), and thallium (Tl) commonly form 3+ ions, but each also forms a series of compounds in which the element exists in the 2+ and 1+ ionic states. Antimony (Sb) and bismuth (Bi) usually exhibit metallic behavior and lose their electrons to form cations.

6.2 Write the chemical symbol (with the proper charge) and give the correct name for the ions formed from each of the following nonmetals.

a. O
b. N
c. I
d. Cl
e. S

BINARY IONIC COMPOUNDS: THEIR CHEMICAL FORMULAS AND NAMES

Knowing how to determine the charges on simple ions formed from metals and nonmetals enables us to predict the chemical formulas of their binary compounds. **Binary compounds** are those formed from two different elements, like LiBr, CaS, and Al_2O_3. Also, because we know how to name metal

IONS IN YOUR DIET

When your doctor says, "You need more calcium in your diet," what does she mean? Should you eat more calcium metal (the pure element)? No. What you need are more calcium *ions* in your diet. The same is true for sodium, iron, potassium, phosphorus, selenium, and so forth (Figure 1). To eat the pure element would be ineffective, dangerous, or possibly fatal.

Most of the metals necessary to good health are ingested simply as ions: iron as Fe^{2+}, magnesium as Mg^{2+}, zinc as Zn^{2+}, manganese as Mn^{2+}, and copper as Cu^{2+}. The nonmetals are used as simple ions, such as chloride (Cl^-) and iodide (I^-), or as polyatomic ions: phosphorus as dihydrogen phosphate ion, $H_2PO_4^-$, sulfur as sulfate ion, SO_4^{2-}, and selenium probably as selenate ion, SeO_4^{2-}.

These ions play specific roles in body functions. For example, sodium and potassium ions preserve body water balance and maintain proper function of the nerves; chloride ion is part of gastric acid, responsible for the digestion of food; calcium and phosphate ions are needed for bone and tooth development and maintenance; and phosphate ion is essential for the production of energy within the body. Iron ions are required for the proper function of hemoglobin, the oxygen-carrying molecule in the blood; iodide ions are needed to produce thyroid hormones, which control growth and regulate heat production in the body; and zinc ions stimulate the growth of taste bud cells to replace those that are lost due to normal wear and tear.

Figure 1
Meat, whole grains, and dark green vegetables are good sources of dietary iron. Iron ions are necessary for the proper function of hemoglobin in the blood.

and nonmetal ions, we can produce the names of these compounds if we are given their chemical formulas.

In an ionic compound, the total number of positive charges must equal the total number of negative charges.

To begin, let's consider common table salt, the binary compound formed from the metal sodium and the nonmetal chlorine. The sodium ion (Group 1A) has a 1+ charge, or Na^+, and the chloride ion (Group 7A) has a 1− charge, or Cl^-. A compound must be electrically neutral; that is, the total number of positive charges and negative charges contributed by all the ions in its formula must add up to zero. If they did not, we would get a shock when we touched our salt shaker! Because a 1+ charge on a sodium ion exactly balances a 1– charge on a chloride ion, we can write the formula for salt as NaCl. Writing NaCl in this way means, in words, "a compound composed of sodium ions and chloride ions in the proportion 1:1 (one-to-one)," or something to that effect. Note that we do not say "sodium and chlorine," but rather we say that they exist as ions in the compound. This will be important later when we discuss the properties of these substances.

Let's take a another example, say the binary compound composed of lithium and sulfur. Lithium is also in Group 1A and thus forms the Li^+ ion. Sulfur, in Group 6A, forms the S^{2-} ion. Putting these ions together and maintaining electrical neutrality for the resulting compound means that we must use two lithium ions for every sulfide ion ($2 \times 1+ = 2+$) to exactly balance the 2– charge of the sulfide ion, and we write the formula as Li_2S. (Recall that we do not write 2LiS. This would be incorrect because the coefficient 2 means that we double the number of formula units, and the ratio of lithium to sulfur (LiS) is different. Writing the subscript 2 following the symbol of an element, such as Li_2, doubles the proportion of that element only.) Thus, the formula Li_2S means "a compound containing lithium ions and sulfide ions in the proportion (ratio) of 2:1." Note also that we do not include the charge on each ion (as a superscript at the upper right of the element's symbol) when we write the entire chemical formula. We reserve this symbolism to talk about the ions individually, when they are separate from one another. The name of the compound we have made is lithium sulfide.

What about the binary compound of calcium and arsenic? Calcium (Group 2A) forms the Ca^{2+} ion, and arsenic (Group 5A) forms the As^{3-} or arsenide ion. The resulting compound, calcium arsenide, has the formula Ca_3As_2. How did we get that? It takes three calcium ions ($3 \times 2+ = 6+$) to balance two arsenide ions ($2 \times 3- = 6-$), and the formula as written means "a compound having calcium ions and arsenide ions in the ratio of three-to-two (3:2)."

A simple shortcut to writing formulas of compounds containing cations and anions of different charges is to simply "crisscross" their numbers. That is, we use the value of the charge on the cation as the subscript for the anion, and the value of the charge on the anion as the subscript for the cation. This is illustrated using the two previous examples:

$Li^{(1)+}$ S^{2-}, crisscrossing numbers $\rightarrow$ $Li_2S_{(1)}$, or Li_2S

Ca^{2+} As^{3-}, crisscrossing numbers $\rightarrow$ Ca_3As_2

Writing the formula for aluminum phosphide is easy. The aluminum ion (Group 3A), Al^{3+}, and the phosphide ion (Group 5A), P^{3-}, have equal but

6.3 Using the Periodic Table, predict the chemical formula of the binary compounds formed from the pairs of elements below. Then name each one.

a. lithium and chlorine
b. aluminum and fluorine
c. calcium and sulfur
d. magnesium and iodine
e. potassium and oxygen
f. sodium and phosphorus

opposite charges, so the formula is simply AlP, "a compound containing aluminum ions and phosphide ions in a proportion of 1:1." (In this instance, crisscrossing numbers results in the subscripts, Al_3P_3, which we simply reduce to the lowest common denominator.)

In the preceding example we wrote the structure of aluminum phosphide simply by having its name. This is easy if we know the names and symbols for the elements. The first part of the name is always the positive ion, and the -ide ending on the negative ion tells us we're dealing with a simple binary compound. Thus, for example, we find aluminum, Al, in Group 3A and phosphorus, P, in Group 5A, determine the charges on their respective ions, and proceed from there. In developing our chemical vocabulary we need to learn to work from the formula of a compound to its name, and also from its name back to its correct formula. Using a similar approach, we can write magnesium fluoride as MgF_2, sodium nitride as Na_3N, and beryllium oxide as BeO.

6.4 Write the chemical formulas for the binary compounds whose names are given below.

a. sodium iodide
b. calcium oxide
c. aluminum oxide
d. potassium sulfide
e. magnesium chloride
f. lithium nitride

POLYATOMIC IONS

Polyatomic ions are ions that contain two or more atoms in a single unit. We will not concern ourselves with their structures here, but we need to learn the formulas of a few of the more common ones. Most of them are anions (■ Table 6.1): the nitrate ion, NO_3^-; the carbonate ion, CO_3^{2-}; the hydrogen carbonate (or bicarbonate) ion, HCO_3^-; the sulfate ion, SO_4^{2-}; the phosphate

■ **TABLE 6.1 SOME COMMON POLYATOMIC IONS**

Name of Polyatomic Ion	Formula
Nitrate ion	NO_3^-
Acetate ion	$C_2H_3O_2^-$
Hydrogen carbonate ion	HCO_3^-
Carbonate ion	CO_3^{2-}
Sulfate ion	SO_4^{2-}
Phosphate ion	PO_4^{3-}
Hydroxide ion	OH^-
Ammonium ion	NH_4^+

ion, PO_4^{3-}; and the acetate ion, $C_2H_3O_2^-$. Note that these ions have endings other than -ide, which tells us they are polyatomic rather than simple. The ending of all of these is -ate, but some polyatomic anions have endings of -ite, such as the sulfite ion, SO_3^{2-}, which is used as a food preservative. One important polyatomic anion, the hydroxide ion, OH^-, has an -ide ending. The polyatomic cation of most importance to us is the ammonium ion, NH_4^+. (Note that the use of the terms cations and anions provides us with a generic way of talking about positive and negative ions, respectively.)

Naming compounds containing polyatomic ions (and figuring out the formulas from their names) is straightforward. Sodium nitrate, for example, has the formula $NaNO_3$ because we are combining a 1+ sodium ion with a 1– nitrate ion. The formula means, "a compound containing sodium ions and nitrate ions in the ratio of 1:1." We must, however, remember that the NO_3 part of the formula is *all one unit* and that we should not call it by some other name when going back to the name from the formula. In a similar fashion we write the formula for calcium sulfate as $CaSO_4$. Here a 2+ calcium ion is paired with a 2– sulfate ion.

6.5 Give the correct name for the following compounds containing polyatomic ions.

a. $NaNO_3$
b. KOH
c. $NaHCO_3$
d. $CaSO_4$
e. $Mg(OH)_2$
f. NH_4Cl

THE IMPORTANCE OF POLYATOMIC IONS

Polyatomic ions have many uses and are important to our lives in many ways. Take phosphates, for example. Phosphate ion is a necessary nutrient for the growth of roots, seeds, and fruit in plants, and so modern agriculture depends on large quantities of phosphate fertilizers. Most of the phosphates produced in this country are used for this purpose. Phosphates also are used in many household and industrial cleaners to soften water and increase the efficiency of the detergents in them. Because they stimulate plant growth, however, phosphates that enter watercourses can produce algal blooms and eutrophication, choking off the oxygen supply for other forms of aquatic life. In some areas the use of phosphates has been restricted or banned because of this problem.

Ammonium ion also is used in fertilizer as a source of nitrogen for plants, essential for the growth of leaves and stems. Ammonium nitrate, NH_4NO_3, is a very good fertilizer because both the ammonium and the nitrate ions can be used by plants. Most nitrates are very soluble in water, and when overapplied as fertilizers, they can penetrate the ground and seep into wells. Too much nitrate ion in water can present a threat to health. Nitrates also are used as explosives.

The sulfites—sulfite ion, SO_3^{2-}, and hydrogen sulfite ion, HSO_3^-—are formed when sulfur dioxide (SO_2) reacts with water. They have been used as food preservatives for more than 2000 years. Many dehydrated fruits are treated with SO_2, and sulfites are commonly used in wines, packaged sauce and gravy mixes, and dried potatoes. Mak-

The rules we have used for writing the formulas of binary compounds of simple metal cations and nonmetal anions also apply to writing the formulas of compounds containing polyatomic ions. Potassium carbonate, for example, is written K_2CO_3. Here, two potassium ions of 1+ charge are required to balance the 2– charge of the carbonate ion. Because polyatomic ions contain elements in groups, we must use parentheses to show when a chemical formula contains more than one of them. Thus, barium nitrate is written $Ba(NO_3)_2$, and aluminum sulfate becomes $Al_2(SO_4)_3$. Recall (Chapter 4) that we used this kind of notation before.

The ammonium ion is treated in a like manner. Ammonium chloride is simply NH_4Cl, ammonium sulfide is written $(NH_4)_2S$, and ammonium phosphate, a compound composed of two polyatomic ions, has the formula

ers of homemade beer and wine often use sodium hydrogen sulfite, $NaHSO_3$, as a sterilant for bottles and equipment because it retards the growth of bacteria but not of yeast. Some people have allergic reactions to sulfites. As a result, the use of sulfites to preserve the fresh appearance of fruits and vegetables in restaurants has recently been banned, and all other foods must be labeled if sulfites have been added to them.

Nitrite ion, NO_2^-, another preservative, is commonly used in processed meats such as ham, hot dogs, and corned beef (Figure 1). Addition of sodium nitrite, $NaNO_2$, to these foods inhibits the growth of the bacteria that cause deadly botulism poisoning. It also is responsible for their pink color and spicy flavor. Used on fresh meats, nitrites help to preserve their fresh-cut redness. In the stomach, however, nitrites can contribute to the formation of compounds called nitrosamines, which are suspected of promoting stomach cancer, so it is wise to eat foods containing nitrites in moderation.

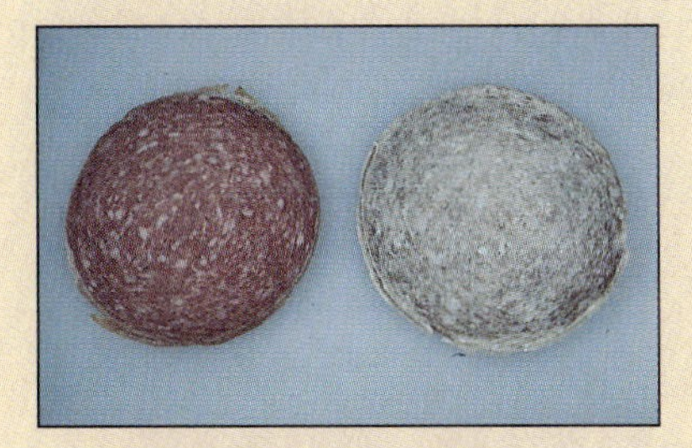

Figure 1
Adding sodium nitrite to processed meats preserves their color and inhibits the growth of the bacteria that cause botulism, a sometimes fatal form of food poisoning. Sodium nitrite was added to the salami on the left, but not to the salami on the right, and both were stored under the same conditions. The salami on the right may be safe to eat, but few people like to eat gray salami.

$(NH_4)_3PO_4$. This last representation means, "a compound containing ammonium ions and phosphate ions in the proportion 3:1."

6.6 Write the correct formulas for the following compounds containing polyatomic ions.

a. potassium hydrogen carbonate
b. ammonium bromide
c. lithium carbonate
d. calcium nitrate
e. magnesium phosphate
f. sodium sulfate

THE IONIC BOND

Electrostatic forces hold positive and negative ions together in a **crystal lattice.**

Ionic substances tend to be crystalline, brittle solids with high melting points. This is because the positive ions are surrounded by a large number of negative ions and the negative ions by a large number of positive ions in a **crystal lattice** having a regular, ordered structure that lends itself well to crystallinity (▶ Figure 6.7). Strong **electrostatic forces** (attraction of positive charges for negative charges) hold the ions together in the lattice, so high temperatures (that is, high energies; see Chapter 10) are required to break the ions apart and make them mobile. The brittleness of ionic substances arises from the regularity of the crystal lattice. Although the electrostatic forces holding the positive and negative ions together are strong, displacement of the lattice by only one ion in any direction causes positive ions to be next to positive ions and negative ions next to negative ions. This results in repulsion between the ions, rather than attraction, and the crystal fractures.

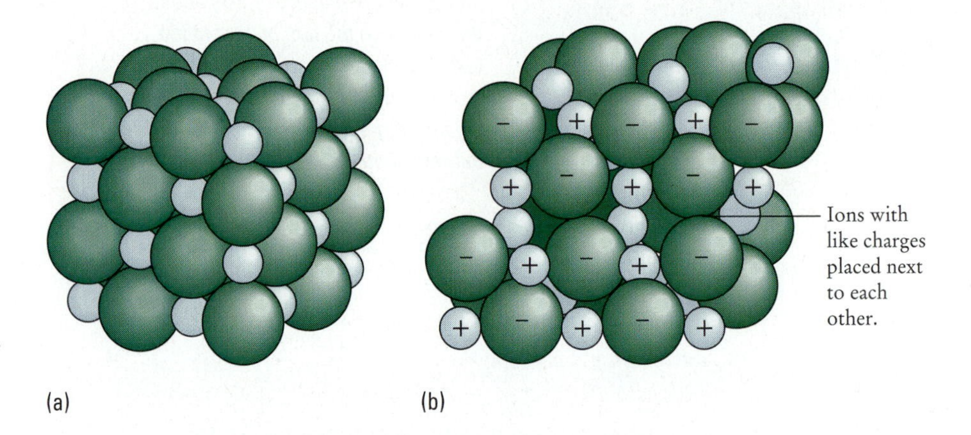

▶ Figure 6.7
An Ionic Crystal Is Hard and Rigid, but Brittle. *(a) Model of sodium chloride crystal lattice. The large green balls represent Cl^- ions; the small, light-colored balls represent Na^+ ions. Each ion is surrounded by ions of opposite charge, and the ions are strongly attracted together. (b) Pushing layers of an ionic crystal past one another places ions with like charges next to each other and the crystal fractures.*

METALS AND NONMETALS: A CHEMICAL DEFINITION

Knowing how the metals and nonmetals react to form simple binary compounds enables us to give them both a chemical definition. Because *metals* are elements whose atoms tend to lose electrons to form positive ions, we can simply define them as such. Likewise, we can define *nonmetals* as elements whose atoms tend to gain electrons to form negative ions. Note that we now are able to recognize as a metal or a nonmetal any element that behaves chemically in the manner we have described.

MOLECULAR COMPOUNDS

A **molecule** is a discrete chemical entity.

Molecular compounds are modular, being composed of small, identical units called molecules. A **molecule** is a discrete chemical entity composed of the number and kinds of atoms contained in its chemical formula. That is, a molecule is a "formula's worth" of atoms bound tightly enough to one another to be recognized as a distinct group.

Ionic substances do not possess this property. There is no molecule of solid NaCl, for example. Instead, its respective ions are packed together in a crystal lattice, each sodium ion surrounded in three dimensions by six chloride ions, and each chloride ion by six sodium ions. No individual sodium ion is associated with a single chloride ion, and it is impossible to find a unique NaCl unit in a crystal of sodium chloride (Figure 6.7a). Likewise, in a crystal lattice of cesium bromide (▶ Figure 6.8a), each cesium ion is surrounded in three dimensions by eight bromide ions, and each bromide ion by eight cesium ions. No single cesium ion can be found paired with a particular bromide ion to make a CsBr unit.

On the other hand, table sugar (sucrose) is a molecular substance composed of discrete $C_{12}H_{22}O_{11}$ units, and each molecule is made up of 12 atoms of carbon, 22 atoms of hydrogen, and 11 atoms of oxygen, all bound together (▶ Figure 6.8b). Over 10^{18} sugar molecules are arranged, like little modules, in a tiny crystal of sugar barely visible to the eye.

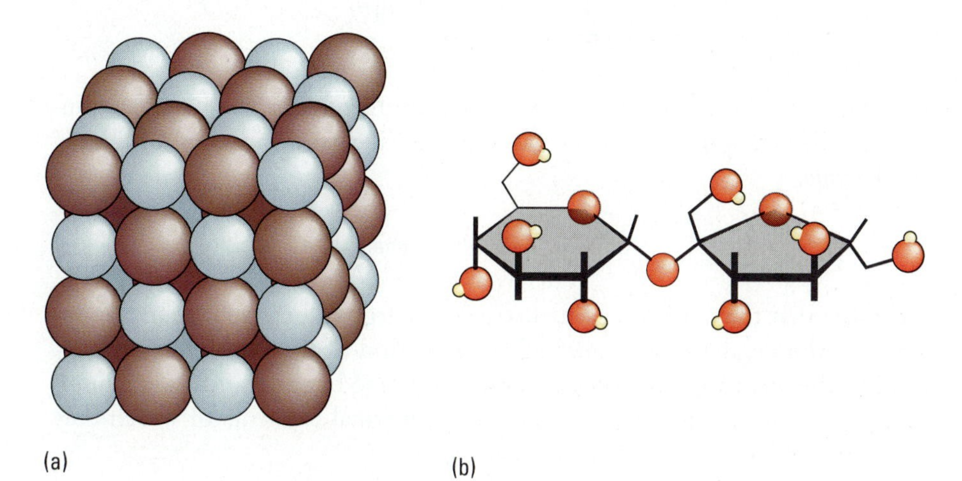

▶ Figure 6.8
Ionic and Molecular Compounds. *(a) In an ionic crystal lattice of cesium bromide, CsBr, there are no recognizable, discrete units of "CsBr." The gray-colored balls represent Cs^{-} ions and the rust-colored balls represent Br^{-} ions. (b) A molecule of sucrose, $C_{12}H_{22}O_{11}$, is comprised of 12 carbon atoms, 22 hydrogen atoms, and 11 oxygen atoms in a single unit.*

MOLECULAR BOND FORMATION: ITS DRIVING FORCE

The atoms of many pure nonmetallic elements commonly are found in pairs, such as H_2, N_2, O_2, and Cl_2. Sometimes they are found in larger aggregates such as P_4 or S_8. These aggregates actually are molecules that are made up of one kind of atom. What causes the atoms to cling together in this manner rather than exist separately? We already know that most elements are found in nature combined with other elements as compounds, because energy is released when combining happens. Compounds usually are more stable energetically than their uncombined elements are, so we might infer that molecules of the nonmetallic elements likewise would be more stable (of lower energy) than their single atoms are.

To visualize this idea, let's consider what it takes to break a hydrogen molecule, H_2, into two individual hydrogen atoms. We find that to accomplish this a considerable input of energy is required. In fact, this reaction is one (of many) that occurs under the high-voltage conditions in a hydrogen discharge tube. Note again that energy is considered to be a reactant in the process.

$$\underset{\substack{\text{Molecule} \\ \text{(reactants)}}}{H_2 + \text{energy}} \longrightarrow \underset{\substack{\text{Atoms} \\ \text{(products)}}}{H + H}$$

If we reverse the preceding process and allow the two hydrogen atoms to combine, forming a hydrogen molecule, energy should be a product of the reaction. In other words, energy is given off when a hydrogen molecule is formed.

$$\underset{\text{Atoms}}{H + H} \longrightarrow \underset{\text{Molecule}}{H_2} + \text{energy}$$

Energy also is released when molecules are formed from dissimilar nonmetals. The reaction of hydrogen gas with chlorine gas, for example, is highly exothermic.

$$H_2 + Cl_2 \longrightarrow 2HCl + \text{energy}$$

Molecules are usually more stable than the separate elements that make them.

We know that the tendency in nature is for energy to be given off in a spontaneous, simple chemical process. The formation of molecules from atoms, just like the formation of ionic substances, tends to be spontaneous, and molecules generally are more stable than the individual elements of which they are comprised.

MOLECULAR ORBITALS

The orbital picture of the atom enables us to explain why energy is released when molecular bonds form. Using the previous example of the hydrogen molecule, let's consider what happens to atomic orbitals when two separate hydrogen atoms are brought together.

In each hydrogen atom a single electron occupies a spherical space, or *s* orbital. This space has the shape and properties of a standing wave, much like the vibration in a string. The standing wave property of each *s* orbital corresponds to a particular "note" on the string. As the hydrogen atoms are brought together, the two *s* orbitals on the respective atoms overlap one another, causing their wave properties to mix, forming a new orbital called a **molecular orbital** (▶ Figure 6.9). This is analogous to using two notes to play a simple chord in music. The molecular orbital is the chord or a new kind of harmony.

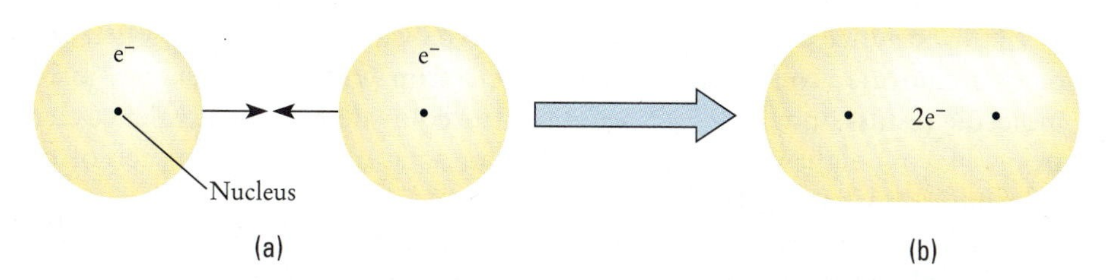

▶ Figure 6.9
Forming a Molecular Orbital. *The atomic orbitals of two hydrogen atoms overlap (a) to form a molecular orbital for the hydrogen molecule (b). The two valence electrons are shared in the space between the two hydrogen nuclei.*

By combining two standing waves (*s* atomic orbitals), each containing one electron, we have, in effect, played two notes together to create a new standing wave (molecular orbital), or chord. This new molecular orbital envelops both nuclei and can accommodate the two electrons so that they reside, most of the time, between the nuclei. The attraction of the two nuclei for the two electrons between them is the "glue" that holds the two atoms (really the two nuclei) together and is the basis for the **molecular bond.** Molecular bonds are also called **covalent bonds,** because the electrons in the bond are shared between the atoms that are bound together.

Molecular bonds are also called **covalent bonds.**

In terms of energy, this state in which a pair of electrons is shared between nonmetal atoms is the more stable. Covalent (molecular) bond formation usually gives off a considerable amount of energy, so these bonds are generally quite strong. (It takes a lot of energy to break them.) However, the attractive forces in bonds of this kind do not pull the nuclei together until they touch, since they are positively charged and repel each other. The attractive force of the electrons for the nuclei is balanced by this repulsion, and so the nuclei are held together but at some distance apart.

VALENCE ELECTRONS IN MOLECULES

Valence electrons are often represented as dots next to the chemical symbol of the element, as follows.

$$H\cdot + H\cdot \longrightarrow H{:}H + \text{energy}$$

Valence electrons in hydrogen atoms (pointing to the dots in H· + H·)

Shared pair of electrons in covalent bond in hydrogen molecule (pointing to H:H)

By combining and sharing their valence electrons, two hydrogen atoms can share a pair of electrons to form a covalent bond between them. (This shared pair of electrons occupies a molecular orbital.) Similarly, combining two chlorine atoms to form a molecule of chlorine also results in a molecular bond containing a *shared pair* of electrons.

:C̤̈l· + ·C̤̈l: ⟶ :C̤̈l:C̤̈l: + energy

Shared electron pair

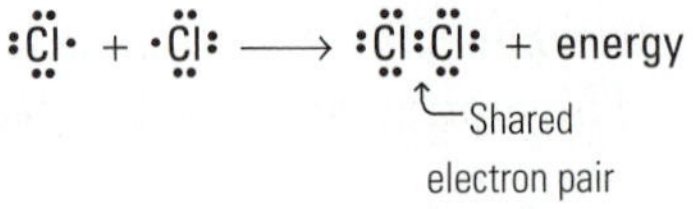

In this instance each chlorine atom has seven valence electrons. When each of two chlorine atoms shares a valence electron with the other, a chlorine molecule results, in which each atom has the equivalent of eight valence electrons. Having eight valence electrons, or an octet of them, gives each of the chlorine atoms a special stability.

The number of valence electrons that a representative element has is equal to its group number.

How do we determine the number of valence electrons an atom of an element has? For a representative element this is easy because the number of valence electrons is equal to that element's group number. Thus, all elements in Group 6A (O, S, Se, and so on) have six valence electrons, those in Group 5A (N, P, As, and so on) have five valence electrons, and those in Group 7A (such as Cl and I) have seven. With the exception of helium, which has two, the noble gases in Group 8A have eight valence electrons (called the noble gas configuration), which we have said is particularly stable.

Sometimes more than one pair of electrons is shared between atoms. Consider the nitrogen atom, which has five valence electrons. When two nitrogen atoms combine to form a nitrogen molecule, N_2, three pairs of electrons are shared and a triple bond is formed.

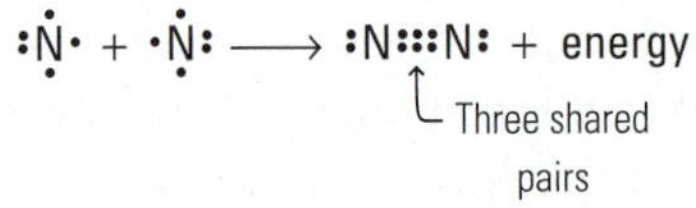

:Ṇ̇· + ·Ṇ̇: ⟶ :N:::N: + energy

Three shared pairs

Each of the bonded nitrogen atoms now is surrounded by a total of eight electrons. How can we say that each nitrogen atom has eight valence electrons? The shared electron pairs are attracted to both of the atoms they join, and each atom is attracted to all of them. Thus, we count shared electrons *twice,* once for each atom.

THE OCTET RULE AND LEWIS FORMULAS

Each of the nonmetal atoms in the previous examples (except hydrogen) has a total of eight valence electrons around it when it is bonded. This is called the **octet rule,** which holds for the bonding in many nonmetals. Specifically, the atoms of these elements (carbon and above in the Periodic Table) will form covalent bonds so as to have a total of eight valence electrons around them. This idea was originated by G. N. Lewis, and structures drawn according to this rule are called *Lewis electron-dot formulas.* They are very useful in explaining and predicting the properties of many molecular substances.

Drawing Lewis (Electron-Dot) Formulas

The rules for drawing Lewis electron-dot structures of molecules are summarized below:

1. All nonmetal atoms, carbon and higher in the Periodic Table, obey the octet rule in forming bonds. Bonded hydrogen is satisfied with only two electrons.
2. Hydrogen atoms usually are bound to the outside of molecules.
3. The central atom in a molecule is the one that is most metallic.
4. The number of valence electrons in a molecule equals the sum of the valence electrons of the atoms that comprise the molecule. (For ions, the total must be adjusted for the charge on the ion.)
5. The Lewis structure must contain all of the valence electrons in the molecule (or ion), and all atoms (except hydrogen) must have a complete octet.
6. It may be necessary to draw multiple bonds to complete the octet of some atoms. Each electron pair in such a bond must be counted twice, once for each atom.

Let's begin by drawing the electron-dot structure for a molecule of hydrogen bromide, HBr, a gas with a stinging, biting odor. It is best to first find the total number of valence electrons contributed by all of the atoms in the molecule. In this case, the hydrogen atom contributes one and the bromine atom contributes seven, for a total of eight. Bonded hydrogen requires only one shared pair of valence electrons, whereas bromine must have an octet. Our Lewis structure, then, must consist of a shared pair of valence electrons located between the hydrogen and bromine atoms and six valence electrons located on the bromine atom. The shared pair is counted twice, once for each atom. In the following, v.e. means valence electron(s).

H:B̈r: (Br with dots above and below)

Shared pair counted for each atom

H = 1 v.e.
Br = 7 v.e.
HBr = 8 v.e. total

Another molecule that serves nicely as an example of writing electron-dot structures is hydrogen sulfide, H_2S, a poisonous gas with the characteristic

6.7 Draw the Lewis (electron-dot) structures for HF, HCl, and HI. How do these structures relate to one another and to that of HBr?

odor of rotten eggs (noticeable at some hot springs). The sulfur atom has six valence electrons (Group 6A) and each hydrogen atom has one, giving a total of eight valence electrons for all the atoms in the molecule. Our problem is, as before, to arrange the atoms and electrons in a Lewis structure that will satisfy the octet rule. Because each hydrogen atom can form only one bond to another atom, we must attach them to the sulfur atom and not to each other. This leaves us with HSH for the arrangement of the atoms, and drawing the Lewis structure gives us

$$\mathrm{H{:}\ddot{\underset{..}{S}}{:}H} \qquad \begin{array}{ll} 2\mathrm{H} & = 2\text{ v.e.} \\ \mathrm{S} & = 6\text{ v.e.} \\ \hline \mathrm{H_2S} & = 8\text{ v.e.} \end{array}$$

Each hydrogen atom is now surrounded by two valence electrons and the sulfur atom by eight. Again, we have counted the shared pairs twice.

6.8 Draw the Lewis structures for H_2O, H_2Se, and H_2Te and compare them with that of H_2S. What kind of pattern do you see? Now try drawing the electron-dot structures for NH_3, PH_3, and AsH_3. Does a pattern persist for this series? What about CH_4 and SiH_4? How does what you observe relate to chemical periodicity?

Next we draw the Lewis structure for a molecule of a binary compound in which both kinds of atoms require an octet of electrons. Carbon and chlorine in CCl_4 make a good example. In this case, the carbon atom brings four valence electrons and the four chlorine atoms bring 28 (4 chlorine atoms × 7 v.e./chlorine atom) making a total of 32 valence electrons that we must accommodate in our structure. Again, the question of arranging the atoms arises. We choose as the central element the atom that is the most metallic. This is the element that (first) is farthest to the left and (second, if necessary) is farthest down in the Periodic Table. Carbon is well to the left of chlorine, so the chlorine atoms surround the carbon atom. Because all atoms in the molecule must have eight valence electrons around them, we write

```
    ••
   :Cl:
 •• •• ••
:Cl:C:Cl:
 •• •• ••
   :Cl:
    ••
```

$$\begin{array}{lcl} C & = & 4 \text{ v.e.} \\ 4Cl & = & 28 \text{ v.e.} \\ \hline CCl_4 & = & 32 \text{ v.e.} \end{array}$$

6.9 Draw the Lewis structures for CF_4, CBr_4, and CCl_4. Again note a family resemblance.

6.10 What is the central atom in each of the following molecules?

a. $AsCl_3$
b. SO_2
c. NF_3
d. H_3PO_4
e. TeF_2
f. NO_2
g. $COCl_2$
h. OF_2

For practice, let's try iodoform, CHI_3, a yellow solid. The total number of valence electrons in the molecule is 26, and carbon is the element leftmost in the Periodic Table, so it is placed in the center. All of the elements except hydrogen need an octet of electrons, so the electron-dot formula is

```
    ••
   :I:
 •• ••
:I:C:H
 •• ••
   :I:
    ••
```

6.11 Draw the Lewis structures for CHF_3, $CHCl_3$, and $CHBr_3$. Also draw them for CH_2Cl_2 and CH_3Br. Now try NF_3, PCl_3, OF_2, and SCl_2. What do you notice about the electron pairs around the central atoms in these last four molecules?

Lewis Structures of More Complex Molecules

To illustrate drawing multiple bonds between atoms we will use carbon dioxide, CO_2, a colorless gas that makes soda fizz and also may contribute to the greenhouse effect. The carbon atom brings four valence electrons to the structure and the two oxygen atoms bring 12, for a total of 16 valence electrons. The carbon atom is the most metallic, so it goes in the center, flanked by the two oxygen atoms. This time, when we try to give every atom an octet of electrons, we find we cannot do so simply by placing one pair of shared electrons between the carbon and each of the oxygen atoms. Instead, we must place two pairs of shared electrons there, forming *double bonds.* Our Lewis structure looks like this:

O::C::O (Double bonds)

$C = 4$ v.e.
$2O = 12$ v.e.
$CO_2 = 16$ v.e.

6.12 Draw the electron-dot structure for CS_2, a poisonous, extremely flammable liquid with a nauseating odor. Also try CH_2O and CH_2S.

Formation of a *triple bond* between two atoms is illustrated by hydrogen cyanide, HCN, a poisonous gas that is very dangerous because it is odorless to most people. The molecule has 10 valence electrons, and it is necessary to place three pairs of shared electrons between the carbon and nitrogen atoms to satisfy the octet rule. Carbon is the most metallic element, so it will be in the center. The Lewis structure is

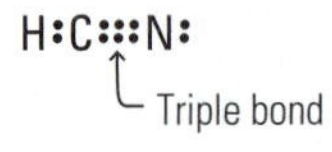

6.13 Draw the Lewis structure of carbon monoxide, CO, a poisonous gas. Also try acetylene, C_2H_2.

Electron-dot structures of more complex molecules, and even polyatomic ions, can be drawn using these rules. The sulfuric acid molecule, H_2SO_4, is a good example. Placing the sulfur atom at the center, with the hydrogen atoms on the outside of the molecule (on the oxygen atoms), and distributing the 32 valence electrons to complete each octet gives the structure,

```
      ..
     :O:
  .. .. ..
H:O:S:O:H
  .. .. ..
     :O:
      ..
```

6.14 Try drawing the electron-dot structures for nitric acid, HNO_3, carbonic acid, H_2CO_3, and phosphoric acid, H_3PO_4.

Polyatomic ions can be represented by Lewis structures. For the sulfate ion, SO_4^{2-}, the total number of valence electrons contributed by the atoms making up the ion is 30, but this number is valid only for a neutral species. To give the ion a 2– charge, we must add two more valence electrons, for a total of 32. With the sulfur atom again in the center, each atom can be given a complete octet in the structure:

```
    ..
   :O:  2-
.. .. ..
:O:S:O:
.. .. ..
   :O:
    ..
```

SO_4	=	30 v.e.
2– charge	=	2 v.e.
SO_4^{2-}	=	32 v.e.

6.15 Try this analysis for the nitrate ion, NO_3^-, the hydrogen carbonate ion, HCO_3^-, the carbonate ion, CO_3^{2-}, and the phosphate ion, PO_4^{3-}.

SHAPES OF MOLECULES: VSEPR

To see how molecular shapes can be explained by the idea of molecular orbitals, let's examine a few simple molecules. A good molecule to start with is methane, CH_4, which we met in a previous study exercise. The methane

LEWIS STRUCTURES AND REALITY

Saturday. Sleep in? No . . . it's your turn to do the recycling. Did you actually dream about electrons last night? Yes, you saw them as little balls dancing around. You'd been studying Lewis formulas, but dots in Lewis formulas haven't yet given you any feeling for chemical interactions in the real world.

You wander, musing, into the bathroom and put toothpaste on your brush. One thinks of the world as made up of many different things, but it isn't, really, it's made of three things: protons, neutrons, and electrons in various arrangements.

Toothbrush in your mouth, you find yourself studying the toothpaste tube: contains fluoride. That's a compound of fluorine, the most reactive element in the universe. Water vapor *burns* in it! But the violent element is tamed in stannous fluoride. Instead of dissolving your head, the fluoride ions in toothpaste gently replace the hydroxide ions in the hydroxyapatite of your tooth enamel, turning it into fluorapatite, which is less reactive with the acidic secretions of bacteria. Thus, the violent element saves you from the dentist's drill.

You rinse out your mouth. Good old water, the very basis of life. You cup your hand under the faucet and let water fill it . . . and suddenly remember the horrendous pictures of the Hindenberg dirigible exploding in the 1930s. Helium wasn't available yet, so they used hydrogen for lighter-than-air craft and $2H_2 + O_2 \rightarrow 2H_2O$ . . . a disaster! Merely a rearrangement of electrons, ten per molecule of water formed.

Electrons are everywhere. Electrons by the trillions making magnets of metal to turn motors, flowing through wires, exciting orbital electrons into giving off photons to make light, the color of which depends on how much energy the emitting electrons have. It's staggering.

Electrons and the differences their numbers and arrangements in atoms make in the real world are still on your mind at the recycling center. When you dump the box of bottles into the bin, they shatter and tinkle (breaking ionic bonds); the plastic bottles bounce into their bin (covalent bonds bend and stretch but don't break easily); the bins

molecule is three-dimensional and has a *tetrahedral geometry.* That is, its four C—H bonds are directed toward the corners of a tetrahedron, with the carbon atom buried in the center of the geometric figure. A *tetrahedron* is a special pyramid having four equilateral triangles for its sides. Another way to visualize this is to look at the molecule from the "outside." Doing so, we see that the hydrogen atoms reside at the corners of the tetrahedron. Figure 6.10 shows this geometry.

for cans have magnets nailed over them so people can separate aluminum cans (an arrangement with electrons in pairs is nonmagnetic) from steel "tin" cans (an arrangement with unpaired electrons is magnetic); and you pour used automotive oil into a tank (electrons arranged in the bonds of weak London forces).

It's a beautiful day, but you decide to do homework after recycling because you want to go out tonight. You're working on Lewis formulas, even enjoying this abstract puzzle with dots, when your mind stumbles over reality again. The dots seem to have nothing to do with reality. The text says H_2S is a bad-smelling toxic gas. OK, you've smelled rotten-egg gas, and it smells bad, and you've heard that the smell of a really rotten egg can make you pass out. That's real. But electrons are all the same. H_2S and H_2O have the same number and arrangement of valence electrons, but you drink one and faint at the mere smell of the other. The sulfur atom has exactly twice the total number of electrons as the oxygen atom and the orbitals are larger. And that's what makes the difference between a poison and a necessity of life!

You're becoming irritated because you want to finish this homework and can't stop your thoughts from wandering. But thoughts often wander into concepts, and it all begins to make sense when you realize that Lewis formulas are models of the results of observations of chemical reactions among millions and billions of atoms. It's the huge numbers of atoms involved that helps you understand. The spot on the desk where your finger is touching is made of billions of atoms, and the repulsive force of their combined electrons for the combined electrons of the billions of atoms in the surface of your finger makes it feel solid.

You extend these numbers to your body. Your being alive is a process of millions of chemical reactions that are rearranging electrons of countless atoms and molecules. You understand how H_2S can be toxic: if you were to replace the electron arrangements in H_2O with those in H_2S in that mass of reactions, they would change some of the billions of electron arrangements and rearrangements that cause you to be alive into some *other* arrangements. And *that* wouldn't be good for you. *That's* what Lewis structures mean in the real world!

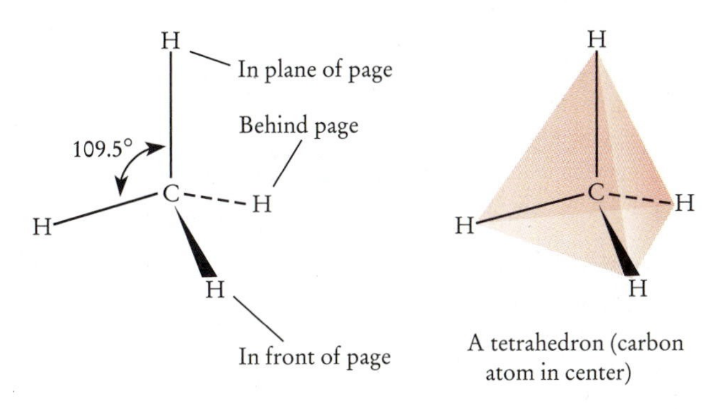

▶ Figure 6.10
Tetrahedral Geometry of Methane.
The hydrogen atoms reside at the corners of the tetrahedron. Hydrogen–carbon–hydrogen bond angles are 109.5°.

The Lewis electron-dot formula for methane is,

$$\begin{array}{c} \text{H} \\ \text{H:}\ddot{\underset{..}{\text{C}}}\text{:H} \\ \text{H} \end{array}$$

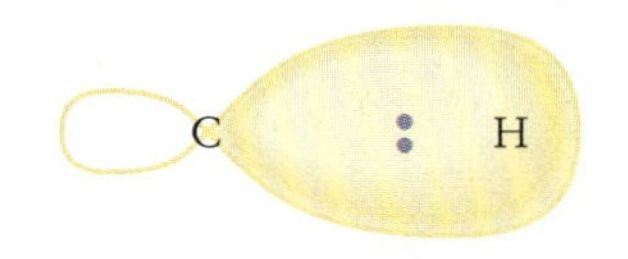

▶ Figure 6.11
A Carbon–Hydrogen Molecular Orbital.

Each C—H bond consists of a molecular orbital that is shaped roughly like a lopsided sausage (▶ Figure 6.11). (The nuclei of the atoms are in the centers of the symbols for the elements.) As in the hydrogen molecule, the two electrons in each C—H molecular orbital reside primarily between the two nuclei. (We say that the electron density, or the probability of finding the electrons, is greatest between the nuclei.)

Covalent bonds around a central atom will orient themselves as far apart as possible.

The electron pairs in the C—H bonds mutually repel each another (like charges repel), forcing the four bonds to orient themselves as far from one other as possible. This is analogous to tying four equally inflated balloons together by their knotted ends, so that they protrude from the same point and press against one another tightly. To relieve the pressure, the balloons will assume the same arrangement in space as the C—H bonds (▶ Figure 6.12). The principle of minimizing the repulsive forces (and thus maximizing the distance) between the electron pairs in bonds around a central atom is called the **valence shell electron pair repulsion** model, or *VSEPR* (pronounced "vesper").

▶ Figure 6.12
Four balloons tied together at their knotted ends assume a tetrahedral arrangement. Four carbon–hydrogen bonds minimize their mutual repulsion by arranging themselves in the same tetrahedral shape.

In this arrangement, the H—C—H bond angles (called tetrahedral angles) are all equal and have a value of 109.5°. Note that the bonds are oriented in three dimensions. If the four bonds were to be oriented in two dimensions (that is, smashed flat), the bond angles would be only 90°, much closer together than in the tetrahedron. Note from ▶ Figure 6.13 that if the molecule, which is three dimensional, were forced to be flattened out into the plane of the page, it would spring back to its tetrahedral shape as soon as it was released (as would the balloon model in Figure 6.12).

The ammonia molecule, NH_3, has a *pyramidal geometry,* in which the nitrogen atom and the three hydrogen atoms reside at the corners of a triangular pyramid, as shown in ▶ Figure 6.14. The H—N—H bond angles are about

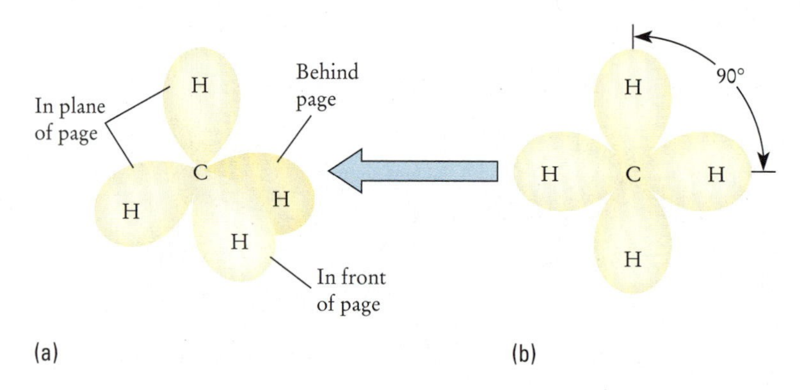

▶ Figure 6.13
Tetrahedral Arrangement of Molecular Orbitals in Methane. *If molecular orbitals in the most stable arrangement (a) are forced closer together, as in (b), electron pair repulsion raises their energy, causing them to spring back to the stable position (a) when application of the force stops.*

107°, slightly less than a tetrahedral angle. VSEPR can also account for the shape of the ammonia molecule.

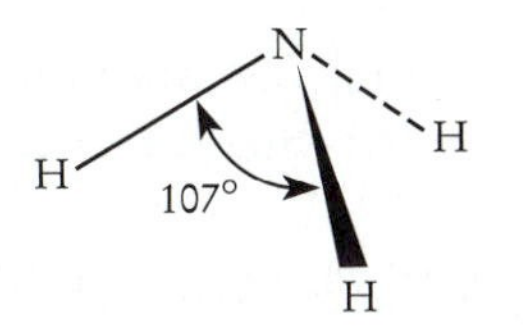

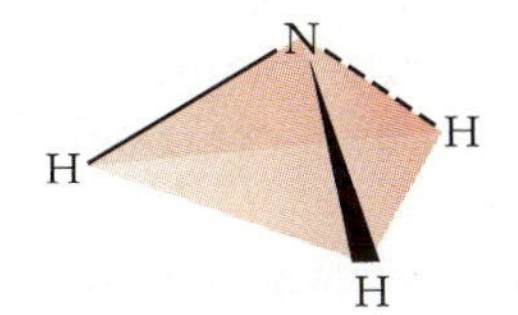

▶ Figure 6.14
The Ammonia Molecule Has a Pyramidal Shape. *The nitrogen atom is at the apex of the pyramid and the hydrogen atoms form the base.*

Note in the electron-dot formula for ammonia that the nitrogen atom has an octet of electrons, but one pair of valence electrons has no hydrogen atom bonded to it.

H
H:N:H — Nonbonded electron pair

This pair of electrons is called a *nonbonded,* or *unshared,* electron pair, as compared with the bonded, or shared, electron pairs in the N—H bonds. The molecular orbital of the N—H bond has roughly the same shape as that for the C—H bond, and so does the orbital that contains the nonbonded electron pair, but the latter orbital is somewhat larger (▶ Figure 6.15).

Even though no hydrogen atom is associated with the nonbonded pair of electrons in ammonia, this pair still takes up space in its molecular orbital and repels the three bonded electron pairs in theirs. Thus, the arrangement of all the electron pairs (bonded and nonbonded taken together) is essentially the same as that in methane: tetrahedral. The shape of the ammonia molecule, however, is different (▶ Figure 6.16). In discussing molecular shape we are talking about the positions of the atoms with respect to one another; we do not consider the nonbonded electron pairs. (We do this because all we can see

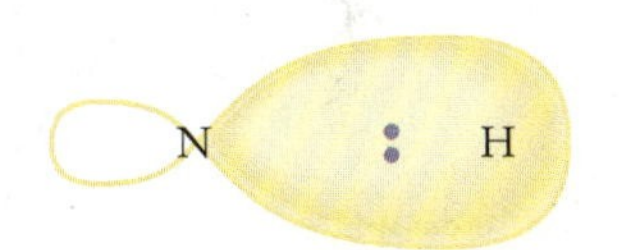

(a)

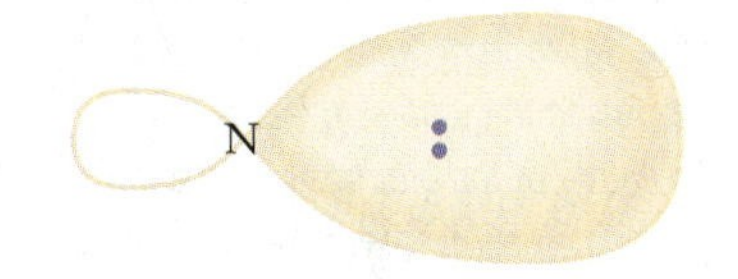

(b)

▶ Figure 6.15
Bonded and Nonbonded Electron Pairs. *(a) The molecular orbital for a bonded electron pair. (b) The molecular orbital for a nonbonded electron pair.*

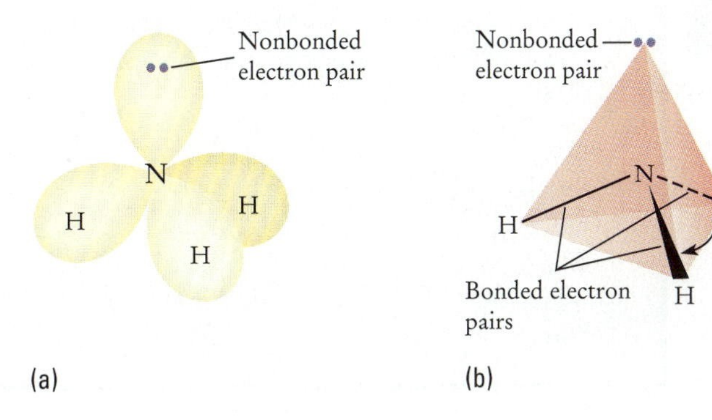

▶ Figure 6.16
VSEPR Analysis for the Ammonia Molecule. *(a) Arrangement of the molecular orbitals in the ammonia molecule. (b) The geometry of the electron pairs around the nitrogen atom in the ammonia molecule is tetrahedral, but the arrangement of the atoms is pyramidal.*

experimentally are the positions of the nuclei of the atoms; the electrons are invisible.) The nitrogen atom and three hydrogen atoms (looking at the molecule from the "outside") describe a triangular pyramid, and so we say the shape of the ammonia molecule is pyramidal. The H—N—H bond angles are about 107°, a slightly compressed tetrahedral angle, because the nonbonded electron pair takes up more room than the bonded pairs and crowds the latter together.

A similar analysis can be made for water, H_2O, which has a bent, or V-shaped, molecule (▸ Figure 6.17). In this case, there are two O—H bonded pairs of electrons and two nonbonded pairs of electrons on the oxygen atom. The shapes and relative sizes of the molecular orbitals that describe each kind of pair are similar to those in ammonia. Arranging the four pairs of electrons (two bonded and two nonbonded) around the central oxygen atom gives the same arrangement as for methane and ammonia—namely, tetrahedral. In this instance, however, the shape of the water molecele can be described as bent or V-shaped because the H—O—H bonds define an angle of about 105°. Although the two nonbonded electron pairs cannot be seen, their presence is felt by their further compression of the H—O—H bond angle.

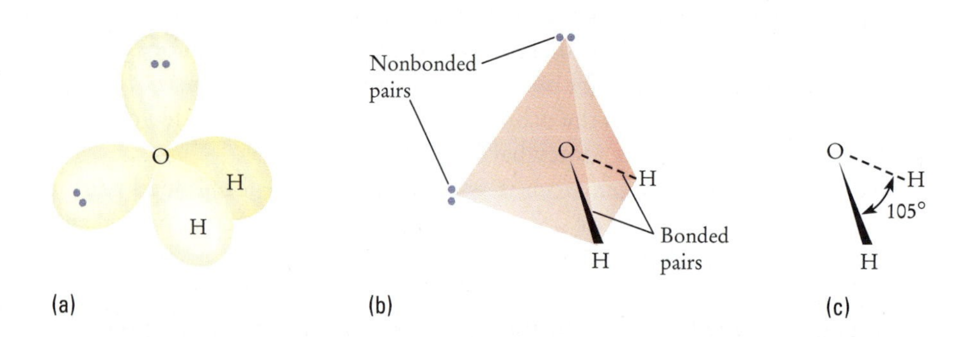

▸ Figure 6.17
VSEPR Analysis for the Water Molecule. *(a) Arrangement of the molecular orbitals in the water molecule. (b) The geometry of the electron pairs in the water molecule is tetrahedral, but the arrangement of the atoms is bent or V-shaped. (c) The water molecule is said to be bent, or V-shaped.*

6.16 Using VSEPR, predict the arrangement of the electron pairs around the central atom in each of the following molecules. You will need to draw the Lewis structures for them first and then determine the three-dimensional arrangement of the electron pairs around the central atom. Do this for:

a. CF_4
b. CH_2Cl_2
c. SiH_4
d. SiF_4
e. NCl_3
f. PCl_3
g. PH_3
h. AsH_3
i. $AsCl_3$
j. H_2S
k. H_2Se
l. OF_2

6.17 Determine the molecular shapes of each of the molecules (a through l) in 6.16. That is, how are the atoms arranged in space around the central atom? Do you see any family resemblances? If so, make some generalizations about what they might be.

OTHER MOLECULAR GEOMETRIES

The electron pairs of some molecules are arranged around the central atom in other than a tetrahedral geometry. The different shapes result when two or three (and sometimes five and six) pairs of electrons exist around the central atom instead of four. We have previously drawn the Lewis structures for some of these in study exercises, so part of our work is already done. Now let's consider formaldehyde gas, H_2CO, a major component in embalming fluid and a reactant used in making plastic foams. This molecule has a triangular shape and is flat: that is, all four atoms lie in the same plane.

Note that its Lewis structure contains a CO double bond,

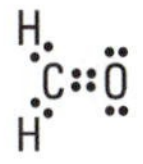

and that both of the electron pairs in that CO bond are pointed in the same direction, toward the oxygen atom. This in effect means that the double bond behaves as if it were one pair of electrons, instead of two, when we use VSEPR. Thus, we have the equivalent of only three electron pairs to orient around the carbon atom. To visualize this we again use the analogy of three balloons fastened together tightly at their knotted ends, so that they tend to spring apart. This time, however, they are as far apart as possible when they are all lying in the same plane (flat surface) and their outer ends are pointing at 120° away from one another. This is a triangular arrangement, and the formaldehyde molecule assumes this same shape (▶ Figure 6.18). Similarly, the HCO and HCH bond angles are all approximately 120°.

All electron pairs in a multiple bond point in the same direction, so they are counted as one when using VSEPR analysis.

6.18 Try this VSEPR analysis with H_2CS, SO_3, CO_3^{2-} (note charge), NO_3^-, and SO_2. Compare the shape of the SO_2 molecule with the arrangement of its electron pairs.

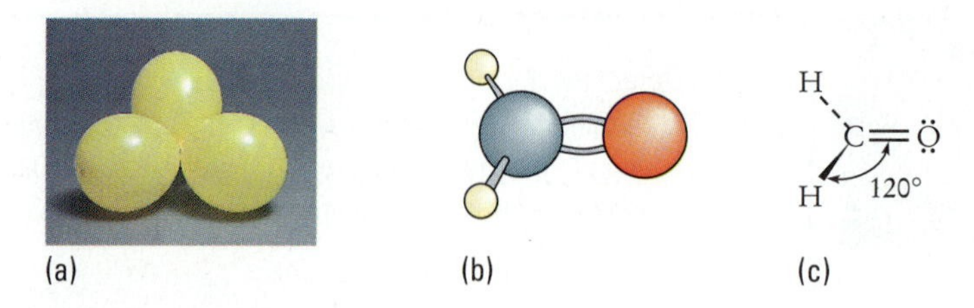

▶ Figure 6.18
VSEPR Analysis for Three Electron Pairs. *(a) Three balloons tied together at their knotted ends assume a triangular shape. (b) The formaldehyde molecule also has a triangular shape. The two bonded pairs of electrons in the double bond have the same direction in space as the electron pair in a single bond would have. (c) Three-dimensional line drawing of the formaldehyde molecule.*

Another molecular geometry can result when multiple bonds are present. This arrangement is *linear,* with the two bonds pointing in exactly opposite directions. Two molecules we have encountered previously exhibit this shape— CO_2 and HCN (▶ Figure 6.19). Their electron-dot formulas are

Ö::C::Ö and H:C:::N:

Carbon dioxide has two CO double bonds, and note again that each of these behaves as if it were only one pair of electrons, not two. This leaves the equivalent of only two bonds (electron pairs, balloons), which orient themselves as far apart as possible. The OCO bond angle is 180°, that of a straight line. The case of HCN is similar, except in this case the molecule has a CN triple bond, which behaves directionally like a single bond. The HC single bond behaves normally; and orienting the equivalent of two bonds (electron pairs, balloons) as far apart as possible results in a linear geometry, with a HCN bond angle of 180°.

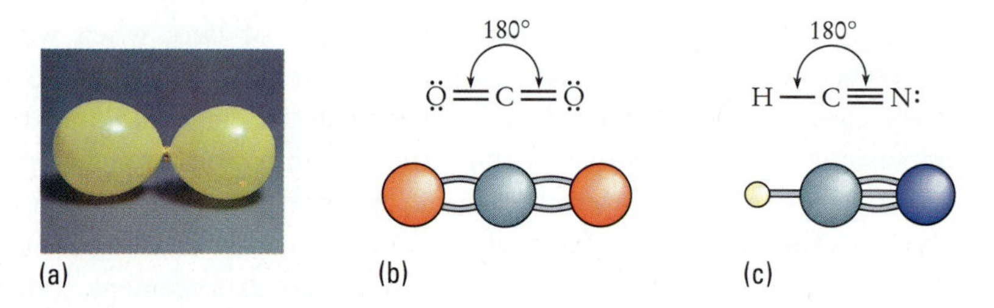

▶ Figure 6.19
VSEPR Analysis for Two Electron Pairs. *(a) Two balloons tied together form a linear arrangement. (b) Carbon dioxide and (c) hydrogen cyanide molecules are linear. Multiple bonds take the same direction that a single bond would.*

6.19 Not many simple molecules have a purely linear geometry. However, try this analysis on HNC, CS_2, COS, and C_2H_2. Note that the last has two carbon atoms, which must be considered separately.

POLAR BONDS

Like the bonds in methane, ammonia, and water, many covalent bonds in molecules join dissimilar atoms. That is, carbon is bonded to hydrogen, oxygen to hydrogen, carbon to chlorine, nitrogen to oxygen, and so forth. The electron pair that holds these dissimilar atoms together is shared by both of them, but because the atoms are different the sharing is unequal. That is, one atom in the pair has the electrons more of the time, on the average, than the other. More correctly, we say that the electron density in the bond is shifted toward one atom in preference to the other.

In molecules like O_2 or Cl_2 the electron pairs joining the atoms are shared equally, and each atom has exactly half of the electron density. In molecules with dissimilar atoms, such as hydrogen chloride gas, HCl, the electron density is shifted toward one atom (in this case the chlorine atom); so one end of the bond (Cl) has more of the electron density around it than the other. When one atom (such as Cl) has a greater affinity for electrons in a covalent bond, we say that it is more electronegative than the other (H). **Electronegativity,** the ability of an atom to attract the electrons in a covalent bond to itself, is greatest for fluorine; and it decreases toward the left of the Periodic Table as well as downward from the upper right-hand corner (▶ Figure 6.20). Thus, oxygen is more electronegative than carbon, chlorine more electronegative than iodine or sulfur, and nitrogen more electronegative than phorphorus. All of the nonmetals are at least slightly more electronegative than hydrogen.

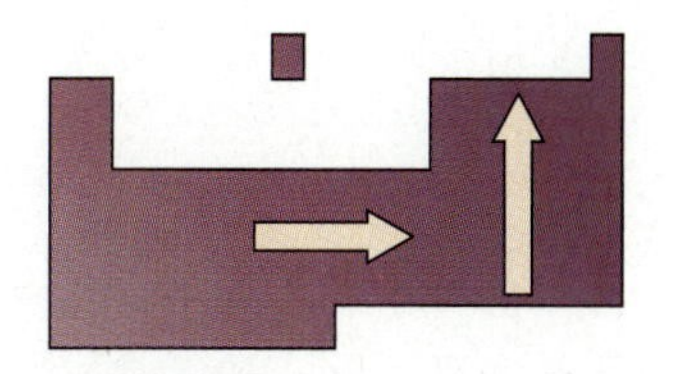

▶ Figure 6.20
Electronegativity Trends in the Periodic Table. *Arrows show the direction of increasing electronegativity (darker shades). Note position and shade of hydrogen in this figure.*

6.20 Which element in each of the following pairs is more electronegative?

a. C or Si
b. As or Br
c. C or O
d. P or Cl
e. N or As
f. O or Se
g. Se or F

6.21 What is the relationship of electronegativity to the rule for finding the central atom when drawing Lewis structures? See rule 3 in the section Drawing Lewis (Electron-Dot) Formulas.

Because electrons are negatively charged and are "closer" to the chlorine atom than to the hydrogen atom in HCl, a partial negative charge (represented by $\delta-$) will reside there. The hydrogen atom, having a smaller than average electron density, develops a partial positive charge (represented by $\delta+$). As a result, the H—Cl bond is polar and is represented as $^{\delta+}$H—Cl$^{\delta-}$.

This partial charge development on two adjacent atoms caused by unequal sharing of the electron pair between them is an example of a *dipole,* and the bond between the two atoms is said to be a *polar covalent bond,* or **polar bond.** In the realm of covalent bonds, polar bonds are the rule rather than the exception. Generally, any covalent bond between two dissimilar nonmetals is polar.

Any covalent bond between two different nonmetals will be **polar** to some extent.

The H—Cl molecule we have used as our example is *diatomic.* That is, it has only two atoms. Because the H—Cl bond is polar, the H—Cl molecule must be polar as well. The entire molecule, then, has a partially positive end (hydrogen) and a partially negative end (chlorine) and thus has an overall dipole. All diatomic molecules with dissimilar atoms (binary molecules), such as CO, NO, and HF, are also polar for this reason. Diatomic molecules that are not binary, such as O_2, N_2, and Br_2, have no dipole and are nonpolar.

IDENTIFYING POLAR MOLECULES

Larger molecules containing three atoms *(triatomic)* or more *(polyatomic)* may or may not be polar, even when they contain polar bonds. To decide whether these molecules are polar we first must know their molecular shape. Then, we must see how their polar bonds are arranged with respect to one another in the molecule as a whole and in three dimensions. If the overall distribution of polar bonds is symmetrical and does not produce "ends" in the molecule, the molecule is nonpolar. If, however, the overall distribution of polar bonds is unsymmetrical and produces an end that has a different kind of atom than another end, then the molecule is polar. (The question is, does the molecule have an endedness to it?) We can use our previous examples of methane, ammonia, and water to illustrate this point.

Polar molecules have an overall dipole.

For the purpose of deciding whether a molecule is polar, we do not have to know the direction of the dipole in each of its polar bonds. All we need to know is the molecule's shape and that all covalent bonds between dissimilar atoms are polar. Consider methane, a tetrahedral molecule (▶ Figure 6.21). The C—H bonds are slightly polar (the negative end is in the direction of the carbon atom). Note that the tetrahedral arrangement of the hydrogen atoms is quite symmetrical, and looking at the molecule from the outside we see no evidence that it is lopsided. That is, there doesn't appear to be a way we can find a positive end and a negative end on the surface of the molecule.

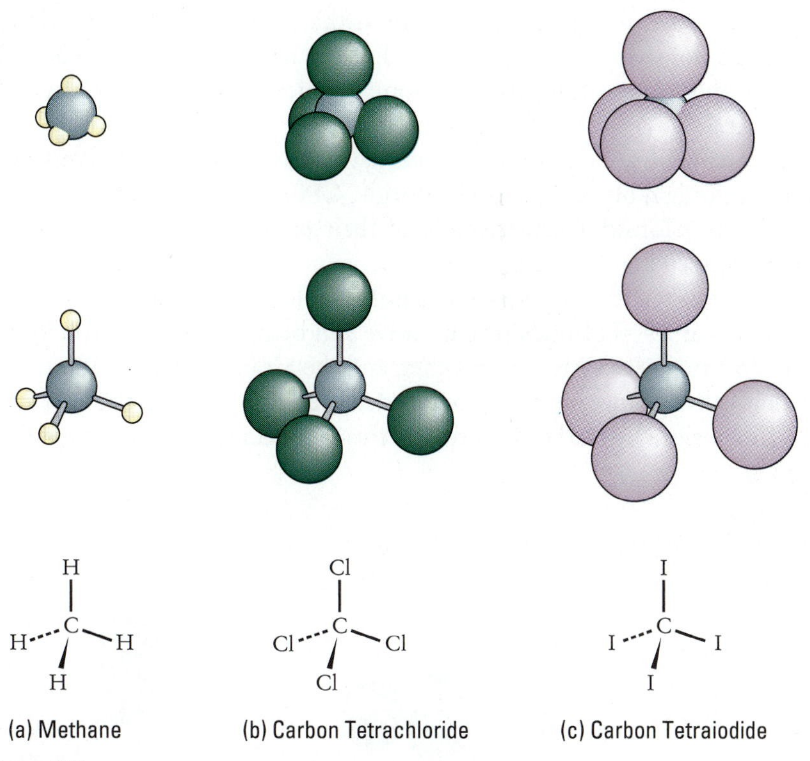

▶ Figure 6.21
Nonpolar Tetrahedral Carbon Compounds. *(Top to bottom) Space-filling models, ball-and-stick models, and three-dimensional line drawings of (a) methane, (b) carbon tetrachloride, and (c) carbon tetraiodide. Each molecule has a symmetrical arrangement of polar bonds*

We could substitute a different kind of atom for all four of the hydrogen atoms in methane and still have a nonpolar molecule. In carbon tetrachloride, CCl_4, for example, the C—Cl bonds are also polar (Cl is more electronegative and is the negative end this time), and the overall molecule is tetrahedral. The molecule again is nonpolar because the tetrahedral arrangement of the chlorine atoms is symmetrical when viewed from the outside.

The ammonia molecule presents us with a different situation. The N—H bonds are polar with the partially negative end at nitrogen, and there is one nonbonded pair of electrons to deal with. Ignoring the electron pair for the moment, we know that the shape of the molecule (arrangement of the atoms) is pyramidal. This results in a definitely lopsided arrangement, with the hydrogen atoms of the three polar bonds toward one end and the nitrogen atom toward the other (▶ Figure 6.22). We immediately suspect that the ammonia molecule is polar. If we choose to visualize the nonbonded pair by attaching it in its proper place to the nitrogen atom, the picture becomes even more unsymmetrical.

Some tetrahedral carbon compounds have an asymmetry analogous to that of the ammonia molecule. Methyl chloride, CH_3Cl, and chloroform, $CHCl_3$, have the same lopsided arrangement of their polar bonds, resulting in their being polar (see Figure 6.22).

The water molecule is very polar.

It is easy to see that water is also a polar molecule. In fact, it is very polar. The two polar O—H bonds (the negative end being oxygen) lie roughly at a tetrahedral angle (105°) to one another, and the arrangement is definitely unsymmetrical. Adding the two nonbonded electron pairs to the picture (in three dimensions) greatly accentuates this imbalance in polarity (▶ Figure 6.23).

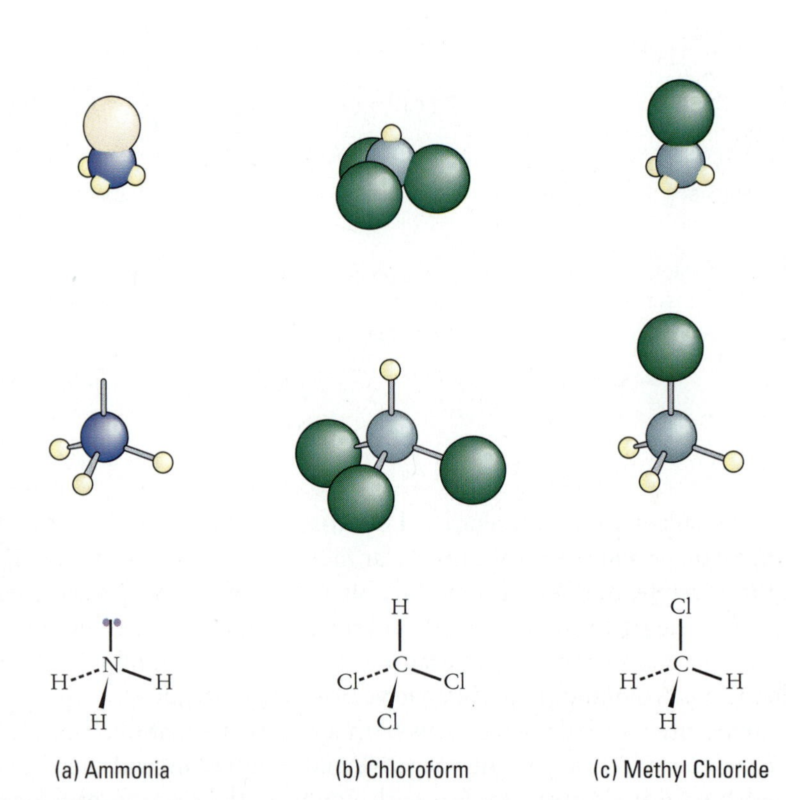

▶ Figure 6.22

Ammonia Molecule and Other Polar Molecules of Like Symmetry. *Space-filling models, ball-and-stick models, and three-dimensional line drawings (top to bottom) show that the arrangement of polar bonds in each molecule is unsymmetrical: (a) ammonia has a hydrogen end and a nitrogen end, (b) chloroform a chlorine end and a hydrogen end, and (c) methyl chloride a hydrogen end and a chlorine end.*

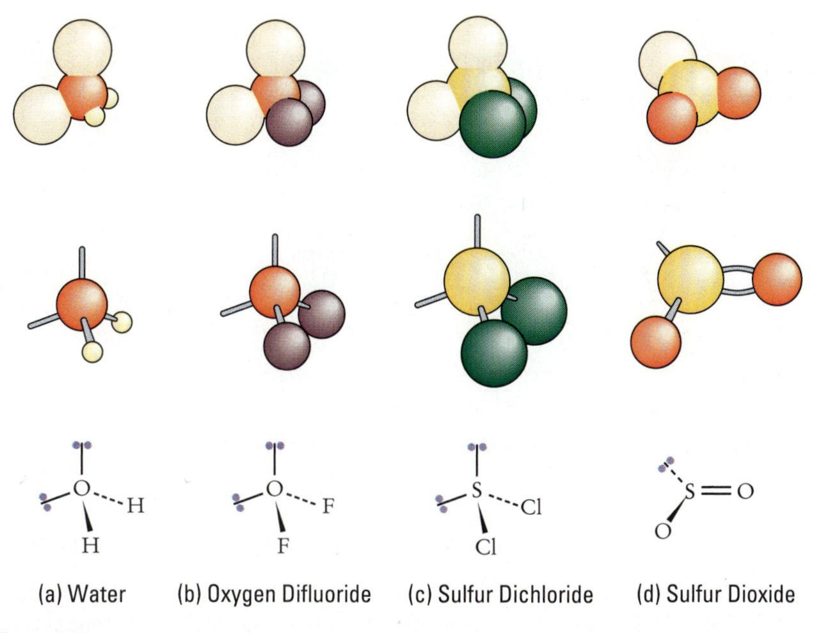

► Figure 6.23

Water and Other Molecules Having V-Shaped Geometries. *Space-filling models, ball-and-stick models, and three-dimensional line drawings (top to bottom) show that all have an unsymmetrical arrangement of polar bonds: (a) water, (b) oxygen difluoride, (c) sulfur dichloride, and (d) sulfur dioxide (triangular electron pair arrangement).*

Other compounds having water's V-shaped geometry are also polar. These include OF_2, SCl_2, and SO_2. The last contains a sulfur–oxygen double bond; and, though bent, its arrangement of electron pairs is triangular instead of tetrahedral. Tetrahedral carbon compounds such as CH_2Cl_2 are also polar because they have an asymmetry analogous to that of water (the C—H bonds take the places of nonbonded electron pairs).

Simple molecules can take geometries other than tetrahedral, pyramidal, and V-shaped. Carbon dioxide, CO_2, is linear (straight) and nonpolar. Hydrogen cyanide, HCN, is also linear but is polar. Formaldehyde, H_2CO, is triangular and polar, while sulfur trioxide, SO_3, is triangular and nonpolar. These examples are shown in ► Figure 6.24.

NAMING BINARY MOLECULAR COMPOUNDS

Binary molecular compounds formed from two different nonmetals are named in a manner similar to that for ionic compounds formed from a metal and a nonmetal. The more metallic element is written first in the chemical

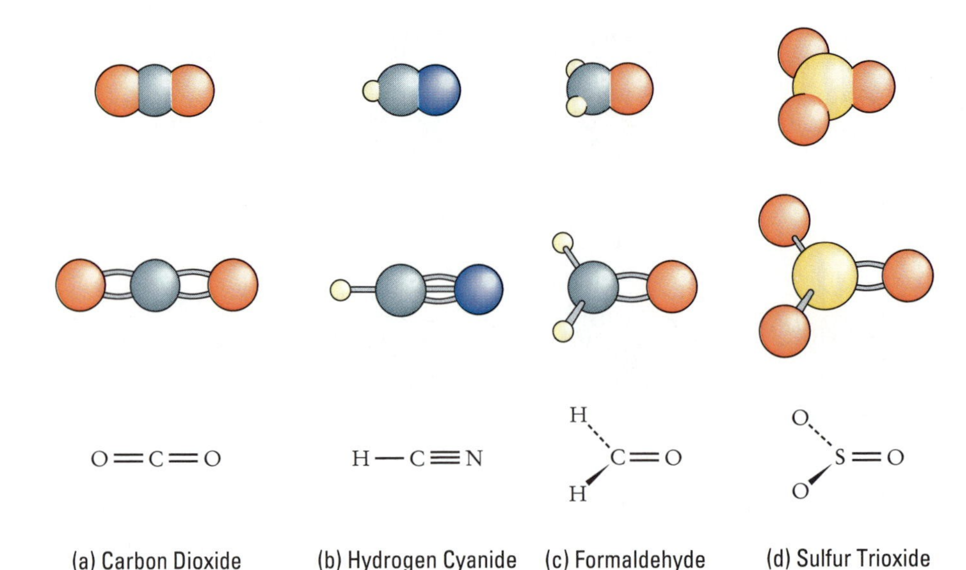

▶ Figure 6.24

Space-Filling Models, Ball-and-Stick Models, and Three-Dimensional Line Drawings of Linear and Triangular Molecules. *Depending upon the arrangement of polar bonds around the central atom, these kinds of molecules can be polar or nonpolar: (a) carbon dioxide is linear, nonpolar; (b) hydrogen cyanide, linear, polar; (c) formaldehyde, triangular, polar; and (d) sulfur trioxide, triangular, nonpolar.*

formula and is named first. The less metallic element is written second and is named as if it were an anion. Thus HF is named hydrogen fluoride. If the relationship of the two elements is within the same group, the element lower in the group is taken first. Thus, a compound of iodine and chlorine is ICl, iodine monochloride.

Often two nonmetals form more than one compound with one another. In this case, Greek prefixes (■ Table 6.2) are used to indicate how many atoms

6.22 Name each of the following molecules:

a. CO
b. NO_2
c. OCl_2
d. S_2Cl_2
e. P_2O_3
f. C_3O_2
g. N_2O_3
h. PCl_5
i. SF_4
j. NF_3

TABLE 6.2 GREEK PREFIXES

Number of Atoms	Greek Prefix
one	mono-
two	di-
three	tri-
four	tetra-
five	penta-
six	hexa-

of each element are present. Thus, CO_2 is carbon dioxide, N_2O is dinitrogen monoxide, SO_3 is sulfur trioxide, and P_2S_5 is diphosphorus pentasulfide. Sometimes the prefix mono- is omitted before the name of the first element. Carbon dioxide and sulfur trioxide, for example, are not named "monocarbon dioxide" and "monosulfur trioxide." In cases like these, mono- is understood. Earlier in this chapter H_2S was named hydrogen sulfide, not "dihydrogen sulfide." Because it is the only compound of hydrogen and sulfur, its common name is not ambiguous. This is also true for H_2Se and H_2Te.

PREDICTING IONIC AND COVALENT CHARACTER

For many binary compounds, we can predict whether or not they are likely be ionic simply by using the Periodic Table, even if we are unfamiliar with the individual elements themselves. Remembering that binary compounds of metals combined with nonmetals tend to be ionic and that those of nonmetals combined with other nonmetals tend to be covalent (molecular), we can look at the Periodic Table to see which elements would be expected to form each of these kinds of compounds.

A binary compound of a metal and a nonmetal is ionic. One of two nonmetals is covalent.

Since metals are grouped to the left of the Zintl border and nonmetals to the upper right of it, we need only determine the relative locations of the pair of elements we want to combine in order to predict whether their binary compound will be ionic or covalent. Combining an element from the left-hand side of the chart with one from the right-hand side should result in an ionic compound. Combining two elements from the right of the Zintl border should produce a molecular compound. (Mixing two different metals usually results in the formation of an alloy, which will not be discussed here.)

As an example, a binary compound formed from strontium (Sr) and iodine (I) should be ionic. Strontium is a metal in Group 2A on the left-hand side of

6.23 Predict whether compounds made up of the following pairs of elements will be ionic or molecular.

a. bromine and fluorine
b. magnesium and oxygen
c. arsenic and chlorine
d. sulfur and zinc
e. oxygen and chlorine
f. potassium and sulfur
g. nitrogen and hydrogen
h. sulfur and carbon
i. iron and oxygen
j. lithium and nitrogen

6.24 Predict whether the following compounds will be ionic or molecular.

a. $CaCl_2$
b. Na_2SO_4
c. NF_3
d. $AlBr_3$
e. NO_2
f. CO_2
g. K_2O
h. P_4O_{10}
i. Li_2Se
j. BrCl
k. CBr_4
l. Na_3N
m. SeO_2

the chart and iodine is a nonmetal in Group 7A on the right-hand side. Similarly, a compound formed from phosphorus (P) and selenium (Se) should be molecular because both of these elements are nonmetals in Groups 5A and 6A, respectively, on the right-hand side of the chart.

BONDING IN METALS

The properties of metals are explained by a different kind of bonding, called the **electron sea model.** The *metallic bond* is thought to involve a three-dimensional lattice of positive metal ions, usually metal atoms without their valence electrons, immersed in a "sea" of negative valence electrons (▶ Figure 6.25). These valence electrons are said to be **delocalized** because they do not belong to any one metal ion, but rather wander freely throughout the sea among all of the ions. This kind of bonding is quite strong because each metal ion is attracted to a large number of electrons, and vice versa. However, the lattice of metal ions is not as rigid as an ionic lattice is. Thus, because metal ions can slide over one another when a force is applied to them, metals are

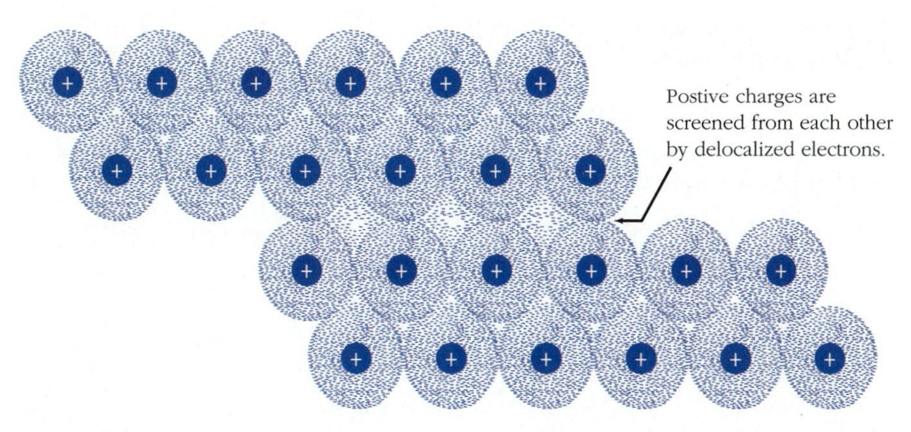
Postive charges are
screened from each other
by delocalized electrons.

▶ Figure 6.25
The Electron Sea Model of Metals. *Positive metal ions are arrayed in a lattice and immersed in a sea of delocalized valence electrons, which account for the electrical conductivity of metals. Metals are malleable and ductile because layers of metal ions can slide past one another, separated by the electrons between them.*

malleable and ductile. Because the delocalized electrons are mobile, they can carry an electric current and conduct heat. These electrons also can occupy a virtually unlimited number of excited state orbitals close to the ground state. Thus, metals can absorb and re-emit light of almost all visible wavelengths, which accounts for their luster.

DID YOU LEARN THIS?

For each of the statements given, fill in the blank with the most appropriate word or phrase from the following list. You *may* use any word or phrase more than once.

octet rule	electronegativity
electrostatic	lattice
molecular orbital	valence electrons
dipole	nonbonded electrons
ionic	tetrahedral
covalent	triangular
metallic	pyramidal
metals	bent
nonmetals	linear
metalloids	delocalized
G. N. Lewis	cations
polar bond	anions

a. _______ A covalent bond formed between dissimilar nonmetals
b. _______ The kind of binary compound that P_2O_5 is
c. _______ Has a partial positive charge at one end and a partial negative charge at the other
d. _______ Positively charged ions
e. _______ A bond in which a sea of electrons surrounds ions
f. _______ The shape of the H_2S molecule
g. _______ The place in which two electrons are found in a covalent bond
h. _______ The spatial arrangement of three electron pairs about a central atom
i. _______ Invented electron-dot formulas
j. _______ The affinity of an atom for electrons in a covalent bond
k. _______ The electrons in an atom that determine how it reacts chemically
l. _______ Describes electrons that account for the electrical conductivity of metals
m. _______ The kinds of bonds that hold atoms together to form molecules
n. _______ An orderly three-dimensional arrangement of ions
o. _______ The kind(s) of forces that hold ionic crystals together
p. _______ Elements that gain electrons to form negative ions

EXERCISES

1. a. The atom of a metal element loses two electrons in a chemical reaction. What is the charge on the metal ion that is formed?
   b. The atom of a metal element loses three electrons in a chemical reaction. What is the charge on the metal ion that is formed?
   c. The atom of a nonmetal element gains one electron in a chemical reaction. What is the charge on the nonmetal ion that is formed?
   d. The atom of a nonmetal element gains three electrons in a chemical reaction. What is the charge on the nonmetal ion that is formed?
2. Write the chemical symbol (with the proper charge) and give the correct name for the ions formed from each of the following metals.

   a. Ga
   b. Na
   c. Ca
   d. Mg

3. Write the chemical symbol (with the proper charge) and give the correct name for the ions formed from each of the following nonmetals.

   a. F
   b. Br
   c. As
   d. P
   e. Se

4. a. A nonmetal X in the representative elements produced an ion X^{3-}. What is its group number?
   b. A nonmetal Z in the representative elements produced an ion Z^{-}. What is its group number?
   c. A metal R in the representative elements produced an ion R^{2+}. What is its group number?
   d. A metal Q in the representative elements produced an ion Q^{3+}. What is its group number?
5. Give the correct chemical symbol (including the correct charge) for each of the following ions.

   a. sulfide ion
   b. barium ion
   c. aluminum ion
   d. sodium ion
   e. fluoride ion
   f. potassium ion
   g. phosphide ion
   h. bromide ion

6. Using the Periodic Table, predict the chemical formula of the binary compounds formed from the following pairs of elements.

   a. beryllium and chlorine
   b. aluminum and sulfur
   c. barium and bromine
   d. magnesium and arsenic
   e. barium and nitrogen
   f. aluminum and nitrogen
   g. magnesium and oxygen
   h. potassium and selenium

7. Name each of the binary compounds formed from the metal and nonmetal pairs in exercise 6.
8. Write the chemical formulas for the following binary compounds.

   a. sodium nitride
   b. aluminum bromide
   c. magnesium nitride
   d. sodium oxide
   e. barium sulfide
   f. aluminum phosphide
   g. beryllium fluoride
   h. calcium arsenide

9. Write the formulas, including correct charges, for the following polyatomic ions. Indicate which are anions and which are cations.

   a. hydroxide ion
   b. ammonium ion
   c. phosphate ion
   d. carbonate ion
   e. nitrate ion
   f. hydrogen carbonate ion
   g. sulfate ion

10. Give the correct name for the following compounds containing polyatomic ions.

    a. $AlPO_4$
    b. $Al(NO_3)_3$
    c. $Ca(OH)_2$
    d. $(NH_4)_2CO_3$
    e. $LiC_2H_3O_2$
    f. $BaCO_3$
    g. Li_3PO_4
    h. $Mg(HCO_3)_2$
    i. K_2SO_4

11. Write the correct formulas for the following compounds containing polyatomic ions.

    a. barium sulfate
    b. calcium carbonate
    c. aluminum nitrate
    d. sodium phosphate
    e. ammonium nitrate
    f. magnesium hydroxide
    g. potassium acetate

12. State the meaning, in words, of each of the following chemical formulas.

    a. KCl
    b. CaF_2
    c. $(NH_4)_2S$
    d. $Al(C_2H_3O_2)_3$
    e. Li_3N
    f. Na_2O

g. $BaSO_4$
h. $Mg_3(PO_4)_2$
i. $NaHCO_3$
j. NH_4NO_3

13. What is a molecule? How does a molecule of vanillin (the flavor in vanilla) differ from the ions in baking soda (sodium hydrogen carbonate, $NaHCO_3$)? How does the Teflon surface, a molecular coating, differ from the aluminum atoms in the frying pan that it coats?
14. Compare the properties of ionic, covalent, and metallic substances with one another. How do the models of the structure for each explain these properties, and how do these models differ from one another?
15. What do the terms diatomic, triatomic, and polyatomic mean when applied to molecules and ions?
16. What keeps the nuclei of two covalently bound atoms from touching one another? What do you suppose determines the length (distance between centers of the nuclei) of a covalent bond?
17. What property of electron pairs is employed when VSEPR analysis is used to determine the arrangement of electron pairs around a central atom? Why must the analysis be done in three dimensions rather than in two?
18. Complete the following chart, which summarizes how to predict molecular shape using VSEPR.

Number of Electron Pairs (Total)*	Number of Unshared Pairs	Geometry of Electron Pairs (All)	Molecular Shape (Geometry)	Bond Angle (Approx.)
4	0			
4	1			
4	2			
3	0			
3	1			
2	0			
2	1			

*Multiple bonds count as one pair.

Once you have completed this chart, you should know how to figure out molecular shapes by determining the total number of electron pairs, the number of nonbonded pairs, and using VSEPR. *Do not memorize this chart!*

19. Decide which of the following molecules are nonpolar or polar.

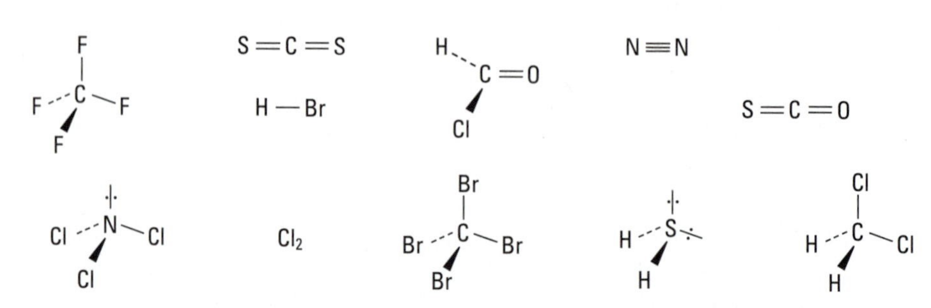

20. Explain how a molecule can have polar bonds but not itself be polar.
21. Predict whether each of the following compounds would be expected to be ionic or molecular.

a. PCl_3
b. $MgSO_4$
c. C_2H_6
d. CBr_4
e. NO_2
f. KI
g. Li_2S
h. CaF_2
i. OF_2
j. $NaNO_3$

22. Predict whether compounds formed from each of the following pairs of elements would be expected to be ionic or molecular.

a. calcium and oxygen
b. nitrogen and chlorine
c. carbon and sulfur
d. oxygen and aluminum
e. arsenic and oxygen
f. arsenic and sodium
g. potassium and iodine
h. nitrogen and sodium
i. sulfur and fluorine
j. barium and bromine

23. Give the correct name for the following binary molecular compounds.

a. N_2O_4
b. H_2S
c. SiO_2
d. As_2O_5
e. OCl_2
f. $SiCl_4$
g. SeO_2
h. SF_4

24. How does the electron sea model account for the malleability of metals?

THINKING IT THROUGH

A Different Kind of Covalent Bond?

We have said that the electrons in a molecular orbital reside, most of the time, between the nuclei of the atoms they bind. Because electrons repel each other, it might seem reasonable that they would try to get as far apart as possible, rather than stay close together. Why would you not expect the electron density in a covalent bond to be found on the opposite sides of the nuclei, away from space between the atoms? (That is, what keeps the electrons in a covalent bond from flying apart?)

Elements That Don't Behave I

The elements beryllium and boron do not obey the octet rule when they form bonds with nonmetals. For example, beryllium forms BeH_2 and $BeCl_2$ whereas boron forms BH_3 and BF_3. Draw electron-dot structures for each of these molecules. What is the number of electrons around the central atom for each of these elements? How stable do you think these compounds would be? BF_3 reacts with HF to form H^+ and BF_4^-. Why do you suppose this happens?

Elements That Don't Behave II

Certain larger nonmetal atoms do not always obey the octet rule when they form bonds, particularly those of P, S, As, and Se. Draw the Lewis structures for SF_4 and PCl_5, and, using the principles of VSEPR, draw each molecule in three dimensions and describe its shape.

ANSWERS TO "DID YOU LEARN THIS?"

a. polar bond
b. covalent
c. dipole
d. cations
e. metallic
f. bent
g. molecular orbital
h. triangular
i. G. N. Lewis
j. electronegativity
k. valence electrons
l. delocalized
m. covalent
n. lattice
o. electrostatic
p. nonmetals

CHEMISTRY AND PLANETARY INTERIORS

Raymond Jeanloz

Geologist

Professor of Geophysics, University of California at Berkeley

B. A. Geology, Amherst College

Ph.D. Geology, Geophysics, California Institute of Technology

I study materials at very high pressures and temperatures—up to millions of atmospheres of pressure and thousands of degrees Celsius in temperature. Scientists have only recently been capable of reaching such extreme conditions in the laboratory, at least in a controlled manner. Why do I do this? To put it simply, I am interested in understanding what happens deep inside planets, and these are the conditions of planetary interiors. The pressure and temperature at the earth's center, for example, are about 3.6 million atmospheres and 6000 to 7000 °C, respectively.

To understand how the planets have evolved over their billions of years of geological history, it is necessary to examine the materials and processes of their interiors. And to do this, one must study what happens in the appropriate conditions.

Such experiments are carried out with lasers and diamonds. Specifically, materials are taken to high pressures by squeezing them between the points of two gem-quality diamonds. We use natural diamonds that are cut the same way as jewelry. Because the points of the diamonds are usually less than 0.5 millimeter across and because pressure is equal to force divided by area, we can achieve pressures of millions of atmospheres by pushing the diamonds together with a modest force.

Why use diamond? It is the strongest material known and can support high pressure at the points without breaking. Because diamond is transparent, we can observe our sample—not just with out eyes, but also with X-rays, lasers, and spectrometers—and thus analyze it in detail at high pressure.

To simulate the temperatures inside planets, we use a powerful laser beam focused right through the diamonds. The material being examined absorbs the continuous (not pulsed) infrared light from the laser, and is heated to thousands of degrees. The temperature is determined by measuring the thermal radiation from the sample, that is, by measuring how red or white hot the sample is. This is the same method used to measure the temperatures of stars from their colors.

Before I describe the insides of planets, I should mention a more basic reason that pressures of millions of atmospheres are interesting. If you calculate the change in the energy of a sample that is caused by such pressures, you obtain a value that is comparable in magnitude to chemical bonding energies. Put another way, the work expended in squeezing the atoms together under a million atmospheres of pressure has a very large effect on the bonding energies. As a result, the nature of bonding—that is, the chemical properties of matter—are greatly altered by pressure.

The underlying reason for these changes is the Pauli principle of quantum mechanics, which states that electrons avoid overlapping each other. As atoms are pushed together, their outer electrons tend to overlap more, to the point that it is easier to change the bonding than to allow greater overlap. This change in bonding occurs when atoms are compressed to pressures of a million atmospheres or more, resulting in major changes in chemical behavior.

Many examples of the effects of pressure on chemistry are now known. For example, xenon, normally an inert element, becomes a metal at pressures of about 1.3 million atmospheres. Similarly, hydrogen has recently been converted into a metal by compression to pressures of about 1.7 to 2.0 million atmospheres. Because hydrogen is the most abundant element in the universe, its metallic form is thought to be the main material in the interiors of giant planets, such as Jupiter and Saturn, and stars. Thus, metallic hydrogen is not just a novel material that is nonexistent at normal pressure, but it is also one of the most important materials for astrophysicists and planetary scientists.

Changes in chemical bonding due to pressure are important inside the earth as well. One way to see this is to think of the deep interior, where the rocky outer shell (mantle and crust) is in contact with the metallic core. Most of the core is made of a liquid iron alloy; this is where the earth's magnetic field is produced.

Rock is composed of oxides, such as the compounds making up ceramics. In fact, rock consists of more than 50% oxygen on an atomic basis. At normal conditions, it is very unlike a metal. At the boundary between the mantle and the core, nearly 3000 km beneath our feet, however, the pressure is over 1.35 million atmospheres, and rock is quite different. We have discovered that oxygen easily combines with iron at high pressures, forming a truly metallic iron-oxide alloy.

When we simulate the conditions existing at the boundary between the mantle and the core, we observe dramatic chemical reactions taking place between liquid iron (or iron alloy)—corelike material—and the high-pressure minerals of the deep mantle. Some of the oxide minerals dissolve into (alloy with) the liquid metal, leaving a heterogeneous mixture of metal alloys and nonmetallic minerals as reaction products.

We expect that the same reactions take place inside the earth, with the rocky mantle dissolving into the liquid metal of the core over geological history. In fact, seismologists, who study the deep interior using the sound waves of earthquakes, find that just such a zone of heterogeneous materials exists at the mantle–core boundary. A simple explanation of why the core is an iron alloy (it does not have exactly the same properties as pure liquid iron at high pressures and temperatures) is that

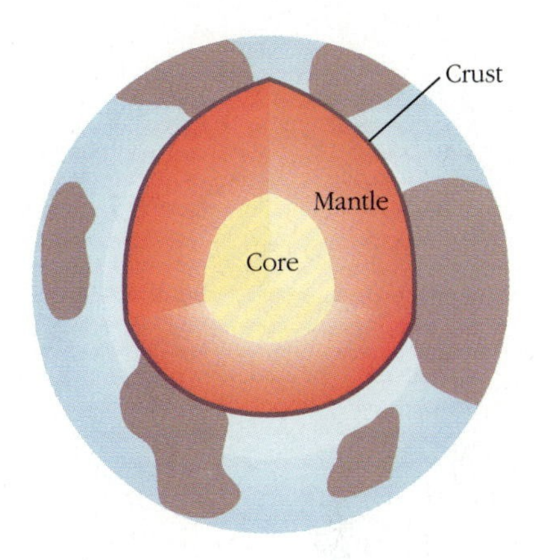

this region has reacted with, and become contaminated by, the mantle above it.

Thus, we may have identified what is the most chemically active region in our planet, the boundary between the mantle and core, and we may have determined one of the main ways in which the earth evolves chemically over geological time.

7 Solutions

Hydrogen bonding is maximized.

Almost all the chemical processes which occur in nature, whether in animal or vegetable organisms, or in the non-living surface of the earth, . . . take place between substances in solution

Wilhelm Ostwald

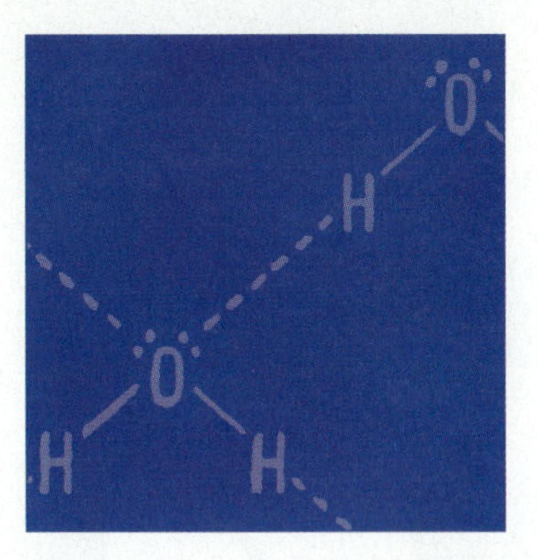

Everyone has used sugar to sweeten tea or coffee or to prepare fruit-flavored drinks. To do this, we add solid, crystalline sugar to water containing the flavored ingredients and stir. Soon the solid sugar disappears, or dissolves, and a uniformly sweetened, homogeneous mixture called a solution results. In a like manner, salt dissolves in water to make salt water, and iodine crystals form a violet-colored solution when swirled in carbon tetrachloride. Nail polish remover can eat holes in, or dissolve, certain fabrics.

SOLVENTS, SOLUTES, AND SOLUTIONS

In the examples just given we observe that a solid substance (sugar, salt, iodine, or fabric) dissolves in a liquid (water, carbon tetrachloride, or nail polish remover). The solid material that dissolves, or "goes into solution," is called the **solute;** the liquid that dissolves it is called the **solvent;** and the homogeneous mixture that results is called the **solution.** Solutions also can be made by dissolving one liquid in another (a martini), a gas in a liquid (soda water), or even a solid in a solid (metal alloys) (▸ Figure 7.1). Air is a solution of gases, primarily oxygen dissolved in nitrogen. Usually, the component that is present in the greater amount is considered to be the solvent.

WHY SUBSTANCES DISSOLVE

What makes salt and sugar dissolve in water, iodine in carbon tetrachloride, and some fabrics in nail polish remover? While we are at it, we also would like to know why salt does *not* dissolve in nail polish remover or gasoline.

For a solid to be dissolved by a solvent (any solvent), the forces that hold the particles together in the solid must be overcome by the attraction of the solvent molecules for the particles. In other words, the solvent molecules must have sufficient interaction to tear an ion or molecule from an edge of the solid and surround it. (▸ Figure 7.2). When surrounded by solvent molecules

(a)

(b)

(c)

▶ Figure 7.1
Solutions Are Homogeneous Mixtures. *Although it contains small quantities of liquid and solid solutes, ginger ale (a) is primarily a solution of carbon dioxide gas dissolved in water. Releasing the pressure on the CO_2 gas causes it to bubble from the solution. Because it contains more sugar than water, chocolate syrup (b) is a solution of a liquid dissolved in a solid. Tap water (c) is a liquid solution of solids and gases. The chrome-plated faucet (also c) is made of brass, an alloy that is a solution of the solids tin and zinc dissolved in solid copper.*

in this manner, the ion or molecule is said to be **solvated.** Thus removed from the bulk of the solid material, each solvated particle *diffuses* (wanders) away from its former neighbors out into the solvent, to become separated from them by many more solvent molecules. As this process continues, the solid dissolves.

INTERMOLECULAR FORCES

Before we can explain how solvation works, we must examine the kinds of attractive forces that hold ions or molecules to one another in solids and liquids. We already know that electrostatic attractive forces account for the

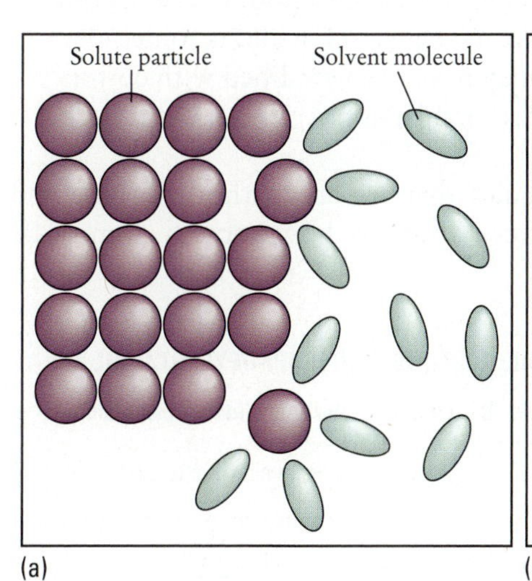

(a)

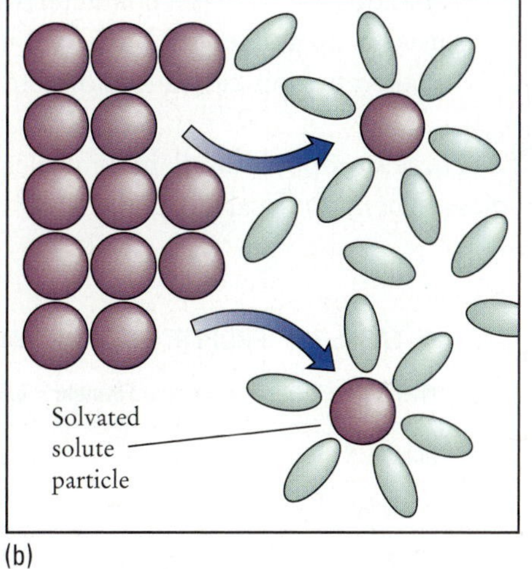

(b)

▶ Figure 7.2
The Solution of a Solid Substance. *(a) Solvent molecules are attracted to solute particles on the edge of a solid. (b) They then pull the solute particles from the solid and surround (solvate) them.*

7.1 From your everyday experience give examples of solutions that are made up of (a) a solid dissolved in a liquid, (b) a liquid dissolved in a liquid, (c) a solid dissolved in a solid, (d) a gas dissolved in a liquid, and (e) a gas dissolved in a gas.

Intermolecular forces determine many of the properties of molecular substances.

properties of ionic compounds such as table salt, NaCl, and Epsom salts, $MgSO_4$. Various other attractive forces, called **intermolecular forces,** hold molecules to one another in the solid or liquid state. These intermolecular forces, whose nature depends on molecular structure, determine many of the properties of molecular substances.

Compounds that are typical of the various intermolecular forces (■ Table 7.1) are the following:

- **Butane** Probably all of us at one time or another have had occasion to use a butane lighter. At the press of a button, butane gas hisses from an orifice in the device and is ignited by a spark. Butane, C_4H_{10} , a flammable and odorless gas at room temperature (b.p. –0.6 °C), is often used as a fuel.
- **Nail polish remover** Although this substance is a mixture, it often contains acetone as a major component. Acetone, C_3H_6O, is a flammable liquid at room temperature (b.p. 56 °C). It has the characteristic odor of lacquer thinner.
- **Rubbing alcohol, or isopropyl alcohol** This flammable liquid, C_3H_8O, boils at about 82 °C, has a characteristic medicinal odor, and is sometimes used as an antiseptic. Your arm may have been swabbed with cotton soaked in isopropyl alcohol before an injection.

These compounds all have molecules that are about the same size and mass. Their physical properties are very different, however, which must be

■ **TABLE 7.1 PROPERTIES OF BUTANE, ACETONE, AND ISOPROPYL ALCOHOL**

Name	Molecular Formula	b.p., °C	Intermolecular Force (kind of molecule)
butane	C_4H_{10}	–0.6	London dispersion (nonpolar)
acetone	C_3H_6O	56	dipole–dipole (polar)
isopropyl alcohol	C_3H_8O	82	hydrogen bond (associated)

due to differences in their molecular structures. Although we shall not dwell on these structures in detail, it is important that we know how they differ (▶ Figure 7.3).

C_4H_{10} O C_3H_7 O—H Hydroxyl group

Butane Acetone Isopropyl alcohol

(a) Nonpolar (b) Polar (c) Associated

▶ **Figure 7.3**
Butane, acetone, and isopropyl alcohol have molecules that are about the same size and mass. (a) The butane molecule is nonpolar, (b) the acetone molecule is polar, (c) and the isopropyl alcohol molecule contains a hydroxyl group, which permits hydrogen bonding.

Beginning our comparison we note that acetone, which has a boiling point that is more than 50 °C higher than that of butane, has a molecule that is polar, whereas that of butane is nonpolar. Butane, like methane, is a member of a family of compounds called *hydrocarbons,* which contain only carbon and hydrogen. Although the carbon–hydrogen bonds in these molecules are slightly polar, the hydrocarbon molecules themselves generally have little or no polarity. The acetone molecule, on the other hand, has a double-bonded electronegative oxygen atom protruding from its middle, making it polar.

Stronger intermolecular attractions require more energy to overcome them.

If we use the rule of thumb that a higher boiling (or melting) temperature means that a greater input of energy is required to separate molecules from one another—as for ions in Chapter 6—we can conclude that polar molecules adhere to one another more tightly (all other things being equal) than do nonpolar molecules. Examining other cases in which we can compare nonpolar and polar molecules in this manner, we find that the intermolecular (between molecules) attraction of polar molecules usually is greater than that of nonpolar molecules, and that polar substances usually have higher boiling or melting points than do nonpolar substances of comparable molecular size.

What causes polar molecules to adhere better to one another? Polar molecules have an unsymmetrical distribution of electron density over their surfaces and have a (partially) positive end and a (partially) negative end, or dipole. Because opposite charges attract one another, the positive end of one molecule is attracted to the negative end of another. The positive end of the latter is, in turn, attracted to the negative end of yet another molecule, and so on, until all of the molecules are attracted to one another by the electrostatic attraction (plus for minus) of their tiny partial charges. This **dipole–dipole attraction** is illustrated in ▶ Figure 7.4. (See also ▶ Figure 7.5.)

Because nonpolar molecules do not possess dipoles and therefore cannot be attracted to one another by dipole–dipole attractions, one might think that they would not stick to one another at all. Yet butane does exist in the liquid

▶ Figure 7.4
Dipole–Dipole Attractive Forces. *(a) The positive end of a dipole is attracted to the negative end of another dipole. (b) Polar acetone molecules are arranged so that their dipoles are attracted to each other in a similar manner.*

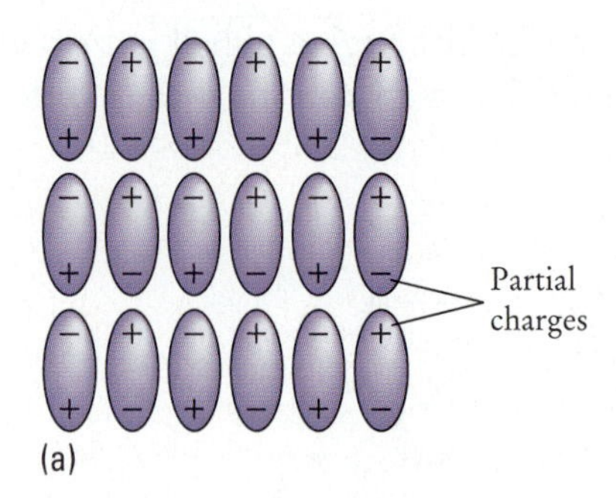

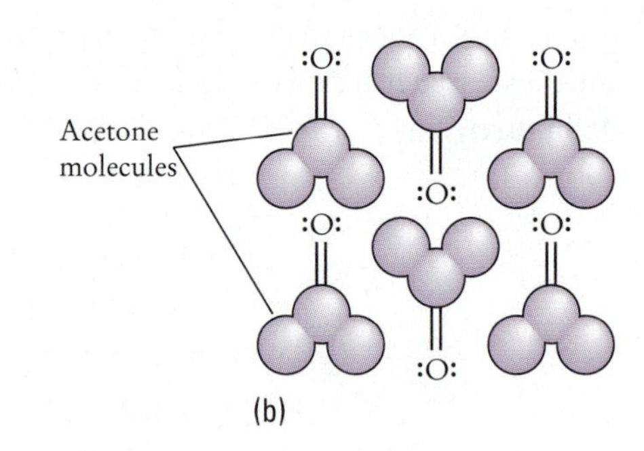

phase when it is slightly below 0 °C or when it is confined at high pressure. Also, carbon dioxide, CO_2, another nonpolar substance, forms a solid (dry ice), which sublimes at –78.5 °C. In fact, many nonpolar substances such as gasoline and paraffin exist in either liquid or solid form at ordinary conditions. We would suspect that the intermolecular attractive forces holding nonpolar molecules together would be very much weaker than dipole–dipole attractions; but what could they be, and how could they work?

▶ Figure 7.5
The dipoles of polar molecules are also attracted to an electric charge brought nearby. The positive ends of the dipoles are attracted to a negative charge, and the negative ends of the dipoles to a positive charge. (a) A stream of water bends toward a charged rod because the dipoles of water molecules are attracted to the charge. (b) A stream of nonpolar carbon tetrachloride (CCl_4) molecules is not deflected by the charged rod. (A dye has been added to both liquids to make them more visible.)

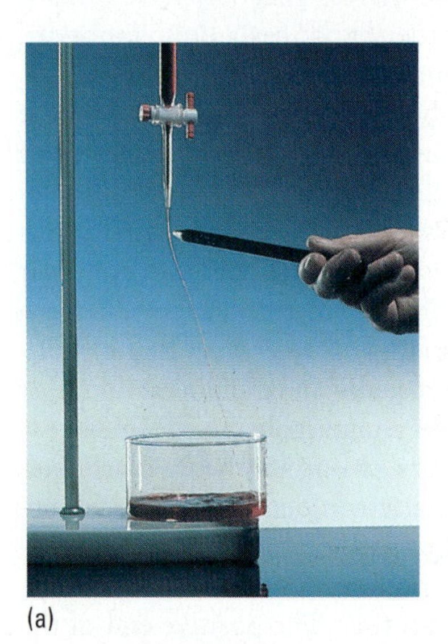
(a)

(b)

Enter the uncertainty principle. Yes, it helps provide the explanation. Our butane molecule is nonpolar because the distribution of electron density on its surface, on the average, is symmetrical. However, at a particular instant we cannot know whether this electron distribution is symmetrical or unsymmetrical. It could be unsymmetrical at that instant; and if so, our butane molecule will have a tiny, fleeting dipole. This tiny, "flickering" dipole is transient; but while it exists, it can interact with a like dipole in a nearby molecule of butane, much like a feeble dipole–dipole attraction. It is also possible for our fleeting dipole to induce (cause the formation of) a dipole in a neighboring, unpolarized molecule of butane and be attracted to it, as in ▸ Figure 7.6. These kinds of fleeting, or flickering, dipole attractions are called **London dispersion forces.** They are much weaker than the dipole–dipole attractions between polar molecules and account for the observation that nonpolar molecules generally have lower melting and boiling points than do polar molecules of comparable size and shape.

London dispersion forces are the weakest of the intermolecular attractions.

Isopropyl alcohol, the third compound in our comparison, boils some 26 °C higher than does acetone. This leads us to infer that its intermolecular forces must be even greater than those of acetone and perhaps even of a different kind. Because isopropyl alcohol molecules are polar, we would expect them to exhibit dipole–dipole attractions, and this is indeed so. But this is not enough. Looking further, we note that the isopropyl alcohol molecule has a unique structural feature. Protruding from its middle is a pair of bonded atoms called the hydroxyl group, or —O—H group; it looks something like part of a water molecule in which one of the hydrogen atoms has been replaced by a hydrocarbon fragment. It is this hydroxyl group that provides an extra attraction between the isopropyl alcohol molecules, an attraction so large that is is actually called a bond. By interacting with each other on nearby isopropyl alcohol molecules, the hydroxyl groups form a bridge between them to form what is called the **hydrogen bond.**

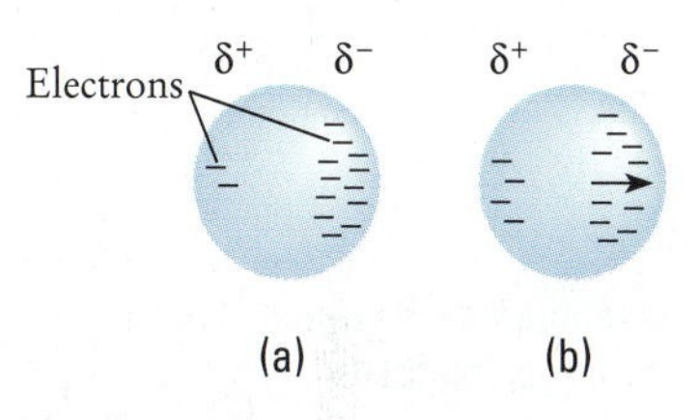

▸ Figure 7.6
London dispersion forces between nonpolar molecules are weak. Transient dipole in one molecule (a) induces a dipole in another molecule (b).

The O—H bond in the hydroxyl group is polarized, $^{\delta-}$O—H$^{\delta+}$. In addition, there are two nonbonded pairs of electrons on the (partially) negatively charged oxygen atom. When two hydroxyl groups are brought close together, the hydrogen atom on one is attracted to the oxygen atom of the other (▸ Figure 7.7). This at first seems like a dipole–dipole attraction, but in this case the hydrogen atom becomes partially attached to the nonbonded pair on the second oxygen atom. The two oxygen atoms are thus joined by a hydrogen bridge, and in the liquid state it is not possible to tell which oxygen atom the hydrogen atom is bound to. The formation of this hydrogen bond releases a fair amount of energy (about one-tenth that of an ordinary covalent bond), and the isopropyl alcohol molecules are held more tightly together than by a simple dipole–dipole attraction. Hydrogen bonding actually occurs among all of the isopropyl alcohol molecules in the liquid, in three dimensions, and hydrogen atoms are actually exchanged from one molecule to another in a continuous, dynamic process.

With the exception of HF, only molecules containing

—OH —NH (with a bond below N) or —NH_2

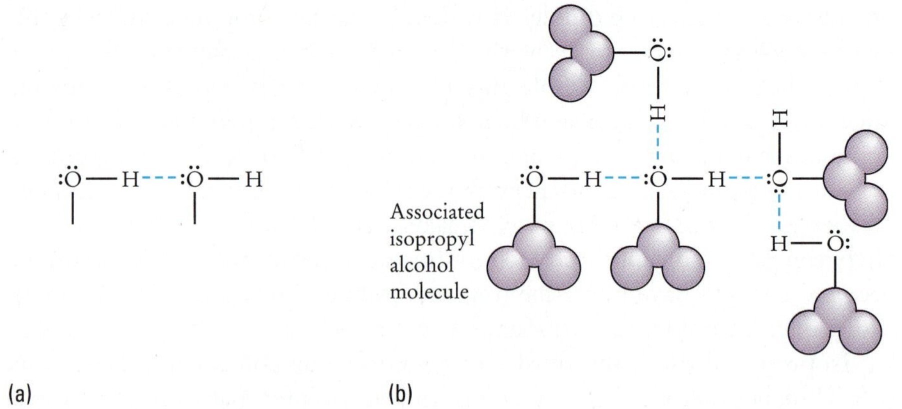

▶ Figure 7.7
The Hydrogen Bond. *(a) The hydrogen atom on one hydroxyl group is attracted to a nonbonded electron pair of a neighboring hydroxyl group. Energy is given off when this happens and a hydrogen bond forms. (b) Hydrogen bonds hold isopropyl alcohol molecules together in three dimensions.*

(a)

Associated isopropyl alcohol molecule

(b)

Associated compounds are hydrogen bonded.

groups hydrogen bond appreciably with one another. Compounds that exhibit hydrogen bonding are said to be **associated.** Because their molecules are held to one another by this extra attractive force, they generally have higher boiling or melting points than do polar compounds composed of molecules of similar shape and size. Overall, then, associated (hydrogen bonded) substances have the strongest intermolecular forces; those in polar compounds are less strong, and those in nonpolar compounds are the weakest.

WATER

Water is a singular natural compound in the universe, and it is questionable whether life as we know it could exist without it. Many of its unique properties derive from its molecular structure, which maximizes hydrogen bonding (▶ Figure 7.8). Having two exchangeable hydrogen atoms (two O—H bonds) and two nonbonded pairs of electrons (to accommodate hydrogen bonds) per molecule. Water promotes the formation of a compact three-dimensional hydrogen-bonded network among molecules in which no opportunities for bonding are wasted. Even liquid water has considerable internal structure.

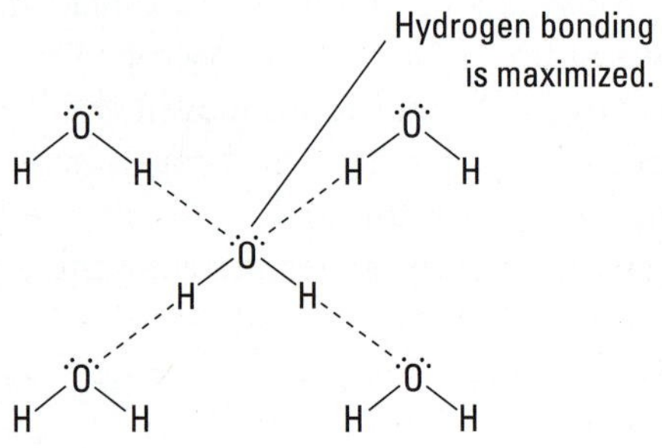

▶ Figure 7.8
Hydrogen Bonding in Water. *The structure of the water molecule maximizes hydrogen bonding, making water highly associated.*

Hydrogen bonding is maximized.

Water, which boils at 100 °C, is remarkable compared with methane, which does not hydrogen bond and boils at –162 °C, and ammonia and HF, which partially hydrogen bond and boil at –33 °C and 19.5 °C, respectively. Because of water's high boiling point, it can exist as a liquid at temperatures commonly found on Earth.

Water is one of only a few compounds that expand when they freeze.

An uncommon property of water is that it expands when it freezes (▶ Figure 7.9). Ice is thus less dense than liquid water and floats on its surface. Most substances contract when they solidify, causing the solid to be more dense than the liquid and sink to the bottom. If ice were more dense than water, aquatic life in temperate and polar climates would be vastly different, because as ice formed, it would sink to the bottom of bodies of water, leaving more water to freeze on the surface. During prolonged cold spells, even very deep lakes would eventually freeze solid, killing practically every living thing. Floating ice insulates the water below it from the frigid air above and usually prevents the entire body of water from freezing. Because much of their habitat is preserved, fish and aquatic life can survive through the winter.

▶ Figure 7.9
A water pipe can burst when the water inside freezes and expands.

The expansion of freezing water is usually detrimental to living cells, which contain water. If they are frozen, this internal water expands, causing the cell membrane to rupture and the cell to die. Frostbite is a painful example of this phenomenon. Freezing water also contributes to the weathering of rocks and concrete; liquid water in small cracks expands with tremendous force when it solidifies, breaking these materials apart.

Although the molecules in liquid water are highly associated, there is some randomness to their hydrogen bonding, which allows them to be fairly close together. When water solidifies, however, these hydrogen bonds become aligned in a more orderly manner, the hydrogen atoms no longer exchange between oxygen atoms, and a rigid structure forms that has considerably more space in it than does liquid water. Ice is actually full of holes. These holes, which look like hexagonal tubes, run parallel through the ice for billions of water molecules (▶ Figure 7.10). This hexagonal pattern also accounts for the shape of snowflakes and ice crystals.

WATER AS A SOLVENT

Water is a good solvent for many kinds of substances.

Water's unique properties make it an excellent solvent for ions, polar molecules, and associated substances. The water molecule is bent, and the resulting unsymmetrical distribution of electron density creates a partially negative end (near the oxygen atom) and a partially positive end (located out in space somewhere between the two hydrogen atoms). (We use *positive end* and *negative end* to simplify the following explanation.)

Because the water molecule has positive and negative ends, water can solvate both positive and negative ions (▶ Figure 7.11). When it solvates a

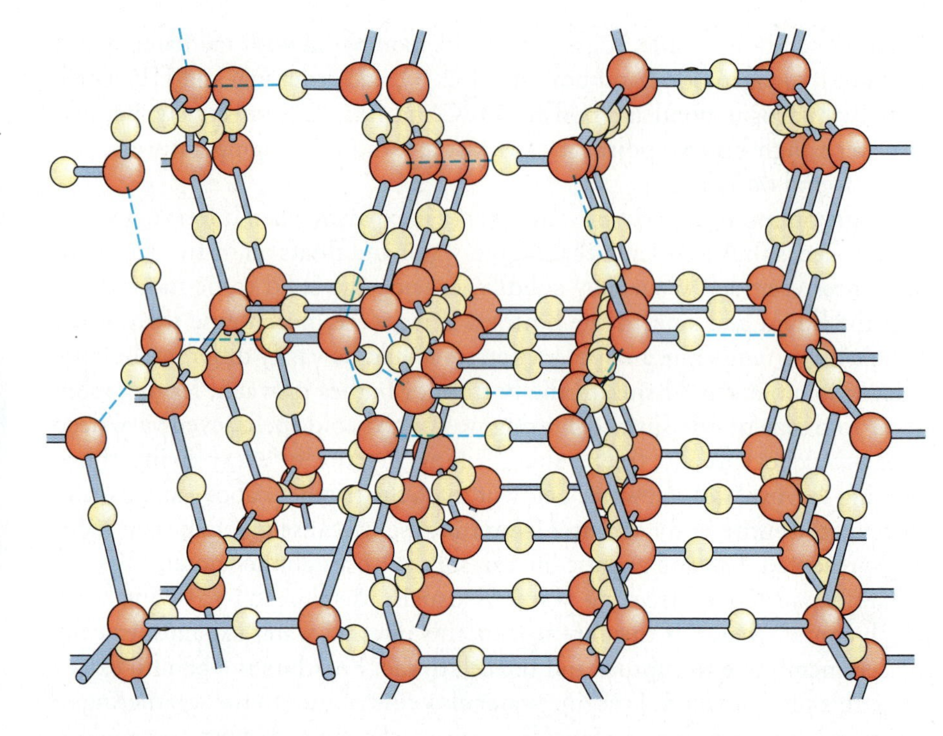

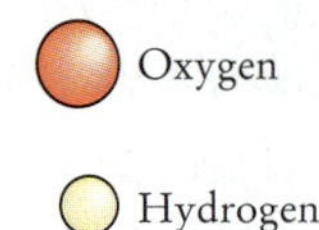

▶ Figure 7.10
When water freezes, all its hydrogen bonds become aligned in a rigid structure that has more space between molecules than liquid water does. The hexagonal tubes that form in ice run for billions of water molecules through the crystal.

7.2 From your everyday experience, what examples of the effects of freezing water have you seen?

positive ion, its negative end (oxygen) is attracted to the positive ion. In fact, the negative ends of *many* water molecules are attracted to the positive ion, resulting in a cluster of water molecules (usually several layers of them). This cluster forms around the ion, pulling it from the crystal lattice and isolating it from ions of opposite charge in the solution. Water solvates negative ions in a similar manner. This time its positive end (near the hydrogen atom) is attracted to the negative ion. Again, many water molecules cluster around the negative ion, pulling it from the crystal lattice and keeping it in solution. In both cases, this is called an **ion–dipole attraction.**

When an ionic substance dissolves in water, its ions **dissociate** (become separated) from one another in solution and usually do not affect each other's behavior to an appreciable extent. For this reason it is common practice to represent this separation of ions in solution by placing a + sign between their

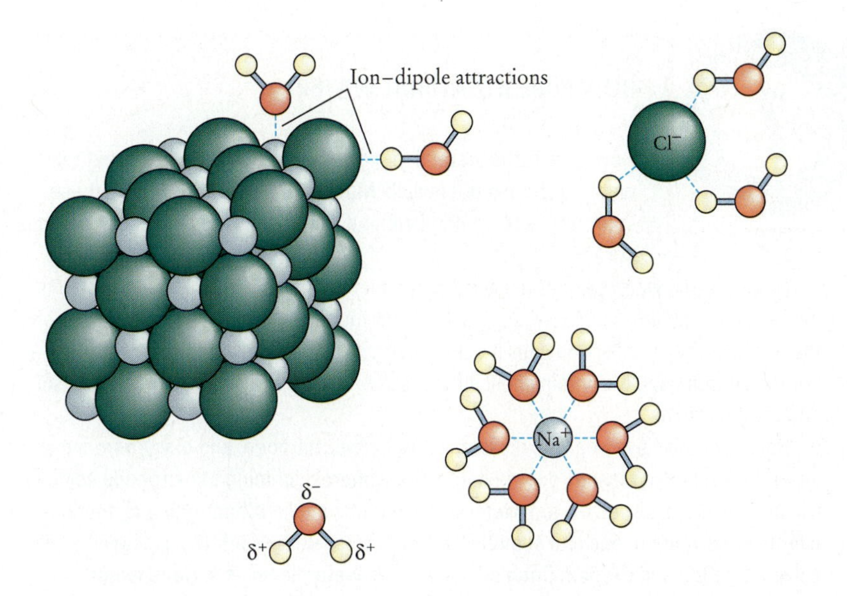

▶ Figure 7.11
Solvation of Ions by Water Molecules. *When an ionic compound dissolves, ionic attractions in the crystal are overcome and replaced by ion–dipole attractions to water molecules.*

symbols when we write chemical equations. For example, the dissociation of lithium chloride in water represented as

$$LiCl(s) \xrightarrow{\text{water}} Li^+ + Cl^-$$

This chemical statement, in English, would read, "Solid lithium chloride when dissolved in water dissociates to produce lithium ions and chloride ions in solution in the ratio of 1:1." In a similar manner we write sodium nitrate ($NaNO_3$, containing the polyatomic nitrate ion) in water solution as

$$NaNO_3(s) \xrightarrow{\text{water}} Na^+ + NO_3^-$$

and magnesium bromide ($MgBr_2$), containing two bromide ions for each magnesium ion, as

$$MgBr_2(s) \xrightarrow{\text{water}} Mg^{2+} + 2Br^-$$

Water also uses its dipole when it solvates polar molecules, but in a slightly different way. Because polar molecules also contain dipoles, the dipole of a water molecule (actually a number of them) can simply align itself with the dipole of the polar molecule it is trying to solvate, in a typical dipole–dipole attraction (▶ Figure 7.12). The negative end of the water molecule is attracted

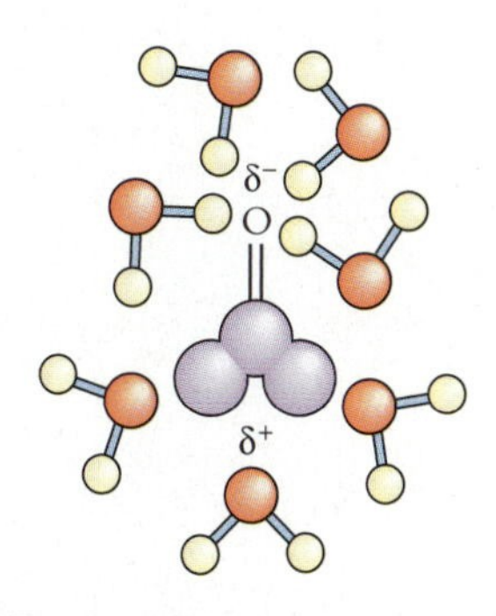

▶ Figure 7.12
In the solvation of a polar molecule (acetone) by water, the primary attractive forces are dipole–dipole interactions.

HOW PURE IS NATURAL WATER?

How pure is the water in a high mountain stream, in natural rainwater, or from a deep well? Much of the time, water from these sources is safe to drink without treatment. Does this mean that it is pure?

Natural fresh waters are actually solutions that contain small quantities of a number of solutes (Figure 1). Water is a good solvent for many substances and will dissolve them from the air, rocks, and soils it comes in contact with in the environment. Also, natural waters usually contain small amounts of undissolved solids, such as dust, soil, and sand particles.

Rainwater dissolves carbon dioxide, sulfur oxides, nitrogen, and oxygen from the air as it falls. It also washes dust from the atmosphere. Lightning can produce some nitric acid, and this dissolves, too. Natural rainwater is acidic (Chapter 8) and contains hydrogen carbonate ions and sometimes nitrate ions (HCO_3^- and NO_3^-), as well as dissolved nitrogen and oxygen. Once on the ground, water flows as surface water through soils and over rocks, dissolving Na^+, K^+, Ca^{2+}, Mg^{2+}, and Fe^{2+} cations and Cl^- and SO_4^{2-} anions.

Percolating downward through the ground, surface water finds its way to underground aquifers where it dissolves more Ca^{2+} and Mg^{2+} ions from the limestone or dolomite formations there. This is the reason that well water often is very hard; it may contain over 200 ppm (200 mg/kg) of dissolved calcium and magnesium compounds. Groundwater in contact with iron-bearing rocks may dissolve even more Fe^{2+} ions. These colorless ions react with oxygen in the air to form rust-colored Fe^{3+} ions, which can badly stain plumbing fixtures.

Natural water is not pure. It is fortunate that this is so because many of the dissolved substances in this water sustain processes necessary for life to exist on this planet. Because water is such a good solvent, however, it also dissolves many harmful things. This makes it easy to pollute, so we must be careful with it.

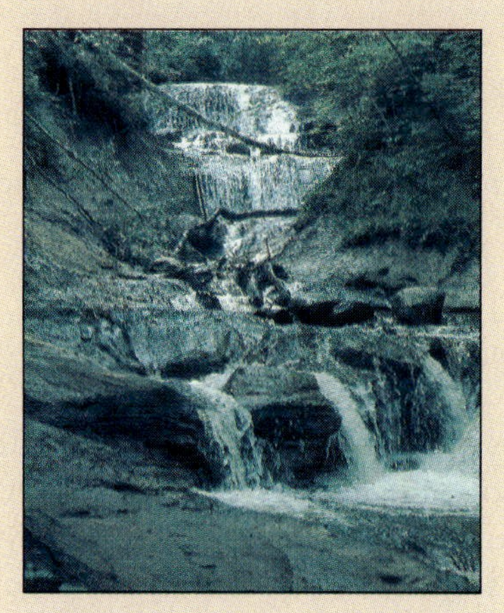

Figure 1
Natural fresh waters are actually solutions of nutrients that support life.

7.3 Write the balanced equation for the dissociation of the following solid ionic substances in water. Then, give a correct English statement for each of the equations you have written.

a. KF
b. $CaBr_2$
c. $NaHCO_3$
d. $MgSO_4$
e. NH_4Cl
f. $Al(NO_3)_3$

to the positive end of the dipole of the polar molecule and vice versa. This attraction is not as strong as an ion–dipole attraction but it is strong enough to enable many purely polar substances to dissolve somewhat in water. For example, highly polar acetone is miscible with water in all proportions. **Miscibility** refers to the solubility of liquids in one another. Even ether and chloroform, which are somewhat less polar than acetone, are slightly soluble in water. Though these two liquids can dissolve in water to a small extent, each will form two distinct layers when shaken with water and allowed to stand. Because of this behavior, ether/water and chloroform/water generally are considered to be *immiscible pairs.*

Because water itself is highly hydrogen-bonded, it is a powerful solvent for almost any substance that is capable of association (hydrogen bonding). Molecules that contain

$$—O—H \qquad —\underset{|}{N}—H \qquad \text{or} \qquad —NH_2$$

groups are often quite soluble in water because these groups can form hydrogen bonds with water molecules, causing the water molecules to cluster around them. The sugar we dissolved in the fruit drink is an excellent example. Sugar (sucrose) molecules have just twelve carbon atoms and are relatively small, yet each contains eight —O—H groups. They literally bristle with —O—H groups on their surfaces. It is not surprising, then, that water molecules surround the sugar molecules in large numbers by hydrogen bonding with these —O—H groups (▶ Figure 7.13). Sugar is so soluble that 85 g of it will remain dissolved in 15 g of water at room temperature! This highly concentrated solution of sugar is called a syrup.

7.4 Why would some sugar syrups be considered solutions of a liquid dissolved in a solid?

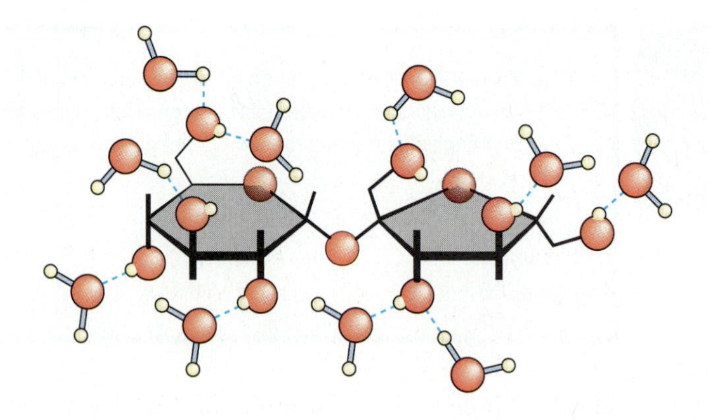

▶ Figure 7.13
Sugar molecules in solution are highly associated with water molecules. About 85 g of sugar can dissolve in only 15 g of water.

Hydrocarbons are compounds that contain only carbon and hydrogen.

Some substances do not readily dissolve in water at all. We have all heard the old adage, "Oil and water don't mix," which is an example of **hydrophobic** behavior (▶ Figure 7.14). Oil is primarily composed of **hydrocarbons,** substances that contain only carbon and hydrogen. Because hydrocarbon molecules are nonpolar and do not have positive and negative ends, there is no way for the polar water molecules to orient around them. Solvation is thus poor. In addition, liquid water itself is highly ordered and structured because it is associated. Placing a nonpolar hydrocarbon molecule in the middle of this structure causes hydrogen bonds to be broken, which requires some kind of energy input. New attractive interactions (ion–dipole, dipole–dipole, or hydrogen bonds) between the water and hydrocarbon molecules are impossible, however, so no energy is released in solvating the hydrocarbon molecule. This overall process thus requires additional energy (is endothermic) and tends not to be spontaneous. We often say that nonpolar molecules are poorly solvated in water because they don't fit the water structure very well.

(a)

(b)

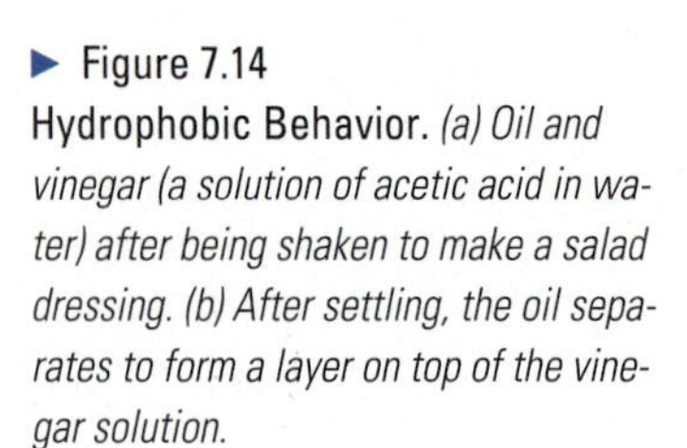

▶ Figure 7.14
Hydrophobic Behavior. *(a) Oil and vinegar (a solution of acetic acid in water) after being shaken to make a salad dressing. (b) After settling, the oil separates to form a layer on top of the vinegar solution.*

NONPOLAR SOLVENTS

But oil and grease do dissolve in gasoline, paint thinner, and cleaning fluid. All these solvents are nonpolar in nature, just like the oil and grease molecules they dissolve. Oil and grease molecules, being nonpolar, are held to one another by London dispersion forces—and so are the molecules in the nonpolar solvents. These same dispersion forces also cause the nonpolar solvent molecules to cluster around the oil or grease molecules, separate them from one another, and carry them into solution. Although these attractive forces between nonpolar solvent molecules and the nonpolar solute molecules are weak, so are the attractive forces between the solute molecules themselves. Thus, relatively little energy is required to separate them from their neighbors in the solid. Also, oil and grease molecules are fairly large and have large surface areas on which the London forces can interact.

POLAR SOLVENTS

Polar solvents that do not hydrogen bond (are not associated) interact with solute particles primarily by dipole–dipole and ion–dipole attractions. The strength of these interactions is greatly influenced by how polar the solvent is. Because the polarities of these solvents range from very polar to slightly polar, depending on molecular structure, their solubility behavior generally lies intermediate between the highly polar associated solvents on one hand and the nonpolar solvents on the other. Some fairly polar solvents dissolve certain ionic substances but not others. Other less polar solvents will not dissolve ionic substances at all, but some of them readily dissolve nonpolar materials.

For example, the acetone in nail polish remover will not dissolve sodium chloride, though sodium iodide, NaI, which has a weaker crystal lattice, is quite soluble in it. Recall that acetone is sufficiently polar to be completely miscible with water. Yet it also will dissolve many greasy substances quite well. Ether, a solvent less polar than acetone, is only slightly soluble in water and will dissolve neither NaCl nor NaI. However, it is a very good solvent for substances of low polarity such as paraffin and wax.

DETERMINING SOLUBILITIES

To predict whether a substance will dissolve in a certain solvent (that is, to find its solubility), it is helpful to use the principle of "like dissolves like." Ionic and polar substances usually (but not always) dissolve in polar and associated solvents. Substances that are themselves associated usually dissolve

"Like dissolves like."

very well in associated solvents. And, of course, nonpolar substances dissolve readily in nonpolar solvents.

Bubble gum (nonpolar) is insoluble in water (polar, associated) but can be removed from clothing readily with a chlorinated hydrocarbon solvent (nonpolar or slightly polar) or lighter fluid (nonpolar, but *flammable*). Contrary to popular belief, sugar (polar, associated) does not dissolve well enough in gasoline (nonpolar) to ruin an automobile engine when a solution of the two is burned in it. Sugar in gas tanks just clogs up fuel lines, mainly by dissolving in the small amount of water usually present in gasoline and making a gooey mess. Neither does salt (ionic) dissolve in gasoline (nonpolar) or nail polish remover (slightly polar). Alcohol (moderately polar but associated) has solvent properties that are intermediate between those of water on one hand and the hydrocarbon solvents on the other. As a result alcohol can dissolve many nonpolar substances as well as ionic and polar ones. An excellent example of this behavior is its use in preparing the antiseptic, tincture of iodine. This is a solution of iodine (I_2, nonpolar) and potassium iodide (KI, ionic) in alcohol. A **tincture** is an alcoholic solution of a medicinal substance.

A **tincture** is a solution of a medicine in alcohol.

Another example is the addition of methyl alcohol (methanol) to gasoline in automobiles to "take water out of the gas" and prevent gasline freeze. Because methyl alcohol is soluble in both water and gasoline, it actually causes the small amount of water present to dissolve in the gasoline; thus, the water cannot settle on the bottom of the tank where it could enter the fuel line to freeze in cold weather.

SOAPS, DETERGENTS, AND MICELLES

The action of soaps and detergents is an interesting demonstration of the like-dissolves-like principle. These compounds usually work by removing—from clothing, dishes, hands, and so on—greasy substances to which other kinds of soil cling. It is usually futile to wash dishes only in water because greasy food substances, which are chiefly nonpolar, will not dissolve in it. The glassware may even be sterile if the wash water was hot enough, but it still will be coated with a film of grease. Somehow, soaps and detergents cause this grease to "dissolve" in water.

Most soaps and detergents consist of strange-looking anions that have a small, negatively charged group of atoms attached to one end of a long hydrocarbon chain (about 12 to 18 carbon atoms in length). Note that even though this chemical species is classified as an *ion,* it has a long *molecular* hydrocarbon tail. A metal ion, usually the Na^+ cation, accompanies it to balance the negative charge (▶ Figure 7.15).

When soaps and detergents are placed in water, they dissociate, just as sodium chloride does. That is, they dissolve by ion–dipole interactions with

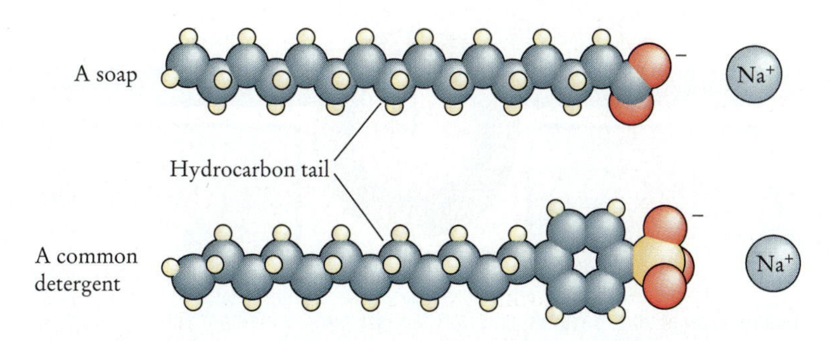

► Figure 7.15
Models of Soap and Detergent Anions.

water molecules, and the ions become separate from one another. The sodium ions, freed from their negative counterparts, actually do nothing except to maintain charge balance, and simply "watch" the anions wash dishes. Ions in solution that "do nothing" are called **spectator ions.**

Spectator ions are ions that do not participate in reactions in solution.

The anion, with its long hydrocarbon tail, has some interesting properties. We would expect its ionic end to be strongly attracted to the water molecules that are solvating it, and indeed it is. It is said to be **hydrophilic,** or water loving. The hydrocarbon tail, being hydrophobic, is repelled by the water molecules. Thus, we get a solvation tug-of-war, with the hydrophilic end being strongly attracted to the solvent and the hydrophobic end being repelled by it. The latter would dissolve more readily in something like itself.

A ready-made solvent for this hydrophobic tail can be found on a greasy plate—namely, the grease itself. One kind of grease should dissolve in another (hydrocarbon in hydrocarbon), so the tail should dissolve in a blob of grease from the plate. However, the ionic end attached to the tail will not dissolve in the grease. This leaves our strange anion in a predicament, with its ionic head dissolved in water and its hydrocarbon tail dissolved in a grease blob.

When we wash something with soap we usually scrub, or agitate, it. Agitation permits a number of these ions to get their heads together, or better, it helps them get their tails together—in a blob of grease. That way they look like they have their heads together, in what is called a micelle.

Here's how it works (► Figure 7.16). When a mixture of grease, water, and soap is agitated, the nonpolar hydrocarbon tails of a host of soap anions dissolve in the grease, but their ionic ends do not. These ionic ends must reside on the *surface* of the grease, covering the surface with negative ionic charge. Agitation breaks this charged grease into tiny globules; and because the surfaces of the globules are of like charge, they are forced to separate from one other by repulsion and can no longer coalesce to reform a larger grease blob. Also, water molecules are "fooled" into thinking the charged globules, called **micelles,** are very large ions. Thus, they are attracted to their surfaces and partially solvate them. Both of these phenomena result in the dispersal of the

NO SOAP

At the laundromat you're standing at the "laundry aids" vending machine trying to decide which slot to put your quarters in. The chemistry course is complicating your life: soap was once just soap to you–or rather, detergent was just "soap" to you, and you'd have dropped in your quarters at random. But now you want to know what's in these boxes.

A bin near the vending machine is full of empty boxes and plastic bottles. You retrieve a bright blue box and look at the list of ingredients on the back. Anionic surfactants. Enzymes: proteases to break down stains such as blood and egg. Sodium perborate and bleach activator: all-fabric oxygen bleach, not as good as chlorine bleach but safer for clothes and colors. Complex sodium phosphates such as tripolyphosphate, $Na_5P_3O_{10}$, called "builders": besides having detergent properties, builders tie up Ca^{2+}, Mg^{2+}, and Fe^{2+} in large water-soluble ions and keep dirt in suspension so it won't redeposit on clothes.

Those are the active ingredients. But then there's sodium sulfate, a processing aid, which adds bulk and acts as a drying agent to keep the powder flowing freely. Sodium silicate: a rust inhibitor to protect the washer. Colorant: presumably to color the detergent, not the clothes. Perfume: to offset the odor of the other ingredients. And finally, fabric whiteners, also known as optical brighteners: colorless dyes that glow in sunlight and fluorescent light. An old jingle pops into your head: "whiter than white!" *That's* what it meant!

A sunshine-yellow box in the bin had contained the same ingredients except bleach. A brilliant red box contained detergent with anionic *and* nonionic surfactants, and the front advertizes NO PERFUME. Otherwise, it's the same as the others.

Most detergents contain anionic surfactants, which work best in hot, soft water and are most effective on oily soils, clay, and mud. Nonionic surfactants work best for oily soils in synthetics at cool washing temperatures. You wonder why all detergents contain the same few surfactants when more than 1200 different ones are available. These few must be optimum in some way. More effective? Less harmful to the environment? Cheaper?

Liquid detergents contain much the same ingredients as powders except for phosphates, which aren't very soluble. For builders, the liquids (and the powders sold in states that ban phosphates in detergents) use old-fashioned washing soda (sodium carbonate), citrates, silicates, and zeolites (claylike minerals that soften water by behaving as ion-exchange resins). Phosphates are an environmental pollutant. They help algae to grow much too well and contribute to the *eutrophication* of lakes and rivers (it means to "enrich with nutrients"). However, phosphates can be recovered from wastewater. Some European countries are recycling them rather than banning them from detergents because only about 20% of the phosphates in wastewater comes from detergent. It's also been found that zeolites may be as hard on the environment as phosphates and are not recoverable.

You pick up a box labeled *Eco-something,* hoping that it's the answer to all your concerns, but the ingredients are the same as the others. Its box says BIODEGRADABLE, but all modern detergents are biodegradable, and this one costs more.

Well, you have to choose one. You reject phosphate but can't avoid the whiter-than-white; you select the nonionic surfactant because your "soils" are just body oils (you're not a muddy child or a mechanic) and you'd rather wash your clothes in warm water.

Now you have to buy a fabric softener because detergents clean the fibers so free of everything that fabrics feel scratchy and can build up a 12,000-volt charge in the dryer, creating a shocking love affair between your socks and underwear. Fabric softeners contain cationic and/or nonionic amine compounds of substituted fatty acids (you'll learn about those later). These waxy substances, related to plain soap, make the fibers stand erect so towels feel fluffy and make fabrics slide smoothly over themselves. It also contains humectants (substances that promote retention of water) so fibers retain enough moisture to dissipate static charge.

When you take your clothes out of the dryer, they're warm and fresh smelling, and you bury your face in them. The smell reminds you of an article about chemists synthesizing various odors: leather, fruits and flowers–and the smell of clothes dried in sunshine. This must be it! And your mood suddenly changes.

Isn't white, white enough? Isn't clean, clean enough? How did it all start? Well, it's the old problem of the chicken and the egg. There's little difference among soaps, or among detergents after the first one. So, to sell their detergent, manufacturers have to make it more convenient or better; and if that isn't possible, at least it should sound, smell, or look different.

Detergent makers spend a half-billion dollars a year advertising slight differences! And once some improvement or difference is advertised, we buy it because if we don't, our appearance may be inferior in some way: ring around the collar or tattle-tale gray. Convenience and the approval of our neighbors are important to us.

Nearly 7×10^9 pounds (how many kilograms is that?) of detergents are sold every year in the United States for $3.8 billion. If complete convenience and whiter than white for 260 million people doesn't harm our environment, then maybe it has to be available in a free society. But you suspect that it is harmful because nothing's harmless if there's enough of it: even too much water is a cataclysm (*clysm,* Greek, meaning "to wash")!

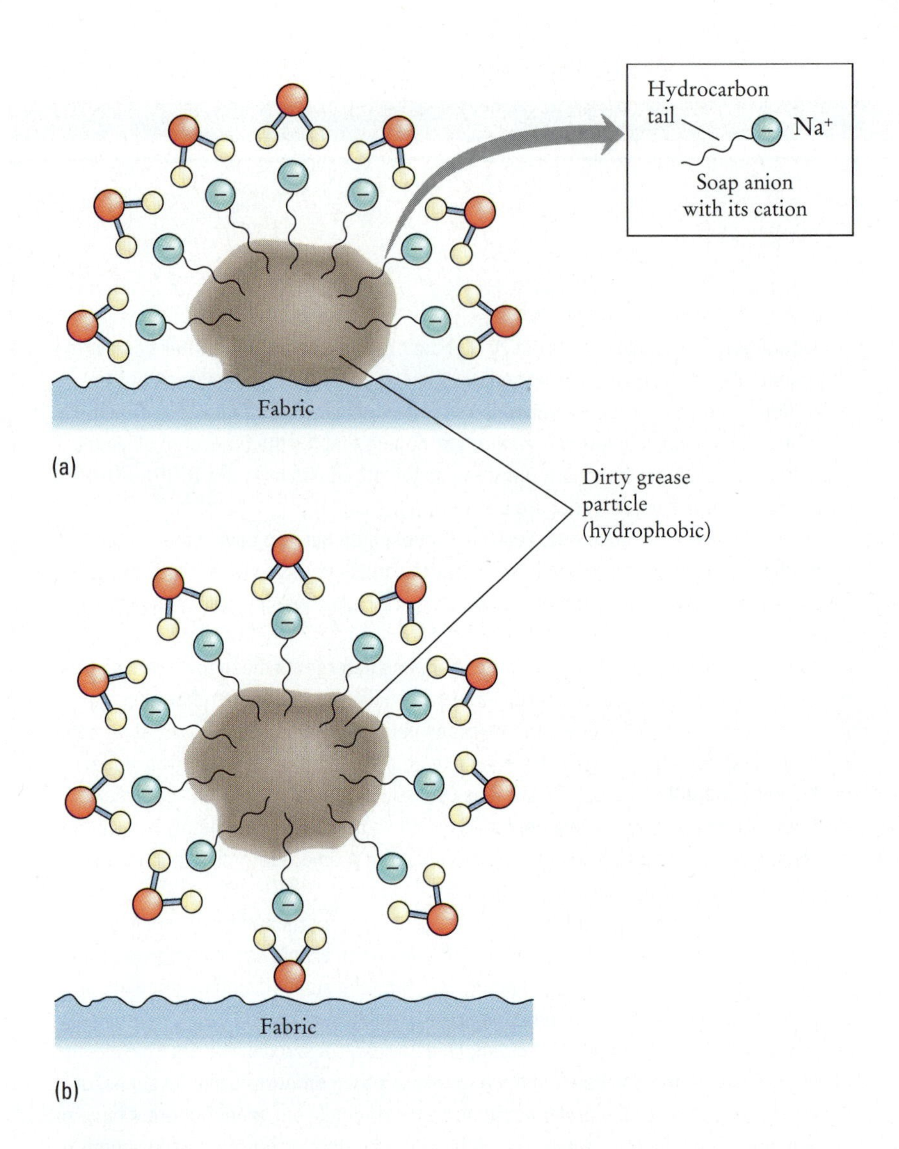

▶ Figure 7.16
Formation of a Micelle. *(a) The hydrophobic ends of soap ions dissolve in a dirty grease particle. Water molecules solvate the ionic ends of the soap ions. (b) A micelle forms, with a grease particle inside. It is hydrophilic and is washed away with the water.*

micelles throughout the water to make an **emulsion** (finely divided droplets of an immiscible liquid suspended in another). The sodium ions, still spectators in the process and doing their job of balancing charge, go down the drain with the wash water.

The chemical difference between soaps and detergents, both also called surfactants, lies primarily in the nature of the ionic group of atoms (a functional group) that is bound to the hydrocarbon tail. Note (see Figure 7.15) that most common *detergents* contain the sulfonate group $R—SO_3^-$ (R means "rest" of ion.) and ordinary *soaps* contain the carboxylate group, $R—CO_2^-$. Although the mode of cleaning (or detergent) action is essentially the same for both, detergents are preferred for use when the water used for washing is "hard." **Hard water** usually contains high concentrations of calcium and magnesium ions, which interfere with the cleansing action of soaps by reducing the soap's solubility in water. Detergents are unaffected by these ions.

Hard water contains Ca^{2+} and Mg^{2+} ions.

The cationic portion of most soaps is usually Na^+ (hard soaps) or K^+ (soft soaps). These *monovalent* (one unit of charge) cations permit their respective soaps to be fairly soluble in water. In hard water the presence of *divalent* Mg^{2+} and Ca^{2+} ions causes the "working" anionic part of the soap *to **precipitate*** (verb), or fall out of solution (▶ Figure 7.17). This solid **precipitate** (noun) is still a soap, but the sodium or potassium ions in the original soap have been replaced by calcium and magnesium ions, rendering it insoluble in water. This material is *soap scum,* which is hard to rinse out of clothes or hair and forms the "ring around the bathtub (▶ Figure 7.18). If more soap is added to the hard water to consume all of the calcium and magnesium ions, the detergent action of the soap can be restored, but this is wasteful.

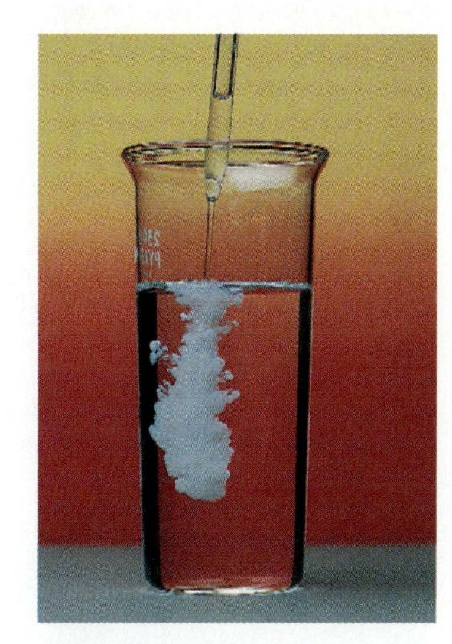

▶ Figure 7.17
The Process of Precipitation. *The white substance that forms is the precipitate.*

$$Ca^{2+} + 2C_{17}H_{35}CO_2^- \longrightarrow Ca(C_{17}H_{35}CO_2)_2(s)$$

In hard water — From dissolved soap — The calcium salt of the soap (the precipitate)

▶ Figure 7.18
The precipitation of the anion of a dissolved soap occurs in hard water. The precipitate is "soap scum." This is a net ionic equation: the spectator ions have been left out. Note that material and charge balance is maintained. See Chapter 8 for an explanation.

SOFT WATER: ION EXCHANGE

Rather than using a detergent, we can use "softened" water with our soaps. Water is softened by replacing its calcium and magnesium ions by sodium ions, which do not precipitate soap. In home water softeners, hard water is circulated through a tank containing an ion exchange resin. A typical *ion exchange resin* consists of very large molecules called polymers (Chapter 12) that have many sulfonate groups attached along their long hydrocarbon chains. These sulfonate groups are anionic, and paired with them are sodium cations (▶ Figure 7.19).

When hard water containing calcium and magnesium ions comes in contact with these sulfonate groups, the water's calcium and magnesium ions, being divalent, selectively adhere to these anionic sites, which sends the sodium

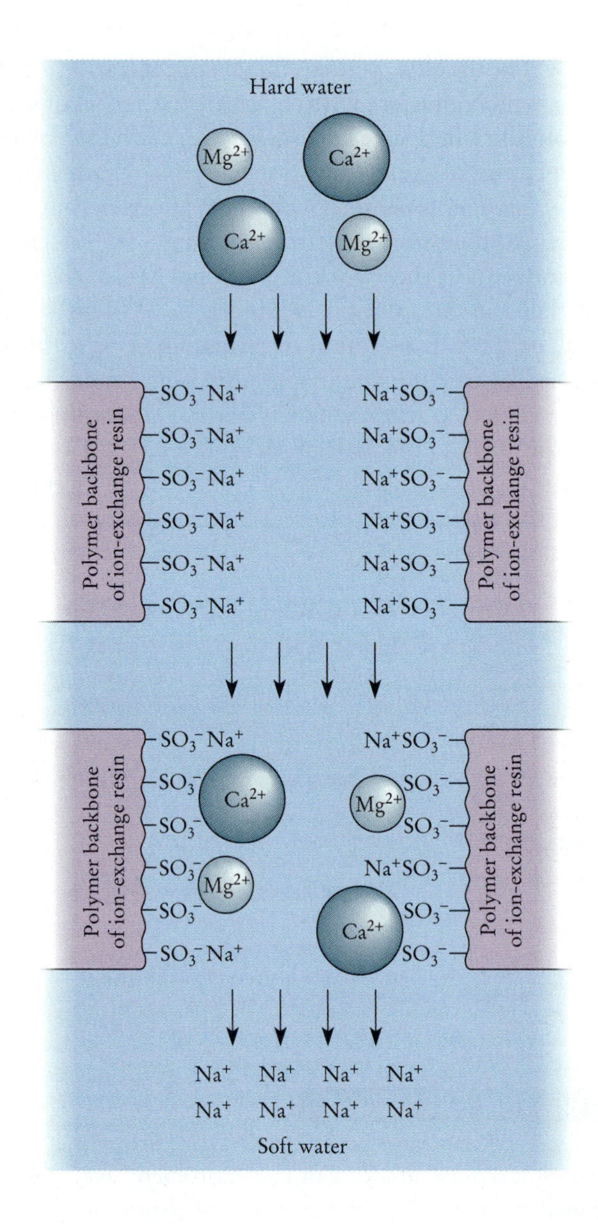

▶ Figure 7.19
Water Softening by Ion Exchange.
When hard water containing calcium and magnesium ions is passed through an ion exchange resin, sodium ions replace the calcium and magnesium ions in the solution, making the water soft.

Soft water usually contains Na^+ ions.

ions into solution. The net effect is to replace each calcium or magnesium ion in the hard water with two sodium ions. Note that this **soft water** is not pure because it still contains ions; the sodium ions and spectator anions, usually HCO_3^- and CO_3^{2-}, that were initially present in the hard water. For this reason, it is not a good idea for persons on low-sodium diets to regularly drink water softened this way.

The calcium and magnesium ions that become bound to the resin remain there until the resin is recharged with sodium ions. This is accomplished by flooding the resin with a *very concentrated* solution of NaCl, which forces Ca^{2+} and Mg^{2+} from the anionic sulfonate sites and replaces them with Na^{+}.

CRYSTALLIZATION

Now that we know what causes substances to dissolve, let's look at the solubility process from a different perspective and ask, "What causes crystals to form?" Even though salt is soluble in water, not all the Earth's salt is found in oceans or brine wells. Much of it is mined in crystalline form in underground salt domes. Sugar is also very soluble in water, but honey and maple syrup often "sugar out" after they stand for a period of time.

We know that when pure substances are melted and then cooled below their melting points, many of them form crystals. However, we are interested here in a different phenomenon—namely, how crystalline substances form from solution. What causes the honey to sugar out? How is crystalline salt obtained from the waters of the Dead Sea or the Great Salt Lake?

This last question gives us a clue to what causes the crystallization of substances from solution. Historically, salt makers have retained briney waters in shallow impoundments and allowed them to evaporate in the hot sun, whereupon they collected the crystalline salt left behind. Salt crystals do not form immediately when this process is initiated, however; only after a number of days, when a substantial amount of water has evaporated, do crystals begin to form. Once crystallization begins, though, it continues until all of the water has disappeared.

How does this process work? We know we must remove much of the water from the brine solution before we begin to obtain salt crystals. Furthermore, when we determine quantitatively how much salt is present in the evaporating brine when crystallization begins, we discover that a given quantity of solution always contains the same amount of salt, as long as the temperature is the same when each measurement is made. It appears that a specific quantity of water can "hold" a certain weight of salt in solution at a given temperature, and no more.

We can test this idea by measuring the amount of salt we are able to dissolve in a given quantity of water at a constant temperature. We begin by adding a small amount of salt to the water and stirring the solution. We see that all the salt dissolves, so we add some more. It, too, dissolves, so we continue the procedure. Eventually, when we add the next portion of salt to the solution, not all of it dissolves and some solid salt remains on the bottom of the container. At this point the water can hold no more salt, and the solution is said to be **saturated** (▶ Figure 7.20).

How much salt does our saturated solution contain? The same amount held by an identical quantity of the brine solution in which salt crystals were

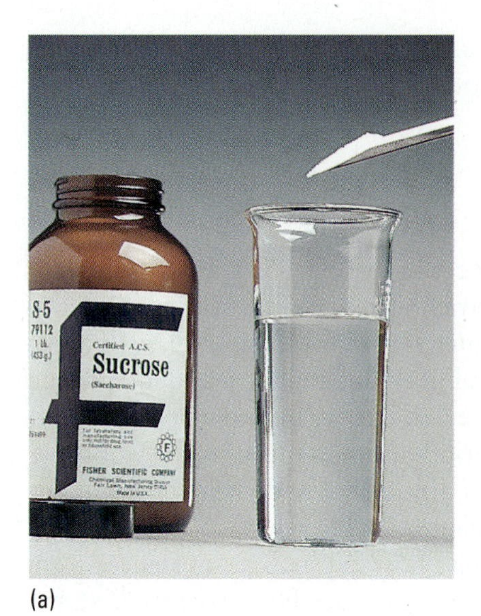

(a)

(b)

▶ Figure 7.20
Saturated and Unsaturated Solutions. *(a) Solid table sugar dissolves in an unsaturated sugar solution. (b) Solid sugar does not dissolve in a saturated sugar solution. It simply falls to the bottom.*

forming. This must mean that the brine solution is saturated, too. Reducing the amount of water in this saturated brine by evaporation, then, causes crystals to precipitate because what water is left can hold only so much salt.

CONCENTRATION

The **concentration** of a solution is independent of its volume.

We note here that a saturated solution of salt in water (or any solute dissolved in any solvent) at a certain temperature has the same concentration of salt, regardless of how much solution there is. In fact, the **concentration** of any solution is independent of its volume. This means that we can express the concentration of a solute as some kind of *ratio* that shows how much solute is dissolved in (per) a given amount of solvent. For example, the maximum solubility of salt in water at room temperature is about 37 g of salt *in* 100 g of water. This means that dissolving 37 g of salt in 100 g of water will produce a saturated solution. No more salt will dissolve at room temperature. A saturated solution of salt also can be made by dissolving 370 g of salt in 1000 g of water or 18.5 g of salt in 50 g of water. As long as the ratio of solute and solvent remains the same, the concentration of the resulting solutions is the same.

The concentration of this saturated salt solution also can be expressed as 27% by mass. Because percent is a ratio (parts per hundred), the 27% value can be obtained by expressing the concentration of salt as 37 g of salt per 137 g of solution (37 g of salt + 100 g of water), or $37\text{g}/137\text{g} \times 100\% = 27\%$.

7.5 Find the concentration in percent by mass for the following solutions.

a. 14.0 g of $NaHCO_3$ in 200. g of solution
b. 250 g of $CaCl_2$ in 1250 g of solution
c. 1.8 g of C_4H_9OH in 90.0 g of solution
d. 22.0 g of KI dissolved in 178 g of water
e. 1.4 g of I_2 dissolved in 138.6 g of CCl_4

7.6 How many grams of solute are there in each of the following solutions?

a. 100. g of a 10.0% solution of KNO_3
b. 750. g of a 20.0% solution of HCl
c. 2.5 kg of a 0.10% solution of $HC_2H_3O_2$

To say that a solution is saturated is specific, because it means that a particular solvent has dissolved all of a given solute that it can hold at a given temperature. To say that a solution is concentrated is ambiguous, however, because concentration is relative. Usually, we consider a **concentrated** solution to be one that contains a large amount of solute in a relatively small amount of solvent. Similarly, a **dilute** solution is considered to be one that contains a small amount of solute. A saturated solution of a certain compound can be either concentrated or dilute, depending on the solubility of that substance in a particular solvent. Also, any solution that contains less than the amount of solute required to make it saturated is said to be **unsaturated.**

CHEMICAL COUNTING: THE MOLE

Chemists often use a very different unit, from those used for mass or volume, to express the *chemical amount* of a compound. This is called the mole (L. *moles;* pile or heap). The mole is a counting unit like a pair, a dozen, a quire (25 sheets), a gross, a ream, or a six-pac or case. In fact, the mole is often called the "chemist's dozen."

Because atoms, ions, and molecules are very tiny, it takes a very large number of them to constitute enough matter for us to accurately determine their mass. For example, a crystal of table sugar having a mass of 1.5 mg contains 2.6×10^{18} molecules! Thus, the mole must be a very large number if it is to be

One **mole** is 6.02×10^{23} of anything.

▶ Figure 7.21
One Mole of Various Substances.
Clockwise from center bottom: copper (Cu, 63.5 g), iron (Fe, 55.8 g), water (H_2O, 18.0 g), sucrose ($C_{12}H_{22}O_{11}$, 342 g), copper sulfate pentahydrate ($CuSO_4 \cdot 5H_2O$, 250 g), mercury (Hg, 201 g). Center: sulfur (S_8, 257 g).

useful for finding the masses of atoms, ions, and molecules in order to count them. This number, called **Avogadro's number,** is 6.02×10^{23}. Thus, a **mole** (or, abbreviated, 1 mol) of iron contains 6.02×10^{23} iron atoms, 1 mol of oxygen contains 6.02×10^{23} O_2 molecules, and 1 mol of sugar contains 6.02×10^{23} sugar molecules, $C_{12}H_{22}O_{11}$.

Why was this number for the mole chosen? It turns out that it corresponds exactly to the formula mass of any pure substance *in grams.* That is, 1 mol of iron atoms has a mass of 55.8 g, 1 mol of oxygen molecules a mass of 32.0 g, and 1 mol of sugar molecules a mass of 342 g (▶ Figure 7.21). **Formula masses** are obtained simply by adding the atomic masses (from the Periodic Table) of *all* of the atoms *in the formula* of a chemical substance and expressing the total in grams. The formula for iron is simply Fe, so the formula mass of iron, in grams, is the same as its atomic mass in grams, 55.8 g Fe. The formula for oxygen is O_2, and each oxygen molecule contains two oxygen atoms. The formula mass for O_2, then, recalling dimensional analysis, is 2 O atoms $\times$ 16.0 g O/O atom = 32.0 g O_2.

Obtaining the formula mass for sugar is an extension of this process. The formula for sugar (sucrose), $C_{12}H_{22}O_{11}$ tells us that the molecule contains 12 carbon atoms, 22 hydrogen atoms, and 11 oxygen atoms. The formula mass, which is the sum of all of the atomic masses, can be set up this way:

$$
\begin{array}{lcl}
12\text{ C atoms} \times 12.0\text{ g C/C atom} & = & 144\text{ g C} \\
22\text{ H atoms} \times 1.0\text{ g H/H atom} & = & 22\text{ g H} \\
11\text{ O atoms} \times 16.0\text{ g O/O atom} & = & 176\text{ g O} \\
\hline
\text{Formula mass (sum of above)} & = & 342\text{ g } C_{12}H_{22}O_{11}
\end{array}
$$

The formula masses of ionic substances can be obtained in the same manner. We recall, for example, that Na_2S contains two sodium ions and one sulfide ion. Because its chemical formula also tells us the proportions of its atoms, we can calculate its formula mass as we did for sugar:

$$
\begin{array}{lcl}
2\text{ Na atoms} \times 23.0\text{ g Na/Na atom} & = & 46.0\text{ g Na} \\
1\text{ S atom} \times 32.1\text{ g S/S atom} & = & 32.1\text{ g S} \\
\hline
\text{Formula mass (sum of above)} & = & 78.1\text{ g } Na_2S
\end{array}
$$

In Chapter 3 we encountered the law of conservation of mass and learned how to balance chemical equations. In writing the balanced equation for the reaction of hydrogen with oxygen to form water, for example, we obtained $2H_2 + O_2 \longrightarrow 2H_2O$. The coefficients meant that two formula units of hydrogen combined with one formula unit of oxygen to produce two formula units of water. At that time we did not say what these "formula units" were.

7.7 The formula mass of any substance can be found in the manner described if its chemical formula is known, regardless of its actual chemical structure (ionic, molecular, or atomic). Find the formula masses for each of the following substances.

a. N_2
b. Br_2
c. Mg
d. Zn
e. NH_3
f. CaO
g. KNO_3
h. C_3H_8O
i. H_2SO_4
j. Al_2S_3
k. C_5H_5N
l. Na_3PO_4
m. $Mg(OH)_2$
n. IBr
o. CCl_4

We can now use moles as our counting unit and say that in this reaction two moles of hydrogen (as molecules) combine with one mole of oxygen (as molecules) to yield two moles of water (as molecules).

Chemical equations are expressed in moles.

Calculating the mass of fractions or multiples of a mole of a substance is straightforward. Because, as we know, one mole is equal to the formula mass in grams, it is a simple matter to multiply this mass by the fraction or

7.8 It also would be correct to say that two molecules of hydrogen combine with one molecule of oxygen to produce two molecules of water. How useful would using molecules be when carrying out experiments in the laboratory?

7.9 Balance the following equations and, using the mole concept, write the English statement that correctly describes each.

a. $H_2 + S \longrightarrow H_2S$
b. $P_4 + O_2 \longrightarrow P_4O_{10}$
c. $H_2 + N_2 \longrightarrow NH_3$
d. $Na + O_2 \longrightarrow Na_2O$
e. $Mg + N_2 \longrightarrow Mg_3N_2$
f. $C_2H_4 + O_2 \longrightarrow CO_2 + H_2O$
g. $C_5H_{10} + O_2 \longrightarrow CO_2 + H_2O$

multiple we are seeking. If we want to know the mass of 3.00 mol of aluminum, we simply multiply the formula mass of Al by 3.00, or 3.00 mol Al × 27.0 g Al/1 mol Al = 81.0 g Al. (Recall that it is possible to use the conversion factor, 27.0 g Al/1 mol Al = 1, because 27.0 g of Al = 1 mol Al.) Likewise, the mass of 0.300 mol of ethanol (ethyl alcohol) is 0.300 mol C_2H_5OH × 46.0 g C_2H_5OH/1 mol C_2H_5OH = 13.8 g C_2H_5OH, and the mass of 0.10 mol of NaI is 0.10 mol NaI × 150 g NaI/1 mol NaI = 15 g NaI.

7.10 Find the mass of each of the following quantities.

a. 2.0 mol I_2
b. 0.30 mol Li
c. 0.75 mol SO_2
d. 5.0 mol NaCl
e. 0.40 mol C_7H_8O

Similarly, the number of moles of a substance can be determined if its mass is known. The same mass-to-moles conversion factor is used except that it is inverted. For example, the number of moles contained in 20.0 g of calcium is 20.0 g Ca × 1 mol Ca/40.1 g Ca = 0.500 mol Ca, the number of moles in 285 g of octane is 285 g C_8H_{18} × 1 mole C_8H_{18}/114 g C_8H_{18} = 2.50 mol C_8H_{18}, and the number of moles of MgO in 4.03 g is 4.03 g MgO × 1 mol MgO/40.3 g MgO = 0.100 mol MgO.

7.11 Find the number of moles contained in the amounts of each of the following substances.

a. 4.5 g Be
b. 56.0 g N_2
c. 4.4 g CO_2
d. 6.35 mg Cu
e. 39.0 g C_6H_6
f. 299 g Li_2O

MOLARITY

Concentration most often is expressed in terms of moles of solute per liter of solution, or **molarity**, *M*. This unit has the advantage that concentrations have the same chemical basis (instead of a weight or volume basis). That is, solutions of the same molarity have the same number moles of solute in a given volume of solution, regardless of composition. In addition, this attribute makes molar solutions convenient for doing solution chemistry because it is possible to exactly "measure out" a desired number of moles of solute simply by measuring out a corresponding volume of solution. Two solutions of different compounds that have the same percentage compositions by mass likely would not have the same number of moles in a certain volume of solution, because usually the formula masses of the two compounds would be different.

In practice, preparing a molar solution involves weighing out the desired mass of solute (which corresponds to the desired number of moles) and placing it in a volumetric flask of suitable size. (A volumetric flask is one calibrated to *contain* a known volume of solution.) Part of the solvent is added to the solute in the flask and the solute is dissolved; then additional solvent is added until the flask is full (▶ Figure 7.22).

Molarity, *M*, is expressed in terms of moles (of solute)/liter of solution, or mol/L. The symbol, *M*, is used interchangeably with mol/L, and both mean the same thing. Finding the molarity of a solution requires that we know the number of moles of solute (which can be calculated from the number of grams of solute) and the total volume of the solution. Note that we do not need to know the total amount of solvent. The molarity of a solution containing 2.0 mol of HCl in 1.0 L of solution, then, will be 2.0 mol/L, or 2.0 *M*. A solution of 0.40 mol of $HC_2H_3O_2$ in 4.0 L of solution will be 0.40 mol/4.0 L = 0.10 mol/L, or 0.10 *M*; and a solution of 0.25 mol of $AgNO_3$ in 500. mL of solution will be 0.25 mol/0.500 L = 0.50 mol/L, or 0.50 *M*.

(a)

(b)

(c)

▶ Figure 7.22 *Preparing a molar solution using a volumetric flask involves three steps: (a) The desired mass of solute is placed in the flask. (b) Some water is added, and the flask is swirled to dissolve all of the solute. (c) The flask is filled to the mark with water, and the contents are mixed thoroughly.*

7.12 Find the molarity, *M,* of the following solutions.

a. 4.0 mol HNO_3 in 2.0 L of solution
b. 0.25 mol NH_3 in 750. mL of solution
c. 1.0 mol $NaC_2H_3O_2$ in 400. mL of solution
d. 0.75 mol CH_3OH in 3.0 L of solution
e. 1.25 mol $CaCl_2$ in 0.25 L of solution

Determining the molarity of a solution containing a given mass of solute in a known volume of solution involves two additional steps. To find the molarity of a solution containing, say, 63.6 g LiCl in 1.50 L solution, it is first necessary to find the formula mass of LiCl, which is 42.4 g/mol. Next, the number of moles of solute must be calculated, in this case

$$63.6 \text{ g LiCl} \times 1 \text{ mol LiCl}/42.4 \text{ g LiCl} = 1.50 \text{ mol LiCl}$$

Finally, the molarity can be determined by dividing the number of moles of LiCl by the volume of solution, as was done in the previous example:

$$1.50 \text{ mol}/1.50 \text{ L} = 1.00 \text{ mol/L} = 1.00\ M$$

7.13 Find the molarity, *M,* of the following solutions.

a. 11.9 g KBr in 0.50 L solution
b. 32.5 g HNO_3 in 2.0 L solution
c. 1.68 g $NaHCO_3$ in 100. mL solution

SUPERSATURATION

It is not uncommon for sugar crystals to appear in honey and certain syrups (they sugar out) when they are allowed to stand for a period of time. The sugar can be made to dissolve simply by gentle warming, but the syrup may sugar out more quickly the next time.

What caused the honey (a mixture of sugars) or syrup to sugar out in the first place? Because these solutions are very sticky (and very sweet), we can assume that they are quite concentrated. Unlike the formation of salt crystals from brine, however, no water can evaporate from honey or syrup to cause crystallization because the lids on the containers are tightly in place. Clearly, some other process is at work. To understand what it is, we look at how syrups are made.

Normally, we make sugar syrups by dissolving a *lot* of sugar in a *small amount* of water. We often mix three or four times the amount of sugar with water in a pan on the stove, but not much sugar dissolves until we heat the mixture and stir for quite a long time—until the mixture begins to boil if we are making a very "heavy" syrup. Usually we add flavorings before we allow our syrup to cool very slowly back to room temperature.

Why must we take all of this trouble to make a syrup? We don't do this to make concentrated, or even saturated, solutions of salt or baking soda ($NaHCO_3$). As a matter of fact, we don't need to do this to make a saturated sugar solution, either. What we have made instead is a solution of sugar that is even more concentrated than one that is saturated. By dissolving the sugar in water in this manner, we have "fooled" the water into holding more sugar than it would ordinarily hold, and have thus made a **supersaturated** solution. Honey is also a supersaturated solution, of the two sugars glucose and fructose. Supersaturated solutions are unstable, and sooner or later form stable saturated solutions by precipitating out the extra dissolved solute as crystals. In some cases, such as with syrups and honey, the supersaturation can persist for a long time before this crystallization begins.

Once a few crystals form, however, the rest follow rapidly. This is because formation of new crystals is usually accelerated if some are already present. In most solutions, solute particles are sufficiently separated to be independent and not well oriented with respect to one another. On the other hand, crystals contain highly ordered arrays of molecules or ions. The presence of even a tiny crystal (crystallite) in a saturated solution provides a template on which the dissolved solute particles can properly align themselves as they coalesce on the tiny crystal and cause it to grow (▶ Figure 7.23). This process of *nucleation* sometimes can be accomplished by the presence of tiny bits of foreign matter (even dust) in a supersaturated solution. For example, crystallization from a supersaturated solution can often be induced simply by scratching the sides or bottom of the solution's container with a glass stirring rod. The crystals begin their growth around the microscopic pieces of glass broken off during the scratching. Likewise, syrups and honey often sugar out after a time because of the presence or introduction of impurities, which can induce crystallization. Heating these sugared solutions dissolves the crystals but does not remove the impurities. Thus, the sugaring process quickly resumes as soon as the solutions cool.

PURIFICATION BY RECRYSTALLIZATION

In making our sugar syrup we heated the mixture of sugar and water to dissolve the sugar. We did this because, as a rule, the solubility of most substances increases with temperature. For example, about 35.7 g of NaCl will

The solubility of many substances in a solvent usually increases with temperature.

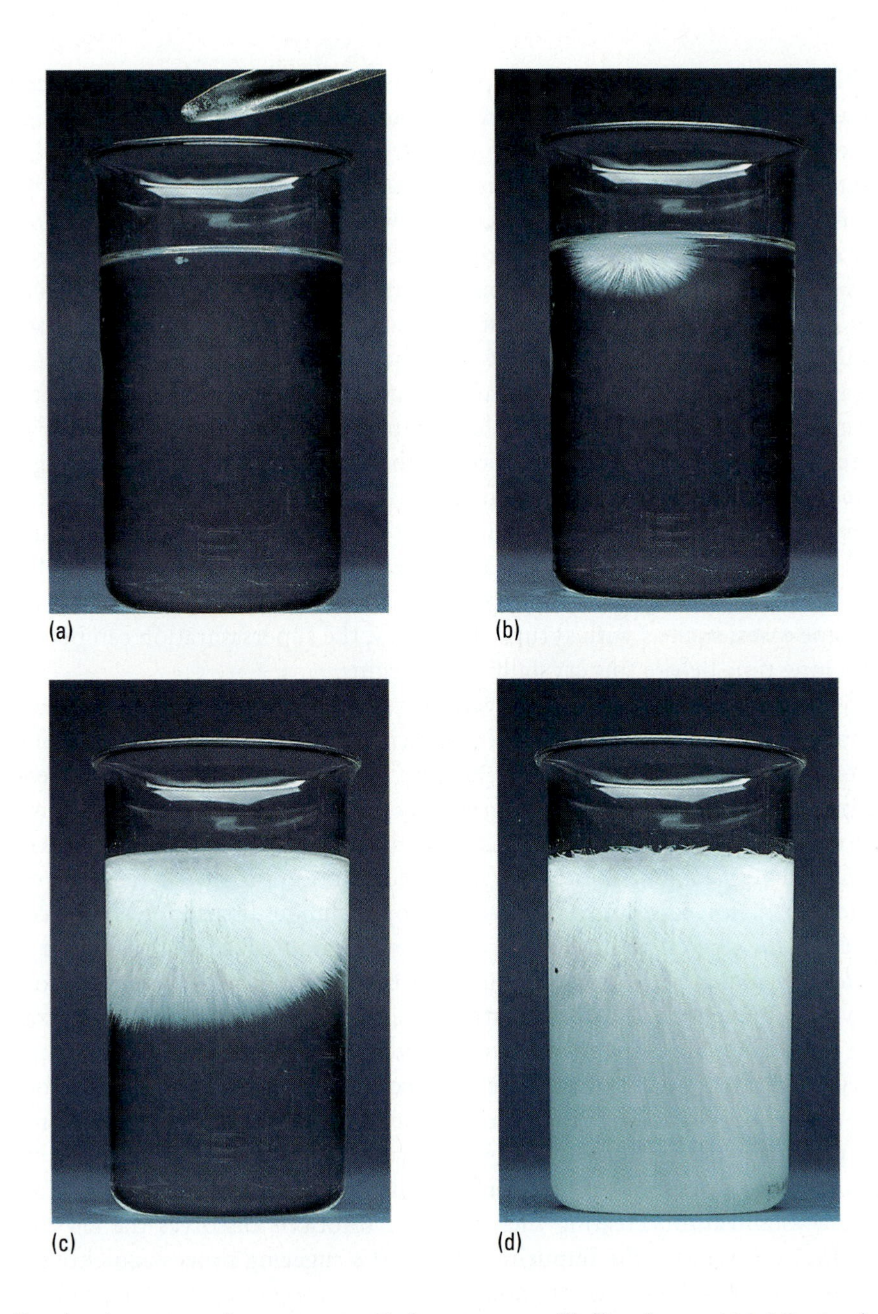

▶ Figure 7.23
Nucleation of a Supersaturated Solution. *(a) A seed crystal of sodium acetate is added to a supersaturated sodium acetate solution. (b) Needlelike crystals of sodium acetate begin to form immediately. (c) . . . and grow very rapidly. (d) All the excess sodium acetate in the supersaturated solution has crystallized. A saturated sodium acetate solution remains between the crystals.*

dissolve in 100 g of water at 0 °C, but 39.1 g will dissolve at 100 °C. At the same respective temperatures, 100 g of water will dissolve 179 g and 487 g of sugar!

Recrystallization is a commonly used means of purifying solid substances.

This increase in solubility with temperature can be used to purify many substances. Called **recrystallization,** the process is one of the most common ways of purifying solid materials, particularly molecular substances. For it to work well the substance to be purified must be more soluble in the hot solvent than in the cold (this is usually true), so that a hot, saturated (or nearly saturated) solution of the substance can be made to form crystals on cooling. In addition, the impurities must be either insoluble in the hot solvent (so they can be filtered from the hot solution) or very soluble in the cold solvent (so they will remain dissolved). For most substances this ideal cannot be met. However, it is usually possible to find conditions that permit partial purification of the desired substance. This means that the process of recrystallization must be repeated several times to yield pure compound. In each repetition, however, some of the material always remains dissolved and is lost with the *filtrate* (the filtered solution). For this reason very pure compounds are often quite expensive to prepare—a large amount of impure substance must be "wasted" in order to obtain the degree of purity desired. (The filtrates usually can be recycled, however, by evaporating the solvent to leave impure material as a solid, which can be repurified.)

REFINING SUGAR

Purification by recrystallization is used in the production of refined sugar from sugar beets or cane. As a part of this process a *slurry* (much like a puree) of the crushed beet or cane pulp in water is filtered to remove all insoluble solid material. The filtrate solution (juice) contains primarily sucrose (table sugar), along with some dissolved ionic compounds, other sugars, and plant by-products. After some of the unwanted impurities are removed chemically, the solution is concentrated by evaporation of the water to make a hot, saturated solution of sugar. (Many of the ionic substances and some of the sugars are in low concentrations and remain dissolved.) Cooling this saturated solution produces crystals of sugar, which are usually brown in color. When removed by filtration, this brown sugar can be sold as is, or it can further be refined by treatment with activated charcoal.

Refined (white) sugar is produced by *decolorizing* the hot, saturated solution of brown sugar with activated charcoal before allowing it to crystallize. *Activated* (or "bone") *charcoal* has a powerful affinity for the colored substances that are found in small quantities in brown sugar. When the finely divided charcoal is added to the hot brown-sugar solution, the molecules of the colored substances stick tightly (a*d*sorb) to the surfaces of its particles. Because the charcoal is completely insoluble in water, it can be filtered from the hot sugar solution, carrying the colored molecules with it. The remaining colorless (decolorized) saturated solution then yields the familiar white sugar crystals on cooling.

DID YOU LEARN THIS?

For each of the statements given fill in the blank with the most appropriate word or phrase from the following list. You may use any word or phrase more than once.

solvent	hydrophilic
solute	associated
solution	spectator ion
dissociation	saturated
solvated	unsaturated
hydrogen bonding	concentrated
dipole–dipole	dilute
ion–dipole	recrystallization
miscibility	mole
micelle	supersaturated
a precipitate	ion exchange
hydrophobic	charcoal

a. ______ A homogeneous mixture
b. ______ Forms when soaps are used in hard water
c. ______ The kinds of attractions that cause ionic substances to dissolve in water
d. ______ Describes a solution that contains very little solute
e. ______ The compound of lesser amount in a solution
f. ______ A process used to soften water
g. ______ The property of one liquid dissolving in another
h. ______ Ions that do not participate in chemical reactions in solution
i. ______ Corresponds to the formula mass of any compound in grams
j. ______ The condition of solute particles in solution
k. ______ The condition of a brine solution in which salt crystals are forming
l. ______ Compounds that undergo hydrogen bonding
m. ______ Used as a decolorizing agent
n. ______ Water loving
o. ______ A microscopic globule with an electrically charged surface
p. ______ The process that occurs when an ionic substance dissolves in water

EXERCISES

1. What is a solvent? A solute? A solution? If two substances dissolve in the one another in all proportions, which is considered to be the solvent? The solute?
2. In terms of attractive forces between particles, what must any solvent be able to do in order to dissolve a solid substance? What does the term *solvated* mean?
3. What kinds of attractive forces hold polar molecules to one another in the solid or liquid phase? Answer this question for nonpolar molecules and for associated molecules. Which of the intermolecular (between molecules) forces is the strongest? The weakest? Given that the molecular mass is the same, which kind of molecular substance would have the lowest boiling point? The highest melting point?
4. Decide which of the following molecules are nonpolar, polar, or associated.

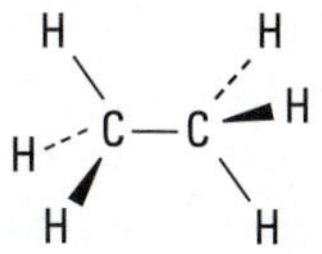

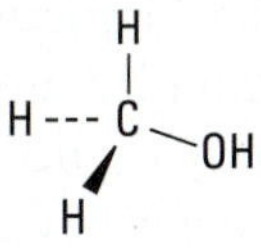

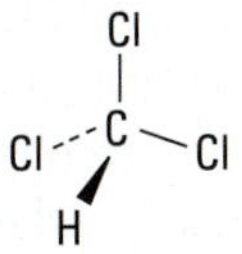

$H_3C{-}CH_3$ (ethane, wedge-dash structure)　　$H_3C{-}OH$ (wedge-dash structure)　　$CHCl_3$ (wedge-dash structure)　　$CH_3CH_2CH_2CH_2OH$

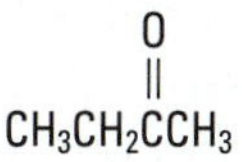

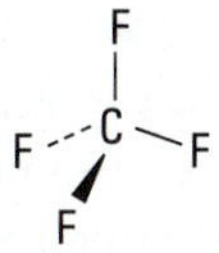

$CH_3CH_2C(=O)CH_3$　　CF_4 (wedge-dash structure)　　CO_2　　HCN　　$CH_2Cl{-}CHCl_2$ (wedge-dash structure)

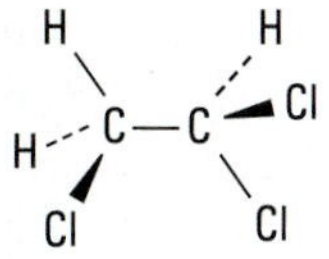

5. Molecules that contain N—H bonds also can hydrogen bond with one another. Show how hydrogen bonding occurs with the following molecules.

$:NH_3$　　$CH_3CH_2CH_2\ddot{N}H_2$

6. Compounds such as dimethyl ether and triethylamine, shown here, are not associated although they are polar. Why is this so?

$CH_3\ddot{\underset{..}{O}}CH_3$
Dimethyl ether

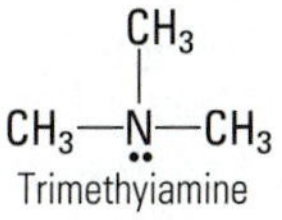

$CH_3{-}\underset{..}{N}(CH_3){-}CH_3$
Trimethyiamine

7. What is it about the structure of the water molecule that accounts for many of its unique properties, such as very high boiling and melting points for its small molecular size? What causes water to expand when it freezes? Why is this attribute of water necessary for life on Earth?
8. What property of water makes it a good solvent for ionic substances? What is the name of the attractive force by which water solvates ions? How does water solvate polar substances? What is the term used to describe this interaction? What kinds of attractive forces are most important when water solvates associated molecules?
9. What is meant by the term *dissociation?* Write equations for the dissociation of the following ionic compounds in water.

   a. KBr d. Li_2S g. $Ca(NO_3)_2$
   b. $MgCl_2$ e. K_2SO_4
   c. NaOH f. Na_3PO_4

10. Write the English statement that corresponds to the equation you have written for each of the compounds in exercise 9.
11. What is the reason that nonpolar substances, such as hydrocarbons and carbon tetrachloride (CCl_4), do not dissolve in water? What is the term used to describe the behavior of nonpolar substances when they interact with water?
12. What intermolecular forces come into play when nonpolar solvents dissolve nonpolar solutes? What is the reason that nonpolar solvents do not dissolve ionic and polar substances very well?
13. What kinds of substances do polar solvents dissolve? What kinds of attractive forces between solvent molecules and solute particles are at work in this process? What structural factor determines, at least to some extent, the kinds of substances these solvents dissolve?
14. What does the term *miscibility* refer to? What do you suppose the term, *100% miscible with one another,* means?
15. What does "Like dissolves like" mean? From your everyday experience list some examples of this relationship.
16. Distinguish between a soap and a common detergent. What is hard water? What property of detergents makes them more desirable for washing in hard water? Write an equation to show the chemical reaction that occurs when a soap is dissolved in hard water. What is the term used to describe the solid product formed?
17. What is soft water? What is an ion exchange resin? What is the chemistry involved when an ion exchange resin is used to soften water?

18. What is a micelle? What properties of soaps and detergents permit them to form micelles from fatty substances? What purpose does agitation of a mixture of a fat and a soap or detergent have in the formation of micelles?
19. What is a saturated solution? When we say that a solution is concentrated or dilute, what do we mean? Why are these two terms ambiguous? When is a solution unsaturated?
20. What is a supersaturated solution? When small crystals of solute (called seed crystals) are added to a supersaturated solution, they begin to grow and new crystals begin to form. What is this process called? What does the seed crystal do to cause crystallization to begin?
21. What kind of a unit is the mole? Why is it used instead of a mass or volume unit when describing the amounts of substances used in chemical reactions and in writing chemical equations? What is Avogadro's number? Why is it so large and why was it chosen instead of some other very large number? (That is, what is it related to that we can measure in the laboratory?)
22. What is molarity? Write down two ways that it is expressed. Give two reasons it is convenient to use molarity when doing solution chemistry instead, say, of using percent by mass to describe solutions.
23. What general property of the solubility of substances is used in purifying substances by recrystallization? Generally, what criteria must be met if purification of a substance by recrystallization from a particular solvent is to be successful? That is, what solubility properties of the solvent and solute (that you want to purify) would you look for? By contrast, what properties should the impurities have? Usually, recrystallization of a substance must be carried out several times to purify it. What is the reason for this? Why is it not possible to recrystallize a quantity of impure compound and recover all of it in pure form? What happens to the portion that is not recovered? How would you recycle it?
24. How can substances be decolorized with activated charcoal? How is the process carried out?

THINKING IT THROUGH

Saturating Solutions

Could a saturated solution be made unsaturated by changing its temperature? Likewise, could an unsaturated solution be made saturated by changing its temperature? Explain. Could a solution be made supersaturated by changing its temperature? What probably would happen?

Ice Water

A glass of water is filled with ice so that the glass is full to the brim and ice floats above the brim of the glass. Will the water overflow when the ice melts? Why?

A Different World

Devise a scene of what the world would be like if water were more dense in the solid state than in the liquid state.

How Do You Dissolve It?

In the following table, common household solvents are listed roughly in increasing order of their polarity, along with pertinent comments about their properties.

COMMON HOUSEHOLD SOLVENTS LISTED IN ORDER OF INCREASING POLARITY

Solvent	Comments
mineral spirits	moderately large hydrocarbon molecules, FLAMMABLE
lighter fluid	small hydrocarbon molecules, EXTREMELY FLAMMABLE
1,1,2-trichlorothane	moderately polar molecule, immiscible with water, POISONOUS but nonflammable
acetone	polar molecule of small size, 100% miscible with water, FLAMMABLE
rubbing alcohol	a mixture of 70% isopropyl alcohol and 30% water by volume, associated, MODERATELY FLAMMABLE
methanol (methyl alcohol, wood alcohol)	small molecule, associated, 100% miscible with water, POISONOUS, FLAMMABLE
water	

Using this information, decide which solvent you could use to dissolve each of the common substances in the following list—either to make a solution of the substance or to remove it from your clothing, a carpet, or other object.

a. $NaHCO_3$
b. rubber cement (long hydrocarbon chains)
c. clean motor oil
d. candle wax
e. fingernail polish (not soluble in water or gasoline)

f. chewing gum
g. honey
h. shortening (long hydrocarbon chains with polar ends)
i. candy
j. pine tree pitch

ANSWERS TO "DID YOU LEARN THIS?"

a. solution
b. a precipitate
c. ion–dipole
d. dilute
e. solute
f. ion exchange
g. miscibility
h. spectator ions
i. mole
j. solvated
k. saturated
l. associated
m. charcoal
n. hydrophilic
o. micelle
p. dissociation

AN ART CONSERVATOR LOOKS AT CHEMISTRY

Lynda A. Zycherman

Conservator of Sculpture

B. A. Art History, City College of the City University of New York

M. A. Art History, and Diploma, Art Conservation, Institute of Fine Arts, New York University

I was nine years old when I discovered archaeology, ancient Egypt, and the Metropolitan Museum of Art in New York City. From then on, I was hooked on art. Almost every Saturday afternoon over the next ten years, I would spend a couple of hours in the museum until I knew the collection by heart. Because of my interest in art, I attended a special high school that offered fine arts as well as a rigorous academic program.

It was a foregone conclusion that I would be an art history major in college, but, of course, I also had to meet science requirements. My father, a chemical engineer, suggested that I take chemistry. After two weeks, the formulas and numbers proved too much for a freshman art historian to handle. I thought I could *never* master the science so I dropped chemistry as fast as I could.

In the middle of my sophomore year, I saw a description of a graduate program in the conservation and restoration of works of art. Fascinating, I thought, but chemistry was a requirement—and not just inorganic, but a year of organic too. I visited the graduate school to see if they were serious about the requirements (they were). I saw laboratories where paintings and sculptures were being treated to mitigate the ravages of time. The conservators were cleaning paintings or treating archaeological artifacts for "bronze disease." Medieval plaques were being examined microscopically for defects in the glassy enamel. Seeing these laboratories, I knew that this was the way I wanted to be involved with art and archaeology.

I was so enthusiastic about applying science to art that I registered for chemistry again, this time with a goal in mind and a will to succeed. Despite my earlier fiasco, I managed to do reasonably well, and later on I was accepted to the graduate program of my choice.

In the graduate program, we studied the materials and techniques of works of art from a chemical standpoint, i.e., the raw materials of a work of art, the way the materials were modified in making the object, and the changes it undergoes in time. Understanding these things can help us to find a treatment that can arrest the deterioration of the object.

As a conservator for sculpture and objects, I am often required to examine archaeological bronzes. When I examine an ancient metal object, say, a weapon, ceremonial vessel, or figurine, in the laboratory, I study the corrosion carefully. With the aid of a simple chemical test, I can take further steps if necessary to arrest its physical deterioration and bring out the beauty that time and burial have concealed.

Bronze is an alloy of copper and tin that was first used around 3000 to 2500 B.C. in Mesopotamia. The alloy was cast into stone or clay molds, producing artifacts whose surfaces were subsequently worked with tools and abrasives to refine them and remove most of the defects produced by the casting.

When buried in damp soil, artifacts made of copper and its alloys lose their metallic appearance and begin to return to their oxides. The oxide layer increases in thickness, and cuprous oxide (Cu_2O) becomes compacted into a purplish red mineral known as cuprite. This in turn may become encrusted with "patina," beautiful basic green or blue carbonates corresponding to the minerals malachite ($CuCO_3 \cdot Cu(OH)_2$) and azurite ($2CuCO_3 \cdot Cu(OH)_2$). Not all green patinas are stable, however. Pale green, powdery spots of corrosion called "bronze disease" may flower up over the surface, disfiguring and weakening the object. The corrosion will continue until the hydrolysis is arrested or the artifact is a pile of green powder.

How does a conservator determine that green spots on a bronze are deleterious copper chloride and not desirable malachite? One simple, accurate way is to perform a chloride test. In a 10-ml graduated cylinder, a small sample of the green powder is dissolved in a weak nitric acid solution. A few drops of silver nitrate solution are added; if chloride is present, it will be precipitated as silver chloride, which appears opalescent when the cylinder is lit from the side in front of a dark background.

Physical and chemical methods have been developed that reach the deep cuprous chloride and render it innocuous without unduly altering the surface appearance of the bronze. In addition, since chloride activity is at a minimum when conditions are dry, objects with bronze disease should be kept in an atmosphere with as low a relative humidity as possible. Another method for stabilizing bronze disease uses benzotriazole (BTA). BTA forms an insoluble complex compound with cupric ions that precipitates over the cuprous chloride and forms a barrier that prevents the ingress of moisture.

Any detailed examination of archaeological artifacts can lead to surprises. While examining a Chinese painted pottery vessel from the Han Dynasty (208 B.C.–A.D. 220), a colleague and I observed blue and purple pigments that, when analyzed by microchemical and instrumental methods, did not correspond to any of the natural mineral pigments known to be in use in China around 200 B.C. Using scanning electron microscopy to identify the elements present (barium, copper, silicon) and optical mineralogy combined with X-ray diffraction to identify the crystalline structure, we identified the blue component as barium copper silicate, $BaCuSi_4O_{10}$, a hitherto unidentified synthetic pigment.

The purple pigment proved to be more elusive. Recently, scientists working in superconductor research synthesized a magenta compound (barium copper silicate, $Ba_3Cu_2Si_6O_{17}$) that proved to be identical to the purple we had found on that pottery.

The conservation and restoration of contemporary works of art presents a different sort of challenge. We can easily determine the technology of modern materials (plastic, Corten steel, acrylic paint) by consulting industry engineers, but no one knows much about the long-term stability of the materials because that is usually not a concern of the industry that produced them.

In the 1920s, when cellulose nitrate was a new plastic resin, some artists used it to create sculptures. We now know that cellulose nitrate is an unstable plastic. It deteriorates by loss of the plasticizer, which migrates out at the surface, followed by the emission of small amounts of nitric acid. The result is severe yellowing, opacification, and fragmentation into small pieces or powder. The nitric acid could also attack any metal fittings, causing the entire sculpture to self-destruct. Cellulose nitrate is also highly flammable and has spontaneously burst into flame in film collections (but, fortunately, never in an art museum.) Although we cannot restore objects made of this material, we can try to arrest the deterioration by storing such works in a stable environment with a relatively low temperature because every 10-degree drop will halve the rate of a chemical reaction.

From the preservation and treatment of a deteriorated artifact to the analysis of ancient technology, I use chemistry in my work every day. If you are a museum visitor, art lover, or collector, you will see the results of chemistry in every work of art and artifact made by humans.

Acids, Bases, and Chemical Equilibrium

All systems that are in a state of chemical equilibrium undergo, as a result of a variation in one of the factors of the equilibrium, a transformation in such a direction that, if that transformation occurred alone, it would lead to a change of opposite sign in the factor considered.

Henri-Louis Le Châtelier, 1888

:ÖH₂ ⟶ :Cl:Al :Cl :Cl

base

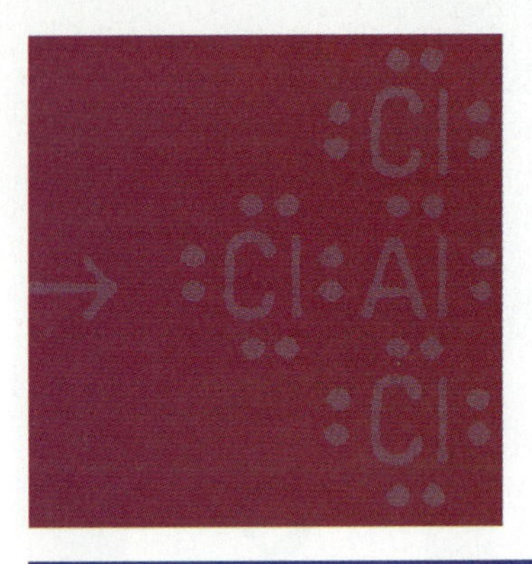

Lemon juice is sour and peach seeds taste bitter. Water holes in Nevada contain undrinkable alkali, and acid rain is harming aquatic life in New York lakes. We use lime to sweeten soil, take antacids to relieve indigestion, and apply deodorants for personal hygiene. Sinkholes swallow cars and houses in Florida (▶ Figure 8.1). All these situations involve the chemistry of acids and bases; which—as one can see—is important to our daily lives.

A DEFINITION OF ACIDS AND BASES

Is there a particular chemical species that produces the sour taste when we eat a pickle or a rhubarb pie? What is it that causes the bitter taste of soap or coffee? When we mix vinegar with baking soda—and if we use just the right amounts—we get lots of bubbles and a solution that is neither sour nor bitter. It appears that the properties of vinegar and baking soda have canceled each other out, or neutralized each other. Are the two chemically related in some way?

The sour substances we have been talking about are acids. **Acids** are chemically reactive substances that not only taste sour but also turn the dye called litmus from blue to red and react with substances called bases, neutralizing their properties (▶ Figure 8.2a). They also react with metals such as iron, zinc, and tin to produce hydrogen gas. The bitter substances are **bases,** also called *alkalis.* They turn red litmus to blue, feel soapy when their solutions are rubbed on our skin, and react with acids to neutralize their properties (▶ Figure 8.2b).

A MODEL OF ACIDS AND BASES

It may not be very satisfying that part of our operational definition of acids and bases describes them in relation to one another. Acids react with bases to

► Figure 8.1
Acid–base chemistry was involved in the formation of this sinkhole on May 8 and 9, 1981, in Winter Park, Florida. The sinkhole, which has a diameter of 100m and a depth of 35m, destroyed a house, numerous cars, and the municipal swimming pool.

neutralize them, and vice versa, yet we haven't said what kinds of chemical species acids and bases are. Because acids are sour, giving almost a "prickly" sensation to the tongue, early chemists thought the particles of acids might have spine-like protrusions on them, which produced this effect. Bases are bitter, sometimes "puckery." Could it be that particles of bases have holes in them that suck in parts of the tongue, and also accept the prickly spines of the acids?

Although this was an interesting idea, a useful model of acids and bases eluded chemists until 1887, when Svante Arrhenius, a Swedish chemist, proposed that in water solution all acids produce *hydrogen ions,* H^+, and all bases produce *hydroxide ions,* OH^-. The **Arrhenius model** met with much resistance and caused considerable controversy in the scientific community; it was not generally accepted until about 1903, when Arrhenius was awarded the Nobel Prize in chemistry.

Acids furnish H^+ in water solution. **Bases** furnish OH^-.

The idea that the acidic properties of aqueous (water) solutions is due only to dissolved hydrogen ions may at first seem strange because many different

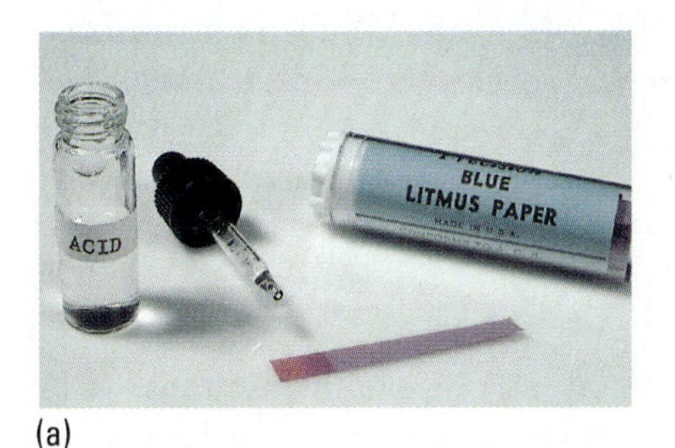

(a)

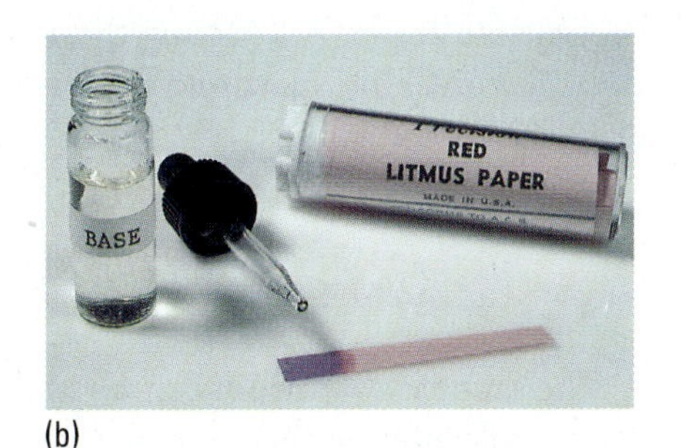

(b)

► Figure 8.2
Testing for Acid or Base. *(a) Blue litmus turns red when moistened with acid. (b) A solution of base turns red litmus blue.*

substances make acidic solutions when dissolved in water. The same can be said for dissolved hydroxide ions being responsible for the properties of aqueous bases. On the other hand, the model is simple: only one chemical species is responsible for acidic properties and one chemical species for basic properties. When we deal with water solutions, we find this always to be true: hydrogen ions (H^+) are ultimately the chemical species that produce acidic properties, and hydroxide ions (OH^-) produce basic properties. When we taste things that are sour, it is the hydrogen ions that we sense on our tongues. Bitter or puckery substances furnish hydroxide ions. Also, hydrogen ions and hydroxide ions react to neutralize each other's properties.

(The idea of keeping a scientific model simple is called the principle of *Ockham's razor.* That is, a scientific theory or model should not be made more complex than is necessary to account for the experimental evidence. Usually, complex models are not used in science if simpler ones will work.)

We must be careful, however, not to extend the Arrhenius model to systems other than water solutions. When we do this, we find that our fairly simple Arrhenius model often does not work. Many chemical reactions behave in ways analogous to those of acid–base reactions in water, and yet it is impossible in these reactions to produce either hydrogen ions or hydroxide ions. Two newer theories, of respectively greater scope and greater power, are now used when the Arrhenius theory becomes inadequate. These are the Brønsted–Lowry and the Lewis theories of acids and bases. For the purposes of our discussion, however, the Arrhenius theory will serve nicely.

ACIDS IN WATER

Substances that are acids, then, must somehow produce hydrogen ions when they are dissolved in water. (See ▶ Figure 8.3 for some familiar acids.) Most of the common acids in their pure form are molecular substances that ionize when placed in water. One of the products of the ionization is always a hydrogen ion, H^+, the other product being an anion. (We recall that all negatively charged ions are called anions.) Let us consider the hypothetical acid molecule, HA. In this acid, H is a hydrogen atom, which will become H^+ when the molecule ionizes in water, and A is the part of the molecule that will become the anion, A^-. Mixing the acid, HA, with water results in its **ionization** (the breaking apart of a neutral chemical species into ions) to form hydrogen ions and anions,

Ionization is the *formation* of ions.

$$HA \xrightarrow{\text{water}} H^+ + A^-$$

The anions (A^-) produced are solvated by ion–dipole interactions with water molecules, in the same manner as are anions from crystalline ionic materials. Solvation of the hydrogen ion is another matter, however, because it bonds tenaciously to water molecules and forms larger ions with the structures H_3O^+, $H_5O_2^+$, and so forth. Each of these chemical species is intensely hydrogen bonded with water, and in fact, each is a part of the water structure. In many discussions of acidity the species H_3O^+, called the *hydronium ion,* is used in place of the hydrogen ion, H^+. We shall stick with the latter for the sake of simplicity. Because both the hydrogen ion and its corresponding anion are solvated and separate from one another in solution, we must again write their symbols with a plus sign between them, for example,

$$HCl\ (g) \xrightarrow{\text{water}} H^+ + Cl^-$$

8.1 Each of the following compounds reacts in water to produce an acidic solution. Write a similar equation for the ionization of $HBr(g)$, $HI(g)$, and $HNO_3(l)$.

BASES IN WATER

When dissolved in water, bases (such as those shown in ▶ Figure 8.4) can form hydroxide ions in two ways. In the first, a metal hydroxide can undergo the process of dissociation in water. A *metal hydroxide* is simply an ionic substance composed of a metal ion and the corresponding number of hydroxide ions required to maintain electrical neutrality. (Recall that the hydroxide ion, OH^-, is a polyatomic ion that always carries a 1– charge and does not break down into smaller pieces.) Thus, sodium hydroxide is NaOH, and lithium

▶ Figure 8.3 (Left)
Some Acidic Substances Commonly Found in the Home.

▶ Figure 8.4 (Right)
Some Basic, or Alkaline, Substances Used in the Home.

hydroxide and potassium hydroxide are LiOH and KOH. Each contains a metal ion and a hydroxide ion in the ratio of 1:1. More complicated hydroxides contain unequal numbers of metal ions and hydroxide ions in order to maintain electrical neutrality. Magnesium hydroxide, $Mg(OH)_2$, which is used in milk of magnesia, contains one magnesium ion, Mg^{2+}, and two hydroxide ions. Similarly, the aluminum hydroxide used in antacids is $Al(OH)_3$.

Dissociation is the *separation* of ions.

When metal hydroxides are dissolved in water, their ions simply dissociate to form separate ions in solution. In the case of sodium hydroxide, for example, a solution of sodium ions and hydroxide ions results.

$$NaOH(s) \xrightarrow{\text{water}} Na^+ + OH^-$$

8.2 Write a similar equation for the dissociation of $KOH(s)$ and $LiOH(s)$.

Hydrolysis is a reaction with water.

A second way that basic substances can produce hydroxide ions in water solution is by **hydrolysis.** In this instance, a molecular or ionic substance which does not contain hydroxide ions itself actually reacts with water molecules to form hydroxide ions. This is usually accomplished by removal of hydrogen from a water molecule. Consider two basic substances, B and B^-, the former molecular and the latter ionic. In reacting with water they both form hydroxide ions:

$$B + H_2O \longrightarrow BH^+ + OH^-$$

and

$$B^- + H_2O \longrightarrow BH + OH^-$$

An example of the first kind of base (which is neutrally charged) is ammonia. Ammonia gas is a molecular substance that dissolves readily in water to make a basic solution by forming the polyatomic ammonium ion, NH_4^+, and a hydroxide ion:

$$NH_3 + H_2O \rightleftarrows NH_4^+ + OH^-$$

Household ammonia is a fairly concentrated solution of this kind, and its cleansing action is due primarily to the presence of the hydroxide ion.

Many compounds related to ammonia also hydrolyze in this manner. One kind is the *alkaloids* (alkali-like compounds). These bitter tasting substances include caffeine ($C_8H_{10}N_4O_2$), found in coffee, tea, and chocolate; quinine ($C_{20}H_{24}N_2O_2$), a drug used to treat malaria (also, the flavor of tonic water); and nicotine ($C_{10}H_{14}N_2$), the potent drug present in tobacco.

A common example of the second kind of base (which is negatively charged) is the hydrogen carbonate, or bicarbonate, ion. This polyatomic ion, HCO_3^-, is found in baking soda, $NaHCO_3$. The hydrogen carbonate ion reacts with water to form carbonic acid, H_2CO_3, and a hydroxide ion:

$$HCO_3^- + H_2O \rightleftharpoons H_2CO_3 + OH^-$$

Note that this equation has been written without a positive ion present to balance the charge on the hydrogen carbonate ion. Of course such an ion, usually Na^+, must be present in solution; but just like the cations in the soap solutions it is a spectator ion and does nothing. An equation like this, which shows only the ions that participate and ignores the spectator ion(s), is called a **net ionic equation.** (The meaning of the double arrows, $\rightleftharpoons$, will be explained later.)

Spectator ions are omitted from a **net ionic equation.**

We already know some other polyatomic ions that hydrolyze in water in the same manner to produce basic solutions. These ions include acetate ($C_2H_3O_2^-$), sulfate (SO_4^{2-}), phosphate (PO_4^{3-}), and carbonate (CO_3^{2-}). The nitrate ion (NO_3^-) does not hydrolyze.

NEUTRALIZATION

We mentioned that when an acid is made to react with a base, the properties of each appear to be neutralized. We neutralize excess stomach acid with milk of magnesia or other antacids, and we use baking soda to keep sourdough biscuits from being too tangy or to take the fire out of a bee sting. A piece of limestone fizzes when a drop of vinegar or muriatic acid is placed on it.

Each of these processes involves the reaction of an acid with a base; and because (in these cases) the reactions occur in aqueous solution, hydrogen ions and hydroxide ions must somehow be reacting with each other. The properties of both ion types disappear, so these ions must be transformed to something else during neutralization. That "something else" is water. Energy (in the form of heat) is evolved, and so this reaction tends to be spontaneous. **Neutralization,** then, is the reaction of a hydrogen ion with a hydroxide ion to produce water:

$$H^+ + OH^- \longrightarrow H_2O + \text{heat}$$

A considerable amount of heat is evolved in neutralization, particularly when concentrated acids and bases are used. For this reason, it is unsafe to mix concentrated acids with concentrated bases, because enough heat may be liberated to cause these solutions to boil or even vaporize explosively. Many drain cleaners contain concentrated alkali, whereas certain toilet bowl preparations and hard water removers are strongly acidic. It can be dangerous to mix the two. To make matters worse, one commercial drain cleaner (available at plumbing shops) is strongly acidic. If you fail to unclog a drain with one of the common alkaline drain cleaners, don't dump the strongly acidic cleaner into the drain on top of it! This reaction can be violent. Always read the labels on household products.

In neutralization, it is often said, an acid reacts with a base to produce a salt plus water. We have accounted for the water part in the preceding equation, but what about the salt? In their respective acidic and basic solutions, both hydrogen ions and hydroxide ions have spectator ions associated with them. Although these spectator ions usually have nothing to do with the neutralization process, they form the basis for the salt that is produced. For example, neutralization of a hydrochloric acid (muriatic acid) solution with a sodium hydroxide (lye) solution yields a solution of sodium chloride (a salt), plus, of course, water:

$$HCl(g) \xrightarrow{\text{water}} \underbrace{H^+ + Cl^-}_{\text{Hydrochloric acid solution}}$$

$$\underbrace{H^+ + \mathbf{Cl^-}}_{\text{Hydrochloric acid solution}} + \underbrace{\mathbf{Na^+} + OH^-}_{\text{Sodium hydroxide solution}} \longrightarrow \underbrace{H_2O}_{\text{Water}} + \underbrace{\mathbf{Na^+} + \mathbf{Cl^-}}_{\text{A salt (sodium chloride solution)}}$$

The Na^+ and Cl^- ions are the spectator ions that form the salt, which can be obtained in pure form by evaporating the water.

A **salt** is an ionic compound.

Generally, the term **salt** means any ionic substance that could be formed from the anion of an acid and the cation of a base. Because in principle almost all ionic substances can be formed in this manner, the terms salt and ionic substance can be used interchangeably. Not all salts (ionic substances), however, are soluble in water. Salts such as silver chloride (AgCl) and barium sulfate ($BaSO_4$) are examples.

Because all neutralizations in water solution involve the reaction of hydrogen ions and hydroxide ions to form water, our antacids, sourdough and soda, and the fizzing of limestone rocks with acid, must all have the same fundamental chemistry. They differ only in the way each produces hydrogen ions or hydroxide ions. For example, several antacids shown in ▶ Figure 8.5

contain magnesium hydroxide $Mg(OH)_2$, which reacts with the excess hydrochloric acid in solution in the stomach. This overall reaction can be written,

$$2H^+ + 2Cl^- + Mg(OH)_2(s) \longrightarrow 2H_2O + Mg^{2+} + 2Cl^-$$

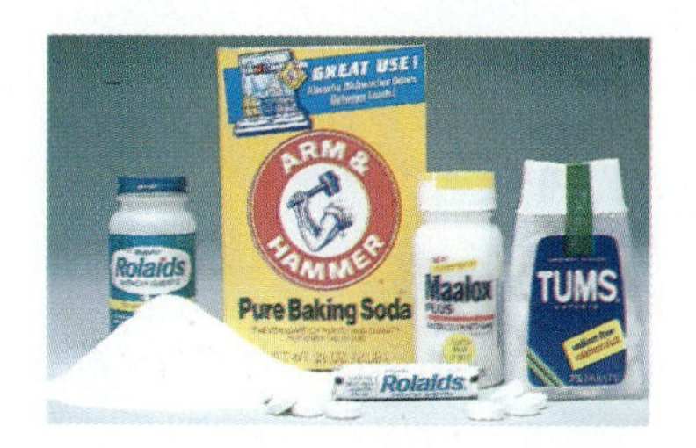

► Figure 8.5
Some Antacids Often Found on the Shelf.

This time, the magnesium and chloride ions are the spectator ions that form the salt, magnesium chloride, in solution. Even though it furnishes hydroxide ions, the magnesium hydroxide is not very soluble in water and does not dissociate very much, so it is written as a solid rather than as ions in solution.

8.3 Another antacid contains aluminum hydroxide, $Al(OH)_3$, which also is not very soluble in water. Write the equation for its reaction with stomach acid.

Sweetening soils by the use of lime (also called quicklime), CaO, is another example of neutralization. In this instance, the calcium oxide reacts with water to produce calcium hydroxide, a source of hydroxide ions:

$$CaO(s) + H_2O \longrightarrow Ca(OH)_2(s)$$

The hydrogen ion in the sour soil then is neutralized by the hydroxide ions in the calcium hydroxide (net ionic equation shown):

$$2H^+ + Ca(OH)_2(s) \longrightarrow Ca^{2+} + 2H_2O$$

Two other important neutralization reactions involve the acid treatment of hydrogen carbonates, such as $NaHCO_3$ (baking soda), and carbonates, such as calcium carbonate (limestone), $CaCO_3$. Both of the polyatomic ions, hydrogen carbonate (also called bicarbonate, HCO_3^-) and carbonate (CO_3^{2-}), react with acids to produce carbon dioxide gas, CO_2. This gas causes the sourdough biscuits to rise and the limestone to fizz. The reactions of carbonates and hydrogen carbonates are so commonplace that it is hard to avoid them in our everyday lives. Their chemistry is part of everything, from soda pop and baking a cake to how rapidly we breathe, how minerals are deposited, and how the greenhouse effect occurs. To understand this important chemistry of the carbonates, however, we must first introduce the concept of chemical equilibrium and explain what the double arrows, $\rightleftarrows$, mean.

NO SWEAT

Getting ready for a Saturday night out, Charley (nature boy) catches you on your way to the shower with your soap and antiperspirant.

"You *use* that stuff?"

"Sure, don't you?" That was a mistake.

"No way! It shuts off your sweat glands."

"I thought that was the idea."

"But sweating's the way you cool off and get rid of excess salt. What's the matter–you don't want to smell like a human being?"

"I don't want to smell like old gym socks . . . ."

"So take a shower every couple of days."

"I usually take a shower every day."

"Every day!" He glances at your soap. "And with alkaline soap! You ruin your skin's natural resistance to bacteria and then spray yourself with stuff that kills them and stops up your sweat glands." He shakes his head hopelessly, and glances at the antiperspirant again. "And you use aerosol–you're breathing the stuff! It's astringent . . . I don't know what that does in your lungs."

You're irritated, mostly because you're guilty of not having thought about any of this. Charley shrugs. "I'm sorry . . . it's none of my business." He walks into his room.

You finish dressing and forget about the problem (another problem!) until the next day. You go to talk to Rhonda, but she's studying and after listening with half an ear, merely points to a book.

We control body temperature by sweating. When we're warm our entire bodies sweat, except our palms and soles, mostly water and NaCl to cool us; when we sweat because we're tense or afraid it's mainly from our palms, soles, and armpits.

We have two kinds of sweat glands. Eccrine ("true") sweat glands, of which about 2×10^6 are distributed over our body surface. The apocrine glands, of which there are many less, are located in armpits and places suspiciously similar to the locations of the scent glands of other animals and absent before adolescence.

CHEMICAL EQUILIBRIUM

In the past we have seen chemical reactions that go only one direction, left to right, from reactants to products. Reactions of this kind are said to be *irreversible*, meaning that they don't go backward. The reaction of hydrogen car-

In fact, pheromones, chemical substances produced by animals to elicit specific responses in other animals of the same species, have been discovered in humans after years of doubt that we produce them. Studies have shown that the menstrual cycles of women living in dormitories tend to become coordinated by chemical signals and that exposure to male pheromones in clinical tests affects the menstrual cycles of some women. It may even be that our odor, like our appearance and behavior, may be attractive or unattractive to another person. Maybe that's what Charley meant by "smelling like a human being." Something else to think about.

In the meantime, the literally nonhuman smells that are so unacceptable are caused by our ever-present skin bacteria, which thrive on the protein secretions of the apocrine glands and which secrete the amines, acroleins, and carboxylic acids that are the olfactory offenders. What Charley meant by our skin's natural resistance to bacteria is that skin tends to be acidic, which is not a good environment for bacteria, and it secretes substances that are poisonous to them.

Deodorants are cosmetics that control the growth of bacteria or mask the odor of their secretions; most deodorants do both with perfumes and triclosan, a broad-spectrum bacteriostat (*stat,* "to hold still" as in *static*). Bacteria are so prevalent on our bodies that they can only be controlled, never eliminated.

Antiperspirants are aluminum chlorohydrates such as $Al(OH)_4Cl_2$. They release aluminum ions ($A1^{3+}$), which denature proteins and contract the ducts that deliver sweat to the skin. No sweat, no odor.

Maybe Charley's right. We worry too much about body odor. The makers of deodorants and antiperspirants seem to have us over a barrel: according to their ads, *everyone* stinks, or will if they don't use their products. You wonder if there's some kind of acid cleanser you could use. But how could you ever be sure it worked? You can ask housemates if you *look* all right when you're going out, but you certainly can't ask anyone if you *smell* all right.

Except, maybe, Charley.

bonate ion with water, however, goes both directions, from reactants to products and once the products are formed, from products to reactants. Reactions of this kind are said to be *reversible* and are indicated by double arrows:

$$HCO_3^- + H_2O \rightleftharpoons H_2CO_3 + OH^-$$

A **chemical equilibrum** is a dynamic process.

The double arrows have special meaning because they designate the reaction to be a dynamic process that is going on all of the time. That is, reactants are forming products and products are forming reactants simultaneously in a process called **chemical equilibrium.** They also mean that, once this equilibrium is "established," the products are forming reactants exactly as fast as the reactants are forming products. Or, both the "forward" and the "reverse" reactions are occurring at the same rate: the reactants are being reformed from products as fast as they are being consumed to form them, and likewise, the products are being formed from reactants as fast as they are consumed to make them (▶ Figure 8.6).

▶ Figure 8.6
How a Chemical Equilibrium Is Established.

$$A + B \longrightarrow C + D$$

Reactants A and B are mixed together and allowed to react to form products C and D.

$$A + B \longleftarrow C + D$$

In a reversible reaction, as soon as products C and D begin to form, they start to react to reform reactants A and B.

$$A + B \longrightarrow C + D$$
$$A + B \longleftarrow C + D$$

When equilibrium is established, both reactions are going at the same rate. Products C and D are being used to form A and B as fast as they are produced. Reactants A and B also are being used as fast as they are formed to give products C and D.

$$A + B \rightleftarrows C + D$$

Dynamic equilibrium is established. Concentrations of reactants and products do not change unless the reaction is disturbed. Reaction appears to be static, but it is not.

POSITION OF EQUILIBRIUM

Although there is constant chemical activity in a chemical equilibrium, the concentrations (or amounts) of both the reactants and the products remain constant, giving an impression that nothing is happening. But something *is* going on, and some slight external force can cause either the forward process or the reverse process to speed up or slow down relative to the other, and can cause the concentrations of reactants and products to change. When this happens, we say that the position of equilibrium has shifted.

In discussing the chemistry of the carbonates we shall want to know where the position of equilibrum lies for the chemical species produced when each of these ions is dissolved in water, and also which factors cause that position of equilibrium to shift in one direction or the other. To say the position of equilibrium lies to the left means that the reactants are in high concentration and the products are in lower concentration, or that the reaction mixture is mainly composed of reactants. When the equilibrium lies to the right, the products are the predominant chemical species present and the reactants are present in a smaller amount. A shift to the right in equilibrium means that some stimulus has caused more products to be formed at the expense of reactants, and a shift to the left indicates that an increase in reactants has occurred with a corresponding decrease in the amount of products (▶ Figure 8.7).

REACTANTS $\rightleftharpoons$ products
Equilibrium lies to the **left.** Mostly **reactants.**

reactants $\rightleftharpoons$ **PRODUCTS**
Equilibrium lies to the **right.** Mostly **products.**

REACTANTS $\rightleftharpoons$ PRODUCTS
Reaction in equilibrium.
reactants $\rightleftharpoons$ **PRODUCTS**
Equilibrium shifted to the **right.** More **products formed.**

REACTANTS $\rightleftharpoons$ PRODUCTS
Reaction in equilibrium.
REACTANTS $\rightleftharpoons$ products
Equilibrium shifted to the **left.** More **reactants formed.**

▶ Figure 8.7
Position of Equilibrium.

WEAK AND STRONG BASES

When the hydrogen carbonate ion hydrolyzes in water, its equilibrium lies to the left. This means that the solution consists mostly of hydrogen carbonate ions and contains very few hydroxide ions and carbonic acid molecules. Such a solution is weakly basic, because the number of hydroxide ions produced by a given number of hydrogen carbonate ions is low. Salts containing the hydrogen carbonate ion, then, are said to be **weak bases** because they hydrolyze only slightly and form weakly basic solutions when dissolved in water. Similarly, ammonia is a weak base because it reacts only slightly with water, producing a low concentration of hydroxide ions (▶ Figure 8.8).

(a) $\mathbf{HCO_3^-} + \mathbf{H_2O} \rightleftharpoons H_2CO_3 + OH^-$

(b) $\mathbf{NH_3} + \mathbf{H_2O} \rightleftharpoons NH_4^+ + OH^-$

▶ Figure 8.8
A Weak Base Hydrolyzes Only Partially in Solution. *(a) Hydrogen carbonate ion is a weak base because it furnishes only a few hydroxide ions by reacting with water. Equilibrium lies mostly to the left. (b) Ammonia is also a weak base because it furnishes just a few hydroxide ions by reacting with water. Again, the equilibrium lies mostly to the left.*

Weak bases react slightly with water to produce weakly basic solutions.

Soap solutions are weakly basic for the same reason. The carboxylate anion, $R—CO_2^-$, hydrolyzes only slightly in water to produce hydroxide ions:

$$\mathbf{R—CO_2^-} + \mathbf{H_2O} \rightleftharpoons R—CO_2H + OH^-$$

In addition to the detergent action of the soap itself, the hydroxide ions have a mild antibacterial effect and also serve to neutralize the acids on skin that contribute to body odor. These acids, volatile molecular compounds with offensive odors, are converted into their salts, which are ionic and nonvolatile. This renders them odorless. When used as a body deodorant, baking soda (sodium hydrogen carbonate) neutralizes these acids in the same manner.

8.4 Write the balanced equilibrium reaction for the hydrolysis of the following anions in water solution to produce hydroxide ions: $C_2H_3O_2^-$, SO_4^{2-}, and PO_4^{3-}. In each case, show an equilibrium reaction that produces only one hydroxide ion. For example, the oxalate ion, $C_2O_4^{2-}$, hydrolyzes as follows:

$$C_2O_4^{2-} + H_2O \rightleftharpoons HC_2O_4^- + OH^-$$

Strong bases completely form hydroxide ions in water solution.

By contrast, a salt (ionic compound) such as sodium hydroxide dissociates so completely that all of its hydroxide ions become available in solution. When we talk about bases being weak or strong, we are not talking about their concentration in solution. Rather, we are concerned with whether their dissociation (or hydrolysis) in water is complete (strong) or only partial (weak). Thus, NaOH is considered to be a **strong base** (▶ Figure 8.9). Some strong and weak bases are listed in ■ Table 8.1.

▶ Figure 8.9
A Strong Base Completely Dissociates in Water. *Sodium hydroxide is a strong base because it dissociates completely to furnish a large number of hydroxide ions.*

$$NaOH(s) \xrightarrow{\text{water}} \mathbf{Na^+} + \mathbf{OH^-}$$

The difference in the behavior of weak and strong bases can be illustrated by comparing baking soda (sodium *hydrogen* carbonate) with sodium hydroxide. Small amounts of baking soda can be safely ingested, and it is used in cooking and as an antacid. Sodium hydroxide is lye. It is used in drain cleaners and for removing the skins from potatoes for processing into French fries and potato chips. It is very dangerous to handle, and severe burns can result if

■ **TABLE 8.1 PARTIAL LISTING OF STRONG AND WEAK ACIDS AND BASES**

Strong Acids	Weak Acids	Very Weak Acids
hydrochloric HCl	acetic $HC_2H_3O_2$	hydrocyanic HCN
nitric HNO_3	citric $HC_6H_7O_6$	boric H_3BO_3
sulfuric H_2SO_4	lactic $HC_3H_5O_3$	
	phosphoric (moderately strong) H_3PO_4	
Strong Bases	**Moderate Bases**	**Weak Bases**
sodium hydroxide $NaOH$	calcium hydroxide $Ca(OH)_2$	magnesia MgO
potassium hydroxide, KOH	sodium carbonate Na_2CO_3	sodium hydrogen carbonate, $NaHCO_3$
lithium hydroxide, $LiOH$	ammonia NH_3	
barium hydroxide $Ba(OH)_2$	trisodium phosphate (TSP) Na_3PO_4	

even a small amount of the solid comes in contact with your skin. Ingesting it can be fatal. Don't even try to taste it!

WEAK AND STRONG ACIDS

Carbonic acid, made simply by dissolving carbon dioxide in water, is a **weak acid.** That is, it ionizes only slightly to form hydrogen ions and hydrogen carbonate ions. The position of this equilibrium lies mostly to the left (less than 1% to the right), so very few hydrogen ions are formed (▶ Figure 8.10). Even so, solutions of carbon dioxide are sufficiently acidic to impart a tart taste to carbonated beverages and to play an important role in the chemistry of groundwater.

Weak acids ionize only a little to form hydrogen ions.

$$H_2CO_3 \rightleftharpoons H^+ + HCO_3^-$$

▶ Figure 8.10
A Weak Acid Ionizes Only Partially in Solution. *Carbonic acid is a weak acid because it furnishes only a few hydrogen ions when it ionizes. Equilibrium lies mostly to the left.*

Natural rainwater, even in the absence of airborne pollutants, is slightly acidic because it contains dissolved CO_2. When this water percolates through limestone underground, it dissolves the limestone according to the following equation,

$$CaCO_3(s) + H_2O + CO_2 \rightleftharpoons Ca^{2+} + 2HCO_3^-$$

thereby forming limestone caves and sinkholes. The stalactites and stalagmites in limestone caves and the mineral deposits around hot springs consist primarily of calcium carbonate, deposited by the reverse of the process (▶ Figure 8.11). These formations occur when groundwater is exposed to the air and CO_2 is lost, which drives the equilibrium to the left.

$$\mathbf{CaCO_3(s) + H_2O + CO_2}\uparrow \rightleftharpoons Ca^{2+} + HCO_3^-$$

The symbol ↑ means that CO_2 is lost from the solution to the atmosphere.

"Scale" in boilers and hot water heaters often forms by this same process. Water containing calcium ions and hydrogen carbonate ions is said to be temporarily hard. Heating decreases the solubility of CO_2 in water and drives it from solution. This removal of a reactant causes the equilibrium to shift to the left (see next section), and $CaCO_3$ precipitates in the pipes or tanks.

Removing a reactant shifts the position of equilibrium to the left.

▶ Figure 8.11
A Lake in a Limestone Cave, Luray Caverns, Virginia. *The stalactites (hanging) and stalagmites (standing) consist primarily of calcium carbonate.*

8.5 In a manner similar to that for carbonic acid, the following weak acids ionize reversibly in water to form hydrogen ion and the corresponding anion. Write an equation for the ionization of each of these in water:

a. acetic acid, in vinegar ($HC_2H_3O_2$)
b. lactic acid, in sour milk, yogurt, sourdough, and sore muscles ($HC_3H_5O_3$)
c. citric acid, in citrus fruits ($HC_6H_7O_7$)
d. butyric acid, in rancid butter ($HC_4H_7O_2$)
e. ascorbic acid, in citrus fruits and rose hips (vitamin C, $HC_6H_7O_6$)
f. oxalic acid, in rhubarb (HC_2HO_4)
g. hydrocyanic acid, from amygdalin found in peach, apricot, and other fruit seeds (HCN)

Strong acids ionize completely.

Strong acids are compounds that ionize completely, or nearly so, to furnish a high concentration of hydrogen ions in solution. Hydrogen chloride (gas) and nitric acid (HNO_3, a liquid) are examples (► Figure 8.12). These molecular compounds ionize so completely in water that for all practical purposes no equilibrium exists: solutions of hydrochloric acid and nitric acid contain only hydrogen ions and chloride or nitrate ions. Some strong and weak acids are listed in Table 8.1.

$$HCl(g) \xrightarrow{\text{water}} \mathbf{H^+ + Cl^-}$$

Hydrochloric acid solution

$$HNO_3(l) \xrightarrow{\text{water}} \mathbf{H^+ + NO_3^-}$$

Nitric acid solution

► Figure 8.12
Strong Acids Are Completely Ionized in Water. *Hydrogen chloride and nitric acid ionize completely in water to furnish a large number of hydrogen ions.*

It is interesting to contrast acetic acid, $HC_2H_3O_2$, the weak acid found in vinegar, with hydrochloric acid, a strong acid found in gastric juice and used in some toilet bowl cleaners and for cleaning masonry. Vinegar contains about 5% acetic acid, which ionizes only a few percent to hydrogen ions and acetate ions, $C_2H_3O_2^-$; see Table 6.1 (polyatomic ions) and 8.5 You Try It in this chapter. We often eat vinegar, with oil and garlic, on our salads. However, to ingest toilet bowl cleaners, which contain 5–8% hydrochloric acid, is *dangerous. Don't try it!* These solutions also can cause severe burns to your skin and eyes, so treat them with respect. (Gastric juice also is a solution of hydrochloric acid, but it is much more dilute, about 0.05%.)

ACIDS AND BASES: OTHER MODELS

The white vapors you see above the open bottles of hydrochloric acid and ammonia solutions in Figure 1 are not "smoke." Hydrogen chloride gas escaping from the bottle of hydrochloric acid is reacting with the ammonia gas from the ammonia bottle to form white, solid ammonium chloride. This compound often constitutes the white film that forms on reagent bottles and windows in chemistry laboratories.

The ammonium chloride that forms in the gas phase,

$$HCl(g) + NH_3(g) \longrightarrow NH_4Cl(s)$$

is a salt, because it contains NH_4^+ and Cl^- ions, and it is the same compound that is produced when a solution of hydrochloric acid is neutralized with an ammonia solution,

$$H^+ + Cl^- + NH_4^+ + OH^- \longrightarrow H_2O + NH_4^+ + Cl^-$$

Recall that a salt is an ionic compound formed from the anion of an acid and the cation of a base in a neutralization reaction. But according to the Arrhenius theory, neutralization is the reaction of H^+ and OH^- to form water. How can this gas-phase reaction be called a neutralization when no H^+ or OH^- ions are present?

These kinds of questions began to puzzle chemists soon after Arrhenius' theory of acids and bases become generally accepted. Many reactions, particularly those in solvents other than water (such as acetone, alcohol, or benzene) and in the gas phase, appeared to exhibit acid–base behavior that did not involve H^+ ions or OH^- ions. Neutralizations seemed to occur because salts were formed, but water was not a product.

Figure 1
Hydrogen chloride gas and ammonia gas react to form the white solid ammonium chloride, which appears as a "vapor" above the bottles.

For example, treatment of solid NaCl with concentrated sulfuric acid (a liquid) produces hydrogen chloride *gas* and solid sodium hydrogen sulfate,

$$H_2SO_4(l) + NaCl(s) \longrightarrow NaHSO_4(s) + HCl(g)$$

Sodium hydrogen sulfate, $NaHSO_4$, is a salt that can be formed by neutralization of a solution of sulfuric acid with sodium hydroxide solution.

A model for this kind of acid–base behavior was proposed independently in 1923 by Johannes Brønsted, a Danish chemist, and Thomas Lowry, an English chemist. In the Brønsted–Lowry model, as it is called, the *transfer* of a proton, H^+, is central to the process. An acid is any species that can donate a proton, and a base is a species that can accept a proton. Neutralization is the transfer of a proton from an acid to a base. Consider the reaction of hydrogen chloride gas with ammonia gas. The hydrogen chloride molecule donates H^+ and the ammonia molecule accepts H^+, producing an ammonium ion and a chloride ion:

transfer of H^+

$$\underset{\text{acid}}{H{-}Cl} + \underset{\text{base}}{NH_3} \longrightarrow [Cl^- \text{ and } NH_4^+] \longrightarrow NH_4Cl(s)$$

Ammonium chloride is the salt formed from proton transfer, which amounts to neutralization.

In the reaction of sulfuric acid with sodium chloride, H_2SO_4 is the proton donor (written HSO_4H for this purpose). The chloride *ion* in NaCl is the proton acceptor and is thus the base (NaCl dissociates in H_2SO_4):

transfer of H^+

$$\underset{\text{acid}}{HSO_4H(l)} + Na^+ + \underset{\text{base}}{Cl^-} \longrightarrow HCl(g) + [Na^+ \text{ and } SO_4H^-]$$

$$\downarrow$$

$$NaHSO_4(s)$$

Aqueous solutions of sodium sulfide, Na_2S, have the unpleasant, rotten-egg smell of H_2S. They are also basic. The Brønsted–Lowry model accounts nicely for what happens here. Like many ionic compounds, sodium sulfide dissociates in water to produce sodium ions and sulfide ions. Subsequently, the sulfide ion hydrolyzes, behaving as a Brønsted–Lowry base that accepts a proton from a water molecule that serves as the acid. The first step in the hydrolysis is essentially irreversible, and the second step involving the hydrogen sulfide ion (which is a weaker base) is an equilibrium. Note that both hydrolysis steps produce hydroxide ions that make the solution basic.

$$Na_2S(s) \xrightarrow{\text{water}} 2Na^+ + S^{2-} \qquad \text{dissociation}$$

ACIDS AND BASES: OTHER MODELS (Continued)

transfer of H^+

$$\underset{\text{acid}}{HOH} + \underset{\text{base}}{S^{2-}} \longrightarrow OH^- + SH^- \qquad \text{hydrolysis}$$

transfer of H^+

$$\underset{\text{acid}}{HOH} + \underset{\text{base}}{SH^-} \rightleftharpoons OH^- + H_2S \qquad \text{hydrolysis}$$

Some of the H_2S escapes the solution as a gas, giving the solution its odor.

Other reactions that appear to be of the acid–base type do not involve a proton transfer, and the Brønsted–Lowry model cannot be used to describe them. For example, gaseous boron trifuoride, BF_3, and solid sodium fluoride react to give solid sodium fluoroborate, $NaBF_4$,

$$NaF(s) + BF_3(g) \longrightarrow NaBF_4(s)$$

This salt can be prepared by neutralization of a solution of fluoroboric acid, HBF_4, with aqueous NaOH solution,

$$\underset{\text{fluoroboric acid solution}}{H^+ + BF_4^-} + Na^+ + OH^- \longrightarrow H_2O + NaBF_4(s)$$

A more general model of acids and bases that explained these observations was proposed by American chemist G. N. Lewis (of Lewis structures) in 1923, the same year that Brønsted and Lowry proposed their model. In the Lewis model, a base is a species that can furnish a *pair of electrons* to form a covalent bond, and an acid is a species that can accept a pair of electrons to form a covalent bond.

Applying this idea to NaF and BF_3, and drawing electron-dot structures of each, we see that the fluoride ion (in NaF) has four pairs of valence electrons, any of which can be used to form a covalent bond. The boron atom in BF_3 is surrounded by only a sextet of electrons: there is a *hole* on the boron atom where another pair of electrons can go to make an octet. Thus, the boron atom is the electron pair acceptor (acid) and the fluoride ion is the electron pair donor (base). When an electron pair from fluoride ion is shared with the boron atom to form the fluoroborate ion, a new B—F bond is formed and the boron atom acquires a complete octet of electrons,

$$\underset{\text{acid}}{BF_3} + \underset{\text{base}}{:\ddot{F}:^-} \longrightarrow BF_4^-$$

acid base new B—F bond

An aluminum compound that was used as an antiperspirant is a hydrate of aluminum chloride, $AlCl_3 \cdot H_2O$, a Lewis acid–base complex formed from aluminum chloride and water. Like the boron atom in BF_3, the aluminum atom in $AlCl_3$ has only a sextet of electrons around it (note the family resemblance) and is thus a Lewis acid. The oxygen atom of the water molecule has two nonbonded pairs of electrons, so water is a Lewis base, and the acid–base reaction is

$$\underset{\text{acid}}{AlCl_3} + \underset{\text{base}}{:\ddot{O}H_2} \longrightarrow Cl_3Al{:}OH_2$$

Al—O Bond formed

Though very effective as an antiperspirant, this compound is no longer used. It is irritating to the underarms and quite destructive of clothing.

Boric acid, H_3BO_3, is a weak acid used in eyewashes. It is not a Brønsted–Lowry acid because it does not ionize in water to furnish H^+ directly. Instead, it furnishes H^+ in solution by behaving as a Lewis acid. Boric acid is often written as $B(OH)_3$ to show its similarity in structure to BF_3. In both, the boron atom has only a sextet of electrons. To complete its octet, the boron atom in boric acid accepts a pair of electrons from a water molecule to form a fourth B—O bond, in the process splitting out H^+ and making the solution acidic.

$$\underset{\text{acid}}{B(OH)_3} + \underset{\text{base}}{:\ddot{O}H_2} \longrightarrow [HO{:}B(OH)_2{:}OH]^- + H^+$$

new B—O bond

Boric acid is a mild antiseptic and once had more than 40 medical uses, most of doubtful effectiveness. Because it is also quite toxic when ingested in small quantities or absorbed through the skin, it should not be used carelessly. Bait made from flour, sugar, a little water, and boric acid makes an excellent poison for roaches and other crawling insects, but it should be kept out of the reach of children and pets.

The acid–base model we use depends on the reaction we want to explain. Although the Arrhenius model accounts for many acid–base reactions in water solution, the Brønsted–Lowry model is more useful for explaining *how* bases hydrolyze in water to furnish OH^- and *how* H_3O^+ ions are produced in water. The Brønsted–Lowry and Lewis models are particularly useful for explaining the reactions of organic molecules (Chapter 11), and they permit us to understand much of the chemistry of other compounds that cannot be explained in other ways. Typically, we use the model that works best to explain the chemistry we want to understand.

THE CARBONATES: SHIFTING THE EQUILIBRIUM

What causes bubbles to form when baking soda is added to sourdough batter or buttermilk biscuit dough? Why does a solution of sodium hydrogen carbonate bubble off CO_2 when acid is mixed with it? Hydrogen carbonate ions in solution hydrolyze to produce a small number of hydroxide ions. These can be neutralized by the hydrogen ions present in the added acid. It might appear that these hydroxide ions would be used up by a small amount of acid, and that addition of more acid would cause the solution to become acidic. However, the hydroxide ions form from hydrogen carbonate ions in a dynamic process, so removal of hydroxide ions by neutralization simply causes more of them to form from the unhydrolyzed hydrogen carbonate ions that remain in solution. As more and more hydrogen ions are added, the formation of "new" hydroxide ions continues as fast as the "old" ones are removed–until all of the hydrogen carbonate ions are used up, no more hydroxide ions can be formed, and neutralization is finally complete. We say that the equilibrium is shifted to the right by addition of the acid (▶ Figure 8.13).

▶ Figure 8.13
Neutralization of a Hydrogen Carbonate Solution with Acid. *Addition of H^+ to a hydrogen carbonate solution in equilibrium removes OH^-, causing the equilibrium to shift to the right and the concentration of H_2CO_3 to increase. The H_2CO_3 is unstable and decomposes to CO_2 gas and water.*

$$\mathbf{HCO_3^- + H_2O} \rightleftharpoons H_2CO_3 + OH^-$$ (undisturbed hydrogen carbonate equilibrium)

$$HCO_3^- + H_2O \rightleftharpoons H_2CO_3 + OH^-$$ (HCO_3^- used up) (hydrogen carbonate equilibrium shifted by addition of acid)

(addition of acid) $$H_2CO_3 + OH^- + H^+ \longrightarrow \mathbf{H_2CO_3 + H_2O}$$ (OH^- from above equation; H^+ added acid)

$$\mathbf{H_2CO_3} \downarrow \mathbf{CO_2(g) + H_2O}$$

Shifting the equilibrium to the right by the irreversible formation of water in the neutralization process simultaneously causes an increase in the concentration of carbonic acid, H_2CO_3, in the solution. Because carbonic acid is only slightly soluble in water, it soon exceeds its solubility limit and is forced to "come out of solution." As it does so, it decomposes into carbon dioxide gas and water, which accounts for the bubbling observed.

Solutions containing the carbonate ion, CO_3^{2-}, behave similarly when neutralized with acid. A solution of sodium carbonate, Na_2CO_3, sometimes called washing soda, is much more basic than a solution of sodium hydrogen carbonate of the same concentration, because the carbonate ion hydrolyzes more completely and furnishes more hydroxide ions in solution.

[Sodium carbonate should not be confused with sodium *hydrogen* carbonate, also called sodium *bicarbonate.* Sodium carbonate is used in dishwasher detergents to produce a strongly alkaline cleaning solution. These solutions are much too *caustic* (alkaline) for hand dishwashing.]

The hydrolysis can be visualized as taking place in two steps, but it should be emphasized that both processes are going on simultaneously and are linked by the amount of HCO_3^- ion present in solutions. The first step is

$$CO_3^{2-} + H_2O \rightleftharpoons \mathbf{HCO_3^-} + \mathbf{OH^-}$$

The equilibrium lies mostly to the right, and the base is fairly strong. The second step is

$$\mathbf{HCO_3^-} + \mathbf{H_2O} \rightleftharpoons H_2CO_3 + OH^-$$

The hydrogen carbonate ion formed in the first step hydrolyzes. Equilibrium lies mostly to the left, and the base is weak.

The chemical species primarily responsible for the increased basicity is the carbonate ion. It hydrolyzes to form a fairly large number of hydroxide ions in solution, along with an equal number of hydrogen carbonate ions. The hydrolysis of the hydrogen carbonate ion in the second step, as we saw before, furnishes very few. Again, all the species in the preceding equations are interconnected by dynamic chemical equilibria, so changing the concentration of one of them changes the concentrations of all of the others. The relationship of carbonate ion to carbonic acid and hydroxide ion can be shown by combining the two hydrolysis steps so that hydrogen carbonate ion, though an intermediate in the reaction, is not shown:

$$CO_3^{2-} + 2H_2O \rightleftharpoons H_2CO_3 + 2OH^-$$

Addition of acid in the form of hydrogen ions, then, will neutralize the hydroxide ions formed in both hydrolysis steps, and both equilibria will be shifted to the right. This means that if enough acid is added, the concentration of H_2CO_3 will again build up sufficiently to cause the evolution of CO_2 from the solution (► Figure 8.14).

$$CO_3^{2-} + 2H_2O \rightleftharpoons H_2CO_3 + 2OH^-$$ (equilibrium forced to right by addition of acid)

$$H_2CO_3 + 2OH^- + 2H^+ \longrightarrow \mathbf{H_2CO_3} + \mathbf{2H_2O}$$

(from above equation) (added acid)

$$\downarrow$$

$$\mathbf{CO_2}(g) + \mathbf{H_2O}$$

► **Figure 8.14**
Neutralization of a Carbonate Solution with Acid. *Hydroxide ion from hydrolysis of carbonate ion is neutralized by H^+, and the concentration of carbonic acid builds up to the saturation point. Carbon dioxide is evolved.*

▶ Figure 8.15
The Reaction of Carbonates with Strong Acids. *Calcium carbonate is insoluble in water. When it reacts with hydrochloric acid, bubbles of carbon dioxide gas escape and the solid dissolves. This reaction is typical of insoluble carbonates.*

This reaction causes the fizzing when a drop of acid is placed on a piece of limestone ($CaCO_3$), or when on antacid that contains calcium carbonate comes in contact with stomach acid (▶ Figure 8.15). Although calcium carbonate is not very soluble in water, it can furnish enough hydroxide ions to react with hydrogen ion, driving the equilibrium in Figure 8.14 to the right. Overall, this (net ionic) reaction can be written.

$$CaCO_3(s) + 2H^+ \longrightarrow Ca^{2+} + H_2CO_3$$
$$\downarrow$$
$$CO_2(g) + H_2O$$

Note that the solid calcium carbonate is converted to an ionic form, which is soluble in water solution. This is why carbonate salts that are insoluble in water will dissolve when they react with acid. Hard-water deposits in sinks, bath tubs, and toilet bowls are composed of insoluble calcium and magnesium carbonates. Many household preparations used to dissolve them contain strong solutions of hydrochloric or phosphoric acid. These should be used with care.

Removing a product shifts the position of equilibrium to the right.

An equilibrium is shifted to the left when a reactant is removed.

When carbonate and hydrogen carbonate solutions are neutralized with acid, the position of equilibrium is shifted to the right by removal of the hydroxide ion. Note that the hydroxide ion appears on the right-hand (product) side of each of the equilibrium equations. The position of equilibrium also can be shifted to the left by removal of a chemical species that appears on the left-hand (reactant) side of the equation. For example, addition of calcium ion (say in calcium chloride solution) will remove the carbonate ion from solution by precipitating it as insoluble calcium carbonate. (The equations in ▶ Figure 8.16 appear to be written backward, but we want to show that the equilibria have moved to the *left* with respect to the processes we already have seen.)

$$H_2O + \mathbf{CaCO_3(s)} + 2Cl^- \longleftarrow Ca^{2+} + 2Cl^- + CO_3^{2-} + H_2O$$

Insoluble — Ions added

$$CO_3^{2-} + H_2O \leftrightarrows HCO_3^- + OH^-$$

(⟵ to above equation)

$$HCO_3^- + H_2O \leftrightarrows H_2CO_3 + OH^-$$
$$\uparrow$$
$$CO_2 + H_2O$$

▶ Figure 8.16
An Equilibrium Is Shifted to the Left by Removal of a Reactant Species. *Addition of calcium ion to a carbonate–bicarbonate system forces equilibria to the left by removing the carbonate ion from solution. Ionic species CO_3^{2-}, HCO_3^-, and OH^-, and molecular CO_2 remain in very low concentrations.*

LE CHÂTELIER'S PRINCIPLE

As a general rule, removal of a product from a reaction in chemical equilibrium will cause more product to be formed at the expense of the reactants.

Likewise, removal of a reactant will cause more reactant to form at the expense of products. This is a manifestation of **Le Châtelier's principle,** which states that when a system in dynamic equilibrium is subjected to some kind of stress, the position of equilibrium will shift to eliminate that stress. In the preceding examples, removal of reactant or product constitutes this added stress.

Another way of stressing the equilibrium is to add extra reactant or product to the reaction mixture. Adding extra reactant causes the formation of more product, and adding extra product forces the equilibrium back to the left, increasing the amount of reactant (▶ Figure 8.17).

Adding more product also moves an equilibrium to the left.

$$CO_3^{2-} + H_2O \rightleftharpoons \mathbf{HCO_3^- + OH^-}$$

$$\mathbf{HCO_3^- + H_2O} \rightleftharpoons H_2CO_3 + OH^-$$

$$\uparrow\downarrow$$

$$CO_2 + H_2O$$

Predominant species in ordinary $NaHCO_3$ solution.

$$\mathbf{CO_3^{2-} + H_2O} \leftrightarrows HCO_3^- + OH^- \quad \text{(extra } OH^- \text{ added)}$$

$$HCO_3^- + H_2O \leftrightarrows H_2CO_3 + OH^-$$

$$\uparrow$$

$$CO_2 + H_2O$$

Predominant species in basic solution

▶ Figure 8.17
Adding Product to a Reaction in Equilibrium Forces it in the Direction of the Reactants. *Addition of OH^- (from NaOH solution) to bicarbonate solution drives equilibrium to the left, forming mostly carbonate ion in solution.*

The extra carbon dioxide placed in the atmosphere when we burn fossil fuels is suspected of contributing to the greenhouse effect and possible global warming (Chapter 13). The oceans help counteract this process by serving as a "sink" for carbon dioxide from the atmosphere. Carbon dioxide is more soluble in seawater, which is mildly basic, than in ordinary terrestrial waters, which are usually slightly acidic. Carbon dioxide dissolves in seawater to form carbonic acid, which is neutralized by the hydroxide ions to form water and hydrogen carbonate ion. The latter reacts with more hydroxide ions to form carbonate ion. Because CO_2 is used up by this process, more of it can dissolve, which removes it from the atmosphere. The equations in Figure 8.17 can be rewritten to summarize what happens:

$$CO_2 + H_2O \rightleftharpoons H_2CO_3 \quad \text{(More } CO_2 \text{ dissolves as } H_2CO_3 \text{ reacts with } OH^-\text{)}$$

$$H_2CO_3 + OH^- \longrightarrow H_2O + HCO_3^-$$

$$HCO_3^- + OH^- \longrightarrow H_2O + CO_3^{2-}$$

Seawater contains dissolved calcium ions, Ca^{2+}, which can react with the carbonate ions produced to form a precipitate of calcium carbonate,

$$Ca^{2+} + CO_3^{2-} \longrightarrow CaCO_3(s)$$

Clams, corals, and oysters use the calcium and carbonate ions in the water to form their shells, which also eventually fall to the ocean floor. Over millions of years, these deposits become limestone.

THE pH ACIDITY SCALE

The **pH acidity scale,** which is based on measurement of the concentration of hydrogen ions in solution, gives us a quantitative way of talking about acidity and basicity. We shall not be concerned here with how pH is measured or calculated, but rather with what it means. For example, knowing that the pH

ACID DEPOSITION: ACID RAIN

When sulfur-containing fossil fuels (primarily coal) are burned to generate electricity and other kinds of energy, the sulfur in these fuels is converted to SO_2 and SO_3. Metal smelters, such as those in which copper, zinc, or lead ores are roasted, also emit sulfur oxides. If these sulfur oxides are not removed from the gaseous effluent, they combine with water in the atmosphere to eventually produce sulfuric acid, H_2SO_4. In addition, various oxides of nitrogen (often referred to as NO_x) are formed in automobile engines and other high-temperature combustion processes. They also react with atmospheric moisture, ultimately to form nitric acid, HNO_3.

These airborne acids find their way into the weather cycles and eventually return to the ground with the rain. Natural rain water is not neutral but has a pH of about 5.6 because it contains dissolved CO_2. Sulfuric and nitric acids are strong acids, and their presence dramatically increases this acidity. The average pH of rainfall in parts of the northeastern United States, neighboring Canada, and northern Europe is now about 4.2. Some values below 2.0 have been recorded, especially in clouds and fog, which tend to concentrate atmospheric acids.

Unfortunately, the weather does not respect boundaries between states or between countries, and effluent from one region often becomes the problem of another. And it is a serious problem. Prevailing weather patterns sweep these "acid precursors" from industries in western Europe northward into Sweden and Norway, and from the midwestern United States across the northeast into Canada. Acid deposition has killed or is killing forests in Germany, Sweden, Norway, Michigan, and the Adirondacks in New York. It also has lowered the pH of numerous lakes in these regions to the point that fish can no longer live.

Acid rain is also corrosive and accelerates rusting and deterioration of metal bridges, machinery, and automobiles. It dissolves metals like copper, lead, and cadmium from particles in industrial smoke and transports their toxic ions to the ground where they contaminate groundwater. It also removes useful plant nutrients from the soil. Throughout Europe and in the eastern United States, acid rain is rapidly destroying

of a lake is now 5.8 and is decreasing tells us that the critters living in it are in big trouble. And if our stomach contents hit a pH of 1.5, it means we'll probably be needing a bottle of antacid soon.

Like all scales that are relative, the pH scale must have a point of reference. Because water is formed in the process of neutralization and is itself neutral, we shouldn't expect it to have any hydrogen ions in it. This means that if we were to measure the concentration of hydrogen ions in pure water, it should be zero, providing us a convenient place to start.

When we actually measure the concentration of H^+ in pure water, however, we find that it is not zero. To be sure, it is very small, about two hydrogen ions per billion water molecules, but it is not zero. Furthermore, we find

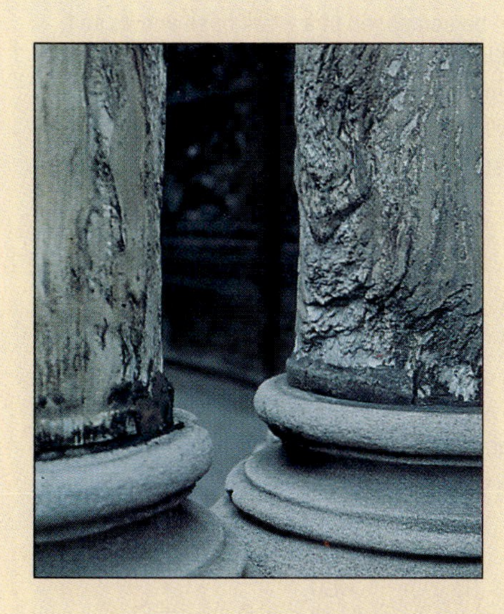

Figure 1
Marble Columns Damaged by Acid Rain.

fine statuary and buildings made of limestone and marble, which are calcium carbonate, $CaCO_3$ (Figure 1). The calcium carbonate is converted into calcium nitrate, which dissolves in water, and into calcium sulfate, which crumbles. Acid fog or smog also causes respiratory distress and other health problems in humans. It is estimated that the damage acid rain causes in the United States alone costs about $5 billion each year.

Not all acid deposition is caused by human activity, however. About 10% of the sulfur in the atmosphere is placed there by natural processes, such as volcanic activity, mineral hot springs, and bacterial activity in soil and water. For example, Mt. Kilauea in Hawaii emits 200 to 300 tons of SO_2 per day, and the 1982 eruption of El Chichón in Mexico blew 300 million tons of sulfur dioxide into the stratosphere. Nitrogen oxides are formed when lighting causes nitrogen and oxygen in the air to combine; and ammonia, NH_3, and amines, RNH_2, enter the atmosphere from bacterial processes on land.

BUFFERED SOLUTIONS

Human blood has a pH of 7.4, a value that usually varies only about 0.1 unit in healthy persons. The body cannot tolerate even very small changes in pH for very long (Figure 1). Below a pH of 7.3 the blood cannot remove carbon dioxide (produced by metabolism) from the cells, and above a pH of 7.7 the blood cannot give up this carbon dioxide to the lungs to be exhaled. A shift in blood pH of about 0.6 unit can prove fatal in just a few seconds.

To maintain this nearly constant pH in the blood the body relies on the chemistry of buffers, solutions that resist changes in the concentration of hydrogen ions. Buffer solutions contain relatively high concentrations of a weak acid and its salt, in roughly equal proportions. An example is a solution of acetic acid, $HC_2H_3O_2$, and sodium acetate, $NaC_2H_3O_2$. Buffer solutions also can be composed of a weak base and its salt, such as ammonia, NH_3, and ammonium chloride, NH_4Cl.

One of several buffer systems present in the blood, the carbonic acid/hydrogen carbonate system serves to illustrate how buffers work. Carbonic acid, formed from dissolved carbon dioxide in the blood, exists in equilibrium with hydrogen ion and hydrogen carbonate ion,

$$\underset{\text{In relatively large supply}}{H_2CO_3} \rightleftharpoons \underset{\text{In low concentration}}{H^+} + \underset{\text{In relatively large supply}}{HCO_3^-}$$

Addition of acid furnishes hydrogen ions, which react with the hydrogen carbonate ions. This uses up HCO_3^- and forms carbonic acid, forcing the equilibrium to the left according to Le Châtelier's principle. However, the concentration of H^+ itself does not change very much;

$$\underset{\text{Increases slightly}}{H_2CO_3} \rightleftharpoons \underset{\text{Concentration changes little}}{H^+} + \underset{\text{Decreases slightly}}{HCO_3^-}$$

The position of equilibrium is shifted slightly to the left by addition of H^+.

Thus, the pH changes only slightly. Similarly, addition of hydroxide ion to the solution neutralizes the hydrogen ions present, forming water and removing the ions from solution. This time, the equilibrium shifts to the right, and the H_2CO_3 ionizes to form more H^+. Again, the concentration of H^+ changes very little, and the pH remains essentially the same.

$$H_2CO_3 \rightleftharpoons H^+ + HCO_3^-$$

Amount changes little

Decreases slightly (ionizes) — Increases slightly

Addition of OH^- removes H^+ (to form water), causing more H_2CO_3 to ionize and replace the H^+ lost. The amount of H^+ changes little.

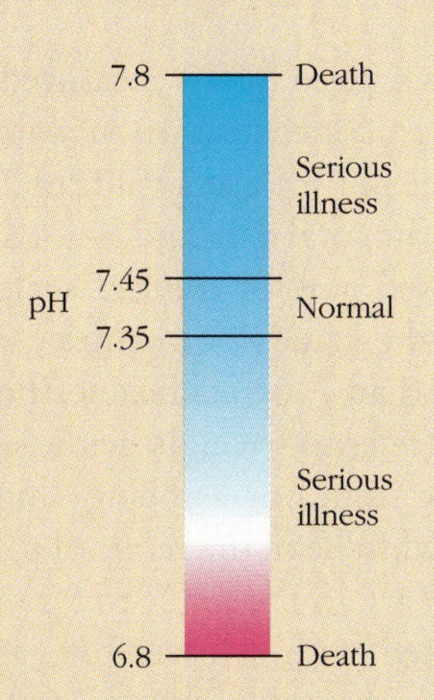

Figure 1
The pH of human blood is maintained within a very narrow range by buffers.

Another buffer system present in the blood uses the dihydrogen phosphate/hydrogen phosphate equilibrium:

$$H_2PO_4^- \rightleftharpoons H^+ + HPO_4^{2-}$$

Dihydrogen phosphate ion — Hydrogen phosphate ion

If acid is added, the H^+ reacts with the HPO_4^{2-} ion, forcing the equilibrium to the left. On the other hand, addition of OH^- uses up H^+ and causes the equilibrium to shift to the right. More $H_2PO_4^-$ ionizes to replace the H^+ lost. Again, the concentration of hydrogen ion changes little in either case.

Buffers are important in other applications. Saliva contains a hydrogen carbonate buffer, along with another system using $—CO_2H$ and $—NH_2$ functional groups, to maintain the pH at a value of about 6.8, almost neutral. Below a pH of 6.0, demineralization of tooth enamel becomes rapid, leading to formation of dental caries (cavities). Seawater is a buffered solution (again a carbonate buffer) with a pH of about 7.8 to 8.3. Salts such as potassium acid tartrate, $KHC_4H_4O_6$, sodium potassium tartrate, $NaKC_4H_4O_6$, and sodium citrate, $NaH_2C_6H_5O_7$, are added to foods to bring their pH to a desired value and maintain it.

that an equal number of hydroxide ions accompany the hydrogen ions, due to the *autoionization* of water.

$$H_2O \rightleftharpoons H^+ + OH^-$$

An acidic solution has more H^+ than OH^-.

This explains why pure water is neither tart nor bitter: it has equal numbers of hydrogen ions and hydroxide ions. A solution that is acidic has an excess of hydrogen ions compared with the number of hydroxide ions it contains, and one that is basic has a surplus of hydroxide ions compared with the number of hydrogen ions present.

A solution with a pH of 7 is neutral.

Because of the way the concentration of hydrogen ions is measured (by the chemical mole), and because the hydrogen ion concentration in water is not zero anyway, it is more convenient to start the pH scale at a value of 7 instead of zero. This number is calculated from experimental data and is not an actual concentration unit. The most commonly used part of the pH scale ranges from a value of 1 (sometimes 0) to 14, with the value 7, or neutral, sitting in the middle. Solutions with pH values lower than 7 are acidic and those with values higher than 7 are basic. Sometimes we refer to strongly acidic solutions as being of "low pH" and to strongly basic solutions as being "high pH." Weakly acidic solutions usually are considered to lie in the range of pH 5 to 7, and those that are weakly basic (or alkaline) are in the pH 7 to 9 range. Regardless, a higher acidity is always associated with a lower pH, and conversely, a lower acidity is associated with a higher pH. The pH scale, along with the pH of some common substances, is shown in ▶ Figure 8.18.

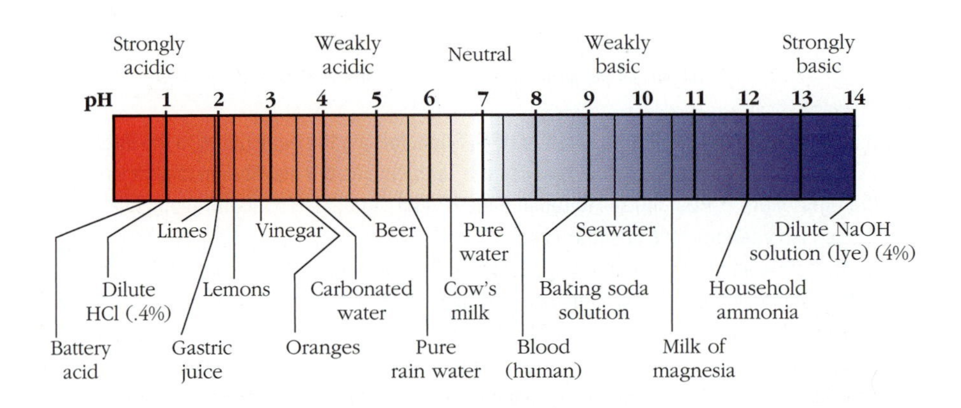

▶ Figure 8.18
Approximate pH Values for Some Common Substances.

A change of one pH unit represents a *tenfold* change in acidity or basicity.

Another feature of the pH scale is that it is not linear, but rather is *logarithmic.* This means that each incremental pH value represents a tenfold increase or decrease in acidity or basicity. Thus, a solution with a pH of 6 has a concentration of hydrogen ions ten times that of one with a pH of 7. A vinegar solution of pH 3 is 100,000 times more acidic than egg white with a pH of 8 (the difference is 5 pH units, or 10^5 in concentration). Similarly, a solution of household ammonia having a pH of 12 is about 1000 times more basic than seawater, which has a pH of about 9 (3 pH units difference, or 10^3 in concentration). Some people can eat lemons with pleasure, whereas others can barely handle the acidity of oranges. Lemon eaters are tough. Note in Figure 8.18 that lemons are over ten times more acidic than oranges.

Because each pH increment represents a factor of 10 in acidity or basicity, even a change of a few tenths of a pH unit is not trivial. A change of this magnitude can indicate that a doubling or tripling (or corresponding decrease) in acidity has occurred. Limes are three to four times more acidic than lemons. The normal pH of human blood lies in the range of 7.3 to 7.5. If an individual's blood pH is found to be very far from this, that person is apt to be severely ill.

The pH of water in most inland lakes is about 6.5 to 7.5. In parts of Europe and the United States, this value has been dropping dramatically in certain lakes exposed to acid rain, often to well below a pH of 6. Some of these lakes already are biologically dead, and in many others the fish populations have seriously declined. A few tenths of a pH unit may not seem significant, but you should be aware that to many delicate biological systems a change of that magnitude can spell disaster.

DID YOU LEARN THIS?

For each of the statements given, fill in the blank with the most appropriate word or phrase from the following list. You may use any word or phrase more than once.

10^2	5	neutralization
10^3	equilibrium	acidic
10^4	salt	basic
10^5	spectator ion	neutral
2	acid	ionized
3	base	litmus
4		

a. _______ The condition of HBr(*g*) in water solution
b. _______ A dynamic process in which reactants are converted to products and products are converted to reactants at the same rate
c. _______ A chemical compound that is formed from the reaction between an acid and a base
d. _______ Having equal numbers of hydrogen ions and hydroxide ions
e. _______ The numerical difference between a pH of 4.0 and a pH of 7.0
f. _______ Ions that maintain charge balance but do not participate in a chemical reaction
g. _______ A substance that turns red litmus blue
h. _______ The reaction of an acid with a base to form water and heat
i. _______ Having an excess of hydroxide ions

EXERCISES

1. What is the chemical species that produces acidic properties in water (aqueous) solution? What species produces basic properties in water solution? What color does each of these species produce when it reacts with litmus dye?
2. Sulfuric acid, H_2SO_4, is a strong acid. Write the balanced equation for its ionization in water, producing just one hydrogen ion.
3. Phosphoric acid, H_3PO_4, is a moderately strong acid. That is, it ionizes in water to produce a large number of hydrogen ions but it does not ionize completely. Write a reasonable equation for this ionization, producing just one hydrogen ion.

4. Both ammonia and the hydrogen carbonate ion react with water to form a basic solution. Write an equation for each of these processes. What do the double arrows between reactants and products mean?
5. In water solution, what is neutralization? What does it always produce? What are spectator ions? Giving a definition based on neutralization, what is a salt? The term *salt* is used to describe a certain class of compounds. What are they?
6. Write equations for the following neutralization reactions in water solution. In each indicate which chemical species is the acid, which is the base, and which are spectator ions. For example, a reaction in which a solution of HI is neutralized by a solution of KOH can be written (remembering that we are dealing with *ions in solution)* as

$$H^+ + I^- + K^+ + OH^- \longrightarrow H_2O + K^+ + I^-$$

where H^+ is the acid, OH^- is the base, and K^+ and I^- are spectator ions.

   a. A solution of HCl is neutralized with a solution of LiOH.
   b. A solution of HNO_3 is neutralized with a solution of NaOH.
   c. A solution of HBr is neutralized with solid $Ca(OH)_2$.
   d. A solution of $HC_2H_3O_2$ is neutralized with a solution of KOH.
   e. Solid calcium carbonate is neutralized by a solution of hydrochloric acid.
   f. The lactic acid ($HC_3H_5O_3$) in sour milk is neutralized by a solution of baking soda.

7. Identify the salts formed in each reaction in exercise 6.
8. What is a reversible reaction? Describe what chemical equilibrium is in terms of the process that is occurring. What happens to the amounts of reactants and products when a reaction is at equilibrium? What can you say about the rates of the forward and reverse reactions at equilibrium? List ways to shift the position of an equilibrium to the right or to the left.
9. Write the equation for the deposition of calcium carbonate from "temporarily hard" water in a hot spring. What drives the position of equilibrium to shift in this instance? How does water become temporarily hard in the first place? Write an equation to show this reaction.

10. What causes solid insoluble carbonates, like $CaCO_3$ and $MgCO_3$, to dissolve when they react with acids? Write an equation using $MgCO_3$ to show how this happens. What kind of chemical species that are soluble in water does the insoluble magnesium carbonate form?
11. There are always hydrogen ions and hydroxide ions in water. What is the relationship between these two ions when a water solution is neutral? Acidic? Basic?
12. Why is the pH acidity scale used? What value on the pH scale is neutral? A solution of boric acid has a pH of 6.8. Is this solution strongly acidic, weakly acidic, neutral, weakly basic, or strongly basic? Answer this question for

    a. a solution of baking soda (pH = 9.0)
    b. a solution of lye (pH = 13.5)
    c. blood (pH = 7.3)
    d. lime juice (pH = 2.2)
    e. milk (pH = 6.5).

13. How much change in acidity does one increment on the pH scale represent? How much more or less acidic is

    a. a solution having a pH of 3.0 compared with a solution having a pH of 5.0?
    b. a solution having a pH of 13.0 compared with one having a pH of 9.0?
    c. a solution having a pH of 4.0 compared with one having a pH of 10.0?

    (This exercise is easy if you express your answers in scientific notation.)

THINKING IT THROUGH

Alkaline Soils

In parts of the western United States, soils are quite alkaline, rather than acidic (sour), and sometimes it is desirable to reduce this alkalinity to make crops grow better. Ammonium nitrate, NH_4NO_3, is a good fertilizer that also can be used to neutralize the basicity of these soils. The ammonium ion is a weak acid when dissolved in water. Write net ionic equations to show (a) how the ammonium ion produces hydrogen ions in solution and (b) how the neutralization of the soils might occur. Would the nitrate ion have any effect on the acidity of the soil? Why or why not?

Bubbles in Lime Water

When you bubble your breath through lime water (using a straw, for example), a white precipitate forms. Lime water is a dilute solution of calcium hydroxide. Calcium hydroxide is not very soluble in water, but a small amount of it will dissociate. Write an equation for this dissociation process. What do you suppose is the precipitate that forms when you bubble your breath through the lime water? What is in your breath that could cause it to form? Write equations to show what happens.

ANSWERS TO "DID YOU LEARN THIS?"

a. ionized
b. equilibrium
c. salt
d. neutral
e. 10^3
f. spectator ions
g. base
h. neutralization
i. basic

A BREWER USES CHEMISTRY

William R. Jenkins

Package Production Manager, Pike Place Brewery

B.A. Chemistry, University of Southern Maine

Working in a microbrewery, I am constantly reminded that chemistry is an inseparable part of the brewing day. From the inspection and analysis of barley and hops when they arrive at the brewery to the finished beer, chemistry and the chemical method must be ever-present to ensure that the beer is as good as it can be. Whereas a larger brewery can usually compensate for undesirable qualities in its product by using flavor additives or stabilizers, microbreweries must rely on quality raw ingredients and a solid understanding of chemistry to ensure that their beer can compete in the marketplace.

Water, of course, is the main ingredient in beers, yet its chemistry is often overlooked as an influence on the finished product. Different styles of beer benefit from the different balances of minerals in the water from which they are brewed. The pale ales of England, for example, could never have achieved their style or popularity if the water in the areas surrounding London had not had an extremely complex mineral composition. Calcium, though having little direct effect on flavor, is important in that it precipitates proteins, clarifying the beer, and helps maintain proper pH balance. Magnesium is a vital cofactor in several enzymatic reactions. And the sulfate complement to these cations imparts a dryness to the beer and accentuates the bitterness of the hops. In contrast to this very hard water is the water from which the classic pilsner style, native to central Europe, is brewed. This beer, with its delicate bitterness and pronounced malt character benefits from water that is relatively free of dissolved minerals.

Many municipal water supplies are chemically treated to meet quality standards for drinking water, and the results are felt in the brewing industry. Heavily chlorinated water produces medicinal flavors and solventlike aromas, and strongly alkaline water may inhibit essential steps in the fermentation cycle. The Seattle area owes a large part of its reputation for quality beers to the fact that its geography and climate minimize the need for most treatment methods. Many breweries adjust the mineral composition of their water to brew a variety of beer styles, and a solid foundation in inorganic chemistry is useful in making such adjustments.

A thorough understanding of biochemistry is also important to the brewer. The production of beer using only barley, hops, water, and yeast is dependent on microbial and enzymatic reactions, procedures far more complex than the simple dissolution of mineral salts in water. Once the brewery receives the malted barley, the mashing procedure begins, which converts the starches in the grain to fermentable sugars. The conditions of the mash greatly influence the character of the beer. Lower mash temperatures, around 140°F (60°C), and a pH near 5.0 result in a more fermentable wort (the term given to beer prior to fermentation) and a drier taste, with correspondingly less body and a decreased malt character. Higher temperatures in the mash kettle, around 158°F (70°C), and a pH closer to 5.7 produce a wort with more complex sugars that are harder for the yeast to ferment. The result is a heavier-bodied beer with a sweeter, more malty taste on the palate. By understanding the chemical processes in the mash tun, the brewer is able to custom craft the beer by the application of chemical principles.

The next step, boiling the wort, also benefits from an understanding of chemistry. Boiling denatures many organic molecules, especially proteins, and causes them to precipitate, clarifying the beer. The hops, the other principal flavoring ingredient, are added during the boil. Because the essential hop resins are only partially soluble in water, a vigorous and lengthy boiling period is required to extract the full flavor of the hops. The heat and agitation produced in the kettle chemically alter the bittering oils and allow them to be assimilated into the beer. Successful fermentation is also dependent on chemical knowledge. Yeast cells need certain narrow ranges of temperature, pH, and nutrient levels to produce the qualities we expect in a fine beer. The presence of amino acids, lipids, and phosphates aids in the growth cycle of yeast, and in the early stages of fermentation the wort must also contain a sufficient level of dissolved oxygen.

Fermentation involves more than the simple conversion of sugars to alcohol and carbon dioxide. Other by-products, though produced in much smaller amounts, greatly influence the flavor of the finished beer; hence, the brewer can carefully craft the beer by manipulating the fermentation conditions. One of the most notorious of these by-products is 2,3-butanedione, or diacetyl, which is produced during the early phase of fermentation in the presence of oxygen and is reduced to ethanol later in the cycle. Diacetyl imparts a strong butterscotch flavor and has a very low flavor threshold, below 0.1 ppm, so its presence can be detected even in very small amounts. Warmer temperatures toward the end of the fermentation cycle aid in the reduction of diacetyl, giving brewers a simple means of controlling its effect.

Fusol oils, alcohols of higher molecular weight, are produced during fermentation. Fusol oils can result in off-flavors or may combine with various organic acids to form esters, which are responsible for unmistakable fruity aromas. Ester formation, however, like many of the reactions producing off-flavors or aromas, can easily be slowed or eliminated by changing conditions under which they are produced. In the case of unwanted esters, simply ensuring that the wort contains a sufficient level of dissolved oxygen prior to fermentation greatly inhibits their production.

Thus, aided by an appreciation of various chemical principles and an understanding of how they relate to the various steps in the brewing process, the brewer is able to fine-tune the beer-making procedure. The brewer's work serves as a reminder that the world of chemistry is not limited to the laboratory or lecture hall—it even has a place beside you at the next barstool.

9

Oxidation and Reduction

What surprised me more than I can well express, was that a candle burned in this air [pure oxygen] with a remarkably brilliant flame.

Joseph Priestley

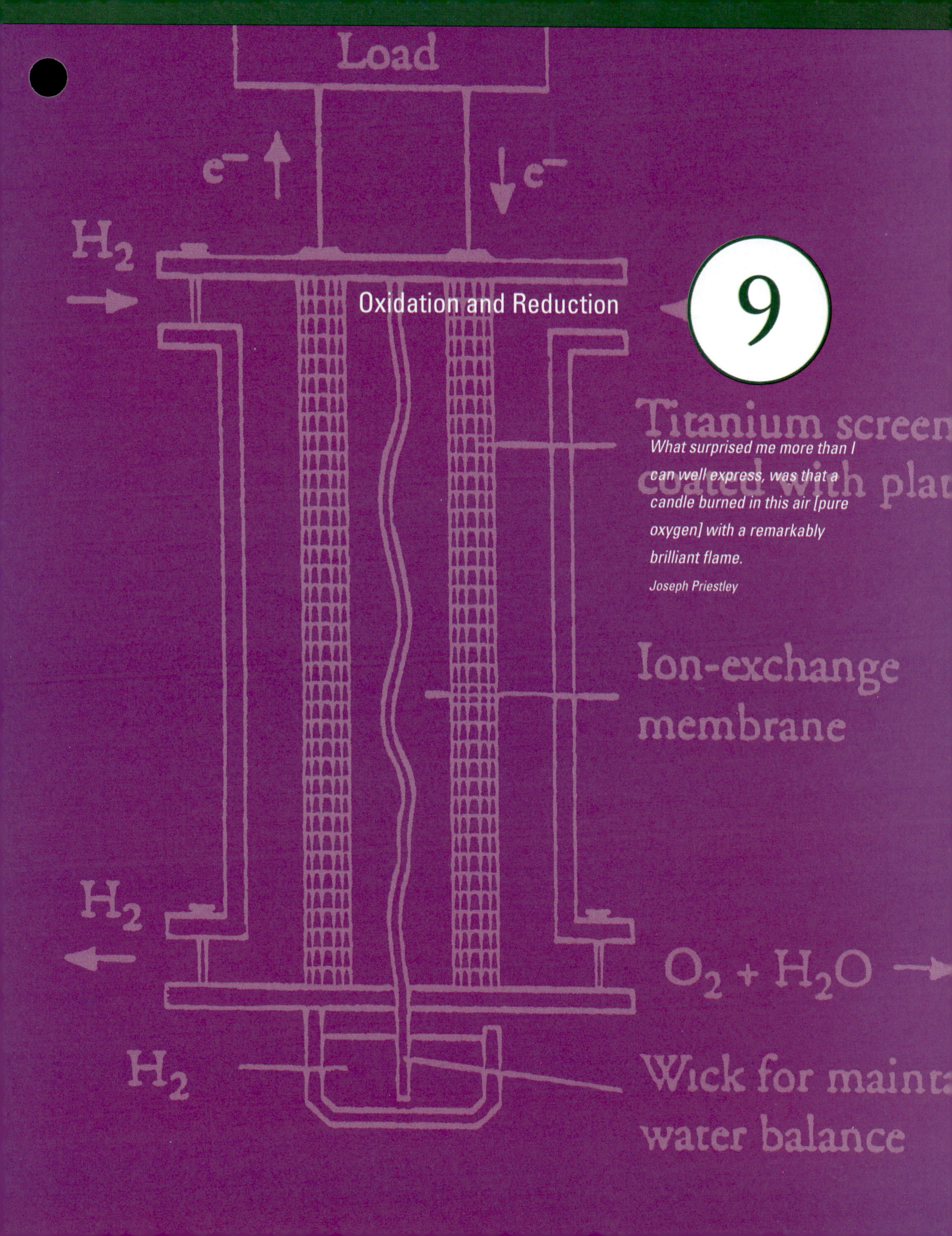

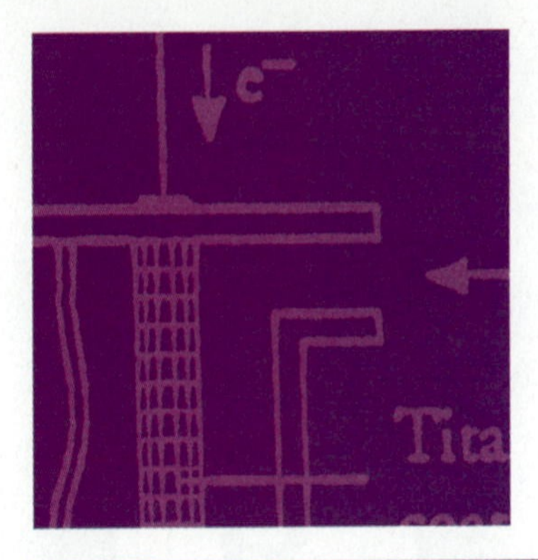

Unprotected iron rusts in the presence of moist air, and a copper wire turns blue-green with corrosion when it comes into contact with battery acid. Batteries start our cars and power innumerable tools, appliances, and gadgets, and fuel cells provide electrical power to operate manned space vehicles. The combustion of natural gas, petroleum, and coal furnishes most of the energy we use in this country today. Our bodies "burn" the food we eat to provide the energy we need to live.

Although it is difficult to "unburn" coal or gasoline, many of these processes can be reversed. For example, many kinds of batteries can be "recharged," and metal ores are routinely smelted and refined to yield the pure metals. In sunlight, plants convert carbon dioxide and water into carbohydrates, using the process of photosynthesis.

OXIDATION

Oxidation can involve the gain of oxygen through a chemical reaction.

The *combustion* of gasoline involves its combination with oxygen from the air, as does the *rusting* of iron (▶ Figure 9.1). This process is called **oxidation.** Originally, it was thought that oxidation is simply the combination of elements or compounds with oxygen, and one way to determine if something is oxidized is to see whether it has gained oxygen in a chemical reaction. For example, iron rusts according to the following equation:

$$4Fe + 3O_2 \xrightarrow{\text{water}} 2Fe_2O_3$$

In this example the iron gains oxygen, so it is oxidized. Likewise, when phosphorus burns in air, it gains oxygen and is oxidized to tetraphosphorus decaoxide,

$$P_4 + 5O_2 \longrightarrow P_4O_{10}$$

▶ Figure 9.1
Rusted Automobiles. *Every year about 20% of the iron and steel produced in the United States is used to replace things that have been ruined by the oxidation of iron to form rust.*

Later, chemists realized that oxidation processes did not necessarily require free oxygen. For example, sulfur deposits near the vents of volcanoes and some hot springs are formed by the reaction of the hydrogen sulfide and sulfur dioxide gases emitted there:

Oxidation can involve the loss of hydrogen atoms from a chemical species.

$$2H_2S(g) + SO_2(g) \longrightarrow 3S(s) + 2H_2O(l)$$

Note that the H_2S loses hydrogen in this process. This is a second way to determine whether something is oxidized: see if it loses hydrogen in a chemical reaction. In this case, H_2S is oxidized to elemental sulfur by SO_2 (▶ Figure 9.2).

9.1 Hydrazine, N_2H_4, is used as a rocket fuel and burns vigorously in pure oxygen to generate a considerable amount of heat:

$$N_2H_4(l) + O_2(g) \longrightarrow N_2(g) + 2H_2O(g) + \text{heat}$$

What chemical species is oxidized? How did you decide?

▶ Figure 9.2
Sulfur Deposits in a Volcanic Crater, Mt. Papandayan, Java. *A redox reaction between the gaseous* H_2S *and* SO_2 *emitted from the vent forms the sulfur:*

$$2H_2S(g) + SO_2(g) \longrightarrow 3S(s) + 2H_2O(l)$$

Sulfur and chlorine react with iron much like oxygen does:

$$2Fe + 3S \xrightarrow{\text{heat}} Fe_2S_3$$

and

$$2Fe + 3Cl_2 \longrightarrow 2FeCl_3$$

In Fe_2S_3 and $FeCl_3$, the iron exists as the Fe^{3+} ion, just like it does in Fe_2O_3, so each reaction has produced iron atoms that have lost three electrons,

$$\cdot\ddot{Fe} \longrightarrow Fe^{3+} + 3e^-$$

(These are binary compounds of a metal and a nonmetal: the oxide and sulfide ions have a 2– charge, and the chloride ion has a 1– charge.) This is typical behavior of metals—namely, they lose their valence electrons to form positive ions. This loss of electrons also is called oxidation. *Oxidation,* now broadly defined, is the loss of one or more electrons from a chemical species.

Oxidation is the loss of electrons from a chemical species.

Many substances, besides metals, undergo oxidation. For example, iodide ion in solution (say, from KI) is easily oxidized to free iodine by the action of chlorine (K^+ is the spectator ion):

$$\underset{\text{colorless}}{2I^-} + \underset{\text{pale yellow-green}}{Cl_2} \longrightarrow \underset{\text{purple}}{I_2} + \underset{\text{colorless}}{2Cl^-}$$

In this case each iodide ion, with eight valence electrons, loses one electron to form an iodine atom with seven valence electrons. Two iodine atoms then combine to form an iodine molecule, sharing a pair of electrons in a covalent bond between them:

$$:\ddot{\underset{..}{I}}:^{-} \longrightarrow :\ddot{\underset{..}{I}}\cdot + e^{-}$$

$$:\ddot{\underset{..}{I}}\cdot + \cdot\ddot{\underset{..}{I}}: \longrightarrow :\ddot{\underset{..}{I}}:\ddot{\underset{..}{I}}:$$

9.2 In each of the following reactions, which substance is oxidized? And which of the preceding methods did you use to determine your answer?

a. $CH_4 + 2O_2 \longrightarrow CO_2 + 2H_2O$
b. $Zn + S \longrightarrow ZnS$
c. $2CH_3CH_2OH + O_2 \xrightarrow{\text{Cu, heat}} 2CH_3CHO + 2H_2O$
d. $Cu + 4H^+ + SO_4^{2-} \longrightarrow Cu^{2+} + SO_2 + 2H_2O$
e. $Si + 2FeO \longrightarrow 2Fe + SiO_2$
f. $2F_2 + 2H_2O \longrightarrow 4HF + O_2$

A popular summer pastime is grilling food over charcoal. Assuming that the charcoal is pure carbon and that there is plenty of air to burn it, we can write its oxidation as:

$$C + O_2 \longrightarrow CO_2$$

If insufficient air is present for complete combustion, carbon monoxide gas, CO, is formed:

$$2C + O_2 \longrightarrow 2CO$$

Note that carbon dioxide is more oxidized than carbon monoxide, since it contains twice as many oxygen atoms per carbon atom. Carbon monoxide can be oxidized to carbon dioxide with the liberation of heat,

$$2CO + O_2 \longrightarrow 2CO_2 + \text{heat}$$

Because electrons are shared in the covalent bonds formed between two nonmetals, like carbon and oxygen, it becomes difficult to see how a nonmetal could have lost electrons when it was oxidized. To simplify matters, we shall consider only those oxidations of nonmetals (except fluorine) that involve oxygen, using the convention that the nonmetal always gives up its electrons to oxygen. That is, we visualize that oxide ions (O^{2-}) are formed in these processes (although they really do not exist in this form in covalent compounds). We think of carbon monoxide, then, as a C^{2+} ion combined with one O^{2-} ion, and carbon dioxide as a C^{4+} ion combined with two O^{2-} ions, even though such combinations do not represent the actual covalent molecule (C^{2+} and C^{4+} ions do *not* exist.)

9.3 In each of the following pairs, which species is more oxidized?

a. SO_2 or SO_3
b. HNO_3 or HNO_2
c. NO_2 or N_2O_3
d. P_2O_3 or P_2O_5
e. H_2O or H_2O_2

REDUCTION

A substance is *reduced* when it gains electrons.

The process of oxidation can never occur by itself. When a substance is oxidized and loses one or more of its electrons, these electrons must have a place to go. This usually means that some other substance must be available to accept them. The substance that accepts these electrons is said to be *reduced.* **Reduction** is associated with the process of gaining electrons.

Consider the burning of magnesium metal in air,

$$2Mg + O_2 \longrightarrow 2MgO$$

Here, each magnesium atom loses two electrons and is oxidized to form a magnesium ion.

$$:Mg \longrightarrow Mg^{2+} + 2e^- \qquad \text{(oxidation)}$$

Nonmetals, we know, like to gain electrons to form negative ions. Each oxygen atom in O_2 can accept two electrons (from the magnesium) to complete its octet:

$$:\dot{\underset{\cdot\cdot}{O}}\cdot \; + 2e^- \longrightarrow \; :\overset{\cdot\cdot}{\underset{\cdot\cdot}{O}}:^{2-} \qquad \text{(reduction)}$$

oxygen atom (6 valence electrons) → oxide ion (8 valence electrons)

Thus, the oxygen atoms are reduced.

In a similar manner, each chlorine atom in Cl_2 needs one more electron to complete its octet and is reduced to form a Cl^- ion when it reacts with Fe:

$$\cdot\ddot{Fe} \longrightarrow Fe^{3+} + 3e^- \qquad \text{(oxidation)}$$

$$:\overset{\cdot\cdot}{\underset{\cdot\cdot}{Cl}}\cdot \; + e^- \longrightarrow \; :\overset{\cdot\cdot}{\underset{\cdot\cdot}{Cl}}:^{-} \qquad \text{(reduction)}$$

chlorine atom (7 valence electrons) → chloride ion (8 valence electrons)

In a previous example, elemental sulfur reacted with iron and was also reduced to sulfide ion. Each sulfur atom gained two electrons in the process:

$$2Fe + 3S \longrightarrow Fe_2S_3$$

and

$$:\dot{\underset{\cdot\cdot}{S}}\cdot \; + 2e^- \longrightarrow \; :\overset{\cdot\cdot}{\underset{\cdot\cdot}{S}}:^{2-}$$

sulfur atom (6 valence electrons) → sulfide ion (8 valence electrons)

Yet sulfur also can be oxidized to form SO_2 and SO_3 when it reacts with oxygen:

$$S + O_2 \longrightarrow SO_2$$

and

$$2S + 3O_2 \longrightarrow 2SO_3$$

This means that we must always consider the *system* in which a reactant participates. In the presence of iron, sulfur is reduced, yet it is oxidized when it is combined with oxygen.

Reduction can involve the gain of hydrogen to a chemical species. . .

There are two other ways to determine whether a substance is reduced. One is to see if it gains hydrogen. For example, in the reaction of hydrogen with bromine,

$$H_2 + Br_2 \longrightarrow 2HBr$$

the bromine gains hydrogen and is reduced. The hydrogen *must* be oxidized, since oxidation and reduction go together. Another way to identify reduction

. . .or the loss of oxygen from a chemical species.

is to see whether a substance loses oxygen. In the preparation of elemental boron, diboron trioxide is heated with elemental magnesium,

$$B_2O_3 + 3Mg \longrightarrow 2B + 3MgO$$

The diboron trioxide loses oxygen and is reduced to elemental boron, while the magnesium gains oxygen and is oxidized.

9.4 We saw that in the reaction by which sulfur is deposited at the vents of volcanoes, hydrogen sulfide is oxidized to elemental sulfur,

$$2H_2S(g) + SO_2(g) \longrightarrow 3S(s) + 2H_2O(l)$$

What happens to the SO_2 in this process? Both the H_2S and SO_2 contain sulfur. In terms of oxidation and reduction, what happens to the sulfur atom in each of these compounds? (Hint: Is the sulfur atom in elemental sulfur more or less oxidized than the one in SO_2? In H_2S?)

Oxidation and reduction must take place at the same time.

Oxidation and reduction must occur simultaneously, and our terminology reflects this. We talk about *oxidation–reduction reactions,* or simply make a contraction of the two terms and say **redox.** There are three ways to tell whether a substance is oxidized or reduced. ▶ Figure 9.3 summarizes them.

9.5 For each of the following reactions, indicate which substance is reduced. Then determine which is oxidized.

a. $2H_2 + O_2 \longrightarrow 2H_2O$
b. $Ca + 2H_2O \longrightarrow Ca(OH)_2 + H_2$
c. $SnO_2 + 2C \longrightarrow Sn + 2CO$
d. $2H_2S + O_2 \longrightarrow 2H_2O + 2S$
e. $2Na + Cl_2 \longrightarrow 2NaCl$
f. $2Br^- + Cl_2 \longrightarrow Br_2 + 2Cl^-$
g. $4CuO + CH_4 \longrightarrow 4Cu + CO_2 + 2H_2O$
h. $3Mg + N_2 \longrightarrow Mg_3N_2$

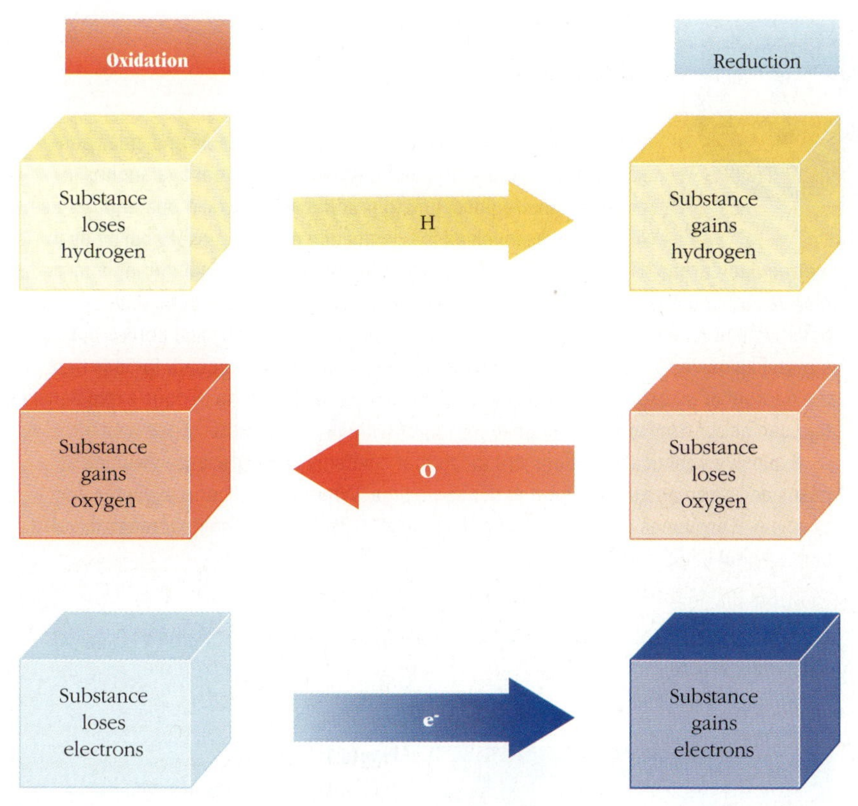

▶ Figure 9.3
Three Ways to Determine Oxidation or Reduction of a Substance.

OXIDIZING AND REDUCING AGENTS

"Agents" cause something to happen. The terms *oxidizing agent* and *reducing agent* derive from this idea. In the smelting of iron ore, for example, iron ore is reduced with carbon in the form of coke (▶ Figure 9.4, page 294).

In talking about redox reactions, we say iron ore *is reduced* to pure iron (Fe) and the carbon *is oxidized* to carbon dioxide. Now, something caused the ore to be reduced to pure metal. That was the carbon; thus, carbon is called the reducing agent. Similarly, since the carbon was oxidized, some other substance had to do it. This was the iron ore, which is called the oxidizing agent.

Agents cause processes to occur. An **oxidizing agent** (oxidizer, oxidant) causes some other species to be oxidized and is itself reduced in the process. A **reducing agent** (reductant) causes some other species to be reduced, and, in turn, is itself oxidized.

EXPLOSIVES

Explosives are used to mine coal, to blast tunnels and clear passageways for highways and railroads, to prevent avalanches in mountainous regions, and to produce fireworks. A spectacular use of explosives involves the demolition of large, unsafe buildings in metropolitan areas (Figure 1). Using their knowledge of structural engineering and explosives technology, highly skilled teams can demolish multistory structures in seconds without damage to other large buildings nearby, often literally just across the street. Explosives are also used for military purposes, but these account for less than ten percent of all explosives manufactured. Closer to home are the serious explosions that can occur from mixtures of gasoline vapor with air, and of natural gas with air.

Explosive chemical reactions are those that suddenly generate large volumes of gases and large amounts of heat at the same time. When nitroglycerin explodes, for example, it produces temperatures of about 5000 °C and a 22,000-fold increase in volume (of gas) in an instant!

$$\underset{\text{nitroglycerin}}{4C_3H_5(ONO_2)_3(l)} \xrightarrow{\text{shock or rapid heat}} 12CO_2(g) + 10H_2O(g) + 6N_2(g) + O_2(g)$$

Figure 1
Old, unsafe buildings are sometimes demolished in seconds with explosives. Highly skilled teams of structural engineers and explosives experts place charges in critical locations to make the building collapse within its foundations.

9.6 Go back to the previous problem (9.5 You Try It) and label the oxidizing agents and reducing agents in each of the reactions.

Explosive mixtures consist of an oxidizing agent (oxidizer) and a reducing agent (fuel) in correct proportion with one another to react instantly and completely. For example, a mixture of methane and air will explode when the methane concentration is 5–13% by volume, but not when it is outside of this range. Mixtures of hydrogen gas and acetylene gas (C_2H_2) with air are much more dangerous: those containing 9–65% hydrogen or 3–73% acetylene by volume will explode.

Because mixtures do not have fixed compositions, making explosives by mixing various amounts of oxidizer and fuel can be used to control the strength of the explosive. Oxidizer and fuel must be in intimate contact with one another, however, if the chemical reaction producing the explosion is to be rapid enough. Explosives with greater shocking power (brisance) are those in which the oxidizer and fuel (reducing agent) are incorporated in the proper proportions in the same compound. This insures not only a constant composition of the explosive but also a faster reaction, because only slight changes in the positions of atoms are necessary to cause a redox reaction to occur. Nitroglycerin (see the preceding equation) is a molecular explosive: part of the molecule is the oxidizer (the $—ONO_2$) and part of it is the fuel. Trinitrotoluene (Chapter 11) is another example of this kind of explosive.

Ammonium nitrate, NH_4NO_3 is a common ionic explosive, which is also used as a fertilizer. In this instance, the ammonium ion is the reducing agent and the nitrate ion is the oxidizer. When heated to more than 300 °C, it decomposes violently as follows:

$$2NH_4NO_3(s) \longrightarrow 2N_2(g) + O_2(g) + 4H_2O(g)$$

Because it is inexpensive, ammonium nitrate is often used in large blasting operations, where it is placed in boreholes and detonated by a small charge of dynamite or other explosive.

Rocket propellants have the same chemistry as explosives, except that the fuel and oxidizer are mixed together at a controlled rate, so that the reaction provides a constant push instead of an explosion. The lunar excursion module of the Apollo space vehicles used liquid dimethylhydrazine [$(CH_3)_2NNH_2$] as a fuel. Liquid dinitrogen tetroxide, N_2O_4, the oxidizer, reacts with this fuel on contact:

$$(CH_3)_2NNH_2(l) + 2N_2O_4(l) \longrightarrow 3N_2(g) + 4H_2O(g) + 2CO_2(g)$$

COMBUSTION, METABOLISM, AND PHOTOSYNTHESIS

The combustion of hydrocarbons derived from petroleum provides most of the energy used for transportation in this country (▶ Figure 9.5). Using isooctane, a component of gasoline, and assuming there is enough oxygen present to completely burn it, we can write a representative equation for the typical redox reaction that occurs in automobile engines:

Combustion is the rapid reaction of fuels with oxygen.

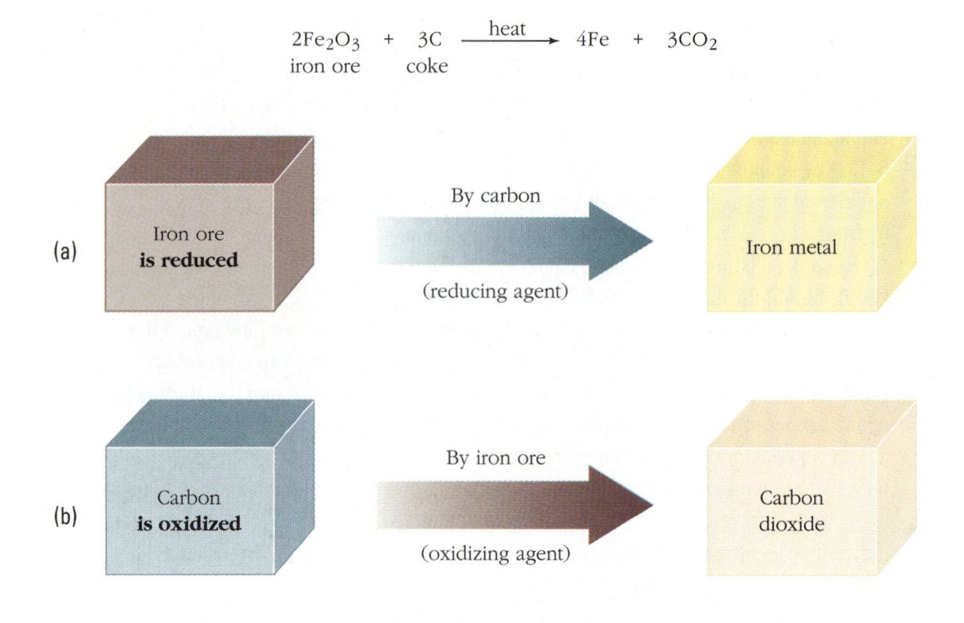

▶ Figure 9.4
Reduction of Iron Ore with Coke. *(a) Carbon* ***reduces*** *iron in ore: reducing agent is C. (b) iron ore* ***oxidizes*** *the carbon: oxidizing agent is Fe_2O_3.*

$$\underset{\text{iso-octane}}{2C_8H_{18}} + 25O_2 \longrightarrow 16CO_2 + 18H_2O + \text{heat}$$

▶ Figure 9.5
A Jet Aircraft Engine. *The combustion of fuels derived from petroleum provides most of the energy for transportation in the United States.*

In this reaction iso-octane is oxidized to form carbon dioxide and water, and oxygen is the oxidizing agent. In turn, oxygen is reduced by iso-octane, the reducing agent. In general, hydrocarbons are oxidized in air to give carbon dioxide and water with the release of heat. If insufficient oxygen is present, deadly carbon monoxide, CO, also forms. Because of the way flames form and propagate (in the open or in an internal combustion engine), a small amount of CO almost always forms, even when enough oxygen is available for combustion. For this reason, it is always wise to provide good ventilation when any combustion process is used in a confined area.

Another important combustion process is the **metabolism** of carbohydrates, which takes place in the mitochondria of cells in many living organisms. In our bodies these carbohydrates are oxidized by the oxygen we breathe, through a series of complicated biochemical steps, to produce carbon dioxide and water (▶ Figure 9.6). A considerable amount of energy is also released in the process.

The name, **carbohydrate,** literally means "hydrate of carbon." Substances like sugars and starches are molecular compounds that contain carbon, hydrogen, and oxygen in roughly the ratio $C_n(H_2O)_n$ (the *n*s are simply integers: 3,4,5,6, etc.). Fruit sugars such as glucose and fructose, for example, have

▶ Figure 9.6
Carbohydrates, Fuel for the Body. *Grains, fruits, and vegetables, which provide complex carbohydrates, are part of a healthful diet.*

the formula $C_6H_{12}O_6$, whereas sucrose (table sugar), maltose (malt sugar), and lactose (milk sugar) have the formula $C_{12}H_{22}O_{11}$.

The overall equation for the metabolism of glucose in the body can be written as follows:

$$\underset{\text{glucose}}{C_6H_{12}O_6} + 6O_2 \xrightarrow{\text{metabolism}} 6CO_2 + 6H_2O + \text{energy}$$

As in combustion, oxygen is the oxidizing agent. The glucose is oxidized and is the reducing agent that reduces the oxygen.

Most combustion processes are difficult to reverse. It is hard to "unburn" gasoline or coal, partly because their combustion processes give off a considerable amount of heat energy. To unburn them would require that this energy be put back in (see Chapter 10). That is, these reverse reactions tend not to be spontaneous. In nature, however, there exists an important reaction that converts carbon dioxide and water, the products of combustion and metabolism, to carbohydrates. This is the familiar process of **photosynthesis,** which occurs in green plants. Using the chlorophyll in their leaves, these plants synthesize carbohydrates from carbon dioxide in the air, using sunlight as the source of energy; oxygen gas is their waste product (▶ Figure 9.7). This redox reaction not only provides food for animals and us, but it also is the only

▶ Figure 9.7
Oxygen bubbles are formed on the aquatic plant, waterweed, as a product of photosynthesis.

Photosynthesis is the reverse of the processes of combustion and metabolism.

REDOX REDUX

Rhonda doesn't waste time getting to her point:

"I've been answering your questions the whole term . . . will you help me with something?" You nod, happy for a chance to repay her for her patience. "I'm going to be a Teaching Assistant in a freshman biology section tomorrow, and I want to give them an idea of how living systems obtain energy. I figure if I can make you understand it, I'll be OK. Will you listen?"

"Sure . . .", not sure her "figuring" is complimentary.

"OK. Have a seat. In a nutshell, plants use sunlight as the source of the energy they need to derive nutrients from water and carbon dioxide, and they release oxygen into the atmosphere as a waste product. Animals derive the nutrients and energy they need from plants and from oxygen in the air, and they release carbon dioxide into the atmosphere as a waste product. In a sense, biology is a huge cycle of redox reactions."

Electrons again. That whole series of reactions could be thought of as electrons being pushed uphill by plants, in the process called *photosynthesis,* and rolled down again by the metabolism of animals. (Plants metabolize, too, but they do it with their own products.) Plants and animals need steady sources of energy to drive the multitude of chemical reactions that constitute living, and the energy-storage "batteries" of both plants and animals are molecules called adenosine triphosphate (ATP), a kind of biological match. When a reaction needs energy, an ATP molecule is "lit." The bonds between phosphate groups take energy to make, and energy is released when they're broken. Forming these phosphate bonds is called phosphorylation: *photophosphorylation* (by sunlight) in plants and *oxidative phosphorylation* (by oxidation of nutrients from plants) in animals.

Photosynthesis takes place in molecules of chlorophyll, the green pigment in plants. It includes a series of light-trapping reactions, a series of energy-controlling reactions, and a series of reactions that make nutrient sugars by combining the products of the reduction of CO_2 with hydrogen ions.

Light trapping. Electrons in molecules of chlorophyll in a leaf are excited to a high-energy state by sunlight, and this energy is passed from electrons in one chlorophyll molecule to electrons in another until it reaches electrons in special chlorophyll molecules called P680 (the wavelength of light they can absorb). These electrons, now in a high-energy state, are taken away from P680 by molecules called electron-acceptors (oxidizing agents); and the P680, now short of electrons, oxidizes water in the leaf.

You interrupt:

"The idea of oxidizing water confuses me! Worse, they even talk about oxidizing oxygen!"

Note that each atom of copper metal gives up two electrons but each silver ion can accept only one. Thus, two silver ions are necessary to accept the electrons from one copper atom. These electrons are transferred directly from the copper metal every time the silver ions come in contact with it.

It is possible to make this reaction produce an electric current. To do this we must prevent the copper metal from coming in direct contact with the silver nitrate solution and at the same time keep electrical contact between two. How can we build a device that will do this?

We know that a metal wire can conduct electricity because metal atoms hold their electrons loosely, enabling them to move about easily from one atom to another. To make an important part of our device we also need to know that when a water solution contains dissolved ions, it, too, can conduct electricity. This is because ions in water solution are mobile; unlike those held rigidly in a solid crystal lattice, they are capable of carrying an electric current because they can migrate about the solution under the influence of an electric field (positive and negative charge) (► Figure 9.9). Salts and molecular compounds, such as hydrogen chloride and sulfuric acid, which dissolve in water to produce mobile ions are called **electrolytes.** Substances that dissolve in water but do not produce ions, for example alcohol and sugar, are called **nonelectrolytes**. Electrolyte solutions provide a way of making a wire to connect two dissimilar solutions that cannot be permitted to mix.

A solution of ions in water conducts electricity.

This all sounds complicated, but our device is actually rather simple to build. First, we take a small sheet of copper and immerse it in a solution of potassium nitrate ($K^+ + NO_3^-$) in a beaker. In a separate beaker containing silver nitrate solution ($Ag^+ + NO_3^-$), we immerse a small sheet of platinum metal. (Platinum metal is inert to most reagents.) The copper and platinum sheets are called **electrodes.**

Because an ordinary metal wire won't work, we shall use a "wire" of mobile ions to make the electrical connection between the solutions in the two beakers. To make this wire, we dissolve some potassium nitrate in hot water ($K^+ + NO_3^-$), stir in some gelatin, pour the solution into a U-shaped glass tube, and allow the mixture to cool and gel. This gel contains potassium ions and nitrate ions dissolved in the water that is held in the gelatin matrix (framework). The ions can move through this matrix, but the water—and hence the other solutions—cannot. We have built a *salt bridge,* which will conduct electric current between the potassium nitrate and silver nitrate solutions without permitting the two to mix. This keeps the Ag^+ ions that are in one beaker away from the copper electrode in the other.

Ions can move through a *salt bridge,* but water cannot.

We're almost there. The silver nitrate solution and copper metal are isolated from one another except for the salt bridge. To find out whether any electron transfer is occurring between them, all we need to do now is join the copper and platinum electrodes with a wire connected to a tiny light bulb, and watch the bulb. If we have hooked everything together properly, the tiny

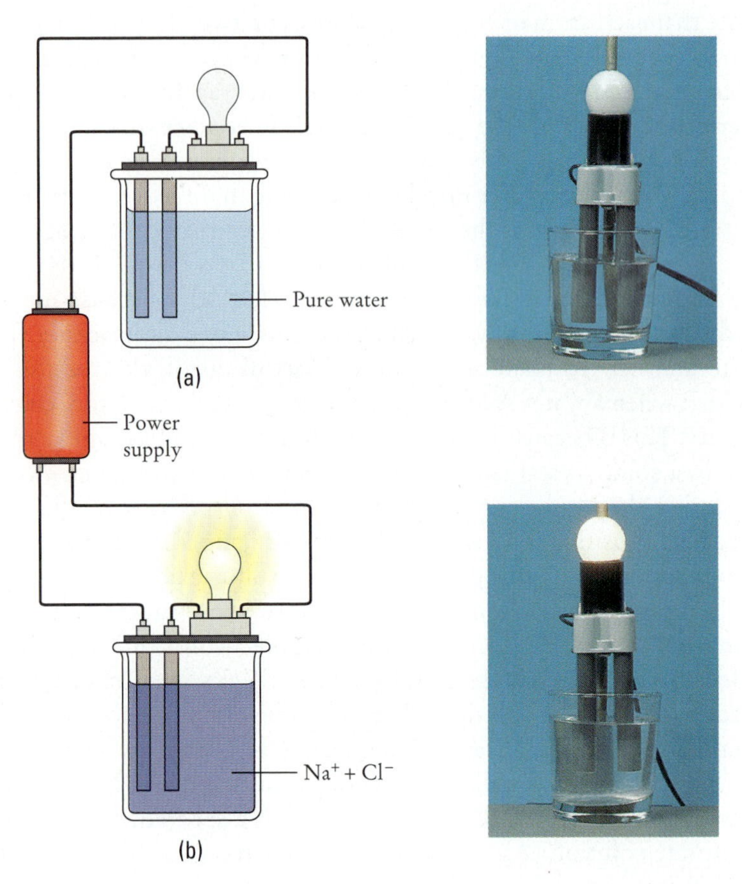

▶ Figure 9.9

Electrolyte Solutions Conduct Electricity. *(a) When the electrodes are placed in pure water, the bulb does not light, indicating the absence of mobile ions. (b) When the electrodes are placed in a solution of sodium chloride, ions conduct an electric current and the bulb lights brightly.*

bulb will glow. If we then place a voltmeter in the circuit instead of the bulb, we can see that the electrons (current) in the wire are flowing from the copper sheet to the platinum sheet. After a period of time we also can see that silver metal is forming on the platinum electrode and that the solution of potassium nitrate in which the copper electrode is immersed is turning blue. This is the same redox reaction we observed for the formation of the silver tree, but now we have made the chemical reaction provide energy by producing an electric current. Such a device is a **voltaic** (or **galvanic**) cell (▶ Figure 9.10). Used to produce and store electrical energy, one or more voltaic cells in a package is commonly referred to as a **battery.** (Our cell is not a particularly powerful

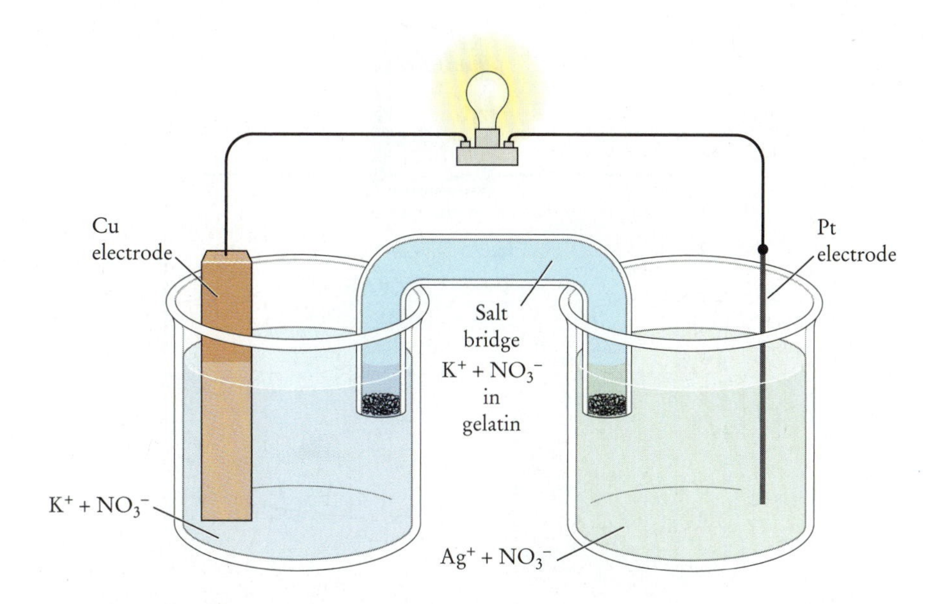

▶ Figure 9.10
Diagram of a Copper–Silver Cell.

one, though, because it only develops about 0.45 volts. By comparison, dry cells provide about 1.5 volts.)

Without the presence of the salt bridge between the two solutions this process would stop almost immediately, because each time a copper atom gave up its two electrons (through the wire and the platinum sheet) to reduce two silver ions, a copper ion, Cu^{2+}, would be left behind in the potassium nitrate solution. Eventually the accumulation of copper ions would cause a strong positive charge to build up there. (Copper ion also causes the blue color to form in the solution.) Similarly, removal of silver ions from the silver nitrate solution would leave behind an excess of nitrate ions, making this solution strongly negative. In this condition, the cell is said to be *polarized.* The salt bridge, being a conduit for ions, permits the excess nitrate ions left behind in the silver solution to migrate to the copper solution, where they can balance the charge on the extra copper ions being formed. Thus, *electrons* flow along the wire between the electrodes while negative *ions* flow between the solutions through the salt bridge, all the while maintaining a balance of electrical charge (▶ Figure 9.11). (Positive ions can also flow through the salt bridge in the opposite direction.)

OTHER KINDS OF CELLS/BATTERIES

The ordinary dry cell, called the zinc–carbon cell, is illustrated in ▶ Figure 9.12. It consists of a zinc can filled with a moist paste made of manganese

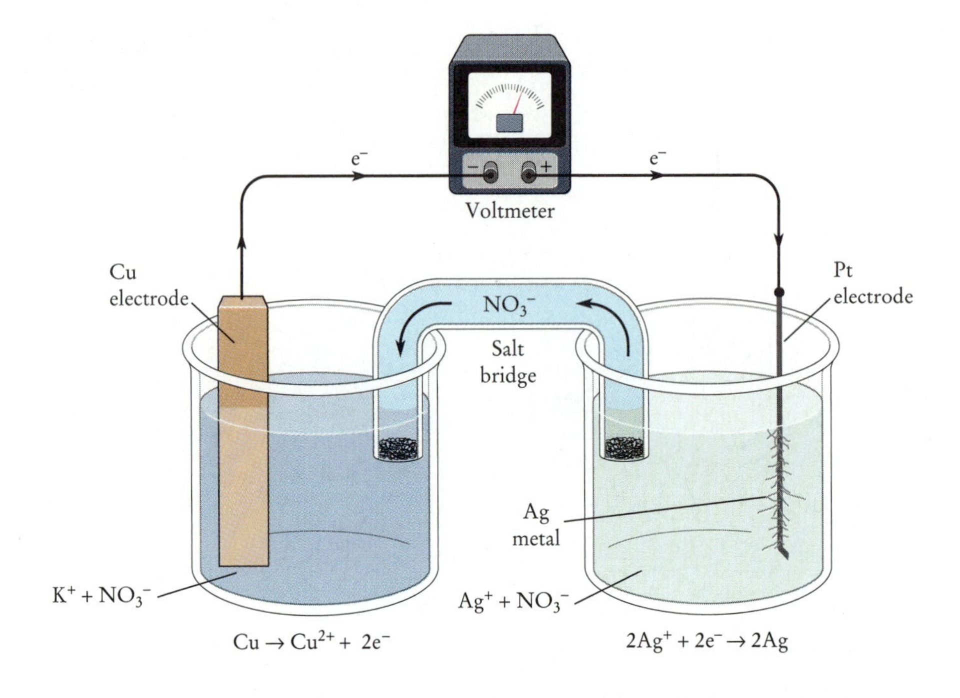

▶ Figure 9.11
Movement of Electrons and Ions in a Copper–Silver Cell.

dioxide, MnO_2, finely divided carbon (carbon black), and electrolytes, $ZnCl_2$, and NH_4Cl. A carbon (graphite) rod in the center serves as the cathode, where reduction of the MnO_2 takes place. When the circuit is completed, the zinc can is oxidized according to the reaction,

$$Zn(s) + 2MnO_2(s) + 2NH_4^+ \longrightarrow Zn^{2+} + Mn_2O_3(s) + 2NH_3 + H_2O(l)$$

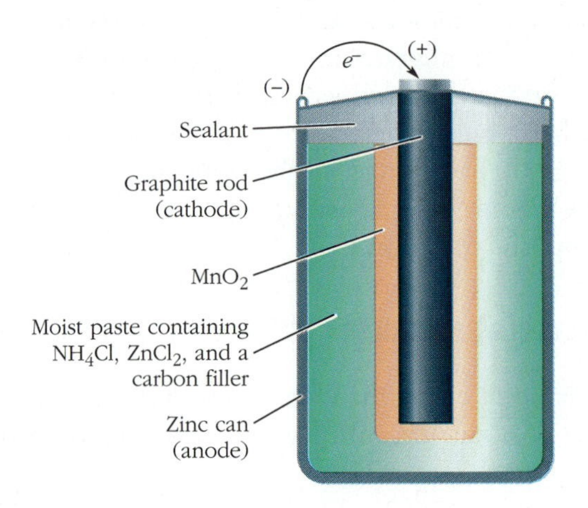

▶ Figure 9.12
A Zinc–Carbon Dry Cell.

The alkaline cell also contains zinc and maganese dioxide, like a dry cell. However, its electrolyte is strongly alkaline potassium hydroxide instead of ammonium chloride (which is slightly acidic). Alkaline cells last longer and have a longer shelf life than do ordinary dry cells, but they cost more. The cell reaction is,

$$Zn(s) + 2MnO_2(s) \longrightarrow ZnO(s) + Mn_2O_3(s)$$

Lead storage batteries consist of alternating plates of lead and lead packed with lead dioxide, PbO_2 (▶ Figure 9.13). The electrolyte is a 30% solution of sulfuric acid ("battery acid"). When the cell is discharged, the lead is oxidized and the lead oxide is reduced:

$$Pb(s) + PbO_2(s) + 2HSO_4^- + 2H^+ \longrightarrow 2PbSO_4(s) + 2H_2O(l)$$

Unlike dry cells and alkaline cells, this cell can be recharged.

A small, rechargeable battery commonly used in cordless power tools and telephones uses the nickel–cadmium cell. It contains cadmium metal, nickel (IV) oxide (NiO_2), and a KOH electrolyte. When this cell is producing electricity, the cadmium is oxidized and the NiO_2 is reduced,

$$Cd(s) + NiO_2(s) + 2H_2O(l) \longrightarrow Cd(OH)_2(s) + Ni(OH)_2(s)$$

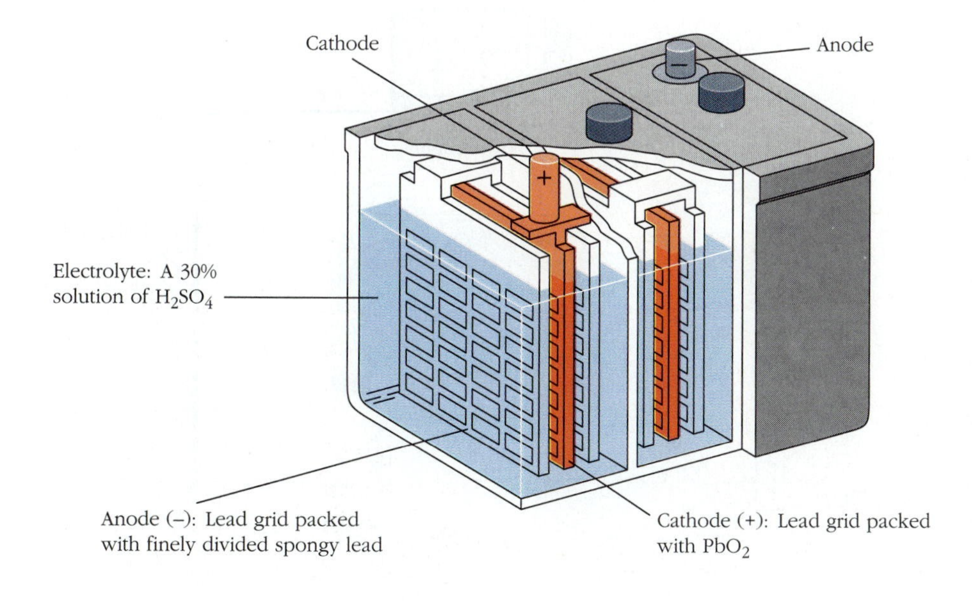

▶ Figure 9.13
A Lead Storage Battery. *The interconnected lead sheets form the anode, and the sheets filled with lead dioxide form the cathode. The electrolyte is a solution of sulfuric acid. A fully charged cell delivers about 2.0 volts. This 6-volt battery consists of three cells connected in series (connections not shown).*

Ni-Cd batteries should be recycled.

Because cadmium and its compounds are poisonous, batteries containing it should be recycled rather than incinerated or discarded when they are no longer rechargeable. Lead storage batteries should also be recycled.

ELECTROLYTIC CELLS

Because batteries convert chemical energy to electrical energy, they all "run down" after a period of use. We can often, but not always, put this energy back into a cell by recharging it—that is, using an external electrical energy source of sufficiently high voltage to force the cell to run backward. In producing energy the cell we made used up its reagents:

$$Cu + 2Ag^+ \longrightarrow Cu^{2+} + 2Ag + \text{energy}$$

To recharge it we must put back this energy, thereby reversing this process and restoring these reagents to their original concentrations (▶ Figure 9.14):

$$Cu^{2+} + 2Ag + \text{energy} \longrightarrow Cu + 2Ag^+$$

The external electrical source must be connected so that electrons are pushed onto the copper electrode, reducing the dissolved copper ion to metallic copper and causing the blue color in the solution to disappear. At the same time, electrons are removed from the platinum electrode and, in turn, from the silver metal that has plated onto it. This oxidizes the silver metal to

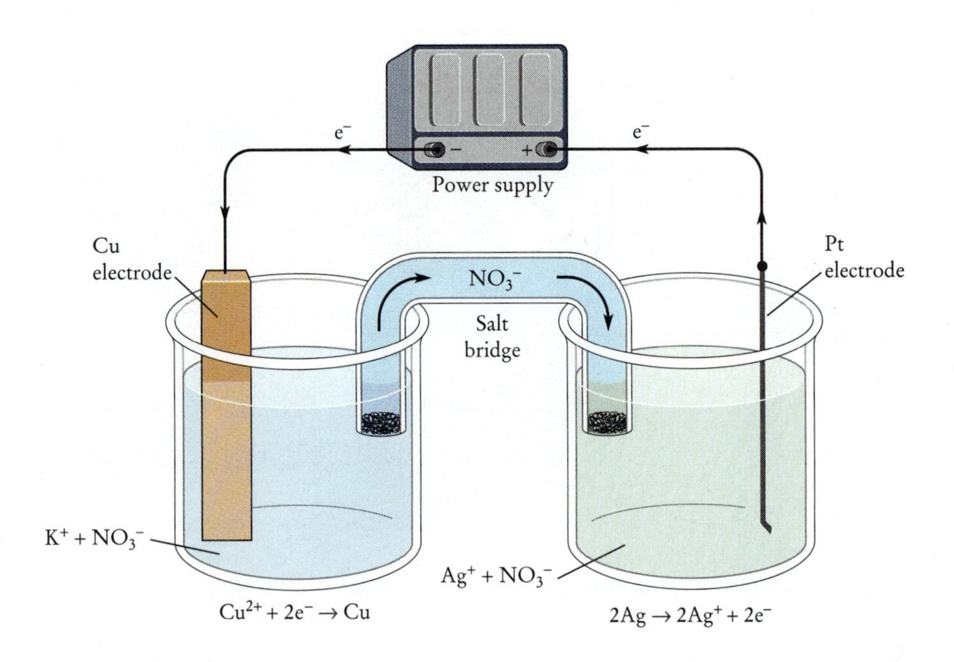

▶ Figure 9.14
Recharging a Copper–Silver Cell.

silver ion, which then goes into solution. Nitrate ions migrate back through the salt bridge from the copper solution to the silver solution, to maintain electrical charge balance. Batteries can thus serve as energy storage devices. This is essentially the way lead storage and nickel–cadmium batteries work as well.

By using an external electrical energy source to reverse the chemical reactions in our copper–silver cell, we have caused it to behave as an **electrolytic cell.** In such a cell, nonspontaneous chemical reactions are made to proceed by the application of electrical energy. Electrolytic cells have many important applications. For example, they are used for *electroplating* one metal on another, such as chromium or tin on steel, or gold on electronic parts and switch contacts (▶ Figure 9.15). Copper, nickel, and zinc metals are purified by a similar process. Many metals, such as aluminum and sodium, are produced from their compounds by a process called *electroreduction. Anodizing* of metals, particularly aluminum, is an electrolytic process that deposits a protective metal oxide coating on the metal surface. Water can be easily decomposed into hydrogen and oxygen by **electrolysis** (cleavage by electricity).

▶ Figure 9.15
Silver-plated Spoons Freshly Removed from the Electroplating Bath.

Chrome plating an automobile part provides an interesting example of electroplating. Because chromium (Cr) metal does not adhere well to steel, the steel part is usually plated first with copper and then with nickel (Ni), before it is electroplated with chromium. In the chrome plating process (▶ Figure 9.16), the part is suspended by a wire (which serves as a conductor of electricity) in a bath containing chromic acid, H_2CrO_4, and a little sulfuric acid, H_2SO_4. Also suspended in the bath, from a separate conductor, are bars of chromium metal. An electric current is passed through the solution such that the part to be plated is given a negative charge and the pieces of chromium metal are given a positive charge. The excess electrons on the part cause the

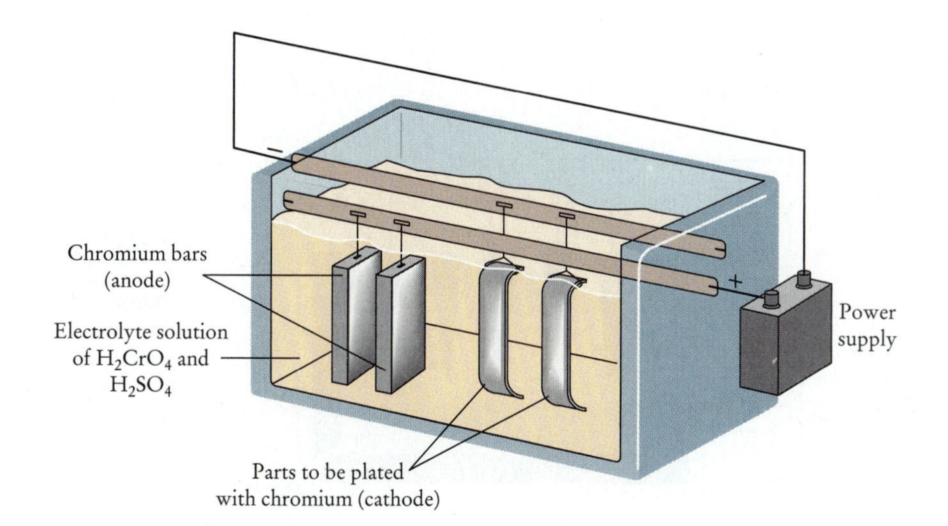

▶ Figure 9.16
A Cell for Electroplating Chromium.

In *any* kind of electrochemical cell, reduction takes place at the **cathode** . . .

. . . and oxidation occurs at the **anode**.

chromium in the chromic acid to be reduced to Cr metal. This electrode, at which reduction takes place, is called the **cathode.** At the positively charged chromium bar, chromium atoms give up their electrons and are oxidized to an ionic form that replenishes the H_2CrO_4 consumed at the cathode. The electrode at which oxidation takes place is called the **anode.** As the electroplating continues, the chromium in the bar (anode) is used up (oxidized) and is deposited (reduced) on the part (cathode) as bright, shiny "chrome plate." The chromic and sulfuric acids ionize sufficiently in water to make a good electrolyte solution.

The most important electroreduction process used today is the production of aluminum metal from its ore, bauxite (primarily aluminum oxide, Al_2O_3, or alumina). Before 1886, when the Hall–Héroult process was developed, aluminum metal was extremely difficult and expensive to obtain because it had to be made by reduction of aluminum chloride with potassium metal. In the Hall–Héroult process (▶ Figure 9.17), the electroreduction takes place in a bath of molten cryolite, a mineral having the formula Na_3AlF_6. Alumina readily dissolves in the cryolite, and the solution serves as the electrolyte. The molten bath, maintained above 660 °C (the melting point of aluminum), is contained in a carbon (graphite) box that also serves as the cathode. Interconnected anodes, also of graphite, are immersed in the hot bath, and an electric current is passed between the them and the box. At the box, which is the negative electrode, ionic aluminum is reduced to aluminum metal and sinks to the bottom of the box where it periodically is drawn off (▶ Figure 9.18). At the interconnected positive electrodes (anodes), oxygen gas is produced from the oxidation of the oxide ions present in the molten cryolite solution. This oxygen then reacts with the carbon in the anodes to produce CO_2. The overall reaction for this process is,

$$2Al_2O_3(l) + 3C(s) \longrightarrow 4Al(l) + 3CO_2(g)$$

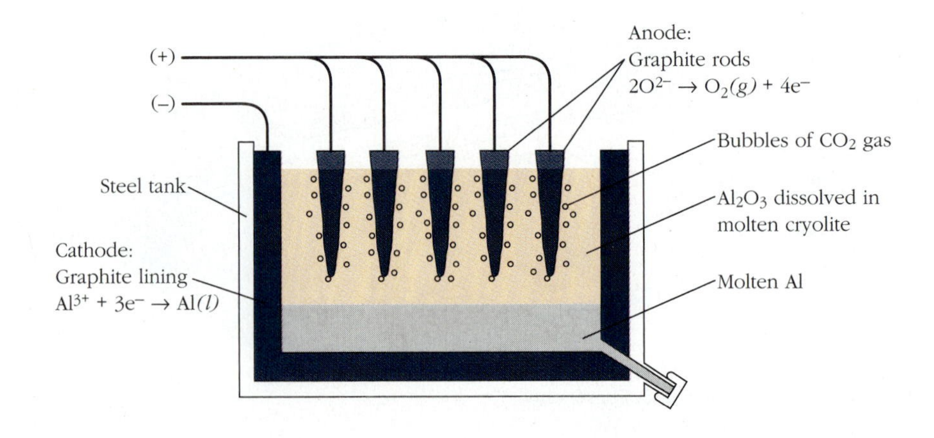

▶ Figure 9.17
A Hall–Héroult Cell for the Electroreduction of Aluminum Oxide in Molten Cryolyte. *The aluminum sinks to the bottom where it is drawn off. The oxygen formed at the anode reacts with the carbon (graphite) electrode forming CO_2.*

Electrolysis of water is a simple means of producing pure hydrogen and pure oxygen gases. Two electrodes, usually made of graphite, are placed in a container of water, along with a small amount of an electrolyte such as sulfuric acid, a molecular substance that dissolves in water to form ions (Chapter 8). (Pure water is a very poor conductor of electricity and thus is a poor electrolyte.)

At the negative electrode (cathode), the hydrogen in the water is reduced to hydrogen gas and is collected there. At the positive anode, oxygen gas is formed by the oxidation of the oxygen in water (▶ Figure 9.19). (The oxygen in water is in its reduced form.)

Care must be exercised to keep the two gases separated because a mixture of the two is highly explosive. Properly controlled, however, the energy released from the combination of hydrogen and oxygen can be used to power rockets and generate electricity in fuel cells.

▶ Figure 9.18
Pouring Molten Aluminum after Electroreduction of Bauxite Ore.

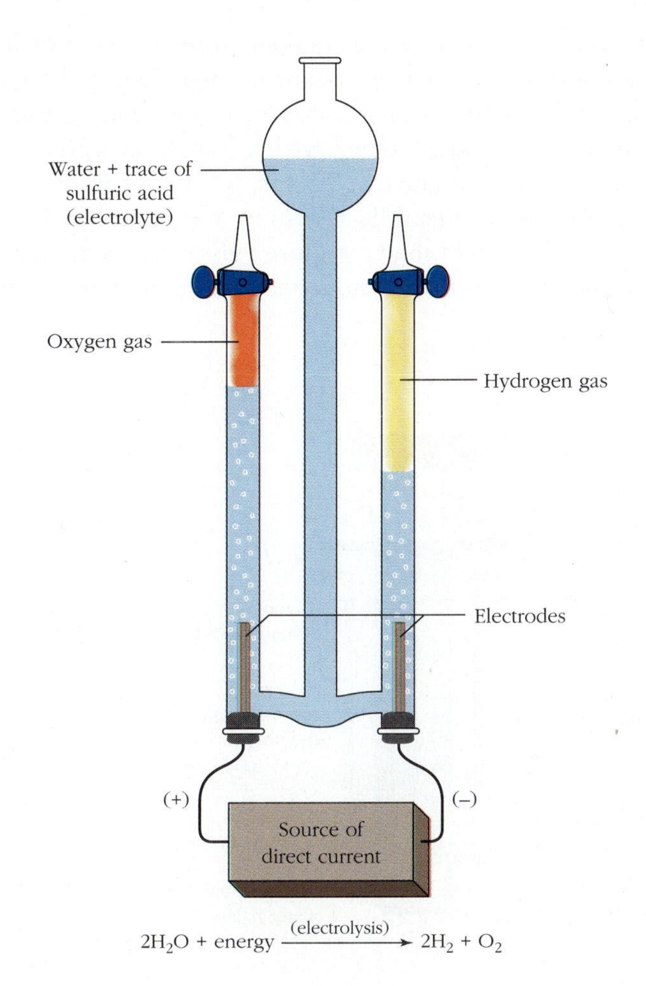

$$2H_2O + \text{energy} \xrightarrow{\text{(electrolysis)}} 2H_2 + O_2$$

▶ Figure 9.19
Electrolysis of Water.

FUEL CELLS

Fuel cells can produce electricity continuously as long as they have a supply of fuel and oxidant.

Fuel cells convert an oxidizable fuel (reducing agent), such as hydrogen or methane, and an oxidizing agent, usually oxygen from the air, directly into electricity. Unlike voltaic cells and batteries, which must be recharged, **fuel cells** continuously produce electricity as long as fuel and oxidant are fed into them. They are also very efficient: about 70% of the energy in the fuel is converted into electrical energy, as compared with about 40% for a very efficient steam-turbine electrical generating plant. ▶ Figure 9.20 shows a hydrogen–oxygen fuel cell used to provide electricity in spacecraft. Note that this cell produces only water as a product and is therefore nonpolluting. This water can be used for drinking during manned space flights. Fuel cells are still too expensive, however, to be used to produce electricity on a large scale.

CORROSION

Corrosion is the unwanted oxidation of metals in the environment.

Most metals tend to oxidize in the natural environment and usually spontaneously form compounds rather than exist in their elemental state for very long. (Recall that this is the reason most of them are found combined in ores rather than in the native state.) When we use metals, however, we want to prevent this process from occurring. That is, we do not want them to **corrode.** Iron is particularly susceptible to rusting, and copper roofs and bronze statues, such as the Statue of Liberty, acquire a green patina in the presence of moist air. Lead compounds that do not conduct electricity form between the

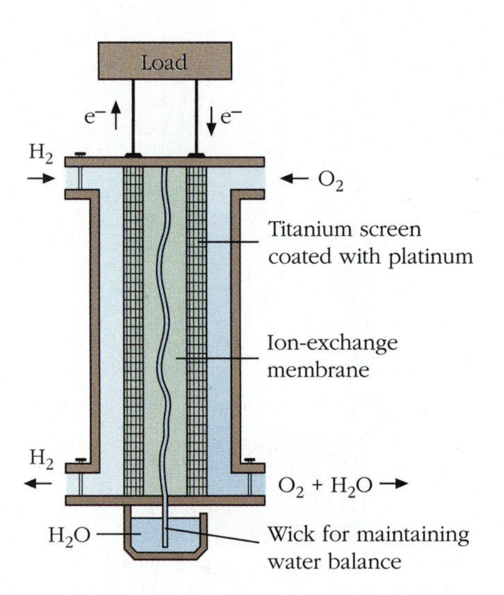

▶ Figure 9.20
Schematic of Single Gemini Hydrogen–Oxygen Fuel Cell.

connections on automobile battery terminals, which are made of lead, sometimes causing symptoms of a dead battery.

Covering a metal surface with paint is one way to prevent corrosion, as long as the paint remains intact. Electroplating with a corrosion-resistant metal, like tin, is another. Sometimes, metals oxidize to form a thin, unbroken layer of oxide on their surfaces that resists further oxidation, a process called **passivation.** The bluing on gunmetal is a form of iron oxide that behaves in this manner, preventing the steel underneath from rusting. Stainless steels contain chromium, which oxidizes to form a similar kind of coating. Aluminum is also passivated. Though easily oxidized, it does not corrode in the environment because it rapidly forms a coat of protective oxide. Aluminum can be oxidized intentionally to protect it by means of an electrochemical process called anodizing. Freshly anodized aluminum can be dyed to produce colored aluminum products.

Stainless steel is not "stainless." It is **passivated** by a protective oxide coating on its surface.

Galvanizing is a process in which iron is given a thin coating of zinc (▶ Figure 9.21). The zinc is more easily oxidized than the iron but reacts with the atmosphere to form a film that protects it. Even if the zinc coating is damaged enough to expose the iron, the iron will not rust. The zinc metal around the damaged area can easily furnish electrons to the exposed iron, and any iron that is oxidized to the ionic state will be immediately reduced back to iron metal before it can migrate from the site:

$$Zn + Fe^{2+} \longrightarrow Zn^{2+} + Fe$$

In this process, the zinc is used up to protect the iron.

The use of *sacrificial anodes* to protect underground storage tanks and pipelines is another application of this principle. Pieces of reactive metals such as magnesium or zinc are attached to the tank or pipe by a wire and buried with it nearby. The active metal provides a supply of electrons to the object (through the conductor) and keeps the iron in a reduced state (cathode). In the process, the active metal is oxidized (anode) and is gradually used up (sacrificed). When consumed, the anode can be more easily and inexpensively replaced than can the object it protects, and more important, leaks of fuel and other substances into the environment are prevented.

▶ Figure 9.21
A sheet of continuously galvanized steel shines as it leaves the cooling chambers at Bethlehem Steel Corporation.

DID YOU LEARN THIS?

For each of the statements given, fill in the blank with the most appropriate word or phrase from the following list. You may use any word or phrase more than once.

electrolytic cell	electrolyte
voltaic cell	nonelectrolyte
fuel cell	photosynthesis
oxidation	oxidizing agent
reduction	reducing agent
corrosion	redox
passivation	bauxite
anode	electrolysis
cathode	salt bridge
galvanizing	

a. _______ Unwanted oxidation of metals in the environment
b. _______ The electrode at which oxidation takes place in an electrochemical cell
c. _______ A device that converts chemical energy into electrical energy as long as reactants are fed into it
d. _______ A substance whose water solution conducts on electric current
e. _______ A device in which nonspontaneous chemical processes are made to proceed by application of an electric current
f. _______ Protection of metals with a thin layer of oxide
g. _______ The kind of chemical reaction that nonmetal elements undergo when they react with metals
h. _______ A natural process which is the reverse of combustion
i. _______ Breaking water into hydrogen and oxygen by use of an electric current
j. _______ Forms the electrical connection between electrolyte solutions in a voltaic cell
k. _______ A name for the oxidation–reduction process
l. _______ Aluminum ore
m. _______ The role of oxygen in metabolism

EXERCISES

1. List three ways to determine whether a substance is oxidized. There are also three ways to decide whether a substance is reduced.

List them. Find several examples in this chapter in which more than one way would have worked.

2. In each of the following reactions, show which chemical species is oxidized and which is reduced. Also, indicate which is the oxidizing agent and which is the reducing agent.

   a. $2Na + 2H_2O \longrightarrow 2Na^+ + 2OH^- + H_2$
   b. $AlCl_3 + 3K \longrightarrow Al + 3KCl$
   c. $CH_4 + Cl_2 \longrightarrow CH_3Cl + HCl$
   d. $(CH_3)_2CHOH + ClO^- \longrightarrow (CH_3)_2C{=}O + H_2O + Cl^-$
   e. $Zn + S \longrightarrow ZnS$
   f. $4NH_3 + 5O_2 \longrightarrow 4NO + 6H_2O$
   g. $P_4O_{10} + 10C \longrightarrow 10CO + P_4$
   h. $2Al + Fe_2O_3 \longrightarrow 2Fe + Al_2O_3$

3. Complete and balance the following combustion reactions. Assume there is enough oxygen for complete combustion. What compound forms if there is not enough oxygen?

   a. $C_2H_6 + O_2 \longrightarrow$
   b. $C_6H_6 + O_2 \longrightarrow$
   c. $C_5H_{12} + O_2 \longrightarrow$
   d. $C_{10}H_{22} + O_2 \longrightarrow$

4. Look through this chapter and make a list of the "fuels" used as examples. What chemical property do they all have in common?
5. Why must the processes of oxidation and reduction occur simultaneously? How does the terminology we use for them reflect this fact?
6. What is a carbohydrate? How are carbohydrates formed? Write the balanced equation for the reaction when ribose, $C_5H_{10}O_5$, is metabolized in the body.
7. What is photosynthesis? What kinds of compounds does it produce? What is its source of energy? Why is the process of photosynthesis so important?
8. What is an electrolyte? A nonelectrolyte? How do electrolytes conduct an electric current? Why doesn't solid NaCl behave as an electrolyte?
9. Which of the following compounds would produce an aqueous (water) solution that would conduct electricity?

   a. $HCl(g)$
   b. $NaOH(s)$
   c. isopropyl alcohol
   d. baking soda
   e. acetone
   f. HNO_3

10. How is a voltaic cell designed to produce electrical energy from a redox reaction? That is, what must be done to make the electron transfer proceed through an external wire? What chemical condition exists when a cell is run down? In terms of energy, what must be done to recharge a cell? What chemical processes occur when a cell is recharged?
11. What would happen if a dry cell had no water in it? Explain.
12. How is an electrolytic cell different from a voltaic cell or a battery? List some uses of electrolytic cells.
13. How does a fuel cell differ from a voltaic cell? How are the two similar?
14. What is corrosion? List some ways that metals can be protected from it. What is passivation?
15. What property of zinc and magnesium makes them useful for protecting iron objects from corrosion? Describe the chemical process by which zinc protects galvanized iron or steel. How is this process similar to the role that sacrificial anodes play in protecting underground storage tanks? Both processes involve the same kind of electrochemical cell. What kind is it? Try to figure out what makes up the salt bridge in each process.

THINKING IT THROUGH

Weak Electrolytes

A solution of acetic acid in water (like vinegar) conducts electricity, but poorly. An aqueous solution of ammonia behaves similarly. These compounds are called weak electrolytes. Write equations that show the behavior of each when dissolved in water. Why do their water solutions conduct electricity only slightly?

Aqua Regia

Gold does not dissolve in either hydrochloric acid or nitric acid separately. However, a mixture of the two acids, called *aqua regia,* readily oxidizes gold and dissolves it according to the equation,

$$Au(s) + 4H^+ + 4Cl^- + NO_3^- \longrightarrow AuCl_4^- + NO(g) + 2H_2O$$

a. How many electrons does a gold atom lose when it is oxidized?
b. The $AuCl_4^-$ ion is quite stable. Which acid is (or furnishes) the oxidizing agent? Why is the other acid necessary to cause the reaction to proceed?

Using Electricity Safely

Water that is pure does not conduct electricity. Yet we know that operating power tools outdoors on a wet day or using an electric appliance while in the bathtub can result in electric shock. Explain how this can happen.

Road Salt

Why does contact with road salt hasten the corrosion and rusting of automobile parts?

Which Is Which?

In an electrolytic cell, the cathode is the negatively charged electrode and the anode has a positive charge. However, in a voltaic cell, the cathode is positively charged and the anode is negatively charged. Explain whether this observation is consistent with the definitions given for *cathode* and *anode.* Can you tell whether an electrode is a cathode or an anode by determining whether it has a positive or negative charge? Explain.

ANSWERS TO "DID YOU LEARN THIS?"

a. corrosion
b. anode
c. fuel cell
d. electrolyte
e. electrolytic cell
f. passivation
g. reduction
h. photosynthesis
i. electrolysis
j. salt bridge
k. redox
l. bauxite
m. oxidizing agent

CHEMISTRY NEEDS OF A SOLAR ENGINEER

Jim Williamson

Solar Engineer, National Renewable Energy Laboratory, *A Division of Midwest Research Institute*

B. S. Mathematics, Montana State University

M. S. Mathematics, University of California, Berkeley

Solar energy is an emerging field that is growing rapidly in the 1990s. Since few universities offer a field of study specifically pointed toward solar technologies, it is one of those fields that you tend to grow into. I started in mathematics for the atomic energy technologies and moved into a specialty energy field of power for satellites and then gradually into solar systems. This pattern is not unusual; many satellites used solar energy when there were very few solar systems on the ground. The common threads of my career have been math and science, with chemistry being particularly significant because of the importance of materials in my fields.

The leading solar technology is photovoltaics, the process of converting sunlight to electricity in solar cells. The reliance of solar technology on chemistry is characterized by this special group of semiconductor materials that nature has blessed with the property of producing electricity when exposed to sunlight. The trick in developing the technology is to select semiconductors that will produce the most electricity at the least cost. Research into materials is conducted at the Solar Energy Research Institute (SERI), in Golden, Colorado. SERI, a national laboratory of the U.S. Department of Energy, is also studying a variety of other solar systems that rely on a basic understanding of chemistry. The chemistry foundation of all research into solar energy technologies cannot be overstated. My high school chemistry teacher told us that "Chemistry is the queen of the sciences because it is the basis for so much of our understanding." To me, that understanding flows from basic chemistry to the interactions of more complex materials to final applications in solar devices. Chemical reactions have always been a major part of my work and have enabled me to take an analytical approach to problem solving. Problem solving in research, though structured and exacting, can still be fun and may include travel to exotic places.

One of my favorite research projects involved the design and operation of a pilot plant for desalting seawater with solar energy. The project was jointly sponsored by the national solar research programs in the United States and Saudi Arabia. Most of the design and engineering was done in the United States, and the plant construction and testing were done on the Red Sea coast near Yanbu, Saudi Arabia. The desalination process used an innovative freezing technique wherein the seawater was frozen into ice crystals and the salt was rinsed off the ice. This process was developed to replace more energy-intensive osmotic membrane separations or conventional distillation techniques. The basic chemistry principle at work is that less energy is required to freeze water than to boil it. The solar technology in the project was an array of large solar thermal collectors with mirrors to concentrate heat into a receiver fluid that operated standard air-conditioning equipment.

The project went smoothly through design, construction, and initial operation. We enjoyed the Red Sea area with its colorful reefs and fishes. The rural community of Yanbu provided a glimpse of Bedouin life-styles and cuisine. Then a problem with the "queen of sciences" occurred.

The individual mirrors of the solar collectors began to turn black almost before our eyes. "Black-lace" degradation of mirrors is very common in old or antique mirrors that do not have adequate backing or protection for the silvering. The black, nonreflective surface occurs when the silver reacts with salts, forming a black silver substance. However, the mirrors for our experiment had been carefully manufactured with a special low-iron glass to improve reflectivity and a thick anticorrosion paint to protect the silvering. The project for making fresh drinking water from the sea quickly turned into a giant experiment in mirror coatings.

We finally identified the source of the corrosive salts as the exhaust plume of a nearby petrochemical plant in combination with the natural salinity of the air near the seacoast. The solution was a thick, resistant epoxy coating that slowed the penetration of the airborne salt into the mirror coatings. Thus, the project was not only successful in demonstrating an innovative method of desalination but also had the unexpected benefit of developing a new mirror protection for the mirror industry. Saudi Arabia has continued to develop desalination plants and now has the largest systems in the world; the plants are the principal sources of drinking water for even the inland cities.

When I first took chemistry in high school and college, I was intrigued by the investigative nature of experimentation. The scientific approach to problem solving that I learned in those classes has been invaluable in my career in solar technology research. In addition, I am always on the lookout for materials problems and reactions that underlie engineering challenges.

Energy and Nonrenewable Natural Resources

You can't get something for nothing. You can't even break even. The more you use, the less you can get from what you have left.

work done = energy available to do work in source − w

G = ΔH − TΔS

rk   energy available   wasted energy

Early in this text we defined energy as "that part of the world which makes things move." We have used ideas such as the energy hill, stability, higher and lower energy, and spontaneity to explain why most of the elements are found in combined form, how electrons are arranged in atoms, what causes chemical bonds to form, and why substances dissolve. However, we have talked little about energy itself.

Energy is an important part of chemistry. In fact, it probably is the most important aspect of physical and biological science because it is the ultimate driving force for all processes in the universe. Things that move are always being pushed by some kind of energy change. Energy manifests itself in motion: we cannot observe energy if it is still.

TYPES OF ENERGY

The different kinds of energy are interconvertible.

There are several kinds of energy, including electrical energy, chemical energy, radiant (electromagnetic) energy, mechanical energy, nuclear energy, and thermal energy, commonly called heat. We have seen **electrical energy,** carried by moving electrons, used to excite atoms in a discharge tube and produce light, which is **radiant energy. Chemical energy** is the energy given off in chemical reactions when stronger bonds are formed at the expense of weaker ones. **Mechanical energy** is produced by moving objects, such as falling water spinning a turbine or a bow shooting an arrow (▶ Figure 10.1). **Nuclear energy** is released when matter is converted to energy in nuclear processes, and **heat** is contained in the motion of the atoms, ions, or molecules that make up a substance. These forms of energy are all interconvertible, such as from electrical to radiant in the discharge tube, from chemical to electrical in a battery, and from thermal to mechanical in a steam engine. Energy is further classified as energy of motion or energy that is stored: *kinetic energy* or *potential energy*, respectively.

▶ Figure 10.1
Windmills in this wind farm convert the mechanical energy in winds to electrical energy.

10.1 Indicate the kind of energy (mechanical, electrical, nuclear, chemical, radiant, or thermal) produced by each source listed (there may be more than one kind).

a. gasoline
b. a lawnmower engine
c. the fission of uranium
d. food
e. lightning
f. a battery
g. the wind
h. the sun
i. a steam radiator
j. the human body
k. a light bulb
l. a waterfall

10.2 Classify each of the following as containing kinetic or potential energy (some could have both).

a. hot water
b. an airplane in flight
c. sunlight
d. a battery
e. a book on a high shelf
f. tides
g. a match
h. natural gas
i. hot soup
j. wood
k. water in a reservoir
l. electricity

(Properly, thermal energy is the energy of motion of the particles in a substance, and heat is the thermal energy that is transferred from one body to another. However, thermal energy is commonly called heat, and so we shall use the term *heat* to mean both.)

TEMPERATURE AND HEAT

We seem to know intuitively that a hotter object somehow contains more energy than a colder one. For example, a liter of boiling water can melt more ice cubes than a liter of lukewarm water can. Yet is temperature a measure of energy? The units of temperature are expressed in "degrees," which mean relative intensity, not amount.

An idea of intensity can be obtained by comparing the sweetness of sugar with that of the artificial sweetener, Nutrasweet. It takes only a small amount of the latter to sweeten a cup of coffee, several hundred times less than the amount of sugar needed to get a corresponding sweetness. The sweetness of

Nutrasweet, then, is very intense: a little goes a long way. Another example of intensity is the body given to perfumes by the addition of tiny amounts of skunk oil or ambergris (a waxlike substance from the intestine of the sperm whale); and another is the pleasant odor of freshly ground, roasted coffee, which is due to the presence of traces of 3-nonenal. Large amounts of these substances have an unbearable odor. So we could say that temperature is a measure of the intensity of heat. The higher the temperature an object has the more heat it contains.

Temperature and heat are different.

But heat is different from temperature. The filament in a light bulb is very hot, about 1650 °C (3000 °F); an electric oven set at 200 °C (390 °F) is much lower in temperature. Which of these devices would we choose to boil a pan full of water? Although the light bulb filament is very hot, using it to heat the water to boiling would take a long time. The oven is cooler, but it will boil the water more quickly. If we had an oven filled with many light bulb filaments instead of glowing Calrod units, our oven would not only be very hot, it would boil the water in an instant.

The amount of heat depends on the amount of matter and the temperature.

The amount of heat an object contains, then, must depend not only on the temperature but also on the amount of matter (mass) that is heated (▶ Figure 10.2). The light bulb filament, though very hot, has a tiny mass and thus contains relatively little heat. The oven, though cooler, is massive in comparison and contains lots of heat. When we pay our electric bill, we are paying for the amount of energy we use. To "burn" a light bulb is inexpensive because it takes relatively little energy, but the electric stove is another matter.

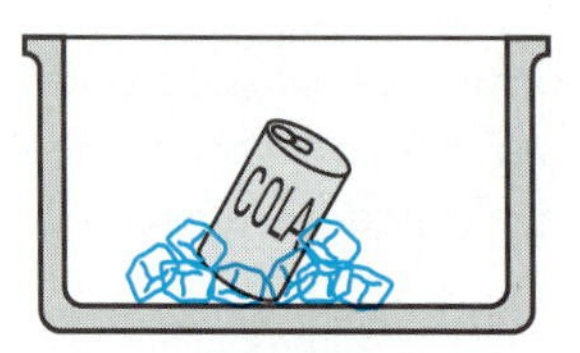

(a) Initial temperature = 25 °C
Cooled to 5 °C

(b) Initial temperature = 25 °C
Cooled to 5 °C

▶ Figure 10.2
The ice must absorb six times as much heat to cool six cans (b) as to cool one can (a), beginning and ending at the same temperatures.

10.3 Which contains more heat?

a. a steam radiator or a hot cup of tea
b. a bucket of ice or the bucket full of cool water
c. a kilogram of molten lead or a kilogram of solid lead
d. a candle flame or a gas burner flame

ENERGY IS CONSERVED

▶ Figure 10.3
Heat flows from the hot burner to the cooler pan and the water in it.

Energy is conserved.

Early scientists who studied heat thought it was an invisible fluid because it seemed to flow from one place to another, and today we still use the term *heat flow* when we talk about it. The study of the movement of heat is called **thermodynamics.** (*Thermo-* = heat; *dynamics* = motion.)

Adding heat to a substance causes its particles (atoms, ions, or molecules) to move faster and become more active. The source of heat for this substance must always have a higher temperature because heat always spreads (flows) from a hotter body to a cooler one, never the other way around (▶ Figure 10.3).

Also, the heat is conserved in the process. That is, the amount of heat received by the cooler body is *exactly equal* to the heat given up by the hotter one. This is called the *law of conservation of energy:* energy can neither be created nor destroyed. This law applies not only to processes that involve the movement of heat; it can be generalized to all other energy conversion processes as well. This general conservation law is called the **first law of thermodynamics.**

(Like the law of conservation of mass introduced in Chapter 4, this energy conservation law is true for all nonnuclear processes. We learned in Chapter 5, however, that matter and energy are interconvertible. In nuclear processes matter can be converted to energy, and energy to matter. Thus, a more general conservation law is that, in the universe, the amount of matter *and* energy remains constant.)

Heat can be transferred from a hotter body to a cooler one in two ways. If the body at higher temperature is in contact with the one at the lower temperature, the motion of the more-active particles at higher temperature is transferred by collision of the particles between the two bodies. As the more-active particles of the hotter body bump into the particles in the cooler body, they cause the particles in the cooler body to move faster. The more-active particles themselves then begin to move more slowly. In this manner, the hotter body cools down and the cooler body warms up until they both reach the same temperature (▶ Figure 10.4). This process is called **conduction.**

If the two bodies are not in contact, heat can be transferred by *radiant heat,* which is simply long-wavelength light (electromagnetic radiation). The radiant heat only heats substances that can absorb it, and this absorption is caused by the particles in the substance *coupling* with the electromagnetic field and absorbing its energy. This absorbed energy then becomes heat in the form of vibrations, rotations, or translations (sideways movements) of the particles. A steam radiator works this way, and so does a campfire. A good example of coupling is the heating of food in a microwave oven. The microwave oven is "tuned" to a wavelength that corresponds to the amount of energy required to spin a water molecule. When the photons of microwave

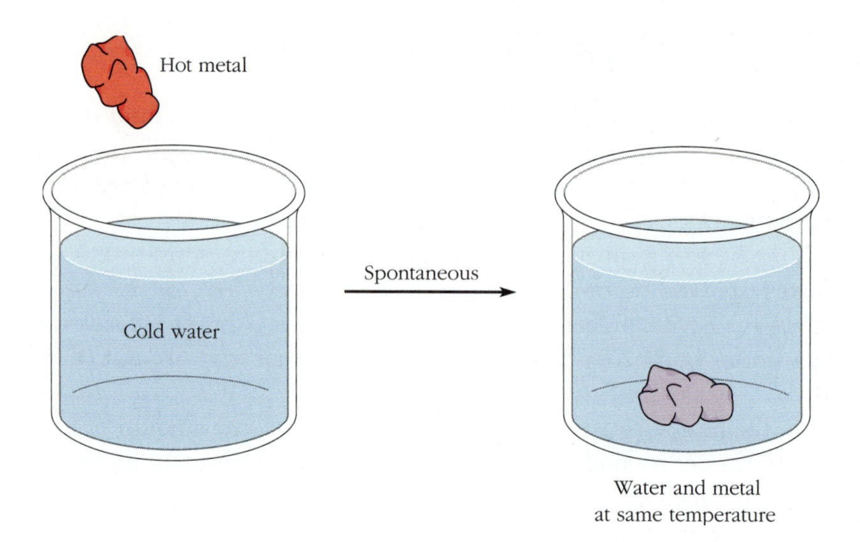

▸ Figure 10.4
The Process of Conduction. *Active atoms in the hot metal give their energy to the less active water molecules by colliding with them, until both substances reach the same temperature.*

radiation strike the water molecules, they are absorbed and their energy causes the water molecules to rotate. This rotational motion is soon converted into vibrational and translational motions by collision of the water molecules with one another, and the substance containing the water "gets hot." Most substances that do not contain water cannot be heated in a microwave oven. For example, popcorn can be popped inside a paper bag in a microwave oven and the bag remains cool. The popcorn contains water and the paper bag does not.

KINETIC ENERGY AND TEMPERATURE

The **kinetic energy** of an object depends on the *square of its velocity.*

The energy of motion of particles, large and small, is called **kinetic energy** (KE). This energy is dependent on the mass *(m)* of the particles and on their velocity *(v):*

$$KE = \frac{1}{2} mv^2 \qquad \text{(Note the } \textit{squaring} \text{ of the velocity.)}$$

The velocity is the more important factor because doubling the velocity of the particles quadruples their kinetic energy. This is one reason that driving a 55 mph speed limit on our highways saves lives. To drive 65 mph instead of 55 mph (an 18% increase in velocity) increases the kinetic energy of a vehicle (and its occupants) by 40%. At 70 mph (a 27% velocity increase), this killing power increases by 62%!

Temperature, which we have described as a measure of intensity of heat, is actually a measure of the average kinetic energy of the particles that make up

10.4 If the ice cubes, liquid cold pack, and air in the freezer all have the same average kinetic energy in their particles, what is it about their chemical structure that causes them to exist in different physical states at the same temperature?

matter. This means the particles of all substances that are at the same temperature have the same average kinetic energy, whether the substances are solids, liquids, or gases at that temperature (▶ Figure 10.5). It at first may seem strange that the particles of ice cubes, the liquid cold pack, and the air in a freezer all have the same average kinetic energy, but this is indeed the case.

ABSOLUTE ZERO

If temperature is a measure of the average kinetic energy of particles in a substance, what would happen if we could cool substances to slow the motion of their particles (atoms, ions, or molecules) until it stopped? The temperature at which this would happen would be very low, and we could never go below it. This is called **absolute zero.** In principle, and in practice, we can never reach this temperature, which is –273.16 °C, because this would violate the

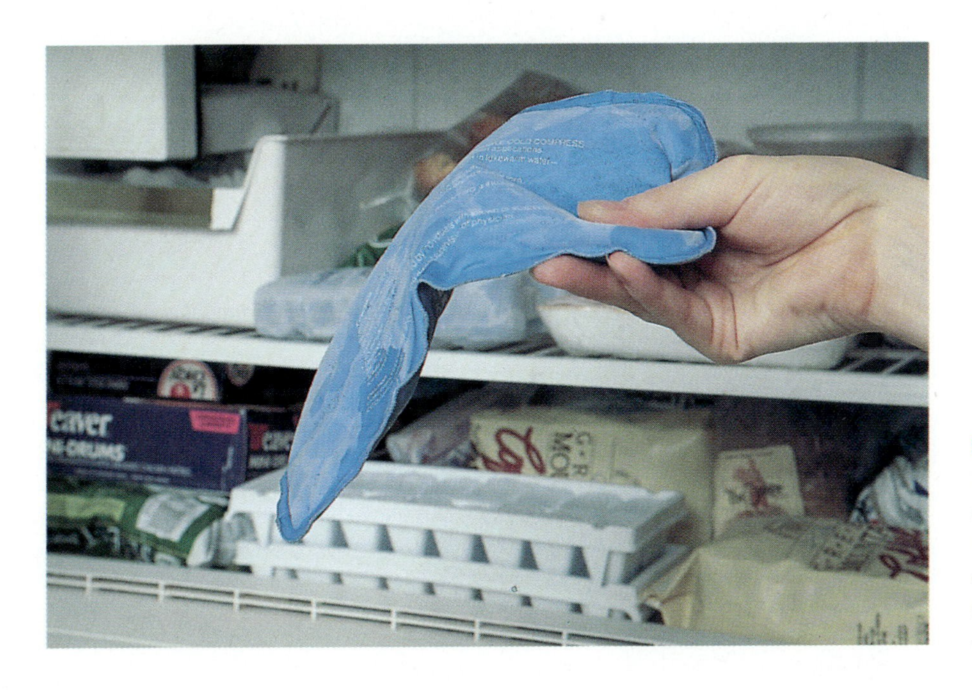

▶ Figure 10.5
The ice cubes, liquid cold pack, and the air in this freezing compartment are at the same temperature. The average kinetic energy of the particles in each (whether solid, liquid, or gas) is also the same.

uncertainty principle. If, at absolute zero, all the particles could be completely stopped, we would know their position and momentum exactly. However, the uncertainty principle states that we cannot know both position and momentum with exactness for very small particles. (This means that when our particles are "stopped" at absolute zero, the limits of error are rather large, and for all practical purposes the particles are actually moving.) In practice, we also find that it would take an infinite amount of energy to cool a substance to absolute zero (say, in a very powerful refrigerator); that is, the colder we get a substance, the more energy it takes to cool it just a little bit more. (We have been able to cool a *few atoms* to 0.0003 K.) Even if we could cool a substance to absolute zero, its particles would not be standing still. The uncertainty principle tells us just how much they would be moving and how much energy they would have. This residual energy of motion of particles at absolute zero is called the **zero-point energy.**

TEMPERATURE SCALES

Various temperature scales commonly in use today are the Fahrenheit scale (°F), the Celsius scale (°C), and the kelvin scale (K, *not* °K) (► Figure 10.6). On the *Fahrenheit scale,* water freezes at 32 °F and boils at 212 °F, and on the *Celsius scale,* at 0 °C and 100 °C. There are 180 degrees between the freezing point and the boiling point of water on the Fahrenheit scale and 100 degrees between these two points on the Celsius scale. Thus, the size of a Celsius de-

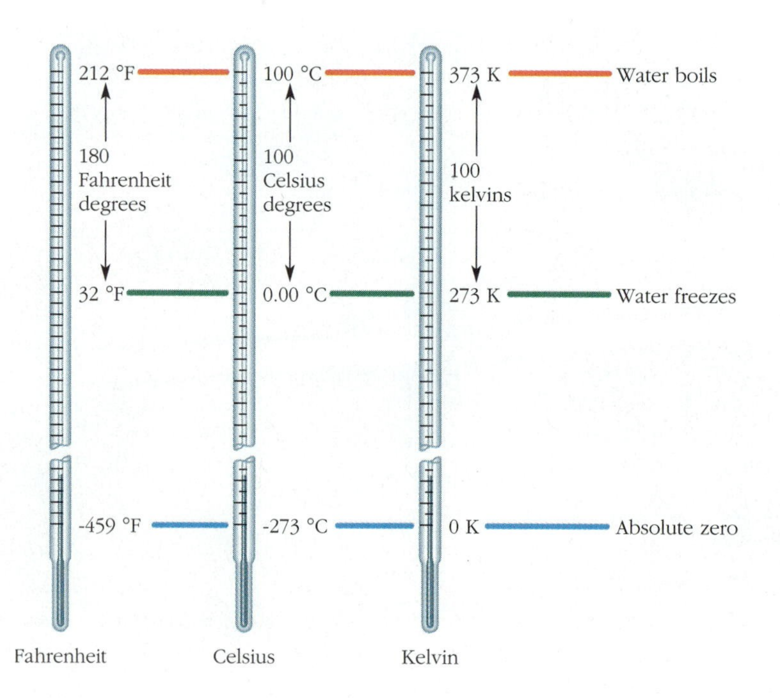

► Figure 10.6
Fahrenheit, Celsius, and Kelvin Temperature Scales. *Absolute zero is the lowest temperature possible.*

10.5 A formula often used to convert from Celsius to Fahrenheit is: $T_F = (9/5)T_C + 32$, where T_F is the Fahrenheit temperature and T_C is the Celsius temperature. Convert the following Celsius temperatures to Fahrenheit temperatures:

a. 100 °C
b. 60 °C
c. –40 °C
d. 5 °C

10.6 The formula commonly used to convert Fahrenheit to Celsius is $T_C = (5/9)(T_F - 32)$. Convert the following Fahrenheit temperatures to Celsius temperatures:

a. 77 °F
b. –13 °F
c. 98.6 °F (normal body temperature)
d. 212 °F
e. 59 °F

gree is 180/100, 9/5, or 1.8 times that of a Fahrenheit degree. The *kelvin scale* is really the Celsius scale adjusted so that zero is at absolute zero. Thus, 0 °C is 273 K, and 0 K is –273 °C (32 °F and –456 °F). The sizes of the Celsius degree and the kelvin are the same.

UNITS OF ENERGY

The amount of heat contained by a substance depends on its temperature multiplied by its mass. The units for heat contain this relationship, although they may not be expressed in mass × temperature directly. Common units of heat are the **calorie** (cal, the amount of heat required to raise the temperature of a gram of water by 1 °C), the *British thermal unit* (Btu, the amount of heat required to raise the temperature of one pound of water by 1 °F), the kilowatt-hour (kwh), the therm (used for natural gas), the joule (J), and the food calorie (Cal). The *Cal* is equal to 1000 cal (1 kcal, or kilocalorie), and in everyday usage we must specify which one we are using. Most chemists have used the calorie (cal) or kilocalorie (kcal), but by international agreement (SI) the **joule,** J, is now the most widely used (1 cal = 4.19 J).

1 *Cal* (food calorie) equals 1000 **calories** (cal).

Because the amount of heat a substance contains is proportional to the product of its temperature and its mass, shouldn't doubling its temperature also double the amount of heat it contains? Not necessarily. This relationship holds only if the temperature is expressed in kelvins, because the zero on this scale is really zero and not some arbitrarily defined temperature. Temperatures on the other scales must be converted to kelvins (or the Fahrenheit equivalent, Rankine degrees) before the amount of heat is proportional to the temperature.

ENERGY CONVERSIONS AND THE SECOND LAW

If we utilized the amount of heat generated by burning a slice of bread (about 100 Cal), how much could we increase the temperature of a liter of water? This is enough heat to raise the water temperature by about 100 °C (180 °F), or enough to heat a liter of ice-cold water to nearly boiling (but *not* boiling). We are assuming here that all of the heat in the bread is conserved when it is transferred to the water and none is lost.

In real-life situations, energy utilization is somewhat different. Let's look at some examples (▶ Figure 10.7).

The *% used* in Figure 10.7 indicates the efficiency of each process. Note that in each case, the source of energy (fuel) contains a certain amount of heat but not all of it is converted into useful work (that is, work we want the energy in the fuel to do). This is a manifestation of the **second law of thermodynamics.** Any time energy is converted from one form to another, some of it cannot do work. In other words, some of the energy is wasted. This is one of the most important laws of physical science. The second law applies not only to thermodynamics but also to natural resource management, living systems and evolution, economics, and social systems.

Some energy is always wasted in an energy-conversion process.

ENERGY AND WORK

Kinetic energy, the energy of motion, is energy in action. We can see what it does and we know how to use it because it is a part of things that move. **Potential energy,** often called stored energy, is different. It is really *potentially* energy, something that has the capacity to be kinetic energy. Potential energy is really not observed until it causes something to move. For example, the energy stored in a car battery does nothing until the starter switch is turned. Coal, petroleum, chemical energy, nuclear energy, and the energy stored in a rock hanging from a cliff are only sources of energy, not energy in themselves. The energy in them must be *converted* before it can become useful.

Work is the expenditure of energy to move mass. This means that for work to be done, kinetic energy in some form must move a mass with some kind of a force through a distance. Sometimes the kind of work being done is not obvious, such as in the case of electrical work; but electrons and magnetic fields are moving in an electric motor.

Work is the expenditure of energy to move mass.

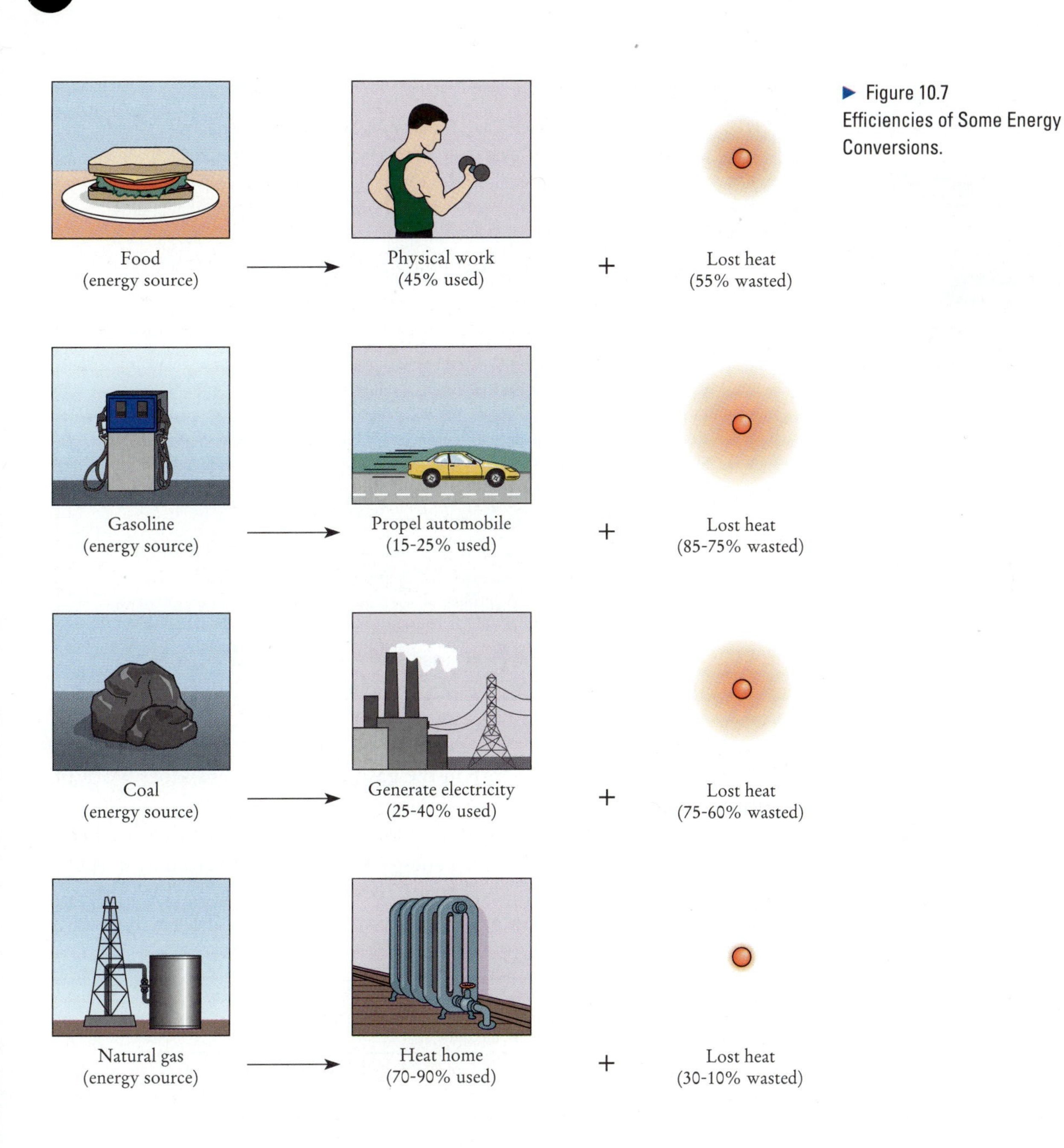

► Figure 10.7
Efficiencies of Some Energy Conversions.

We usually think in terms of using an energy source to do work for us, such as cook our food, power a radio or computer, or propel an automobile (► Figure 10.8). This energy source, say a gallon of gasoline, is a system that contains a certain amount of potential energy, which can be converted into kinetic energy, which then can do work to move a car. In this instance, work

▶ Figure 10.8
A ski lift does work on skiers when taking them to the top of the mountain, "charging" them for a downhill run.

10.7 What kind(s) of energy conversions (mechanical to electrical, electrical to radiant, mechanical to mechanical, and so on) take place in each of the following? (Note that some have several steps.) In which step(s) is potential energy being converted to kinetic energy? In which step(s) is kinetic energy being converted to potential energy?

a. A battery produces electricity.
b. A battery produces electricity that runs a small electric motor.
c. Gasoline is burned in air.
d. Gasoline is burned in an internal combustion engine to turn a wheel.
e. Water behind a dam is used to drive a turbine to produce electricity.
f. Electricity is used to run an electric motor to pump water into a tank in a water tower.
g. Wind is used to generate electricity to charge a battery.
h. Sunlight is used to power a calculator.
i. Sunlight warms a black Laborador retriever on a cold winter day.
j. A bowstring is pulled, then released to shoot an arrow.

is done *by* the system. The energy in the gasoline (the system) does work in moving the car (▶ Figure 10.9a).

Work can also be done *on* a system. Consider a car battery that has been discharged. In this state the potential energy that can be converted to start the car is insufficient. Connecting the battery to a battery charger, which uses an external electrical energy source of higher potential energy (PE), will restore

▶ Figure 10.9
Energy Conversion Processes. *(a) An exergonic process. The energy in gasoline does work moving a car. (b) An endergonic process. Work is done on a battery when it is recharged.*

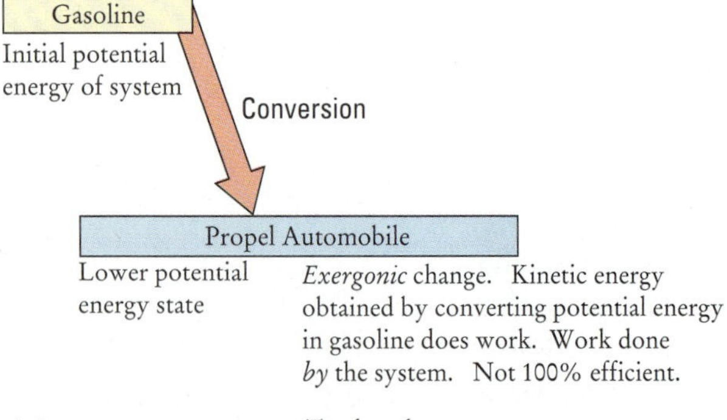

Higher potential energy state

Charged battery

Endergonic change. Work done *on* the system. Kinetic energy to do this is obtained from external, more intense source. (Electrical source of higher potential energy.) Must put in more energy than just potential energy difference between initial and higher state. Not 100% efficient.

Conversion

Dead battery

Initial potential energy of system

Tends not to be spontaneous.

(b)

10.8 In each of the following situations, work is done by or on something. In each example, indicate whether work is being done by or on the object written in bold type.

a. A man pushes a **lawnmower.**
b. Gasoline is pumped by an **electric motor.**
c. **Hot expanding gases** spin a turbine.
d. A **bungee cord** is stretched by a falling bungee jumper.
e. An electric motor turns when **electrons** flow through it.
f. Passengers are lifted in a **hot-air balloon.**
g. A **woman** rows a boat upstream.
h. A man pumps **water** from a well.
i. A woman lifts a **baby** into her arms.
j. A kayaker floats with the **water** downstream.
k. A **battery** produces sound from a portable radio.
l. Electric current causes **fluorescent lights** to glow.
m. Falling water turns a **generator** to produce electricity.

it to full charge, a higher potential energy state, capable of turning the engine. In this case, the electrical energy source—one of higher potential—does work *on* the battery to recharge it (Figure 10.9b).

The term **exergonic** refers to work done by a system. Similar to exothermic processes (Chapter 3), exergonic processes tend to be spontaneous. **Endergonic** processes, as when work is done on a system, tend not to be spontaneous. The terms exergonic and endergonic refer to energy in *any* form, not just heat, that is given off or taken up by a system, respectively. Note the difference between heat and work. Heat is thermal energy, the kinetic energy of particles in a substance. Work is the expenditure of energy to move mass.

The term *stability* can be used to refer to either heat (or other form of energy) or work. Any energy state that can spontaneously change (or tends to change) to a lower energy state is said to be unstable. An energy state that cannot change to a lower energy state is said to be stable.

POWER

The earliest sources of energy, other than animals or humans, were moving water and air. The waterwheel—as a means of harnessing the kinetic energy of moving water to grind grain, pump water, and eventually do mechanical tasks—has been used in some form since recorded history began (▶ Figure 10.10). Windmills in Holland were used to pump water from behind dikes in order to uncover fertile land. The problem with these devices was that they

Power is the amount of energy converted in a certain period of time.

▶ Figure 10.10
Waterwheels convert energy slowly, and thus are of low power.

were of low power. They could do a lot of work, but they could not do it very fast. **Power** is the rate of energy conversion; the amount of energy used or produced in a given amount of time. Units of power are, for example, horsepower, watt, kilowatt, Btu/h, and therm/day.

To produce more power, it was necessary to convert some form of high-intensity heat to mechanical motion. In the steam engine this was done by burning wood or coal. The steam engine generated power more conveniently than did the waterwheel; it could be built anywhere and could be used to power steamships and trains. In large measure the Industrial Revolution was made possible because of the availability of the power (the ability to convert energy *rapidly*) produced by steam engines.

The steam engine had a number of disadvantages, though. It needed a continual source of water, it was heavy for the amount of power it produced (had a low power-to-weight ratio), it was slow to warm up, and its energy source (primarily coal) was inconvenient to use. The internal combustion engine, developed in the late 1800s, had a higher power-to-weight ratio, gave instant power, needed no continual source of water, and its energy source (petroleum) was concentrated (high intensity), portable, and cheap (▶ Figure 10.11). Steam engines are seldom seen today; instead, more efficient and powerful steam turbines are used to generate huge amounts of electricity and propel ships at sea.

GETTING WORK DONE (ENERGY IS WASTED)

The amount of work obtained is always less than the amount of energy expended to do it.

Energy is conserved, but work is not. The amount of work that can be obtained from an energy source is always less than the amount of energy originally available in it, and some energy, as heat, is always wasted in the conversion process. This is a consequence of the second law of thermodynamics. The work actually done is equal to the amount of energy available to do work in the source minus the energy wasted in the conversion process:

$$\text{work done} = \begin{matrix}\text{energy available}\\ \text{to do work in source}\end{matrix} - \text{wasted heat}$$

This same relationship between work, the energy available to do work, and the energy wasted can be expressed mathematically as follows:

$$\underset{\text{work}}{\Delta G} = \underset{\text{energy available}}{\Delta H} - \underset{\text{wasted energy}}{T\Delta S}$$

▶ Figure 10.11
A Radial Aircraft Engine. *Internal combustion engines are powerful, portable, and inexpensive to operate.*

The term, ΔG, is called the change in **free energy** (actual work) and ΔH is the change in **enthalpy** (change in heat available). [The symbol Δ (delta)

10.9 Where does the waste energy go in each of the following devices?

a. a lightbulb
b. an electric motor
c. a tennis player
d. a hacksaw cutting metal
e. an electric refrigerator
f. an automobile
g. a stereo amplifier
h. a cannon
i. a washing machine

means *change in.*] Of importance is the $T\Delta S$ term (T in kelvins). This represents the energy that cannot do work, the energy that is wasted. This energy is not destroyed, but rather it takes the form of dissipated, diffuse, and generally useless low-potential-energy heat. It has been said that "low-quality heat is the garbage dump of energy."

ENTROPY

Entropy is a measure of disorder.

The change in **entropy,** ΔS, is a measure of the amount of randomness or disorder caused by a conversion of energy from one form to another when work is done. A state of high disorder is one of high (or positive) entropy, and a state of low disorder (or increased order) is one of low (or negative) entropy. The tendency in nature is toward a state of greater disorder, and changes in this direction, called **entropic** changes, tend to be spontaneous (► Figure 10.12). Changes in the direction of greater order are said to be **negentropic** and tend not to be spontaneous.

Energy is required to create order.

Creation of order requires the use of energy. This energy must come from an external source that is in a state of greater order (lower entropy) than the state that is to be created. Creation of order in one place causes a larger increase in disorder somewhere else.

The entropy change depends on the amount of energy converted and the change in temperature that occurs during the conversion. This relationship can be expressed as follows:

$$\Delta S = \frac{q}{\Delta T}$$

(a)

(b)

► Figure 10.12
The most probable state is one of disorder. *(a) A drop of dye is released at the bottom of a beaker filled with water. (b) With no mixing the dye spontaneously spreads out (diffuses) and becomes distributed throughout the liquid.*

where S = the entropy, q = the amount of heat, and T = the absolute temperature. Note that the change in entropy is much smaller when the change in temperature (ΔT) is large. This means that less energy is wasted (the process is more efficient) when an energy conversion process starts at a very high temperature and ends at a very low one. To accomplish this, however, a source of high potential energy must be available to generate high temperatures.

IMPLICATIONS OF THE SECOND LAW

Some energy is always wasted (not destroyed) when work is done in energy conversion processes. This means not only that we will be unable to break even in getting work from an energy source, but also that we must always come out behind. This does not mean, however, that entropy cannot be decreased (order created) in a system and that potential energy cannot be increased within it. Given a suitable source of higher potential energy (and greater order) from an external source, it is perfectly possible to do so. We do it all the time, for example, when we mine ores, refine them, and use the metals to build automobiles (► Figure 10.13). It does mean, though, that the tendency for these events to occur is not very probable and that they are not likely to happen on their own.

10.10 Which of the following in each pair has greater entropy?

a. popcorn packaging material that is in a box or that is loose in the room
b. a used car lot or an automobile salvage yard
c. sulfur dioxide emissions in the air or at the stack
d. iron in an ore or in a tool
e. a shirt or a bolt of cloth
f. pure water or polluted water
g. a neat house or a messy one
h. a lump of coal and cold water or hot water and ashes
i. water vapor or ice
j. a landfill or a supermarket

▶ Figure 10.13
Creation of order requires the use of energy. *Iron and steel have much greater order than the ores from which they are obtained.*

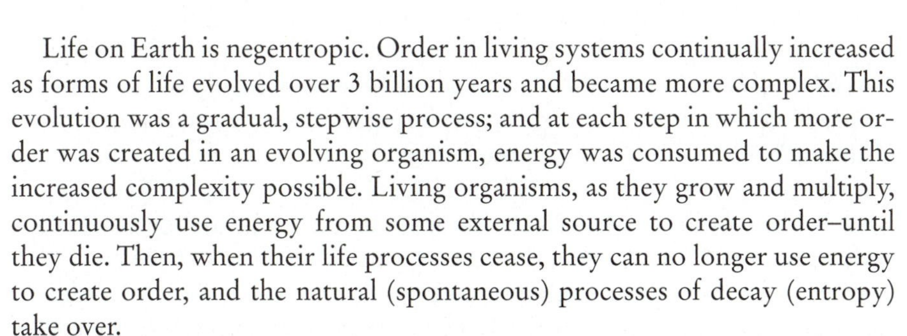

Life on Earth is negentropic. Order in living systems continually increased as forms of life evolved over 3 billion years and became more complex. This evolution was a gradual, stepwise process; and at each step in which more order was created in an evolving organism, energy was consumed to make the increased complexity possible. Living organisms, as they grow and multiply, continuously use energy from some external source to create order–until they die. Then, when their life processes cease, they can no longer use energy to create order, and the natural (spontaneous) processes of decay (entropy) take over.

The sun is the source for the creation of order on the earth.

The source of energy (and therefore order) that living organisms use for creating order is ultimately the sun. The sun is a giant nuclear reactor, which is becoming more disordered (is running down) as it gives off its energy in all directions in space. Part of this energy falls on the earth and is used to support complex and highly ordered forms of life, which can evolve and flourish using the sun's order as a source for creation of their own unique structures.

An implication of the second law is that the world is constantly running down, and that all of the energy (in the universe) will eventually be at one, very low, potential energy state, a state at which it can no longer do work. It has been estimated that when this happens, all things will be at a temperature of about 4.2 K (barely above absolute zero), and everything will be at a state of total disorder. Clausius first stated this idea: namely, that the entropy (disorder) of the world is always increasing and will eventually approach a maximum value. This time is a long way off, however, and we need not worry about it too much: the sun is expected to run down in 4 billion years or so.

Increase in entropy gives us a sense of time.

It has been said that "entropy is time's arrow." Fires burn. Mountains erode. Wood rots. We are born, we grow, and we age. We cannot unscramble an egg, nor can we undo an earthquake. The fact that entropy is always increasing, that things are running down, gives us a sense of time in a forward direction.

EFFICIENCY OF ENERGY CONVERSION PROCESSES

When we carry out an energy conversion process, we would like to get as much work (free energy) out of our energy source (enthalpy) as possible and to minimize the amount of energy that is wasted (due to entropy). That is, we want our process to be as efficient as possible. **Efficiency** is the amount of work done divided by the amount of converted energy times 100 percent, or

$$\% \text{ efficiency} = \frac{\text{amount of work done}}{\text{amount of energy converted}} \times 100\%$$

Because the energy wasted is $T\Delta S$, making the entropy change, ΔS, as small as possible will minimize energy waste and enable more of the energy in our source to do work. Making a process less entropic is always a challenge, and a number of basic approaches can be taken. One way already mentioned is to have a large temperature change occur during the energy conversion process. This requires the use of a high-potential-energy source to drive it. Coal-fired electric power plants, for example, are most efficient when the steam generated in their boilers is fed to the turbines at very high temperatures (and hence pressures). In addition, the turbines are designed to use the energy released from the cooling steam until it exits as warm water. To achieve the desired high temperatures in the boilers, it is necessary to pulverize the coal to the consistency of talcum powder and blow it, mixed with forced air, through large nozzles, which produces a very hot flame.

A second important way of increasing efficiency is to reduce friction, which takes many familiar forms. A wagon with wheels is more efficient than a sled for moving heavy objects over bare ground. A barge moving over water is even better. To "grease the skids" we use lubricants of all kinds, and without them our machines would soon fail. We streamline our cars, boats, and aircraft to reduce their resistance to air or water, so that they will go faster and use less fuel. Although entropy manifests itself in other ways, it is fair to say that in many respects entropy is friction, the friction of doing business energywise. Perpetual motion machines cannot work (and cannot do work) because entropy robs some of the energy required to move them. If no additional energy is put into them, they run down.

The more power used to do work, the more energy is wasted.

A third way to increase efficiency is to slow down energy conversion processes and use energy at a lower rate. This means using energy under conditions of low power, which is contrary to our habitual way of doing things. The human body is a relatively efficient energy converter (about 45%), but it can only do work at a fairly slow pace. Early humans, who quickly learned that there were faster (and easier) ways of getting work done, soon began to use the greater power of draft animals and (when they could get away with it) the power of large numbers of other human beings as slaves. Today our work is done rapidly by machines driven by powerful engines, and each American now uses daily an amount of energy that is equivalent to the work of about 120 slaves. Unfortunately, we pay a price for having our work done faster, because the more power we use to do a particular amount of work, the more energy we waste.

▶ Figure 10.14
Voyager, the first aircraft to be flown around the world without refueling had an average speed of only 120 mph. By flying slowly at low power, it could use its 1200-gallon fuel supply very efficiently.

In principle, we could obtain a maximum efficiency from an energy conversion process by making it proceed infinitely slowly. (As long as work is done, however, the efficiency can never be 100%.) This, of course, is impractical, but the idea can be illustrated in practical application. *Voyager*, the first aircraft to be flown around the world without being refueled, was a high-tech, ultralight aircraft carrying two persons (▶ Figure 10.14). Weighing only 460 pounds empty, it consumed 1200 gallons (3.5 tons) of fuel and traveled at an average speed of about 120 mph. By contrast, a much heavier jet fighter aircraft with a crew of two, flying at more than the speed of sound (700 mph), has a range of only a few hundred miles on the same energy supply.

Several other strategies can be used to minimize the consequences of the second law. One that is often not recognized is to reduce the number of energy transfers in the chain (▶ Figure 10.15, page 338). Generally, the fewer energy conversions used to accomplish a desired result the better the efficiency of the process. For example, from the standpoint of efficiency it makes no sense to burn natural gas (a **primary source** of energy) to generate electricity and transmit that electrical energy over long distances for the purpose of heating homes. Generating the electricity (a **secondary source** of energy), at an overall efficiency of at most 40%, involves four energy conversions: converting the chemical energy in the natural gas into heat, transferring that heat to water to make high-pressure steam, transferring the kinetic energy in the steam to mechanical (rotary) energy in a turbine, and finally, converting that mechanical energy to electrical energy. Transmission losses due to electrical resistance and transformer inefficiencies (electrical "friction") can waste another 10% of the energy in the fuel. Generating heat from electricity is almost 100% efficient (little work is done), but overall about 70% of the energy in the natural gas is wasted. Or, the overall efficiency is about 30%.

Fossil fuels are **primary sources** of energy.

Electricity is a **secondary source** of energy.

Burning the natural gas directly for home heating is much more efficient. Gas can be transmitted long distances by pipeline at low energy cost, and with the new high-tech gas furnaces it is possible to heat homes with efficiencies of 70–90%.

Although the use of electric automobiles may help reduce air pollution in certain urban areas, it will not conserve much energy. Some batteries approach 90% efficiency; but when they run down, they have to be recharged—

PERPETUAL MOTION MACHINES: Wouldn't It Be Great?

Wouldn't it be great if we could use the energy in water stored behind a dam to generate electricity that could be used to pump the water back into the reservoir and still leave us some electricity to use? Wouldn't it be great if we could use a battery to run an electric car whose motor was connected to a generator that would recharge the battery?

As early as the beginning of the sixteenth century Leonardo da Vinci wrote why such a machine—what we call a perpetual motion machine—wouldn't work, and scientists have known for centuries from an accumulation of evidence that building one is impossible. But even in the early 1900s, when the laws of thermodynamics were firmly established, many people still believed in perpetual motion machines, at least enough of them to provide the promoters of such schemes with an audience (and sometimes a living).

Usually, an inventor would claim to have a discovered a way to convert an energy source (preferably a newly discovered or mysterious one, like "electricity," "radium," or an "etheric force") to another kind of energy (commonly mechanical or electrical) that would be used to fully restore the original energy source. The machine would be a closed energy system (so no energy could escape), and therefore the machine would be self-sufficient and continue to run, indefinitely. Because a machine that would sustain its own motion wasn't much good except as a curiosity, it was often claimed that the device would also produce some extra energy that could be drawn off for other purposes.

Of course, these ideas run headlong into the laws of thermodynamics. Any time a given amount of energy is converted from one form to another, entropy makes some of it incapable of performing work, and each successive conversion of that energy renders more of it useless. Even if all of the original energy could somehow be kept inside a perpetual motion machine, after this energy had gone through a number of conversion cycles it would have no potential left to do work. In any machine, enough work has to be done to overcome internal friction (which can never be zero), so eventually any machine will run down. And of course, a machine that generates more energy than it takes to run it is also impossible. Energy cannot be created from nothing.

Some interesting machines have been designed to try to beat the laws of thermodynamics. One was called the Garabed, designed by Garabed Giragossian in 1917. The Garabed consisted of a large flywheel, which was started spinning by hand with pulleys and then was kept in motion by a small electric motor of 1/20th horsepower. To stop the wheel required 10 horsepower, and the idea was to use this large amount of "free energy" to do some work. Unfortunately, Giragossian had confused energy with power. To obtain the 10 horsepower, the flywheel had to be stopped rapidly, using almost instantly all of the energy that had gradually been stored from running the little motor.

Another idea was to fit a bicycle with a piston that would produce compressed air each time the person riding the bicycle hit a bump. This compressed air would then drive a motor that would turn the wheels. Thus, the rougher the ride, the faster the bicycle would go using the energy produced by bumping along. Think of the ride one would get on a bumpy road if it really worked!

And then there was an engine that worked, but not very well. It was the caloric engine, invented in 1853 by John Ericksson, who had built the first screw-propelled steamship and later built the famous iron-clad warship of the Civil War, the Union's *Monitor.* Ericksson understood that *caloric* (an eighteenth-century term for heat) was wasted in engines, but he believed that it could be recycled and used repeatedly without losing its ability to do work. (Thermodynamics was an infant science then, and few people understood it.) His huge engine used heated air to drive four cylinders, each 14 feet in diameter, displacing among them a total of 3700 cubic feet (6.4×10^6 cubic inches)! The hot air exhausted from the cylinders was passed through a "regenerator," a mattress-sized box stuffed with fine wires that was supposed to trap the leftover caloric and recycle it by passing it back to the heated air entering the cylinders.

The caloric engine actually propelled a ship (named the *Ericksson*), but it produced far less power than steam engines of its day, and it was no more efficient. Thus, it was never used to generate large amounts of power. However, very small (fractional-horsepower) hot-air engines (without the regenerator) later found widespread use. Quiet, safe, inexpensive, easy to operate, and able to use any kind of fuel, they pumped water and ran small machinery of all kinds from the 1860s into the early twentieth century.

Some perpetual motion machines actually ran, and a few did a little work. But they all shared something in common: a hidden, external source of energy, such as compressed air or a tiny electric motor, to keep them going.

It is interesting that although the United States Patent Office began to discourage patent applications for perpetual motion devices in the 1890s, it did not totally refuse to consider them until 1918, long after the scientists there knew better. Even today, inventors occasionally try to patent perpetual motion machines. As late as 1985, the Patent Office was sued for immediately rejecting a patent application for a motor that would deliver 111% efficiency.

Wouldn't that be great?

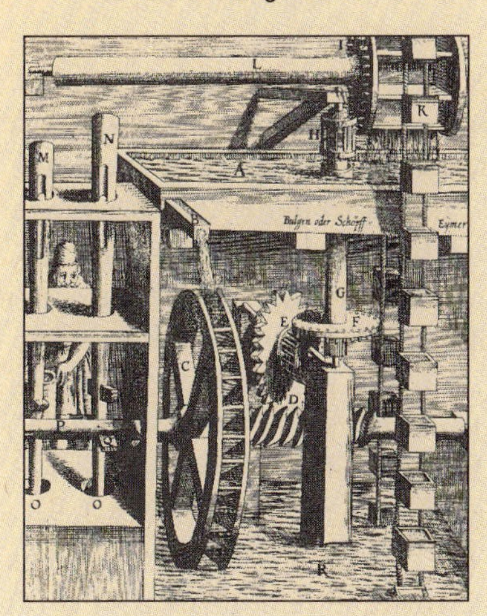

Figure 1
Drawing of a Perpetual Motion Machine, Published in Nurnberg, Germany in 1673. *Water from an upper reservoir (A) turns a water wheel (C) attached to a shaft (P). The rotating shaft drives grain crushers (M, N) and a series of gears and shafts (D through L), which lift water in buckets (K) from a lower reservoir (R) to replenish the water in the upper reservoir (A).*

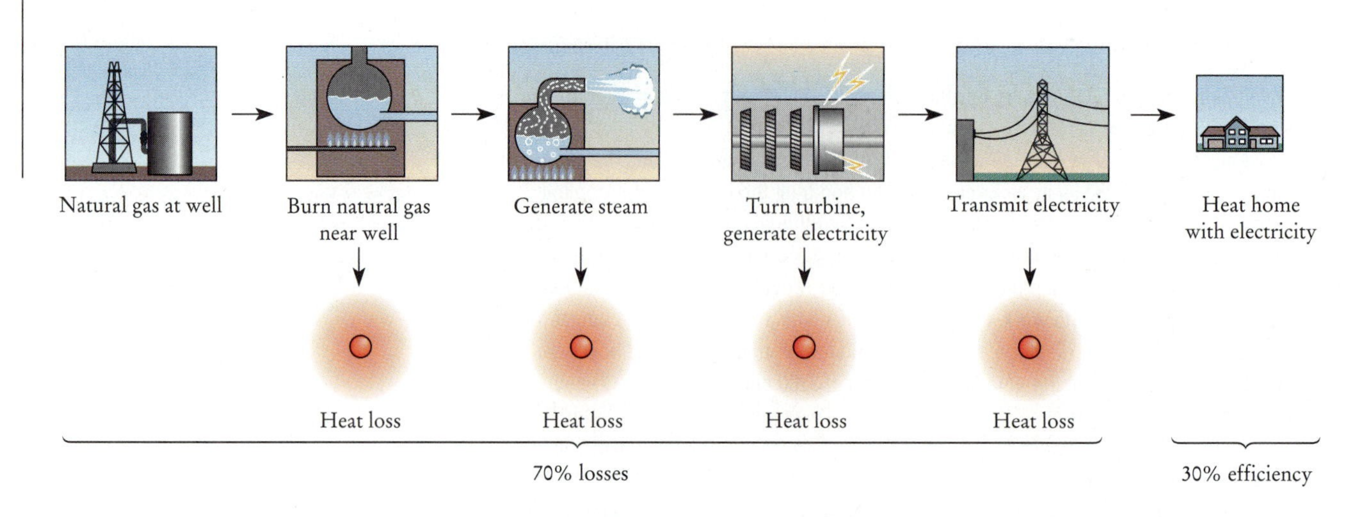

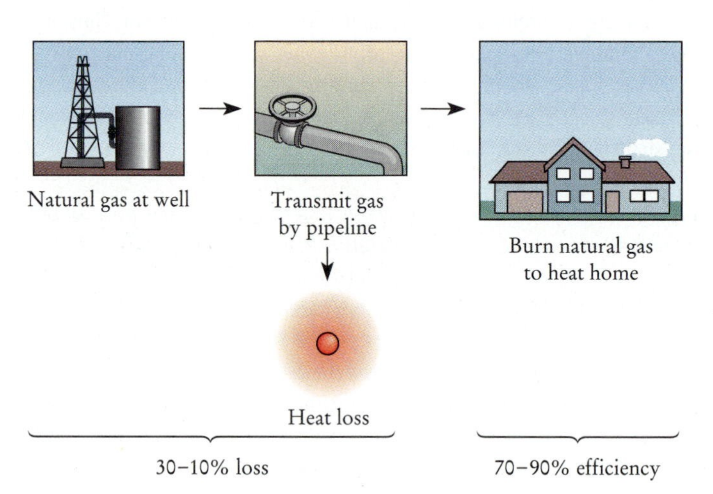

▶ Figure 10.15
Minimizing the number of energy transfers in a chain increases efficiency.

using electricity. Energy losses are also associated with recharging the battery and running the electric motor in the car, reducing the overall efficiency of the vehicle, based on a primary source of energy, to under 20%. A modern, gasoline-powered automobile can do as well or better.

Another way to minimize energy waste is to match the temperature of the energy conversion with the temperature of its intended use. If a home is to be heated to a temperature of 20 °C (68 °F), it is not necessary to burn coal at 1650 °C (3000 °F) in a power plant, or even natural gas at 550 °C (1020 °F) in a furnace, to do so. These high-intensity fuels should be reserved for uses in which their ability to produce high temperatures really is necessary. Low-intensity solar energy is a better alternative for heating homes and buildings in climates where there is adequate sunshine. In some remote areas, it is often cheaper to install photovoltaic collectors if only a small amount of power is needed (such as for pumping water), than it is to run power lines. Because solar energy is so diffuse, however, it is not an attractive source of high-intensity electric power on a large scale.

Finally, like the penny saved that is the penny earned, the calorie or Btu that is conserved is an energy source. Mass transportation is more energy efficient than using automobiles, but little effort has been made in this country to develop high-speed rail or efficient bus service. Recycling an aluminum can requires less than 10% of the energy necessary to make one by refining bauxite ore, and more than half of the aluminum produced in the United States is now recycled (▶ Figure 10.16). This represents a large energy saving, because refining aluminum uses almost 2% of the energy converted in this country each year.

▶ Figure 10.16
Making an aluminum can from recycled aluminum cans uses less than 10% of the energy needed to make one from aluminum ore.

DISCOVERY AND UTILIZATION OF NONRENEWABLE NATURAL RESOURCES

In the early 1970s we experienced our first energy crisis in this country, and an energy shortage was foremost in our thinking. The burning of Kuwaiti oil fields by Iraq in 1991 reminded us of our dependence on foreign oil. Yet today, long lines of cars at gasoline stations no longer occur, and the price of gasoline is lower than it was in 1973. Is there really an energy shortage? Or is there a conspiracy among the countries and corporations that produce energy to make us think so?

In a very strict sense there is no energy shortage because energy is neither created nor destroyed in an energy conversion process (first law of thermodynamics). However, there *is* a worldwide shortage of readily available energy in high potential-energy states. To create the high standard of living to which we have become accustomed, we have consumed high-intensity energy sources at a prodigious rate. Though only 6% of the world's population, we in the United States use 30% of the world's energy. We use tremendous amounts of power, heedlessly converting our sources of high-intensity energy (high potential-energy sources) to waste heat that is dissipated, diffuse, and of such low potential energy that it is useless for doing work. This waste heat is often a pollutant in the environment, and it cannot be reused.

Why is there a shortage of sources of high-intensity energy? Can't we just drill for more oil or mine more coal? What about nuclear energy? There's lots of that if we can solve the environmental problems associated with it. And why not tap the energy in the wind, the ocean tides, or use geothermal steam?

Energy resources are classified as being renewable or nonrenewable. **Renewable energy sources** ultimately come from the sun, as sunlight falling on the earth, and include tides, wind and water power, solar energy, and wood

Renewable energy sources ultimately come from the sun.

SILICON MAGIC

Charley shows everyone his solar battery charger.

"We leave it here in the sun, and we'll always have fresh batteries. It'll save us money eventually and we won't be throwing so much weird stuff into the landfill."

"It uses special batteries, doesn't it?"

"Yeah. They cost more and we have to buy twice the number we need because recharging's slow, but they can be recharged almost indefinitely."

Scientists would like to find a way to imitate photosynthesis. They haven't succeeded, but they have learned how to generate electricity with sunlight. It's about electrons again! You know that electrons can be separated from their atoms in chemical reactions, and you've learned to be interested in what the resulting ion is like and how it behaves.

But, as you probably know from experience, electrons can also be removed from their atoms by sliding leather shoes over a carpet and tumbling clean, dry clothes in a dryer. It can also be done by moving a conductor in the magnetic field of a generator. When an electron moves out of its place between two atoms in a crystal, it leaves a positive charge behind it. If the electron and positive charge can be separated—for instance, by a magnetic field that forces them to move apart—then you can generate useful electricity. The electrons will flow in a circuit made of some conductor.

Pure silicon has an orderly crystalline structure like diamond, with every atom bonded covalently to four others, and as a result it's nearly an insulator at room temperature. Light of certain wavelengths in the infrared (including about half the wavelengths in sunlight) excites the electrons in silicon so that they move from their positions in bonds between silicon atoms, leaving behind regions of positive charge called "holes." In pure silicon, the holes are immediately filled by other excited electrons, which leave other holes. Imagine a vast chinese-checkers board, vibrating enough to jar marbles from the holes but not enough to keep them from catching in other holes for a moment before being shaken out again. The marbles move about randomly, leaving and filling holes. If you tilt the board slightly, you'll see the marbles flow slowly downward from hole to hole, and if you imagine that the marbles falling off the bottom fly back to the top, you'll see a current of marbles flowing down the board. Semiconductors can be "tilted," in a manner of speaking, by adding tiny amounts of another substance (about 1 atom in 6×10^6 atoms of silicon), a treatment called "doping," and they can be "vibrated" by sunlight.

Phosphorus atoms have five valence electrons; so when they're bonded to silicon, each phosphorus atom has a left-over electron that's free to wander. Thus, a phosphorus-doped silicon crystal is *n*egative, making an *n*-type semiconductor. Boron has only three valence electrons, so each boron atom is short one electron when bonded to silicon, leaving free *p*ositive charges (holes). Thus, boron-doped silicon is a *p*-type semiconductor.

Photovoltaic cells are made by placing a layer of *n*-type crystal on top of a layer of *p*-type crystal. In your imagination, hold the semiconductor sandwich horizontally and look at it edge-on. The boundary between the layers is called a *p-n* junction, and when the junction is made, the electrons and holes of the two layers are attracted to each other.

Now, shine some mental sunlight on the *n*-layer and you'll see excited valence electrons move out of their positions in the bonds between silicon atoms. If you connect the two layers of the sandwich with a loop of conducting wire, the excited electrons will flow from the *n*-layer into the wire, around the loop of wire, and into the *p*-layer where the holes are. The *p*-layer now has extra electrons, and the *n*-layer has a shortage of them, so the electrons flow across the *p-n* junction back into the *n*-layer, completing the circuit and making an electric current flow. If you cut the wire and interpose a battery, the current will charge the battery.

The current flows as long as light shines on the solar cells. Because one solar cell supplies only a very small amount of electricity, in solar collectors many cells are connected in a way that adds their currents together.

But when the sun goes down, the process reverses, and the photovoltaic cells turn into infrared radiators, using the charge on your battery. Fortunately, you can install a device called a diode to shut off this nightly back flow.

The integrated circuit chips used in computers and various control devices are similar to photovoltaic cells except that they're used to control rather then create electric currents. Millions of circuit-control devices—the *p-n* junction and diode among others—can be built onto minute chips of semiconductor using various kinds, amounts, and arrangements of dopants.

You're all lazing in the sun watching the solar collector expectantly. A hover-fly, attracted by the blue crystals for its own reasons, investigates the collector for a few moments. Otherwise, nothing visible happens. Charley puts your thoughts into words:

"I wish it would hum or something; it's hard to believe anything's happening."

and fuels derived from plants (photosynthesis). These are limited in availability, however, because only a certain amount of sunlight falls on the earth's surface each day. For this reason, the amount of power we can obtain from these sources is limited, and they generally provide energy only at a relatively low rate. Our demands for energy far exceed this. As a modern society, even a very frugal one, we cannot "live in equilibrium with the sun," although we should use this energy source much more than we presently do.

Coal, natural gas, and petroleum are **nonrenewable energy sources.**

To meet our demands we now rely heavily on the most common **nonrenewable energy sources,** the fossil fuels: coal, natural gas, and petroleum. (Tar sands and oil shales are also fossil fuels but are not yet developed as a resource.) Not so obvious as nonrenewable, due to their immense potential availability, are nuclear and geothermal energy sources. However, like the fossil fuels, they are finite in quantity; only a certain amount of each is available on the earth. When any one of these resources is used up, *no more of it can be obtained.*

The boom-and-bust saga of the Gold Rush days in the United States during the late 1800s is a familiar story. Many times a prospector made a strike in virgin territory and staked his claim. Soon, many fortune seekers flocked to the area, and more discoveries were made. Fervent mining activity resulted in a rapid increase in gold production. A boom was underway.

Within a short time, however, new deposits of gold became hard to find. Despite this, more and more gold could be produced because it was still plentiful and easy to mine at existing claims. Yet the amount of gold really was limited, and eventually it became difficult to get what was left. Production soon peaked and then declined rapidly as the claims became played out. The bust had come. Gold finally became so scarce that it was no longer worth the trouble to find it, and the miners moved on, seeking richer lodes. Now, a century later, little easily obtained gold is left, and only a few gold mines are active in the United States today. Even the gold being mined in South Africa, although still found in reasonably rich deposits, is mined with considerable difficulty several thousands of feet underground.

The discovery and use (recovery) of the fossil fuels have followed a similar pattern over the course of the past century (somewhat longer in the case of coal). In fact, these same trends can be shown for many other nonrenewable resources, such as metal ores and other minerals of value to us.

When we first discover nonrenewable resources, they usually have already been concentrated by forces of nature into deposits that can be easily exploited. In this concentrated state they are ordered and of low entropy. Little energy and other resources (like steel for mining equipment or drills) are required to obtain them. Demand for a resource drives its recovery, which in turn drives exploration to find more of it. For a while, additional easily recovered deposits are found; but eventually, like gold, the deposits become scarcer, deeper, less accessible, and generally less concentrated and of poorer quality. What is left is more disordered and in a high entropy state. To recover these scattered resources (create order) requires a much greater expenditure, and waste, of energy than before. Thus, as a resource plays out, more and more energy and other resources are required to obtain diminishing amounts of it. Eventually, it becomes too expensive, or too wasteful of critical resources, to continue. (Demand usually remains high or even continues to grow after the resource starts to become scarce, however, and a shortage often results.)

IT TAKES ENERGY TO OBTAIN ENERGY

Energy must be expended to obtain energy, or any other natural resource. Energy is required to find, extract, process, and transport the resource to its place of use, and the scarcer (more entropic) the energy resource, the more energy is required. This energy cost must be subtracted from the energy value of the resource before it is converted to do work, where conversion losses due to entropy subsequently take their toll.

Energy is required to obtain *any* natural resource.

$$\text{net work obtained} = \text{energy in resource} - \text{energy lost in recovery} - \text{energy wasted in conversion (second law)}$$

Energy losses in recovery are often now surprisingly large. In 1910, it required less than 10% of the energy in a barrel of U.S. crude oil to produce a barrel of it. It is estimated that by the year 2005 the amount of energy needed to obtain a barrel of U.S. crude will be equivalent to that in an entire barrel. Oil shale is also a high-entropy source of energy. To produce a barrel of crude oil from oil shale using today's technology requires the energy contained in almost a barrel of oil.

At these efficiencies, it makes little sense to burn petroleum for fuel unless it is absolutely necessary. Rather, the hydrocarbons in it should be used to make petrochemicals, organic compounds that often are not as easily or efficiently obtained from other sources (Chapter 11). Petrochemicals are necessary for the manufacture of a host of products, like plastics, rubber, detergents, medicines, antifreeze, and fibers, which in the long run are more useful and valuable to us than the energy supplied from the combustion of hydrocarbons.

Mendeleev is reported to have said that burning petroleum is like fueling a stove with bank notes.

Because it is ultimately a product of photosynthesis, the food we eat is usually considered to be a renewable source of energy that comes from the sun. Closer examination, however, shows that modern agriculture and food production are heavily energy intensive, particularly of petroleum (▶ Figure 10.17). Fertilizer is energy. Tractors, made of steel, require fuel. Steel must be

Modern agriculture is energy intensive.

► Figure 10.17
Modern agriculture uses large amounts of energy to power machinery and make fertilizers.

made with fire. Irrigation water must be pumped. Food must be processed and packaged, refrigerated, transported, and cooked. In this country, 8 to 10 Cal of energy are required to put 1 Cal of food on the table. In so-called primitive societies, 1 Cal of human work often produces 5 Cal of food.

Because modern agricultural methods are so energy intensive, using them to grow crops such as grain, corn, or sugar to make ethanol for use as a motor fuel (gasohol) is a poor thermodynamic investment. Often, more energy is required to produce the alcohol than is obtained from burning it. Worse, food—which is of low entropy (highly ordered and complex) and valuable for human consumption—is converted into a small molecule (ethanol, Chapter 11) and by-products, both having higher entropy and little nutritional value.

ENERGY AND METALS

Metallic ores can be classified according to the energy required to extract metals from them. A *class-I ore* is like a placer deposit, in which a mineral containing a certain metal is distributed throughout sand, gravel, or other loose material. Gold in the Gold Rush days was often found this way, and many streams in the western United States still bear the scars of placer and dredge mining. A *class-II ore* is hard rock that contains a discrete mineral that can be physically separated from the rock by crushing and grinding. The zinc ore *sphalerite* (ZnS), the iron ore *hematite* (Fe_2O_3), and the copper ores *chalcocite* (Cu_2S) and *chalcopyrite* ($CuFeS_2$) are obtained in this way. A *class-III ore* contains small amounts (less than 0.1%) of the desired metal (usually as the ion) dissolved in a solid solution of hard rock. Extraction of the metal requires chemical separation.

About 10^5 Btu per ton of ore processed is needed to extract a metal from a class-I ore. This is the energy contained in one gallon of gasoline. To produce a metal from a class-II ore requires, on the average, ten times more energy, or 10^6 Btu/ton of ore. Class-III ores are highly entropic and require another ten times more energy than do class-II ores, or 10^7 Btu/ton. Although there is a large supply of metals that could be obtained from class-III ores, 100 gallons of gasoline per ton of ore is a prohibitive energy expense. If we were to produce an annual supply of some metals (such as copper) from ores of this kind, we would use more energy than is presently converted in the United States each year.

Most of the class-I ore deposits have been depleted, and we are mining those in class II. These, too, are becoming more entropic and of lower quality. For example, in the early 1940s much of the copper ore mined in this country contained more than 3% copper. Today it contains less than 0.5%. To make a copper concentrate (25–30% Cu) that is suitable for smelting requires about eight times as much energy today as it did in 1940.

10.11 However, in Nevada, class-III gold ores are being mined profitably. Why do you think this is possible? (These ores also contain some silver.)

OUR ENERGY TODAY

SOURCES OF ENERGY

The only source of energy available to early humans was the food they ate, which they obtained by gathering and hunting. This amounted to some 2000–3000 kcal/day (Cal/day) for each person, about the energy consumption of a 100-watt light bulb. When they discovered fire about 100,000 years ago, our ancestors gained the use of an additional 5000 kcal/day with which to keep warm and cook food. By about 5000 B.C. they had domesticated draft animals for farming and transportation, adding another 4000 kcal/day to their energy inventory. At the time of Christ, some 12,000 to 20,000 kcal/day were available per capita, six to ten times the amount needed to survive. For the most part, this energy was renewable: photosynthesis, which uses sunlight, provided food for humans and work animals, and wood for fires.

Ancient societies lived in equilibrium with the sun.

Coal was first used as a fuel around A.D. 1300 in Europe, where it became a primary source of energy by the end of the sixteenth century. It also provided much of the fuel for the steam engines of the Industrial Revolution, which by the late 1800s had made 75,000 kcal available to each person daily.

In the United States, where there were vast stands of virgin forests, wood remained the primary source of energy until about 1885, when it was overtaken by coal. By 1915, coal accounted for 75% of the energy converted here; oil and natural gas combined contributed only about 12%, and the use of wood had declined to about 10%. Today the blend of energy sources we depend on is very different. In 1990, 23.5% of our energy was supplied by coal, 65.0% by petroleum and natural gas, and the remainder mostly by nuclear and hydroelectric power. Almost 96% of the energy we convert today comes from nonrenewable sources, and very little from the sunshine our ancestors depended on. ▶ Figure 10.18 illustrates how U.S. energy use has changed in representative years since 1850.

Almost all the energy used in the United States today comes from nonrenewable sources.

USE OF ENERGY

The rate at which the total energy consumption in the United States has grown is dramatic. In 1850, about 2.4 quadrillion Btu (2.4×10^{15} Btu) of energy was converted. By 1900, this had almost quadrupled to 9.6 quadrillion Btu, and from then until 1955 it quadrupled again to about 40 quadrillion Btu. In 1990, U.S. energy consumption was 81.3 quadrillion Btu, more than double the amount in 1955. During that period, our population increased from about 165 million to 249 million, and our per capita use of energy increased from 162,000 kcal/day to 226,000 kcal/day, 113 times that needed for subsistence. Total U.S. annual energy consumption increased by a factor of almost twelve during the century between 1890 and 1990. ▶ Figures 10.19 and ▶ 10.20 illustrate respectively the total energy converted and the per capita use of energy in the United States for representative years since 1850.

How much is one quadrillion Btu (called a *quad*)? It represents the amount of petroleum that the United States imports in 24 days, or the amount of gasoline we use in 26 days. In effect, we run our gasoline-powered cars and trucks on the petroleum we import. In 1991, 40% of our crude oil was imported, which constitutes about 16% of all the energy we consumed. (We imported a little energy from other sources and exported some coal, in almost equal amounts.)

In effect, all of our gasoline-powered vehicles run on imported oil.

▶ Figure 10.21 shows the use of energy in the United States by sector in 1991, based on primary sources (coal, oil, and natural gas). These percentages include the amount of primary energy required to generate the electricity used by each sector. In 1991, 36.7% of the total energy we converted was used to generate electricity at an overall efficiency of 29.4% delivered to the consumer. ▶ Figure 10.22 shows the percentage distribution of primary sources used to generate that electricity. Note that more than half of this energy was provided by coal, and that over one-fifth came from nuclear energy. Overall, nuclear energy accounted for about 8.1% of all energy converted in the United States in 1991.

In 1991, about three-eighths of the energy consumed in the United States was used to generate electricity.

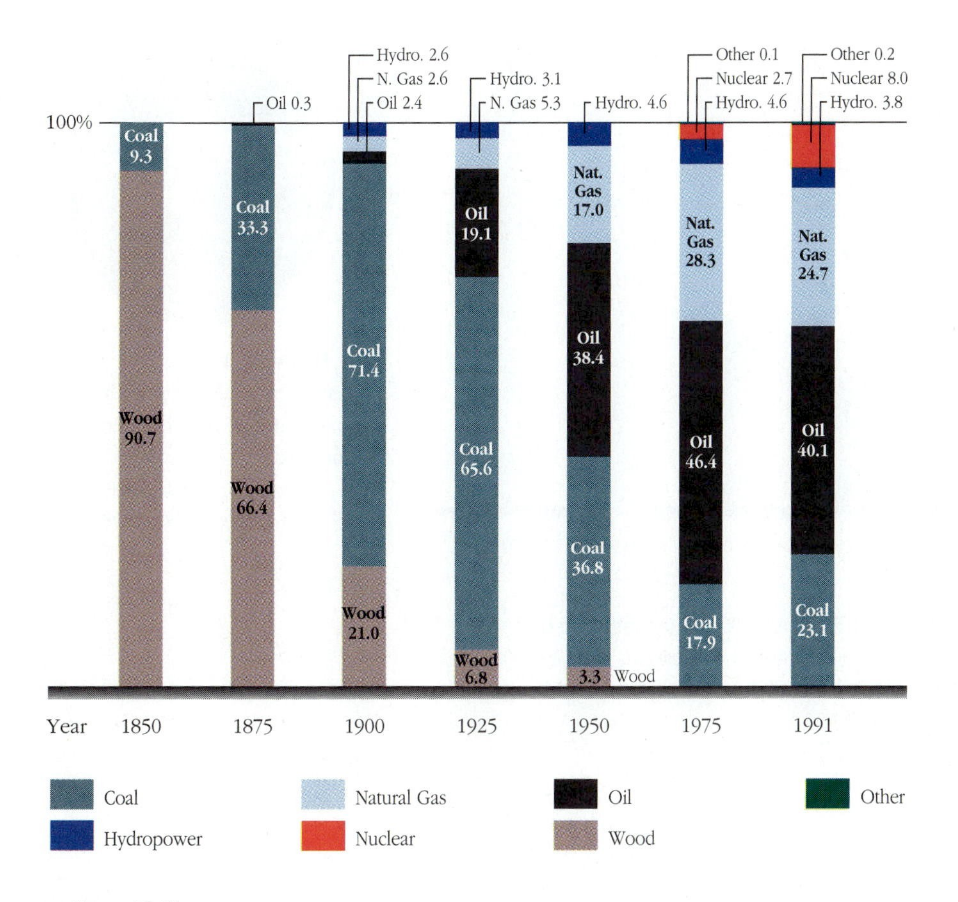

▶ Figure 10.18

Historical Energy Conversion in the United States (Percent by Source).

Sources: Data from Sam H. Schurr and Bruce C. Netschert, Energy in the American Economy, 1850-1975, *1960: Baltimore, The Johns Hopkins Press; and* Annual Energy Review, 1991. *Energy Information Administration.*

FOSSIL FUEL RESERVES

Oil reserves in the United States amounted to 26.3 billion barrels (1991), which is only 2.6% of the 999.2 billion barrels estimated to remain in the world. At our present rate of production, our reserves will last us about 10 years. Two-thirds of the known world oil reserves are in the Middle East (▶ Figure 10.23, page 350). Mexico and Venezuela, a little closer to home, have about 11%. Nonetheless, it is disconcerting that in a few years we shall become almost entirely dependent on imported petroleum. Also worrisome is the fact that, at the present rate of world oil consumption, *all* reserves will be depleted in less than 50 years.

Two-thirds of the known reserves of petroleum are in the Middle East.

The scenario for natural gas is no brighter. U.S. reserves comprise about 4.0% of the world total, or about 170 trillion cubic feet. These supplies will

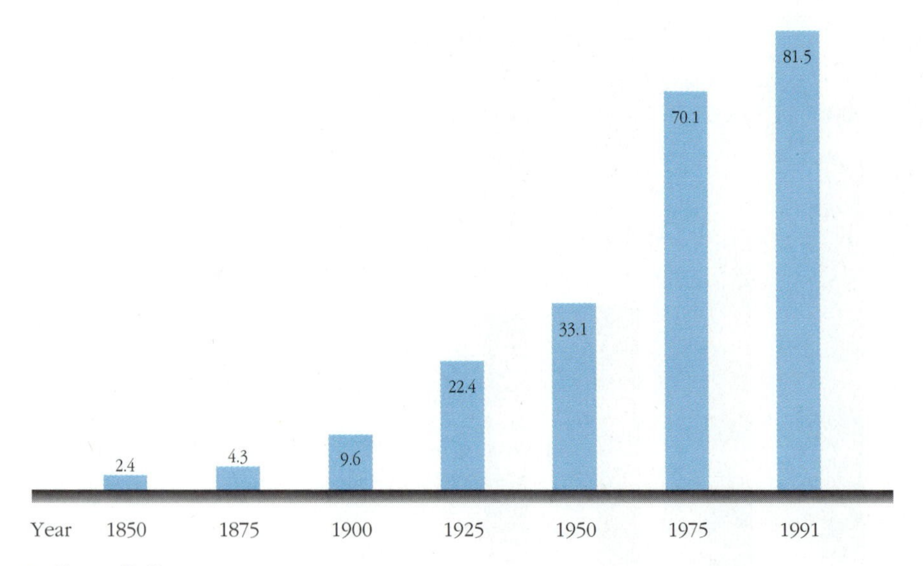

▶ Figure 10.19
Total Energy Conversion in the United States, All Sources 1850–1991 (quadrillion Btu).

Sources: Data from Sam H. Schurr and Bruce C. Netschert, Energy in the American Economy, 1850-1975, *1960: Baltimore, The Johns Hopkins Press; and* Annual Energy Review, 1991. *Energy Information Administration.*

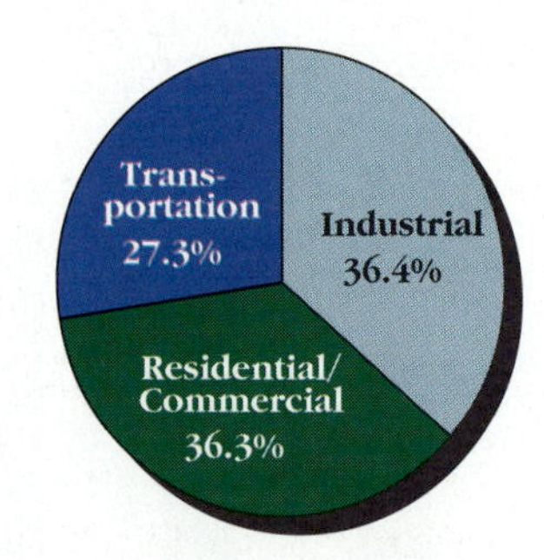

▶ Figure 10.21
Distribution of Energy Use in the United States, 1991, from Primary Sources.

Source: Data from Annual Energy Review, 1991. *Energy Information Administration.*

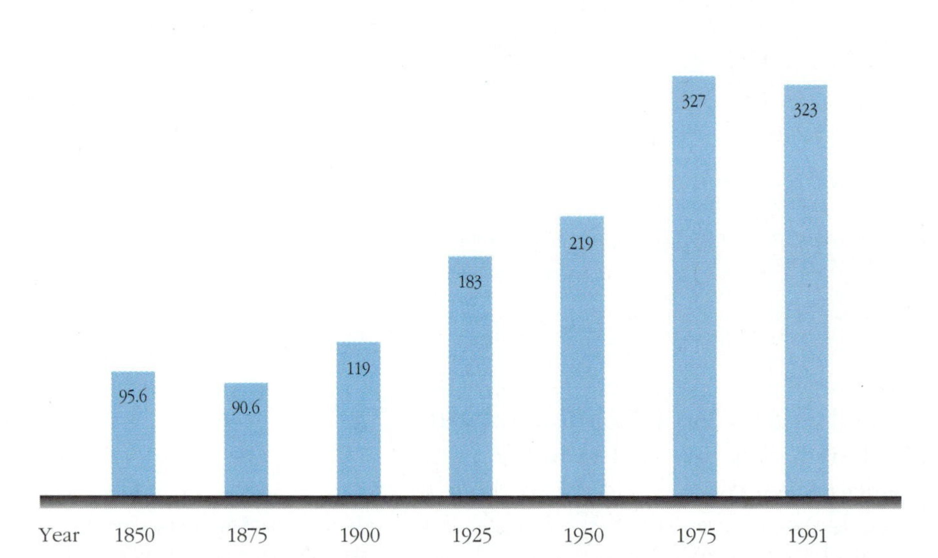

▶ Figure 10.20
Per Capita Energy Consumption in the United States, All Sources 1850–1991 (million Btu). *Between 1850 and 1885, per capita consumption of energy varied between 96.9 million Btu (1855) and 90.2 million Btu (1865), and then it started upward.*

Sources: Data from Sam H. Schurr and Bruce C. Netschert, Energy in the American Economy, 1850-1975, *1960: Baltimore, The Johns Hopkins Press; and* Annual Energy Review, 1991. *Energy Information Administration.*

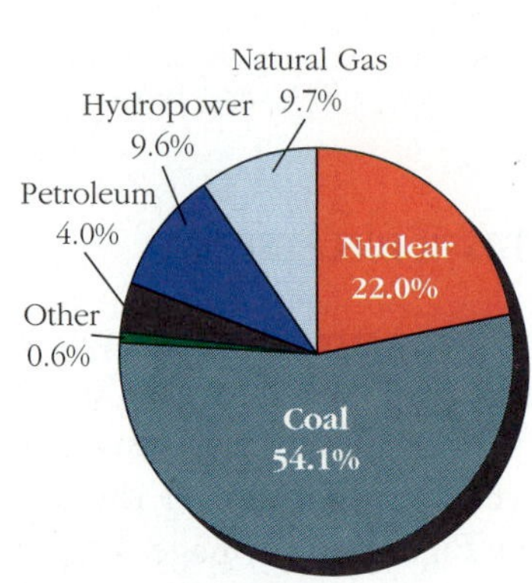

▶ Figure 10.22
Sources of Energy for Generation of Electricity in the United States, 1991.

Source: Data from Annual Energy Review, 1991. *Energy Information Administration.*

last less than 10 years if we continue production at today's rate. Most of the world's reserves of natural gas are in the former Soviet Union and Iran (▶ Figure 10.24, page 351). The technology has been developed to transport natural gas in liquefied form, called *liquefied natural gas (LNG),* so it can now be shipped long distances without pipelines. Still, at the present world rate of consumption, all reserves will be depleted in about 55 years.

The United States has a large amount of coal, about 23% of the world's reserves and sufficient to last us about 300 years at the rate we presently consume it (▶ Figure 10.25, page 352). If we were to replace the energy we obtain from oil and natural gas with that from coal, our coal reserves would be depleted in less than 80 years. On a global scale, coal should last about 230 years at the rate it is now consumed. ▶ Figure 10.26 (page 353) shows the distribution of coal reserves in the world.

The United States has large reserves of coal.

The previous estimates of the amount of time required to deplete fossil fuel resources were based on *present* U.S. and world consumption rates. However, these rates have been increasing and are likely to continue to do so—rapidly. What is driving this growth in energy consumption?

Today's world population of 5.5 billion is projected to increase by 1 billion in the next 10 years, double to 11 billion 40 years from now, and reach 12 billion by the year 2050. Ninety percent of this growth is expected to occur in developing nations. The U.S. population is projected to increase more slowly, from about 259 million today to 392 million in 2050. About three-fourths of the world's energy is used by industrialized nations having only 25% of the world's population. As the developing nations aspire to a better standard of living, they will strive to increase their per capita consumption of energy. These developments, combined with population growth in industrialized nations where energy consumption is already high, can be expected to place considerable pressure on world energy reserves.

The world population is expected to double by the year 2035.

THE FUTURE OF NONRENEWABLE ENERGY

Discovery of new sources of oil in the United States has peaked, and production from our wells may already have begun its decline. On a global scale this picture isn't much brighter. New sources of oil on the earth are scarcer and very much harder to reach than they were in 1960, and it is doubtful that many new major oil discoveries will be made. World oil production is expected to peak early in the next century, probably within 20 to 30 years. Even if the reserves of oil in the world were to be double those presently known, severe oil shortages would be postponed by only about a decade at the present increasing world rate of oil consumption. In your lifetime, petroleum and natural gas will become scarce and expensive.

Petroleum will become scarce and expensive.

DID YOU KNOW ?

THE OPTION OF NUCLEAR POWER

Because our supplies of fossil fuels are limited, and much of our high-intensity energy is derived from them, many people view nuclear energy as an attractive substitute. The energy released by nuclear fission is enormous: the fission of 1 g of uranium-235 releases the same amount of energy as burning 13.7 barrels of crude oil or about 3 tons of coal. There are over 100 nuclear power plants in the United States, which in 1991 generated 22% of our electricity, slightly more than 8% of the energy we consumed that year. These reactors are fueled by uranium-235, and most of them share common features in their design.

Natural uranium contains mostly uranium-238, which is not fissionable. Uranium-235, the fissionable isotope, is only about 0.7% of ordinary uranium; it must be enriched to a concentration of about 3% before it can be used as fuel in a nuclear fission reactor. This concentration of uranium-235 is much too low to produce a nuclear explosion, and a nuclear reactor using this kind of fuel *cannot* become an atomic bomb. For use in a reactor, the enriched uranium is fashioned into fuel rods, which are inserted into the core of the reactor (Figure 1). These fuel rods contain UO_2 pellets encased in a zirconium jacket about 1 cm in diameter and 4 m long.

The fission of uranium-235 nuclei produces mostly fast neutrons, which have too much energy to be captured by another nucleus of uranium-235 and support a chain reaction. To slow these neutrons, substances called moderators are placed in the core, typically water or carbon (graphite). Most reactors in the United States are of the light-water (H_2O) type, but a few use heavy water (D_2O; D is ${}^{2}_{1}H$). Graphite, commonly used in reactors in the former Soviet Union, is combustible and was responsible for the fire that burned ten days during the Chernobyl disaster in 1986.

For a reactor to produce energy safely, the rate of the fission process must be strictly controlled. Just enough slow neutrons should be produced to make the chain reaction self-sustaining, but not so many that too much heat is liberated and a meltdown of the core occurs. Control rods made of boron or cadmium, elements that absorb slow neutrons, are placed between the fuel rods in the core. These rods are

Figure 1
Fuel rods are loaded into the core of a nuclear reactor. The long, thin metal tubes are filled with UO_2 pellets.

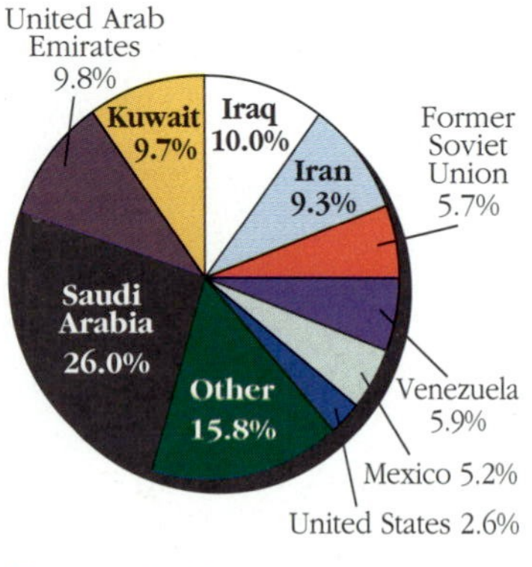

▶ Figure 10.23
Distribution of Known World Reserves of Petroleum, 1991. *The world's petroleum reserves in 1991 were 999.2 billion barrels.*

Source: Data from Annual Energy Review, 1991. *Energy Information Administration.*

raised or lowered as needed to obtain the right number of neutrons, and lowering the rods completely stops the chain reaction and shuts down the reactor.

In light-water reactors, the water that is used as a moderator is also used to cool the core. Maintained under high pressure to keep it from boiling, this water absorbs the heat produced by the nuclear chain reaction and transfers it from the core to a heat exchanger that produces steam (Figure 2). Usually, this steam is used to generate electricity, but it can also be used to power turbines in nuclear submarines.

Although many countries are building additional nuclear power plants to supply their energy needs, construction of new reactors in the United States has all but ceased due to public concerns about their safety and cost, and about the waste they produce. Nuclear reactors are complex, and any errors in their design, construction, and operation can lead to serious consequences. For example, a typical coal-fired plant has about 4000 valves, but a nuclear facility the same size may have 40,000, ten times as many. A stuck valve started the chain of events that led to the incident at Three Mile Island in 1979.

Frightening though the Three Mile Island event was, it produced no known adverse health effects in persons living in the area, primarily because the reactor was housed in a containment building made of concrete. Such a structure is designed to hold any radioactive materials that escape from the reactor and prevent them from entering the

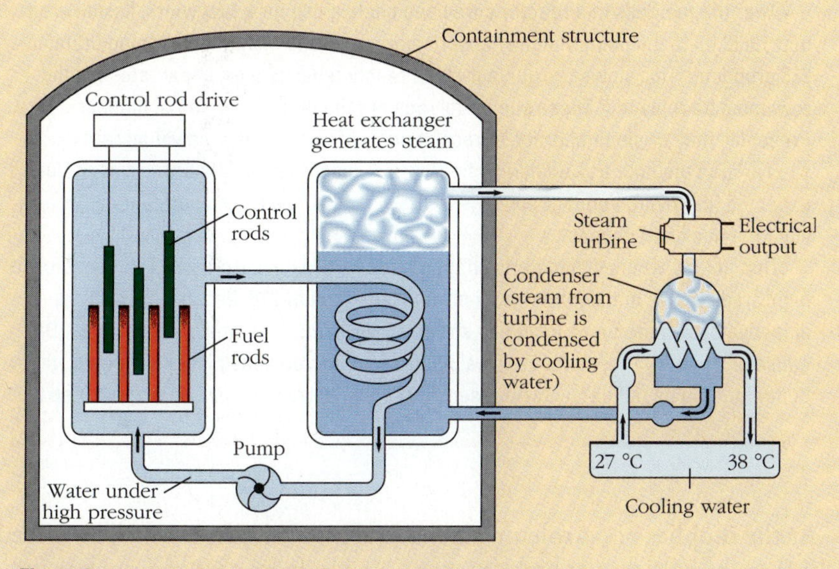

Figure 2
A Light-Water Reactor.
Nuclear fission produces heat, which is used to form steam. The steam is used to drive a turbine and generate electricity. Note that the reactor is isolated from the environment by the containment structure, and that the steam that drives the turbine does not come in contact with the reactor core.

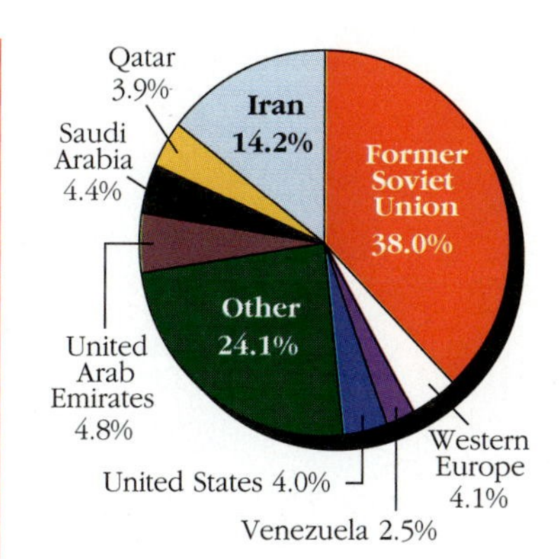

► Figure 10.24
Distribution of Known World Reserves of Natural Gas, 1991. *The world's natural gas reserves in 1991 were 4213 trillion cubic feet.*

Source: Data from Annual Energy Review, 1991. *Energy Information Administration.*

THE OPTION OF NUCLEAR POWER (Continued)

environment. The radiation that leaked from the containment building into the environment is estimated to have exposed residents nearby to a dosage equivalent to that of four chest X-rays. By contrast, the disaster at Chernobyl in the Ukraine directly caused 31 deaths and the hospitalization of hundreds, and contaminated a large area with radioactive debris. It has been projected that persons who live in the fallout area will suffer a much higher-than-normal incidence of cancer in the future. Unlike the reactor at Three Mile Island, the reactor at Chernobyl was not housed in a containment structure. Had it been, the disaster would have been lessened considerably or possibly prevented.

A new generation of reactors is being designed that promises to be much simpler and safer than the reactors now in use. The large number of pumps and valves will be reduced by using natural forces such as convection and gravity. For example, should a reactor core overheat, huge amounts of water stored in tanks overhead will be released to flood it. The new reactors will be about half as large as those today and will be standardized in design, construction, and operation. This will simplify maintenance and repair, and make it easier to train operators, thus reducing the possibility of human error.

After the fuel rods in a reactor core have been in use for a few years, fission products build up to the point that they absorb more neutrons than the remaining uranium-235 produces. This slows the nuclear chain reaction below a practical rate, and the rods must be replaced. These spent fuel rods are the primary source of nuclear waste. Presently, they are in temporary storage in pools of water at the individual reactor sites around the country, awaiting a final resting place. The amount of this spent fuel is expected to grow to about 44,000 tons by the year 2000. So far, no permanent disposal site for it has been chosen.

That no one wants this material in their back yard is not surprising. The spent fuel is highly radioactive and contains not only unused uranium-235, but also over fifty radioactive fission products, including strontium-90, iodine-131, and plutonium-239. Because of its family resemblance to calcium, strontium replaces calcium in bones, thus concentrating its radioactivity near sensitive bone marrow. Iodine is concentrated by

► Figure 10.25
The United States has a large amount of coal, but serious environmental problems are associated with using it.

As the supplies of petroleum and natural gas dwindle, we shall be forced to rely more heavily on the high-intensity energy from coal and nuclear fission. Although more plentiful, the latter sources present serious environmental problems, and to use them safely will be expensive in terms of both energy and other resources. (Nuclear fission is more heavily subsidized by other sources of energy than is commonly realized.) Besides, like petroleum, coal and nuclear fission are nonrenewable. We should use them wisely now to ensure a reasonable quality of life for those who follow us in history.

the thyroid gland, and radioactive iodine there can increase the likelihood of thyroid cancer. Plutonium-239 is a fissionable metal that is used in fuel rods for reactors and in nuclear weapons. It is also carcinogenic and highly toxic: inhalation of microgram quantities can be fatal. Its half life, the amount of time required for one-half of its radioactivity to disappear, is 24,000 years. Because it takes about 10 half lives for a radioactive isotope to decay to harmless levels, plutonium-239 must be isolated from the environment for about 240,000 years.

Any method for storing nuclear wastes must meet the following conditions. It must:

1. safely isolate the waste for 250,000 years
2. provide safety from sabotage or accidental entry
3. provide safety from natural disasters (floods, landslides, tornadoes, hurricanes, etc.)
4. protect any nearby natural resources from contamination
5. be placed in a geologically stable area (no earthquakes, volcanic activity, changes in groundwater level, etc.)
6. provide methods of fail-safe handling and transport of the waste

These criteria give us reason to reflect on the validity of our scientific models, in particular how well they allow us to make predictions. Can we use our models of the weather to predict the likelihood of floods in a disposal area 1000 years from now? How probable is an earthquake at a given site within 10,000 years, or a volcanic eruption within 50,000? Do we know enough about human behavior to design security systems that cannot be breached? Will our theories of metallurgy and chemistry allow us to build containers for waste that will not leak 5000 years from now? An attractive way to stabilize radioactive waste and decrease its mobility in the environment is to vitrify it (make it into a glass). Can we confidently predict that this glass will withstand intense radiation for thousands of years without decomposing?

Whether or not we choose nuclear energy as an option for the future, we must deal with the nuclear waste we have already created. To continue to demand scientific certainty before we act to properly store it is to ask the impossible. We *do* know many ways to safely deal with nuclear waste, and although there is some uncertainty about *all* of them for the long term, not using our knowledge now will undoubtedly make the situation worse later.

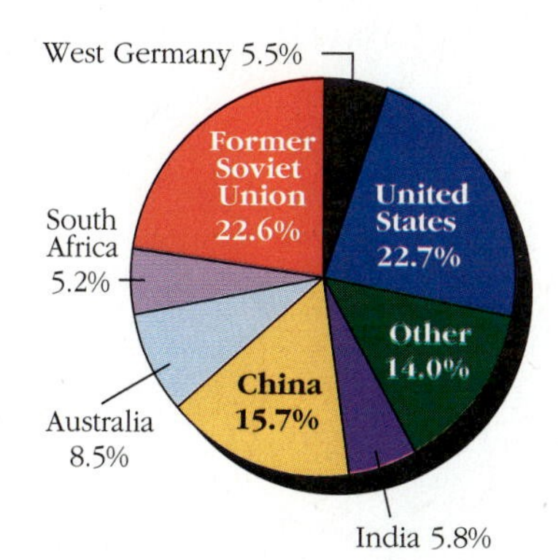

▶ Figure 10.26
Distribution of Known World Reserves of Coal, 1990. *World coal reserves in 1990 were 1,167,346 million tons.*

Source: Data from Annual Energy Review, 1991. *Energy Information Administration.*

Nuclear fusion, which could provide a virtually unlimited supply of clean energy, has been heralded as a final solution to our energy problems. It is not. Waste heat, an unavoidable consequence of the second law, has to go somewhere. Though not useful to do work, it becomes thermal pollution, which does considerable damage to streams and lakes whose water is used for cooling industrial processes and electrical power plants. We cannot continue to convert stored energy—from any source—at the present growth rate for very far into the future without running the risk of dangerously heating our

Waste heat becomes thermal pollution.

environment. It has been estimated that human activity now generates waste heat equivalent to about 0.005% of the earth's daily solar heat input. At the present rate of growth, this amount could increase tenfold to 0.05% in a little more than 50 years. This might not cause problems, but another tenfold increase, to 0.5%, would cause about a 1.5 °C increase in the surface temperature of the earth. Atmospheric warming like this could cause serious changes in global climate.

SOLVING THE ENERGY/NATURAL RESOURCE PROBLEM

Resource depletion is an increase in entropy.

The energy/natural resource problem (the two are inseparable) is not one of technology. Rather, it is a social problem. We must realize that our resources are finite and that the consequences of the second law are inevitable: resource depletion is an increase in entropy. We cannot stop waste completely, nor can we stop using nonrenewable resources, but we can employ strategies that will enable us to forestall serious shortages far into the future. Some of these will involve dramatic changes in the way we live:

- We need to conserve our resources. It has been estimated that about 80% of the energy converted in this country is wasted. About half of this is entropically unavoidable, but much of the other half could be reduced or eliminated by using known conservation practices. We make mineral resources, such as metals, less entropic by expending energy in mining them, refining them, and turning them into useful products. Once we use these products, however, we often discard them in junk yards and landfills, only to have the resource become entropic again. This practice must stop. Goods must be built to last, maintained conscientiously, and recycled when worn out.
- We need to halt population growth in the entire world. The finite resources of this planet simply cannot support more people. On a sustained basis (over several centuries) it is likely that the planet can support only about one-half the present world population. This is assuming a society that maximizes use of renewable energy resources, is highly resource efficient, and at the same time maintains a reasonable quality of life. If we do not seriously address human population, our efforts to solve our energy, resource, and environmental problems will be in vain.
- We need to maximize our use of alternative energy sources, such as solar, wind, and geothermal, consistent with good environmental and thermodynamic practice. In particular, technologies that use renew-

able energy from the sun (solar, wind, and vegetation, for example) should be developed fully.

Hydropower, a high-intensity source of renewable solar energy, is already largely developed in this country and accounts for about 10% of our annual supply of electricity (about 4% of our total energy use). Most of the other solar energy sources are of low intensity, not suitable for generating power on a large scale. However, it has been estimated that the end use of over half of the energy converted in this country is for low-temperature heat. If so, solar energy (including wind and biomass) is an excellent energy-quality match for applications such as heating homes and domestic hot water (▶ Figure 10.27), providing electricity for local or regional use, and using waste vegetable matter to produce ethanol or methane for transportation fuels on a small scale.

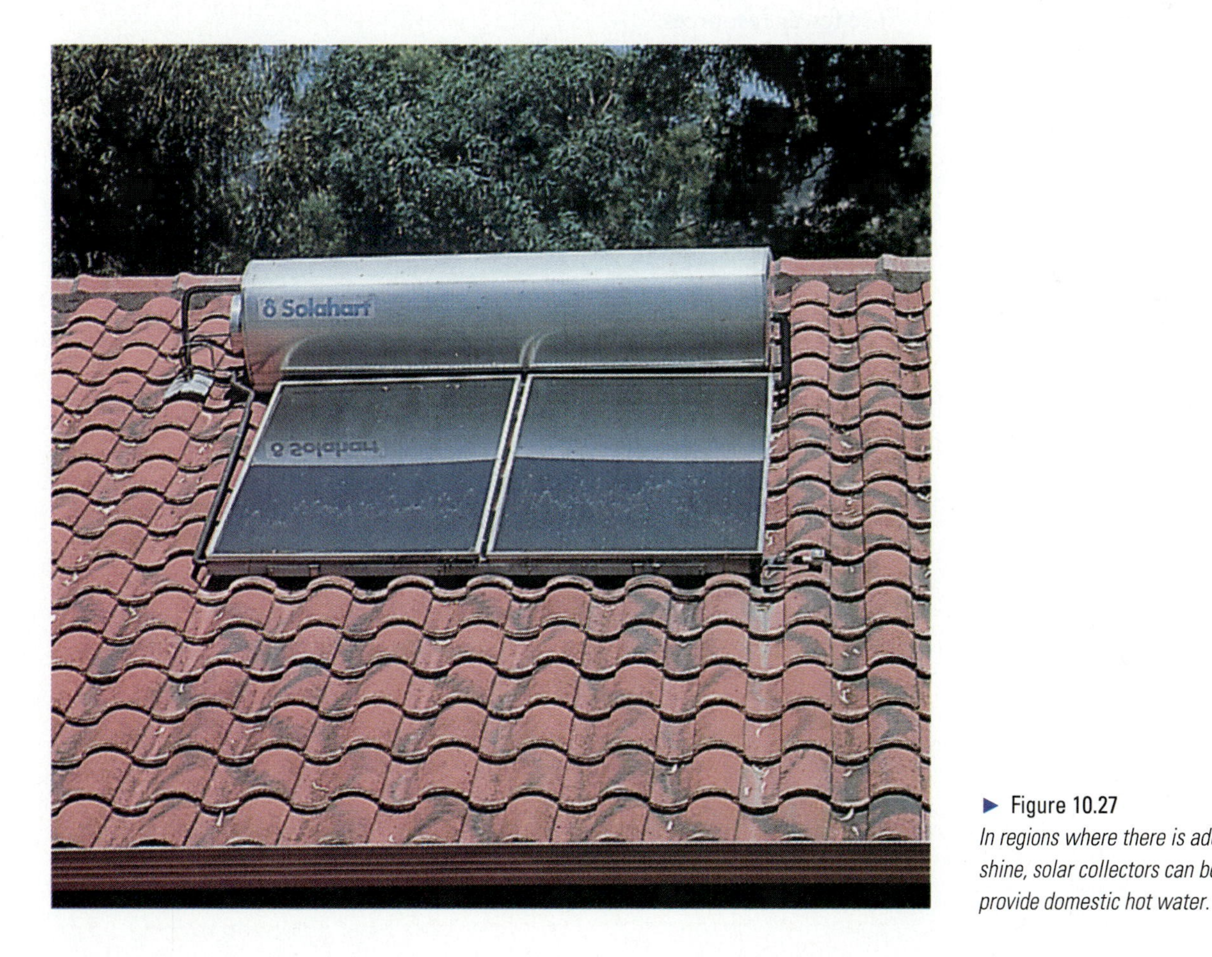

▶ Figure 10.27
In regions where there is adequate sunshine, solar collectors can be used to provide domestic hot water.

- We need to adjust our economy to a no-growth, and eventually a "negative-growth" environment. The economic model of supply and demand does not work when there is a huge demand but little or no supply. Under these conditions, resources are allocated by decree or by war.
- We need to change our lifestyles. Small savings add up to large ones over time. We can walk more, ride a bicycle, car pool, and use mass transportation when possible. We can turn down the thermostat in our homes in winter and turn it up in the summer. We should insist that goods be made to last and that what they are made of can be recycled. Convenience foods waste energy and resources, as do throw-away containers and unnecessary packaging. Most of us have material possessions far in excess of our subsistence needs. This situation is likely to change as energy and other natural resources become scarcer and more expensive. To forestall a shortage of essentials, we can consume less now and instead pursue activities that use fewer resources.

DID YOU LEARN THIS?

For each of the statements given, fill in the blank with the most appropriate word or phrase from the following list. You may use any word or phrase more than once.

negentropic	temperature	endergonic
entropy	joule	endothermic
heat	zero-point energy	nonspontaneous
exergonic	potential energy	less stable
exothermic	efficiency	kinetic energy
spontaneous	power	watt
more stable	work	absolute zero

a. _______ Describes a process that does work
b. _______ A measure of the average kinetic energy of the particles in a substance
c. _______ The temperature at which all motion of particles should theoretically cease
d. _______ A process in which order is created from disorder
e. _______ "Stored energy"
f. _______ The use of energy to move mass
g. _______ The amount of work obtained from an energy conversion process
h. _______ A unit of power
i. _______ A unit of energy
j. _______ Describes a process in which heat is absorbed by a system
k. _______ Energy of motion predicted by the uncertainty principle
l. _______ The tendency of a process toward greater disorder
m. _______ The rate at which work is done

EXERCISES

1. What is temperature? What is the difference between temperature and heat? What is the relationship between temperature and the motion of particles of a substance?
2. State the first law of thermodynamics. What does it tell us about the nature of our energy resources? About thermal pollution? What other conservation law(s) do you know?
3. What is the difference between conduction of heat and radiant energy? What happens when radiant energy is absorbed by a substance?

4. What is kinetic energy? Why is it much more likely that injuries will result from an automobile collision at 60 mph than from a collision at 40 mph?
5. In terms of kinetic energy, what is temperature? What can be said about the kinetic energies of substances that are maintained at the same temperature, regardless of their physical state?
6. In principle, what is absolute zero? Why is it impossible to reach this temperature in practice? What is zero-point energy? What property of tiny particles makes zero-point energy possible?
7. Compare the Celsius and kelvin scales of temperature. What similarity do they have? What difference? Why is the kelvin scale considered to be "absolute" and why is its abbreviation written simply K and not °K?
8. What is the difference in size between a Fahrenheit degree and a Celsius degree? What is the difference in the zero temperature of the two scales?
9. The amount of heat contained by a substance depends on three factors. Two of these have been discussed in this chapter. What are they? What do you suppose the third one might be? (Hint: it is an intrinsic property.)
10. List some common units of heat. When is the amount of heat contained by an object proportional to its temperature? Why can't other temperature scales be used?
11. State the second law of thermodynamics. Heat is considered to be the kind of energy that flows from a hotter body to a colder one. What other kinds of energy are there?
12. Distinguish between energy and work. Why is it more appropriate to use the term *energy conversion* than the term *energy use.*
13. What do the terms *endergonic* and *exergonic* mean? How do they differ from endothermic and exothermic? Which tend to be spontaneous? Which tend to lead to a less stable state?
14. Distinguish between work and power. Why is it often desirable to produce greater power? List some units of power.
15. What is entropy a measure of? What entropy state is the most probable? In terms of energy, what is necessary to create a more ordered state from one of less order? How can this be possible? (That is, what must occur somewhere else?)
16. What is efficiency in an energy conversion process? List strategies that can be used to increase the efficiency of an energy conversion process.
17. If energy is not destroyed in an energy conversion process what happens to it?
18. Explain why it takes energy to get energy and other nonrenewable resources. What other factor, in addition to the losses due to en-

tropy, serves to determine the amount of work that can be obtained from an energy source (net work)? What does the second law tell us about the amount of energy used in recovery of nonrenewable natural resources as they become scarcer and harder to find?

THINKING IT THROUGH

Efficient Agriculture?
It has been said that modern agriculture is highly efficient, in that one person working in the field can feed eighty or more persons. How has this been made possible? From a thermodynamic standpoint, how efficient is modern agriculture?

Fuels from Biomass
Why do you suppose emphasis was placed on the use of *waste* vegetable matter (biomass) to produce transportation fuels?

Electric Automobiles
Although electric automobiles presently are no more efficient in the use of primary energy sources than are gasoline and diesel-powered vehicles, they are being developed (and in California, mandated) for use because they "transfer tailpipe emissions to the power plant." In terms of thermodynamic principles, what advantages does this strategy have in controlling air pollution caused by automobile exhaust? What social or ethical considerations might have to be made about this kind of emissions control?

Would Consumption Double?
Assuming that the rate and distribution of energy use in the world remained the same as they are today, would a doubling of the world's population cause a doubling in the rate of the world's energy consumption? Explain.

ANSWERS TO "DID YOU LEARN THIS?"

a. exergonic
b. temperature
c. absolute zero
d. negentropic
e. potential energy
f. work
g. efficiency
h. watt
i. joule
j. endothermic
k. zero-point energy
l. spontaneous
m. power

WHAT IS SCIENCE FOR?

Henry Linschitz

Professor of Chemistry, Brandeis University

B. S. Chemistry, City College of the City University of New York

Ph.D. Physical Chemistry, Duke University

Early on the morning of July 16, 1945, I found myself, together with a few others, on a ridge in the desert of southern New Mexico. We were staring into the predawn darkness at a point about 15 miles away where the first atomic bomb waited to be tested. The previous afternoon, high up on a steel tower, I had participated in the assembly of that bomb, connecting firing cables to the detonators set in the bomb casing. Later, someone would write that on that morning "the sun rose twice"—and indeed the first "sun" was far brighter than the sun itself. A dazzling whiteness lit up the entire desert. As the light faded and we dared put aside our protective blue glasses, a strange purple glow appeared—an enormous volume of ionized nitrogen shining in the sky above the vaporized tower. Then the blast reached us with an echoing roar that filled the vast space around us. The sun finally rose on an immense, looming column of smoke and dust, the giant "mushroom" that appeared again over Japan and since then has appeared many hundreds of times at test sites in Nevada, central Siberia, the Soviet Arctic, western China, and islands in the South Seas.

We were there as engineers and weaponeers, but in fact most members of the research staff at Los Alamos were scientists, both in outlook and training. We had been brought together in that isolated place by the pressures of war. The average age of the staff was only about 27. Most were American, but there was also an English delegation as well as a critically important group of European refugees, driven from their universities by Fascist governments.

As for myself, I grew up in New York City and at an early age was fascinated by science. New York's wonderful libraries and schools, as well as an encouraging family, made it possible to develop this interest. At City College of New York, I was tempted to major in history, but the experience of the Depression and the need for "practical" studies, as well as my constant bent toward science, kept me in chemistry. Graduate work in physical chemistry at Duke University immediately followed my bachelor's degree. However, I felt a strong obligation to contribute to the war against Hitler and interrupted my doctorate work to take a post in a government agency (NDRC) laboratory, where I worked on explosives and new weapons (the "bazooka"). After some months, I was recruited to join the Los Alamos laboratory as an explosives "expert," which gives some idea of the state of our knowledge at that time. After the war, like most of the staff, I returned to academic life, doing research and teaching at Chicago, Syracuse, and Brandeis in areas (spectroscopy, photochemistry) far removed from detonations.

The development of atomic bombs required an immense scientific and industrial effort. The plutonium that exploded in the desert was a new element, made by processes hitherto unknown in a coordinated plan that involved hundreds of thousands of people and the building of completely new cities in Tennessee, Washington, and New Mexico. Months before the desert test, we had lined up at Los Alamos to gaze in awe at the first shipment of plutonium, a few milligrams, that had been isolated by remote-controlled chemical manipulations of intensely radioactive material at the first nuclear reactors built on earth. The cost was roughly one billion dollars. The world stock of separated plutonium is now about two hundred *tons,* an indication of the scale of the arms race. The total number of nuclear warheads is about 50,000—far more than the number of possible targets.

Even prior to the surrender of Japan, the danger of such an arms race was recognized by scientists working on the Manhattan Project. In an extraordinary, spontaneous movement immediately after the war, groups organized at the various research sites around the country and soon came together to form the Federation of Atomic Scientists (FAS), which is still active today. Our function was educational and political: to explain the essential, harsh facts about nuclear war, and to work to control that threat. It was an amateur but intense effort. Scientists spoke everywhere to interested and concerned church and school groups. FAS set up an office, manned by volunteers, in Washington, D.C. On my way back to Duke to finish my thesis, I stopped off in Washington to help send out literature and carry out informal lobbying. We at least succeed in persuading Congress to establish a civilian, not a military, agency (the Atomic Energy Commission) to foster and regulate nuclear research and development. A worldwide petition by scientists protesting atmospheric contamination caused by bomb tests led to a treaty banning such tests (bombs are now detonated underground). For organizing this protest, Linus Pauling won a Nobel Peace Prize to add to his earlier award for chemistry. Nevertheless, the huge stock of bombs now distributed around the earth indicates how much is left to do.

Now, as we near the end of the twentieth century, the problem of war, including war with nuclear, chemical, and biological weapons, emphasizes our obligation to think not only about what science *is,* but what science is *for.*

The United States ranks first among all nations in military strength and expenditures, but is fourteenth in life expectancy, and twenty-eighth in infant mortality (data from *World Watch,* 1993). Third World nations expend their limited resources on arms, rather than on the desperate needs of their people. The disastrous impact of unregulated technology on the environment is familiar to us all and indeed is touched upon at many places in this book. What does all this say about our values or even our common sense?

Science offers the technical feasibility of profoundly bettering human life. Chemistry in particular provides a host of new materials, embodying new properties. These materials range from fibers to computer chips and include medicines based on a deeper understanding of life processes. It provides means of improving the food supply and controlling the environment. Chemistry is a science in which intellectual satisfaction can be joined with social and economic benefits. This is what science is *for.*

11 The Chemistry of Carbon: Organic Chemistry

"I fell into a reverie and lo! the atoms were gambolling before my eyes. . . I saw how, frequently, two smaller ones united to form a pair, how a larger one embraced two smaller ones; how still larger ones kept hold of three or even four of the smaller. . . I saw how the larger ones formed a chain. . ."

August Kekulé, 1890

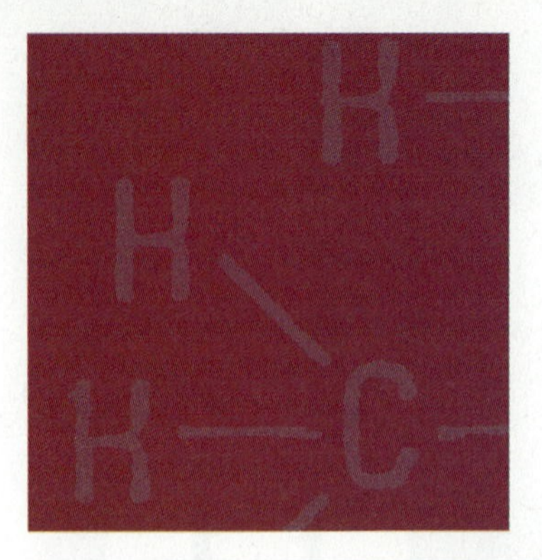

Approximately 11 million chemical compounds are presently known. Of these, about 10.5 million are compounds of carbon, and each year about six hundred thousand new ones are discovered or made in the laboratory. These compounds are immensely important to our lives. For example, they make up food, fibers, wood, our bodies, medicines, viruses, plastics, paints, cosmetics, refrigerants, and adhesives. The combustion of the carbon compounds in coal and petroleum provides most of the energy we use in this country.

THE TERM *ORGANIC*

Early chemists used the term *organic* to describe compounds of carbon because they observed that these substances originated from living or once-living matter. Fats were obtained from animals, vinegar and alcohol from fermented fruits, camphor from Asian camphor trees, and sugar from cane. Inorganic compounds, such as salt and lime, came from nonliving matter. The observation that inorganic substances could be formed by heating organic substances, but not the reverse, led to the theory of vitalism. In this model, living things possessed a mysterious vital force with which they could make organic compounds but which was outside of human control. It was thus believed impossible to make organic compounds from inorganic substances in the laboratory.

The theory of vitalism was weakened in 1828, when the German chemist Friedrich Wöhler heated a mixture of potassium cyanate and ammonium chloride, two inorganic compounds, and produced urea, an organic compound found in urine:

$$\underset{\substack{\text{potassium}\\\text{cyanate}\\\text{(inorganic)}}}{KOCN} + \underset{\substack{\text{ammonium}\\\text{chloride}\\\text{(inorganic)}}}{NH_4Cl} \longrightarrow \underset{\substack{\text{urea}\\\\\text{(organic)}}}{CH_4N_2O} + KCl$$

Wöhler's demonstration that an organic compound could be produced from inorganic compounds stimulated chemists to look for other examples, and soon a number were found. By about 1850, the idea of a vital force had died out, but the term *organic* has persisted. Today, **organic chemistry** is the chemistry of carbon compounds. Many of these are of living origin, but many are also made in the laboratory.

THE BONDING OF CARBON

Carbon forms millions of compounds because carbon atoms can form strong covalent bonds with one another, over and over, to form long chains in a process called **catenation.** These chains can be hundreds, even thousands, of carbon atoms in length. No other element does this to such an extent. Also, because each carbon atom can form up to four bonds, any carbon atom in a chain can have additional carbon atoms bonded to it, causing branching, or the formation of side chains. Other possibilities are the formation of rings rather than chains and the formation of double and triple bonds between carbon atoms. Because carbon forms stable bonds with most nonmetals, in addition to hydrogen, these *heteroatoms* can be incorporated into the molecules as well. Finally, these structural features can be combined; for example, a ring with a chain. Thus, the possibilities for forming different kinds of compounds are limitless (▶ Figure 11.1).

Carbon atoms bond with each other to form long chains and many other structures.

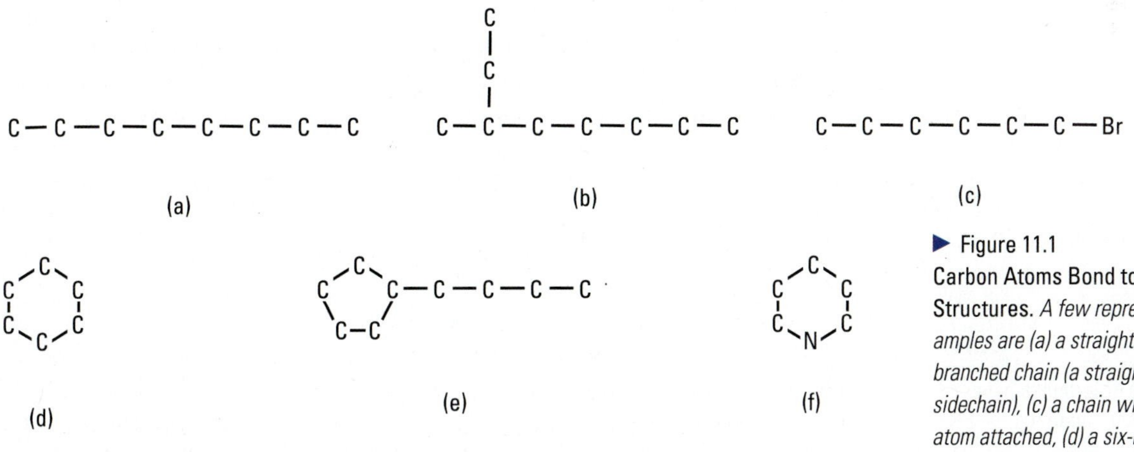

C—C—C—C—C—C—C—C

(a)

C—C—C—C—C—C—C (with a C—C side chain on the second carbon)

(b)

C—C—C—C—C—C—Br

(c)

(d)

(e)

(f)

▶ Figure 11.1
Carbon Atoms Bond to Create Many Structures. *A few representative examples are (a) a straight chain, (b) a branched chain (a straight chain with a sidechain), (c) a chain with a nonmetal atom attached, (d) a six-membered ring, (e) a five-membered ring with a sidechain, and (f) a six-membered ring containing a heteroatom, in this case a nitrogen atom. Hydrogen atoms bonded to carbon atoms have been omitted to emphasize the carbon* skeletons.

CLASSIFICATION OF ORGANIC COMPOUNDS

Although there are many different organic compounds, it is possible to systematically classify and name them all. There are two main classes of organic compounds, hydrocarbons and derivatives of hydrocarbons. **Hydrocarbons** contain *only* carbon and hydrogen (hence their name), whereas **hydrocarbon derivatives** also contain nonmetallic elements such as oxygen, nitrogen, sulfur, chlorine, and the like. (Some contain metals such as lithium, sodium, or iron, and are called *organometallic compounds.*)

Hydrocarbons are further classified as being saturated, unsaturated, cyclic, or aromatic, whereas derivatives of hydrocarbons usually are classified by the kind of functional group they contain: aldehydes, esters, amines, and so on. A **functional group** is an atom or group of atoms present in a molecule that exhibits a particular kind of chemical behavior. The hydroxyl group, —O—H, which we encountered in the isopropyl alcohol molecule in Chapter 7, is an example. We shall have more to say about functional groups later.

SATURATED HYDROCARBONS: ALKANES

Saturated hydrocarbons contain only single bonds.

The term **saturated** applied to hydrocarbons means that they contain only single bonds. They also have the highest possible hydrogen-to-carbon ratio. The simplest of these compounds are called the **alkanes,** and they have the general formula C_nH_{2n+2}, where $n = 1, 2, 3, 4, 5, 6$, and so on. Methane, which we know from Chapter 6, is the first in this series and has the formula CH_4, where $n = 1$. Butane, which we saw in Chapter 7, is the fourth member and has the formula C_4H_{10}, where n = 4. ■ Table 11.1 lists the first ten alkanes. You should learn their names and their chemical formulas because they form the basis for naming many organic compounds.

Note that in Table 11.1 a *condensed structural formula* is given. This is one of several ways that structures of carbon compounds can be represented, and it is important to pay particular attention to which one is used because each has a different level of meaning. Methane, for example, can be written

CH_4, H:C:H (with H above and below), H—C—H (with H above and below), and [wedge-dash structure]

expanded formula

Different kinds of formulas have different levels of meaning.

The first, CH_4, is simply its *molecular formula,* which tells us the ratio of carbon to hydrogen atoms in the molecule. Next is its Lewis electron-dot formula. In the third (the *expanded,* or *Kekulé structure*), the electron-dots have been replaced with lines that mean the same thing: a pair of bonded electrons between carbon and hydrogen. These two representations tell us that a carbon atom is bonded to four hydrogen atoms each through single bonds, and we could get the impression from them that the methane molecule is flat,

■ **TABLE 11.1 THE FIRST TEN ALKANES**

IUPAC Name	Molecular Formula	Condensed Structural Formula	Boiling Point, °C
methane	CH_4	CH_4	−162
ethane	C_2H_6	CH_3CH_3	−88.5
propane	C_3H_8	$CH_3CH_2CH_3$	−42
butane	C_4H_{10}	$CH_3(CH_2)_2CH_3$	−0.6
pentane	C_5H_{12}	$CH_3(CH_2)_3CH_3$	36
hexane	C_6H_{14}	$CH_3(CH_2)_4CH_3$	69
heptane	C_7H_{16}	$CH_3(CH_2)_5CH_3$	98
octane	C_8H_{18}	$CH_3(CH_2)_6CH_3$	126
nonane	C_9H_{20}	$CH_3(CH_2)_7CH_3$	151
decane	$C_{10}H_{22}$	$CH_3(CH_2)_8CH_3$	174

with bond angles of 90°. But we already know that it is tetrahedral, with 109.5° bond angles, as shown in the last representation.

Which structural representation do we use? It depends on what we are trying to communicate. If we need only the information a molecular formula can provide, say for balancing an equation, we use CH_4. This representation would do us no good, however, if we needed to show the bond angles in the molecule or talk about it in three dimensions. The electron-dot and expanded (Kekulé) representations are useful when we need to talk about breaking and forming certain bonds in the molecule, to show chain branching in the larger alkanes, and to indicate where functional groups are located in hydrocarbon derivatives. Thus, we must keep in mind what we want to communicate or visualize when we decide which representation to use.

Ethane (C_2H_6) and propane (C_3H_8) can be written:

CH_3CH_3,
ethane

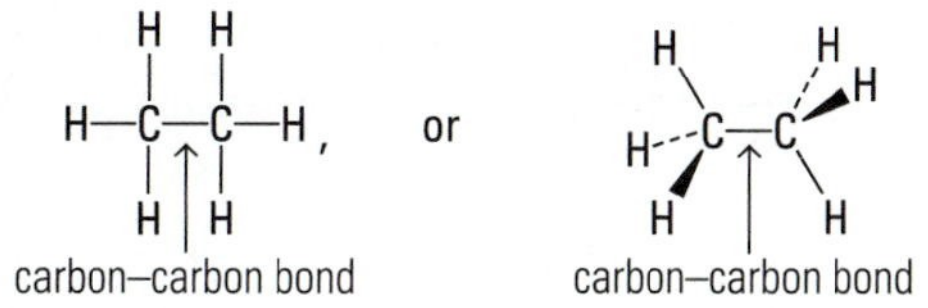

$CH_3CH_2CH_3$,
propane

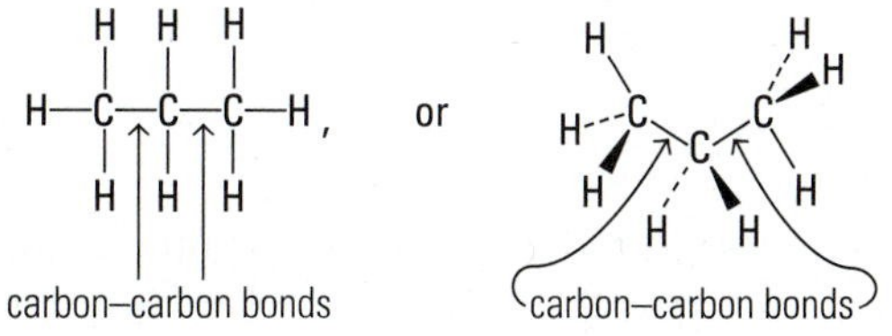

The two carbon atoms in ethane are joined by a carbon–carbon single bond; in propane there are two carbon–carbon single bonds. Each carbon atom must have four bonds. In ethane the remaining six bonds go to hydrogen atoms (three for each carbon atom), and in propane the remaining eight bonds must be to hydrogen (three for the two end carbon atoms and two for the central one). Note that *each* carbon atom in these molecules has a tetrahedral geometry. As the number of carbon atoms joined together increases, these tetrahedra cause the chain to form a zigzag arrangement. This is evident in propane, but is more pronounced in the larger alkanes, such as pentane and octane (▶ Figures 11.2 and 11.3).

$CH_3CH_2CH_2CH_2CH_3$

(a)

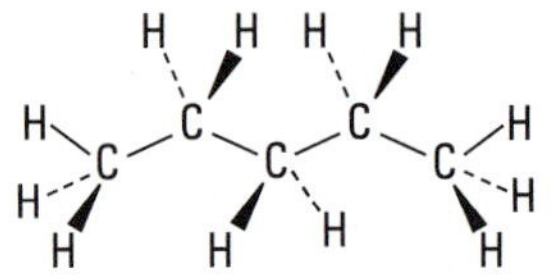

(b)

$CH_3CH_2CH_2CH_2CH_2CH_2CH_2CH_3$

(c)

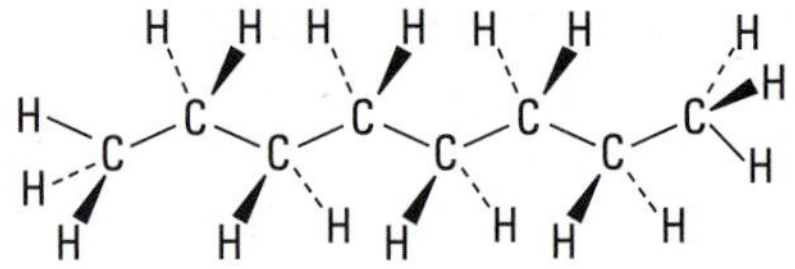

(d)

▶ Figure 11.2
Straight-chain hydrocarbons have a zigzag structure. Condensed formulas of straight-chain hydrocarbons pentane (a) and octane (c) do not show the actual zigzag arrangement of carbon atoms in their respective chains (b,d).

11.1 Draw the condensed, expanded (Kekulé), and three-dimensional zigzag formulas for butane.

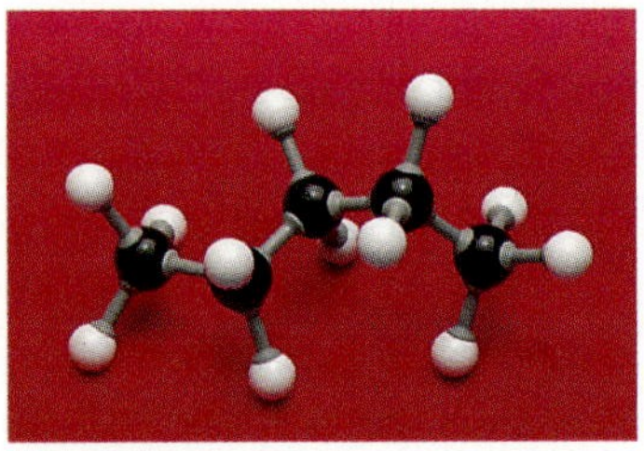

Another feature of the longer alkane molecules is that they are flexible. Little energy is required to cause rotation about carbon–carbon single bonds, making it possible for the chains to assume a number of shapes, such as stretched out (▶ Figure 11.4a) or twisted up (▶ Figure 11.4b). Waxes, for example, work this way. Paste wax is made up of polar functional groups at-

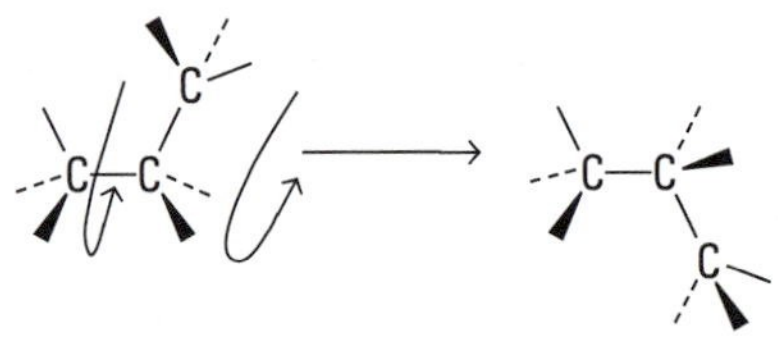

▶ Figure 11.3
A Model of Pentane Showing the Zigzag Arrangement of Tetrahedral Carbon Atoms.

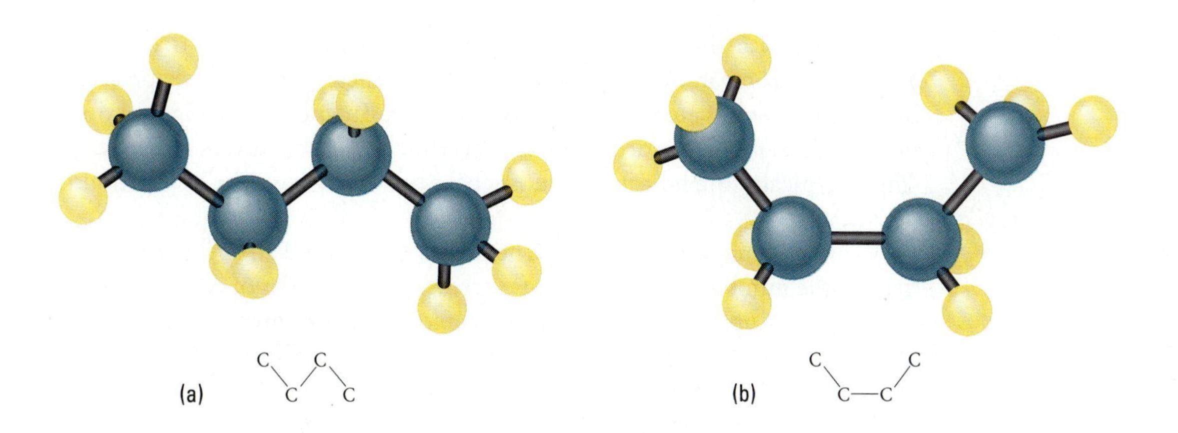

▶ Figure 11.4
Alkane molecules are flexible. Little energy is required to cause rotation about the middle carbon–carbon bond in the stretched-out form of butane (a) to produce a more twisted arrangement (b). Rotations about bonds like this allow alkane molecules to assume a twisted-up shape.

tached to long hydrocarbon chains that are randomly twisted and coiled when the wax is first applied to a surface. Polishing warms the wax and causes these hydrocarbon chains to straighten and line up beside each another. This increases the London attractive forces between the chains, and the wax hardens and produces a shine.

STRUCTURAL ISOMERS

There are two compounds that have the molecular formula C_4H_{10}, one boiling at –0.6 °C and the other at –10 °C. Because these butanes have the same molecular formula but different properties, they must differ in the way their atoms are arranged. Compounds of this kind are called **isomers.** The structural formula of one isomer is easy to imagine: namely, four carbon atoms bonded to one another in a straight chain,

Isomers have the same molecular formula but different arrangements of their atoms.

$$CH_3CH_2CH_2CH_3 \quad \text{or} \quad \begin{array}{ccccccccc} & H & & H & & H & & H & \\ & | & & | & & | & & | & \\ H- & C & - & C & - & C & - & C & -H \\ & | & & | & & | & & | & \\ & H & & H & & H & & H & \end{array}$$

This is the isomer that boils at –0.6 °C; it is called butane. The other isomer, boiling at –10 °C, is called isobutane; it has only three carbon atoms in its chain, and its fourth carbon atom forms a branch from the center carbon atom of that chain,

$$\begin{array}{c} CH_3CHCH_3 \\ | \\ CH_3 \end{array} \quad \text{or} \quad \begin{array}{ccccccc} & H & & H & & H & \\ & | & & | & & | & \\ H- & C & - & C & - & C & -H \\ & | & & | & & | & \\ & H & & | & & H & \\ & & H- & C & -H & & \\ & & & | & & & \\ & & & H & & & \end{array}$$

▶ Figure 11.5 shows models of butane and isobutane. Note that because the carbon atoms in each have a tetrahedral geometry, butane is shaped like a hot dog and isobutane is more nearly spherical. Butane has the higher boiling point because it has more surface over which London forces can interact. This is typical: isomers with the most chain branching have the lowest boiling points.

A similar situation exists for the pentanes, C_5H_{12}, and all of the larger alkanes. As the number of carbon atoms increases, the number of isomers that are possible goes up dramatically. For example, there are three isomers of pentane, five of hexane, 18 of octane, and 75 of decane. Eicosane, $C_{20}H_{42}$, has 366,319 possible isomers!

The three isomeric pentanes are illustrated in ▶ Figures 11.6a, b, c. Note that pentane has a straight chain and isopentane and neopentane have branched chains. A *straight-chain hydrocarbon,* regardless of length, (for example, dodecane, ▶ Figure 11.6d) has all of its carbon atoms lined up in sequence and has no carbon branches. (It does, however, have the zigzag structure due to tetrahedral carbon atoms and may be coiled or twisted.) *Branched-chain hydrocarbons* have one or more carbon atoms (branches) attached to a larger, straight chain. Iso-octane (▶ Figure 11.6e) has three branches.

11.2 Identify the following as straight-chain or branched-chain hydrocarbons.

a. $CH_3CH_2CHCH_2CH_3$ (with CH_3 attached to the middle CH)

b. $CH_3—CH—C—CH_3$ (with CH_3 on the CH, and CH_3 above and CH_3 below the C)

c. $CH_3CH_2CH_2CH_2CH_2CH_2CH_2$ (with CH_3 attached to the last CH_2)

d. $CH_2—CH_2$ (with CH_3 attached below each CH_2)

e. $CH_2—CH_2—CHCH_3$ (with CH_3 attached above the first CH_2 and above the CH)

f. $CH_3CH_2—C—C—CH_3$ (first C bearing CH_3 above and CH_3 below; second C bearing CH_2CH_3 above and CH_3 below)

11.3 Draw all of the possible isomers of C_6H_{14}

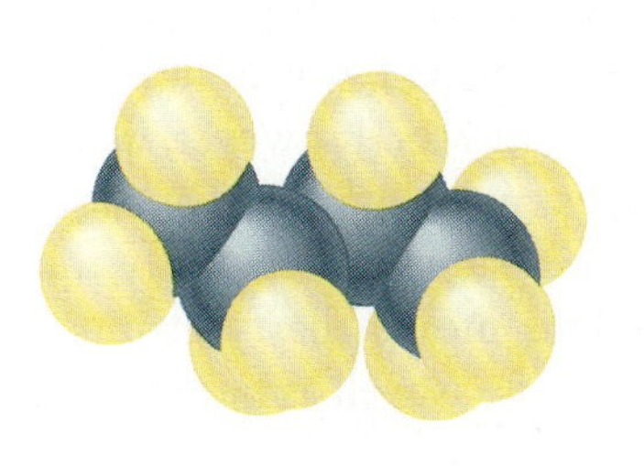

$CH_3—CH_2—CH_2—CH_3$

(a) Butane

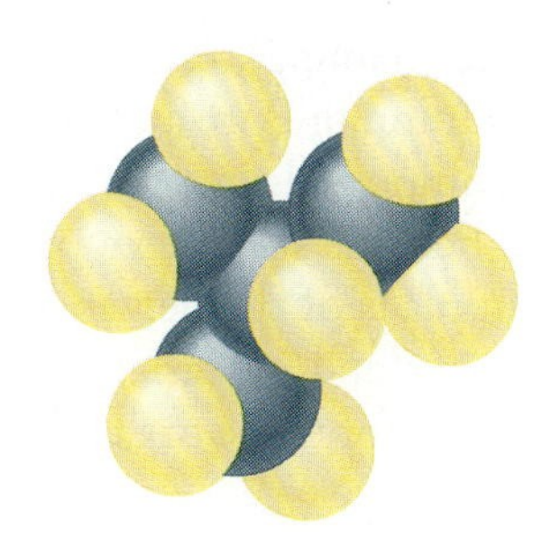

```
CH3—CH—CH3
     |
    CH3
```

(b) Isobutane

▶ Figure 11.5
Space-filling Models of Butane and Isobutane. *The butane model (a) is less spherical and less compact than the branched isobutane model (b). The butane molecule has a greater surface area on which London forces can act and has the higher boiling point.*

(a)

```
                          H   H   H   H   H
                          |   |   |   |   |
CH3CH2CH2CH2CH3  or  H—C—C—C—C—C—H
                          |   |   |   |   |
                          H   H   H   H   H
```

pentane (straight chain, b.p. 36 °C)

(b)

```
                           H   H   H   H
                           |   |   |   |
CH3CHCH2CH3   or      H—C—C—C—C—H
    |                      |   |   |   |
    CH3                    H   C   H   H
                              /|\
                             H H H
```

isopentane (branched chain, b.p. 28 °C)

(c)

```
                              H
                              |
   CH3                     H—C—H
    |                   H     |     H
 CH3CCH3     or       H—C—C—C—H
    |                   H     |     H
   CH3                     H—C—H
                              |
                              H
```

neopentane (branched chain, b.p. 9.5 °C)

(d)

$CH_3CH_2CH_2CH_2CH_2CH_2CH_2CH_2CH_2CH_2CH_2CH_3$

dodecane ($C_{12}H_{26}$), a straight-chain hydrocarbon

(e)

```
   CH3  CH3
    |    |
 CH3CCH2CHCH3
    |
   CH3
```

isooctane (C_8H_{18}), a branched-chain hydrocarbon

▶ Figure 11.6
Straight-Chain and Branched Hydrocarbons. *Pentane (a) and dodecane (d) have straight chains. Isopentane (b), neopentane (c), and iso-octane (e) are branched. Note that for the isometric pentanes, boiling points decrease as chain branching increases.*

NAMING ALKANES

Many organic compounds have two or more names, their common name(s) and their systematic name. Common names are not useful for systematically classifying the millions of organic compounds that are known, so a system of nomenclature (naming), called the **IUPAC system** (abbreviation for the International Union of Pure and Applied Chemistry), was designed for naming all organic compounds in a unified, consistent way. However, many common names have become so much a part of organic chemistry that they remain in use, often in place of the newer, systematic names. (Frequently the common name is shorter and/or easier to use.) This means that you must learn the common names for some organic compounds, as well as be able to write their systematic names.

You will need to know a few IUPAC rules to name simple alkanes, but these rules also apply to naming all of the other organic compounds you will encounter. Learn them well because they will form the basis for systematically naming other organic molecules later on:

1. An alkane (straight-chain or branched) takes its name from the longest, continuous carbon chain that it contains. This chain, called the *parent chain,* is named for the alkane having the same number of carbon atoms. (The first ten alkanes were listed in Table 11.1.) Consider the following hydrocarbon, for example,

$$\begin{array}{c} \quad\;\; CH_3 \\ \quad\;\; | \\ CH_3CH_2CHCH_2CH_2CH_3 \end{array}$$

   It contains three continuous carbon chains, which are shown in bold as follows:

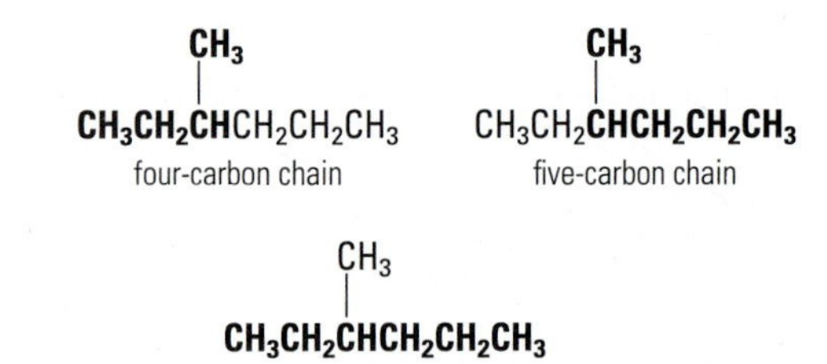

   The longest continuous chain in this instance has six carbon atoms in it. The alkane having six carbons is hexane, so the parent chain (and thus the molecule itself) is named as a hexane. Note that we proceeded along the molecule in a number of directions to find our longest chain. It does not matter which way we go, as long as we follow a *continuous* chain of carbon atoms. Our object is to find the longest one, regardless of how the molecule is presented on paper. Our molecule could have been drawn, for example, like this:

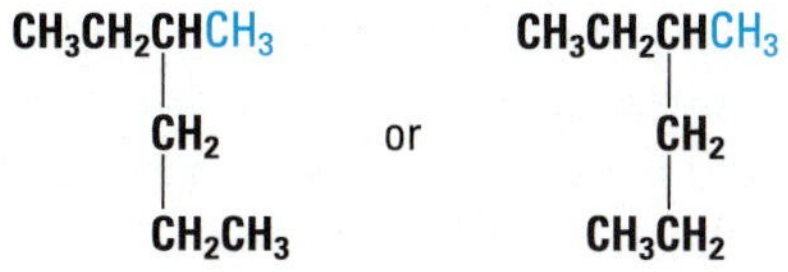

$$\begin{array}{c} CH_3CH_2CHCH_3 \\ | \\ CH_2 \\ | \\ CH_2CH_3 \end{array} \quad \text{or} \quad \begin{array}{c} CH_3CH_2CHCH_3 \\ | \\ CH_2 \\ | \\ CH_3CH_2 \end{array}$$

The parent chain still is six carbons long and is named hexane. We can stand the molecule on end or twist it around (it's flexible) like we've done here and change nothing. What about the —CH_3 group that is a branch off the parent chain? It is called a substituent and must be dealt with in the next steps.

2. Identify substituents and name them. A *substituent* is any atom or group of atoms, excluding hydrogen, that is *not a part* of the parent chain. In alkanes, these groups are also called branches. The —CH_3 group is a substituent called the meth*yl* group. Other substituents you should know are the ethyl group, the propyl group, and the isopropyl group. Their structures are as follows:

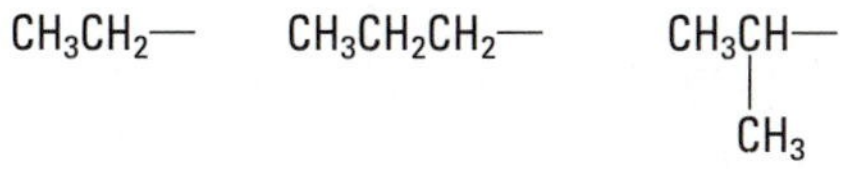

$$\underset{\text{ethyl group}}{CH_3CH_2—} \qquad \underset{\text{propyl group}}{CH_3CH_2CH_2—} \qquad \underset{\text{isopropyl group}}{\begin{array}{c} CH_3CH— \\ | \\ CH_3 \end{array}}$$

These substituent groups do not exist by themselves. The *-yl* ending on their names indicates that they must always be attached to something else, in this case the parent chain. They are attached to it at the carbon atom having the unspecified bond (indicated by —).

3. Number the carbon atoms on the parent chain so as to minimize the number(s) of the substituent(s) attached to it. (Or, number from the end closest to the first branch.) In our example the hexane parent chain could be numbered from either end:

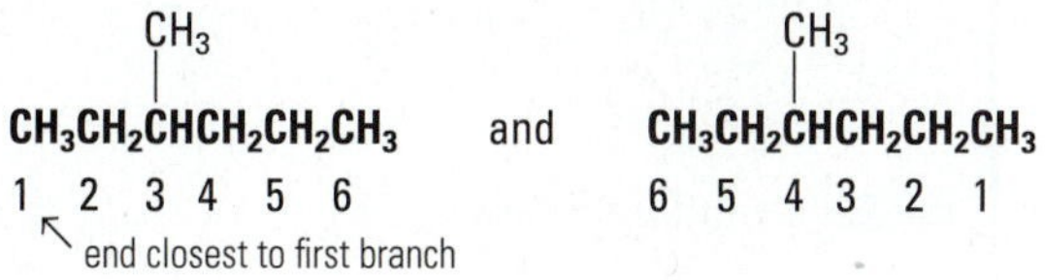

$$\begin{array}{c} CH_3 \\ | \\ CH_3CH_2CHCH_2CH_2CH_3 \\ 1\ \ 2\ \ 3\ 4\ \ 5\ \ 6 \end{array} \quad \text{and} \quad \begin{array}{c} CH_3 \\ | \\ CH_3CH_2CHCH_2CH_2CH_3 \\ 6\ \ 5\ \ 4\ 3\ \ 2\ \ 1 \end{array}$$

↖ end closest to first branch

In the first case, the methyl group is attached to the carbon atom bearing the number 3, and in the second it is attached to the carbon atom having the number 4. The lower of these two numbers is 3, so the methyl group takes this number. Numbering the chain from the end closest to the first branch gives the same result: the methyl group takes the number of carbon 3.

4. Write down the name of the compound. The name of the parent chain is like a surname; it goes at the end of the name, like Wiegand. Thus, the last name of our compound is *hexane.* Although Wiegand is not a very common name, there are quite a few of us, such as Ted, Lois, Danika, and Fred. The family of hexanes has a much larger number of members, and we give them first names in a similar way. We do this by naming each substituent attached to the parent chain and giving each substituent a number to indicate which carbon atom it is attached to. Our example has only one substituent, a methyl group, and it is on the third carbon atom of the chain. We give our hexane a first name as follows:

3-methylhexane

Note that the first and last names are run together, with no space between them. The number indicating the position of the methyl group is separated by a hyphen (-) from the word. As a rule, numbers are separated from words with hyphens (on both sides of the number, if necessary), and numbers are separated from each other by commas (,).

11.4 Be able to provide the IUPAC name for the following compounds before going on:

a. $CH_3CH_2CHCH_3$
   (branch on the 3rd carbon: CH_3)

b. $CH_2CH_2CH_2CHCH_2$
   (above CH: CH_3; below first CH_2: CH_3; below last CH_2: CH_3)

c. $CH_2CH_2CHCH_2CH_2CH_2CH_3$
   (first CH_2 bears CH_2–CH_3; CH bears CH with two CH_3 groups)

d. $CH_3CH_2CHCH_2CH_2CH_3$
   (branch: CH_2CH_3)

e. $CH_3CH_2CH_2CHCH_2CH_2CH_2CH_3$
   (branch: CH_2CH_3)

f. $CH_3CH_2CH_2CH_2CH_2CHCH_2CH_3$
   (branch: CH_2CH_3)

CYCLOALKANES

Cycloalkanes exist in rings instead of chains.

Some alkanes do not exist as chains, but rather as rings of carbon atoms. Their general molecular formulas are C_nH_{2n} instead of C_nH_{2n+2}. They contain two fewer hydrogen atoms because an additional carbon–carbon bond is required to form a ring of carbon atoms:

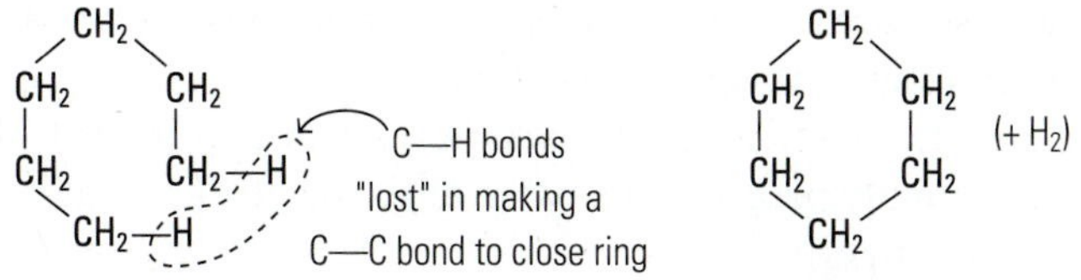

CH_2 CH_2 CH_2 CH_2 CH_2—H CH_2—H

C—H bonds "lost" in making a C—C bond to close ring

hexane (alkane, C_6H_{14})

CH_2 CH_2 CH_2 CH_2 CH_2 CH_2 (+ H_2)

cyclohexane (cycloalkane, C_6H_{12})

The **cycloalkanes** are named simply by placing the prefix *cyclo-* in front of the name of the alkane having the same number of carbon atoms. The simplest cycloalkane is cyclopropane, having a three-carbon ring. It is sometimes used as an anesthetic that allows the patient to remain concious during surgery. Some common cycloalkanes are listed in ■ Table 11.2. Rings having

■ **TABLE 11.2 STRUCTURAL FORMULAS AND SYMBOLS FOR COMMON CYCLOALKANES**

Name	Structural Formula	Abbreviated Formula
cyclopropane	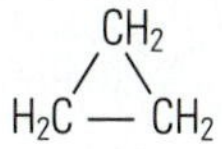 CH_2 / H_2C — CH_2	 △
cyclobutane	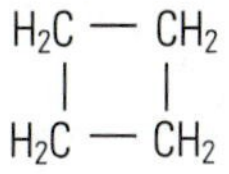 H_2C — CH_2 / H_2C — CH_2	□
cyclopentane	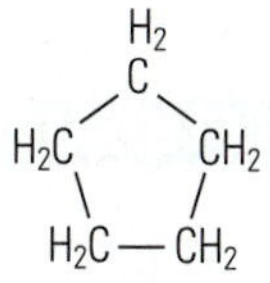 H_2C / H_2C CH_2 / H_2C—CH_2	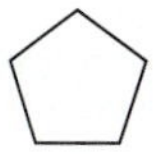 ⬠
cyclohexane	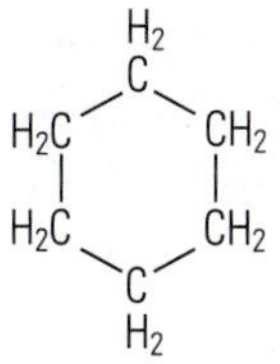 H_2C / H_2C CH_2 / H_2C CH_2 / CH_2	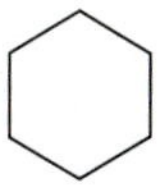 ⬡

Source: Seager/Slaubaugh, Chemistry for Today: General, Organic, and Biochemistry, 2/e, *© 1994 by West Publishing Company. Reprinted with permission.*

more than eight carbon atoms exist but are not common. A special abbreviation, or shorthand, is used to represent the cycloalkanes. Cyclopropane is represented simply by a triangle, cyclobutane by a square, cyclopentane by a pentagon, and so forth. You should keep in mind that there is a —CH_2— group *in* each corner of the geometric figure.

Naming simple substituted cycloalkanes is straightforward: the name of the substituent simply precedes the name of the cycloalkane, as it does for the straight-chain alkanes:

CH_3 — methylcyclobutane

$CH_2CH_2CH_3$ — propylcyclohexane

11.5 Give the correct IUPAC name for each of the following:

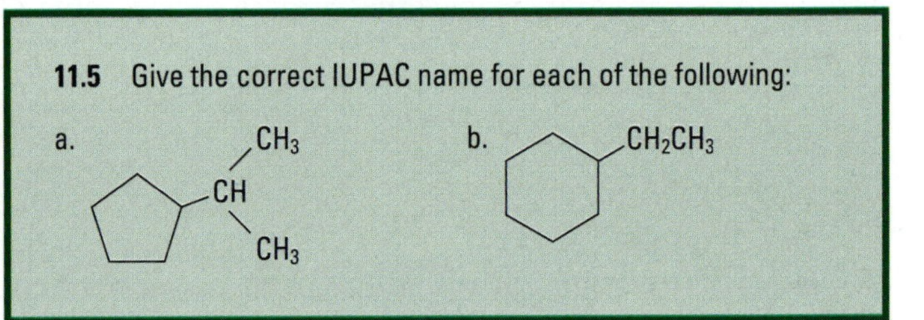

11.6 Write the correct structure from each of the following IUPAC names.

a. ethylcyclopropane b. propylcyclobutane

ALKENES AND ALKYNES: UNSATURATED HYDROCARBONS

Alkenes and **alkynes** are **unsaturated** hydrocarbons.

Alkenes contain carbon–carbon double bonds and have the molecular formula C_nH_{2n}. **Alkynes** have carbon–carbon triple bonds and their molecular formula is C_nH_{2n-2}. The simplest alkene is ethene, $H_2C{=}CH_2$ (common name, ethylene). Its molecule is planar: all six of its atoms lie in the same plane.

$$H_2C{=}CH_2$$

11.7 Draw the Lewis electron-dot structure for ethene and use VSEPR to show that this is so. You will need to treat each carbon atom individually. What are the approximate bond angles in ethene?

Ethyne (common name, acetylene), C_2H_2, is the simplest alkyne. Its molecule is linear: all four of its atoms lie on a line.

$$H—C{\equiv}C—H$$

11.8 Use electron-dot formulas and VSEPR to show that ethyne is linear. What are the bond angles in ethyne?

Note that the formulas of alkenes and alkynes contain two and four hydrogen atoms fewer, respectively, than do alkanes containing the same number of carbon atoms (C_nH_{2n+2}). Alkenes and alkynes are said to be **unsaturated** because their double and triple bonds will react with hydrogen gas to produce alkanes. This reaction, called **hydrogenation,** is usually carried out in the presence of nickel, platinum, or palladium metals, which act as catalysts. A **catalyst** is a substance that alters the rate of a reaction but is not itself consumed in the process. Thus, only a small amount is needed because it can be used over and over. Alkanes, which are saturated, do not react with hydrogen under the same conditions.

A **catalyst** alters the rate of a reaction but is not used up in it.

$$\underset{\text{1-butene}}{H_2C{=}CHCH_2CH_3} + H_2 \xrightarrow{\text{nickel}} \underset{\text{butane}}{H_3C—CH_2CH_2CH_3}$$

$$\underset{\text{2-pentyne}}{CH_3C{\equiv}CCH_2CH_2CH_3} + 2H_2 \xrightarrow{\text{platinum}} \underset{\text{pentane}}{H_3CCH_2—CH_2CH_2CH_3}$$

$$\underset{\text{heptane}}{CH_3CH_2CH_2CH_2CH_2CH_2CH_3} + H_2 \xrightarrow{\text{nickel}} \text{no reaction}$$

NAMING ALKENES AND ALKYNES

Naming the alkenes and alkynes follows naturally from the rules for naming alkanes. (1) This time, the parent chain is the longest continuous carbon chain that *contains the double or triple bond.* The ending of the name of the parent chain is changed from *-ane* to *-ene* if an alkene or to *-yne* if an alkyne. (2) The chain is numbered so as to give the first carbon atom of the double or triple bond as low a number as possible. (3) Substituents (if any) are named and numbered just as they are for alkanes. Two examples have been given previously; 1-butene and 2-pentyne. Their names were derived as follows:

4 3 2 1 $H_2C{=}CHCH_2CH_3$ 1 2 3 4	Chain has four carbon atoms. It is a butene. Number of first carbon atom of double bond is 1. (If numbered the other way, it would be 4.) Name of compound is 1-butene.
5 4 32 1 $CH_3C{\equiv}CCH_2CH_3$ 1 2 34 5	Chain has five carbon atoms. It is a pentyne. Number of first carbon atom of triple bond is 2. (If numbered the other way, it would be 3.) Name of compound is 2-pentyne.

11.9 Name the following alkenes and alkynes.

a. $CH_3CH{=}CHCH_3$
b. $CH_3CH_2CH_2CH_2C{\equiv}CCH_3$
c. $CH_3C{\equiv}CH$
d. $CH_3CH_2CH{=}CHCH_2CH_2CH_2CH_3$

11.10 Write the correct structure given the IUPAC name.

a. 3-hexyne
b. 1-nonene
c. 2-pentene
d. cyclohexene

AROMATIC HYDROCARBONS

One class of cyclic hydrocarbons has rings made up of alternating single and double bonds. The simplest of these is benzene, C_6H_6, which has the structure shown in ▶ Figure 11.7.

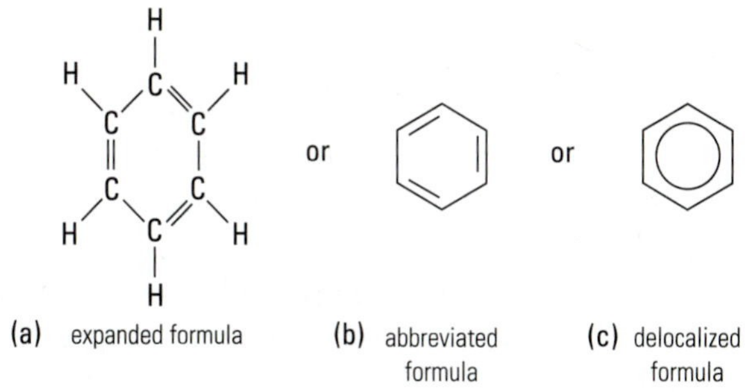

(a) expanded formula (b) abbreviated formula (c) delocalized formula

▶ **Figure 11.7**
Representations of the Benzene Molecule. *Expanded formula (a) and abbreviated formula (b) have rings made up of alternating single and double bonds. Delocalized formula (c) is used to indicate aromatic properties of benzene.*

Although its structure resembles that of an alkene, benzene and compounds like it do not behave chemically like alkenes do. In fact, compared with alkenes, benzene is not very reactive. For example, the catalytic hydrogenation of alkenes discussed earlier proceeds at room temperature and at moderate pressures. Hydrogenation of benzene requires the use of very high temperatures and extreme pressures.

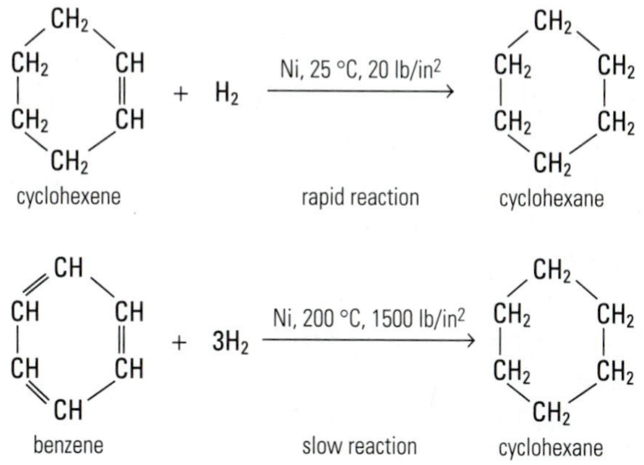

Compared with alkenes, benzene is not very reactive.

$$\text{cyclohexene} + H_2 \xrightarrow{\text{Ni, 25 °C, 20 lb/in}^2} \text{cyclohexane} \quad \text{(rapid reaction)}$$

$$\text{benzene} + 3H_2 \xrightarrow{\text{Ni, 200 °C, 1500 lb/in}^2} \text{cyclohexane} \quad \text{(slow reaction)}$$

(A word is in order about what is written over the arrow in reactions involving organic compounds. This space will be used to indicate reaction conditions: temperature, pressure, catalyst, solvent, special techniques, and so on. As in the past, reactants will always be specified before the arrow and products after it.)

Hydrogenation of alkenes, such as cyclohexene, is an exothermic process, and a considerable amount of heat is evolved from the reaction. The hydrogenation of benzene, though exothermic once it is made to proceed, gives off

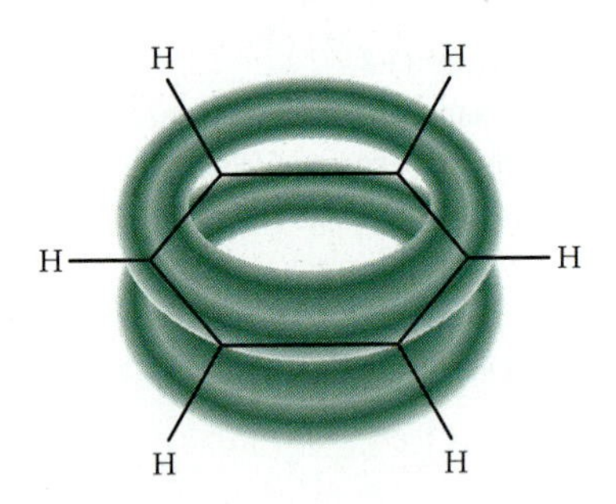

▶ Figure 11.8
One of Three Molecular Orbitals Used as a Model to Explain the Shape and Special Stability of the Benzene Molecule. *The two valence electrons occupying this orbital can move freely about both rings (they are delocalized) and form a very strong bond. The other two molecular orbitals are similar.*

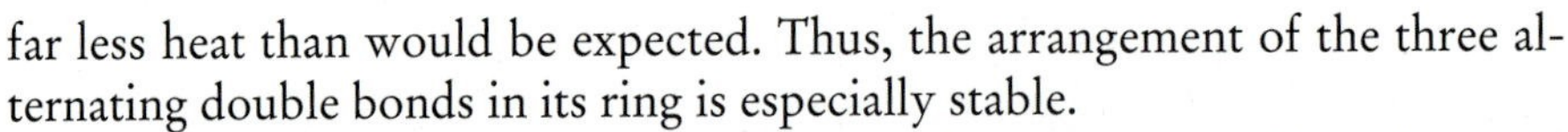

far less heat than would be expected. Thus, the arrangement of the three alternating double bonds in its ring is especially stable.

The benzene molecule is planar. That is, all twelve atoms lie in the same plane. Also, all carbon–carbon bond lengths are equal (the carbon atoms lie in the corners of a regular hexagon), and all bond angles are 120°. The model that explains this molecular shape and special stability requires a new kind of bond. ▶ Figure 11.8 shows one of the molecular orbitals that forms these bonds (there are three). It consists of two doughnut-shaped regions in space, above and below the plane of the six-carbon ring, which are occupied by one pair of valence electrons. The electrons are free to move about both of these doughnuts and are not fixed between any two carbon atoms. Delocalized in this manner, they form a very strong bond between all of the carbon atoms in the ring, which is very hard to break. (We have encountered delocalized electrons before in the electron sea model for the metallic bond.) The molecular orbitals that form the two other bonds (each containing two valence electrons) are similar in shape, except that their doughnuts have "bites" taken out of them at two places across the ring from one another.

Cyclic compounds that possess this extra-stable system of apparent alternating double bonds are said to be **aromatic,** a term used in organic chemistry to express this special property. The delocalized representation of benzene in Figure 11.7c is often used to show this attribute, but it does not reveal that there are actually six valence electrons in the aromatic system.

Many aromatic compounds are important in our daily lives, especially benzene, which is used as a solvent and motor fuel additive. But the primary use of benzene is as a starting material for making a host of compounds we use every day, including pharmaceuticals, fibers and plastics, disinfectants, explosives, dyes, preservatives, and detergents. Benzene can cause suppression of the immune system in humans, however, and long-term workplace exposure to it can make some persons more susceptible to leukemia and to infectious diseases. Some examples of aromatic compounds and their applications are given in ■ Table 11.3. (Rather than memorize the detailed structures of these compounds, look instead at the similarities—and differences—in their structural features.)

▶ Figure 11.9
Petroleum is the primary source of alkanes and cycloalkanes. It supplies 40% of the energy requirements in the United States.

SOURCES OF HYDROCARBONS

The primary source of methane is natural gas, which also contains smaller amounts of ethane, propane, butane, pentane, and helium. Petroleum, a complex mixture of hydrocarbons, is the major source of alkanes and cycloalkanes (▶ Figure 11.9). These are separated from one another by a process called *fractional distillation,* or **fractionation.** The boiling points of hydrocarbons increase as the number of carbon atoms in their molecules increases.

■ TABLE 11.3 SOME AROMATIC COMPOUNDS

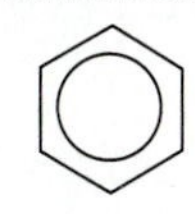

benzene
(octane booster, solvent, chemical feedstock)

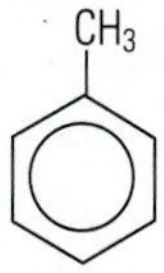

toluene
(solvent, octane booster, explosives)

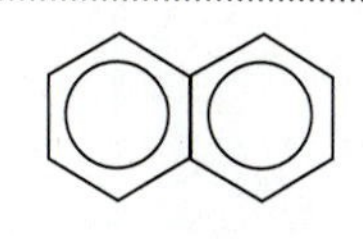

naphthalene
(moth repellant, intermediate for making dyes)

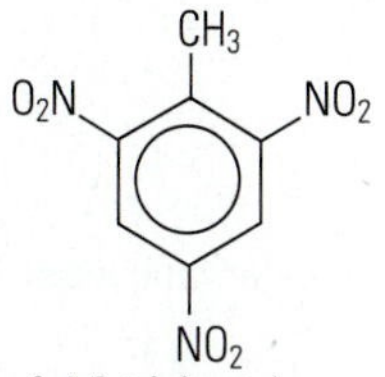

2,4,6-trinitrotoluene
(explosive)

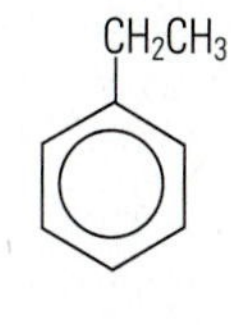

ethylbenzene
(making polystyrene plastics)

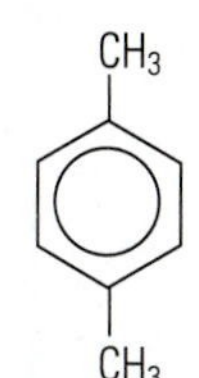

para-xylene
(solvent, making polyster fibers and films)

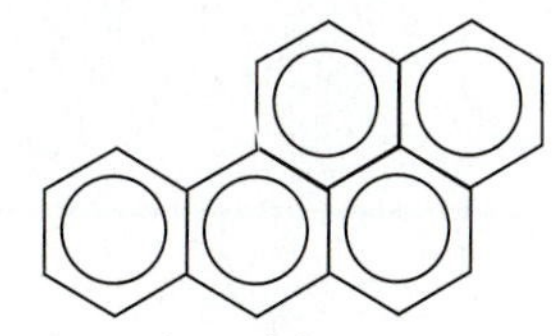

benzo(α)pyrene
(carcinogen found in coal, tobacco smoke, charcoal broiled foods)

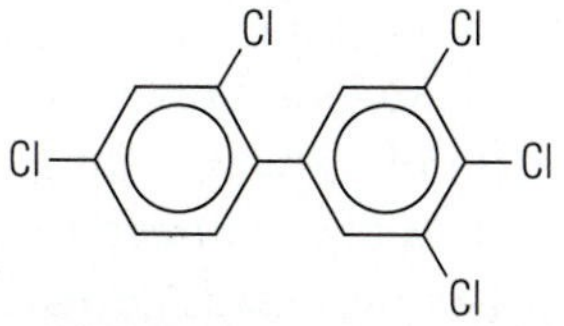

a PCB (polychlorinated biphenyl)
(transformer coolant and insulator; suspected of causing birth defects in humans)

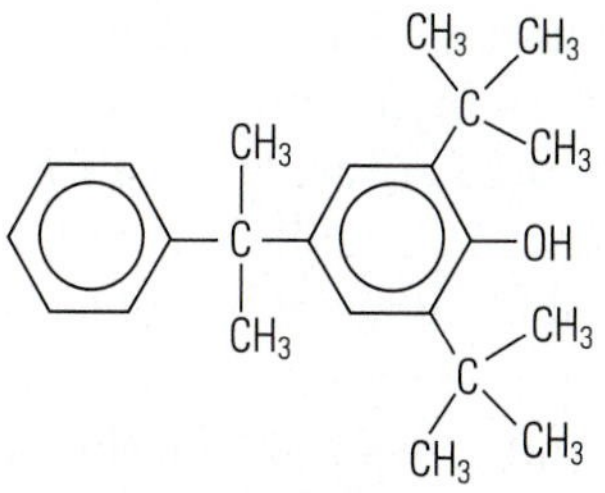

MON-0585
(nontoxic, biodegradable larvicide that is selective against mosquito larvae)

11.11 Circle the aromatic portion(s) of each of the following molecules.

H_2N—(benzene ring)—SO_2NH_2

a. sulfanilamide (antibiotic)

OCH_3, CH_3O, CH_3O, O, C—N, O

b. trimetozine (a sedative)

O, —O—CH_2—C—NH, S, CH_3, CH_3, N, O, CO_2H

c. penicillin V (an antibiotic)

d. fluorene

N, $NHCH_3$, Cl, N, O

e. librium (a tranquilizer)

Methane, ethane, propane, and butane are gases at room temperature; pentane, hexane, and heptane are liquids with low boiling points; and, on the other extreme, triacontane, $CH_3(CH_2)_{28}CH_3$, is a waxy solid melting at about 66 °C. Candles are made of a mixture of similar high-melting alkanes.

Fractionation separates compounds on the basis of their boiling points.

A diagram of the operation of a fractionating tower is shown in ▶ Figure 11.10. Crude oil is preheated to high temperature in a furnace called a pipe still and then is pumped to the fractionating tower. Entering at the bottom of the tower, which is hot at the bottom and cool at the top, the hot oil vaporizes and the vapors begin to rise. The hydrocarbon vapors that have the lowest boiling point (the low boiling fractions) rise to the top of the tower before they condense, but those having higher boiling points condense on their way up. Hydrocarbons with very high boiling points (high boiling fractions) remain near the bottom. Located at various levels in the tower are trays, which permit the different fractions to be collected and drawn off as they condense. ■ Table 11.4 shows typical fractions of hydrocarbons that are obtained from crude oil.

Often the gasoline fraction, about 20% of the hydrocarbons in crude oil, is too small to satisfy the demand for motor fuel. More gasoline can be made by breaking the larger hydrocarbon molecules in the kerosene/fuel oil fraction

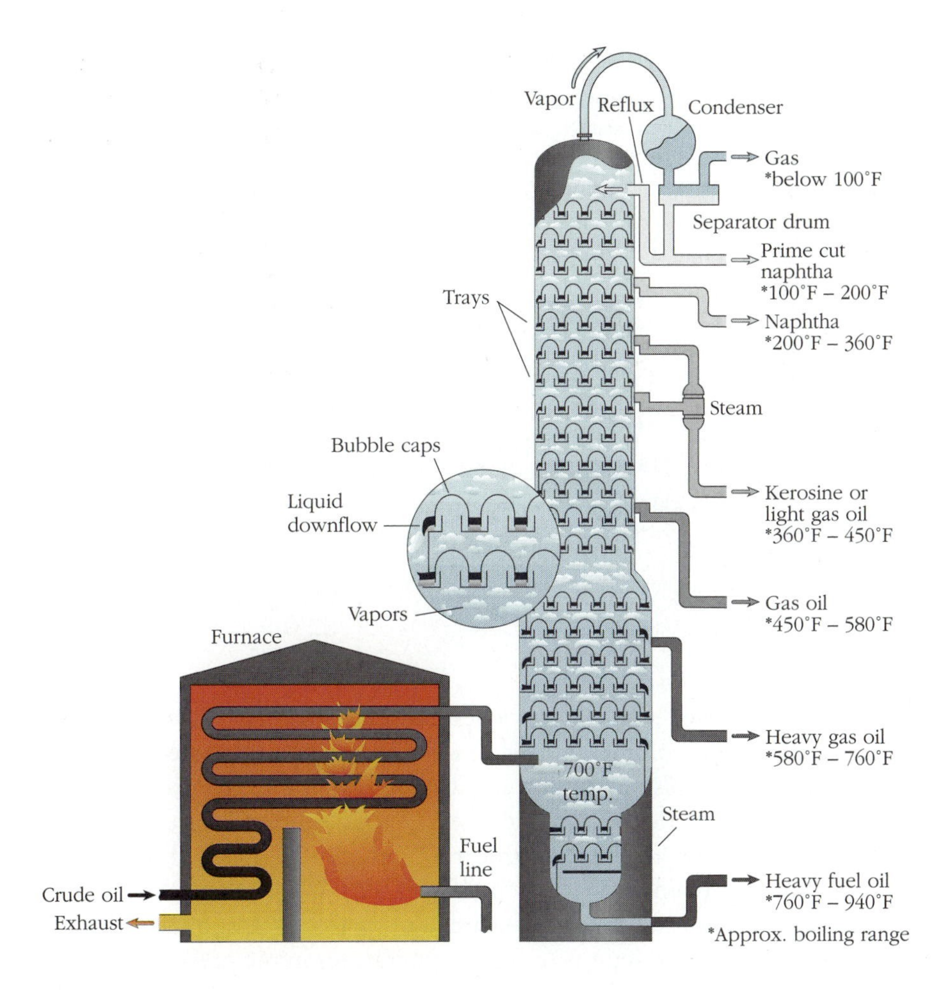

▶ Figure 11.10
A Fractionating Tower in an Oil Refinery. *Fractions are separated according to boiling point. The lower boiling fractions rise to the top of the tower whereas the higher boiling fractions are collected below.*

into smaller ones, a process called **catalytic cracking** (▶ Figures 11.11 and 11.12). Overall, about half of each barrel of crude oil is used as gasoline.

Petroleum also is a major source of cycloalkanes and alkenes. Ethylene, the most important organic compound industrially, is obtained by catalytic cracking of the larger hydrocarbons found in petroleum. Its major uses are to make plastics, antifreeze, and detergents. Acetylene, the most important alkyne, is made by treating calcium carbide with water or by partial oxidation of methane.

$$CaC_2 + 2H_2O \longrightarrow HC{\equiv}CH + Ca(OH)_2$$

$$6CH_4 + O_2 \xrightarrow{1500\ ^\circ C} 2HC{\equiv}CH + 2CO + 10H_2$$

▶ Figure 11.11
An Oil Refinery. *A catalytic cracking unit is shown at the right.*

TABLE 11.4 MAJOR FRACTIONS IN CRUDE OIL

Fraction	Number of Carbon Atoms*	Boiling Range*(°C)	Uses
Gas	1-4	–162-30	Fuel gas; starting material for plastics manufacture
Petroleum ether	5-6	30-60	Solvent, gasoline additives
Gasoline	5-12	40-200	Gasoline
Kerosene	11-16	175-275	Diesel fuel, jet fuel, heating oil
Heating oil	15-18	275-375	Industrial heating
Lubricating oil	17-24	Over 350	Lubricants
Paraffin	20 and up	Solid residue	Candles, toiletries, wax paper
Asphalt	30 and up	Solid residue	Road surfacing

*The exact ranges may vary.

Source: Radel/Navidi, Chemistry, 2/e, *© 1994 by West Publishing Company. Reprinted with permission.*

Because acetylene produces a very hot flame (about 3000 °C) when burned in pure oxygen, it is commonly used in welding and in cutting steel (▶ Figure 11.13). Acetylene is also an important starting material in the production of vinyl and acrylic plastics, resins, and fibers (see polymers, Chapter 12).

Benzene and other aromatic hydrocarbons are present in small amounts in crude oil but are obtained primarily by heating coal in the absence of air. Benzene and toluene are also made by heating the cyclohexane and methylcyclohexanes in petroleum fractions to high temperatures in the presence of a catalyst.

$$\underset{\text{methylcyclohexane}}{\text{(CH}_3\text{–CH, 5 CH}_2\text{ ring)}} \xrightarrow[300\ \text{lb/in}^2]{\text{Mo}_2\text{O}_3,\ 560\ ^\circ\text{C}} \underset{\text{toluene}}{\text{(CH}_3\text{–C, 5 CH ring)}} + 3H_2$$

REACTIONS OF HYDROCARBONS

Although alkanes are not very reactive, they combine with oxygen in the air, in a process called combustion, to produce a considerable amount of heat. This reaction provides the energy to heat our homes (natural gas and heating

$$C_{16}H_{34} \xrightarrow{\text{zeolite catalyst, high temperature}} \text{mixture of alkanes and alkenes}$$

SOME OF THE PRODUCTS

Alkanes		Alkenes
C_8H_{18}	+	C_8H_{16}
$C_{10}H_{22}$	+	C_6H_{12}
C_6H_{14}	+	$C_{10}H_{20}$
C_9H_{20}	+	C_7H_{14}
C_7H_{16}	+	C_9H_{18}
$C_{11}H_{24}$	+	C_5H_{10}
C_5H_{12}	+	$C_{11}H_{22}$
⋮		⋮

▶ Figure 11.12
Catalytic Cracking of Hydrocarbons. *In this example, hexadecane ($C_{16}H_{34}$) is broken into smaller alkanes and alkenes. Note that each alkane formed is accompanied by an alkene whose formula has the same number of atoms as the remainder of the hexadecane molecule. (That is, the formulas of the alkane and its corresponding alkene add up to $C_{16}H_{34}$.)*

oil), cook our food (natural gas and propane), and power most of our transportation (gasoline, deisel, and jet fuel). When enough oxygen is present, *all* hydrocarbons (saturated, unsaturated, and aromatic) undergo combustion to produce carbon dioxide, water, and heat:

$$\underset{\text{methane}}{CH_4} + 2O_2 \longrightarrow CO_2 + 2H_2O + \text{heat}$$

$$\underset{\text{octane}}{2C_8H_{18}} + 25O_2 \longrightarrow 16CO_2 + 18H_2O + \text{heat}$$

$$\underset{\text{benzene}}{2C_6H_6} + 15O_2 \longrightarrow 12CO_2 + 6H_2O + \text{heat}$$

▶ Figure 11.13
Temperatures on the order of 3000 °C can be attained when acetylene is burned in pure oxygen. Here, an ironworker uses an oxyacetylene torch to cut steel.

If insufficient oxygen is available, combustion is incomplete and carbon monoxide, an odorless and poisonous gas, forms as well. For this reason it is important to provide adequate ventilation when burning any hydrocarbon fuel in an enclosed area.

$$\underset{\text{propane}}{C_3H_8} + 4O_2 \longrightarrow CO_2 + 2CO + 4H_2O + \text{heat}$$

Incomplete combustion of hydrocarbons produces poisonous carbon monoxide.

The double and triple bonds in alkenes and alkynes, respectively, are reactive functional groups. The catalytic hydrogenation of these bonds described previously has a common use. Liquid fats, such as soybean, cottonseed, and

11.12 Write the balanced equation for the complete combustion of each the following hydrocarbons. All you need in order to do this is the molecular formula. Compounds having the same molecular formula produce the same balanced equation regardless of their structure. For example, note pairs a, f and e, i in this list.

a. pentane
b. propene
c. toluene
d. naphthalene
e. 3-methylheptane
f. neopentane
g. 1-butyne
h. ethylcyclobutane
i. iso-octane
j. cyclohexene

Monounsaturated and **polyunsaturated** fats have double bonds in their hydrocarbon chains.

canola oils, are **monounsaturated** or **polyunsaturated.** That is, they contain long hydrocarbon chains having one carbon–carbon double bond (monounsaturated) or more than one carbon–carbon double bond (polyunsaturated). Saturated fats, which have no double bonds in their hydrocarbon chains, tend to be solids instead of liquids (▶ Figure 11.14). Hydrogenation or partial hydrogenation of liquid fats can be used to make fats that melt at a desired temperature for use in a particular processed food. For example, partially hydrogenated *squeezeable* margarines are thicker than nonhydrogenated cooking oils. *Tub* margarines are somewhat more hydrogenated but still are soft enough to be spreadable, whereas *stick* margarines are the hardest and most saturated.

The hydrogenation of alkenes is an example of an **addition reaction.** This kind of reaction, in which a double or triple bond becomes saturated, is typical of alkenes and alkyes. Addition of water to ethylene (hydration) results in a reaction that produces ethanol (common name, ethyl alcohol), an important solvent and starting material for other industrial chemicals,

$$H_2C{=}CH_2 + \mathbf{H}{-}OH \xrightarrow{\text{acid catalyst}} \underset{\mathbf{H}}{\underset{|}{CH_2}}{-}CH_2{-}OH$$

ethanol (ethyl alcohol)

11.13 A *desert chocolate bar* has been marketed that is not supposed to melt at summer temperatures. What do you suppose it contains to give it these properties?

▶ Figure 11.14
A Comparison of Saturated and Unsaturated Fats. *The more unsaturated a fat, the more liquid it is. The liquid vegetable oil in the pan contains mostly monounsaturated and polyunsaturated fats. Although the chicken fat (yellow) and the lard (white) contain about the same amount of monounsaturated fat, the chicken fat contains more polyunsaturated fat whereas the lard contains more saturated fat.*

Alkenes and alkynes undergo **addition reactions**.

An addition reaction of a different kind can be used to make antifreeze (common name, ethylene glycol) from ethylene. Here, hydrogen peroxide, H_2O_2, in the presence of an acid adds two OH groups to the double bond to form a glycol,

$$H_2C{=}CH_2 + \mathbf{HO}{-}\mathit{OH} \xrightarrow{\text{acid, then water}} H_2C(\mathbf{OH}){-}CH_2{-}\mathit{OH}$$

ethylene glycol
"permanent antifreeze"
(a glycol)

The ethylene glycol in antifreeze is very poisonous. It is particularly dangerous because it has a sweet taste that appeals to children and pets. However, propylene glycol, made in much the same manner from propene, is much less toxic and is used as an antifreeze for water systems in recreational vehicles, as a preservative for certain foods (wines, for example), and as a solvent in numerous pharmaceuticals and cosmetics.

$$CH_3\underset{\displaystyle OH}{\underset{|}{C}}H—CH_2—OH$$

propylene glycol (common name)

Aromatic hydrocarbons undergo **substitution reactions**.

Because their aromatic bonds have a special stability, aromatic hydrocarbons such as benzene and toluene undergo **substitution reactions** instead of addition reactions. In these reactions a hydrogen atom on the ring is replaced by another group. For example, benzene reacts with chlorine gas in the presence of an iron catalyst to produce chlorobenzene and hydrogen chloride. Allowed to proceed, this reaction continues by substitution of a second chlorine atom on the ring.

$$C_6H_6 + Cl_2 \xrightarrow{Fe} C_6H_5Cl + HCl(g)$$

benzene chlorobenzene

$$C_6H_5Cl + Cl_2 \xrightarrow{Fe} C_6H_4Cl_2 + HCl(g)$$

chlorobenzene *para*-dichlorobenzene (moth repellant)

Aromatic substitution reactions have many uses. For example, treatment of toluene with a mixture of concentrated nitric and sulfuric acids under *carefully controlled* conditions produces trinitrotoluene, TNT, an explosive.

$$C_6H_5CH_3 + 3HNO_3 \xrightarrow{H_2SO_4} CH_3C_6H_2(NO_2)_3 + 3H_2O$$

toluene trinitrotoluene (TNT)

The reaction of dodecylbenzene with SO_3 yields dodecylbenzenesulfonic acid. This is neutralized with NaOH to make sodium dodecylbenzenesulfonate, a commonly used detergent.

$$C_6H_5\text{-}CH(CH_3)(CH_2)_9CH_3 + SO_3 \xrightarrow{H_2SO_4} HO_3S\text{-}C_6H_4\text{-}CH(CH_3)(CH_2)_9CH_3$$

$CH_3(CH_2)_9CHCH_3$ a docecylbenzene; $CH_3(CH_2)_9CHCH_3$ a dodecylbenzenesulfonic acid

$$HO_3S\text{-}C_6H_4\text{-}CH(CH_3)(CH_2)_9CH_3 + Na^+ + OH^- \longrightarrow Na^+\,{}^-O_3S\text{-}C_6H_4\text{-}CH(CH_3)(CH_2)_9CH_3 + H_2O$$

$CH_3(CH_2)_9CHCH_3$ a sodium dodecylbenzenesulfonate

11.14 Write the product of each of the following reactions.

a. $CH_3CH{=}CHCH_3 + H_2O \xrightarrow{\text{acid catalyst}}$

b. $CH_3CH{=}CHCH_2CH_3 + H_2 \xrightarrow{Ni}$

c. $CH_3CH_2C{\equiv}CH + 2H_2 \xrightarrow{Pt}$

d. $CH_3CH_2CH{=}CH_2 + H_2O_2 \xrightarrow{\text{acid, then water}}$

e. benzene + $Br_2 \xrightarrow{Fe}$
(Bromine acts like chlorine. Note family resemblance.)

f. benzene + $SO_3 \xrightarrow{H_2SO_4}$

g. *para*-xylene + $HNO_3 \xrightarrow{H_2SO_4}$

h. *para*-xylene + $Cl_2 \xrightarrow{Fe}$

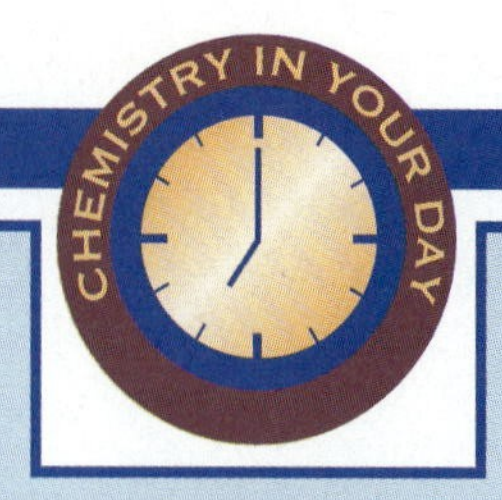

FOOD?

Breakfast. You're all around the kitchen table at the same time–a rare event–and you're all on Charley's case.

"What kind of cereal is that, Charley?"

"I don't know . . . some kind of natural stuff . . . Heart-of-Oat or something like that." Jeanne picks up the box and reads the back.

"Have you read the label on the back?"

"No, but it's all natural and I like it." Jeanne smiles at him impishly.

"You like sugar! Look, it's the second ingredient listed, and they have to list ingredients from highest percentage to least." Rhonda looks at the box.

"'No artificial preservatives.' They're right. Sugar's a *natural* preservative, which is why people preserve fruit in it."

"This one has aspartame, Charley." Carl's reading another box. "Try it. I hear it's 160 times as sweet as sugar."

"No thanks. I don't eat any more chemicals than I have to. Anyway, I don't see how anything can be sweeter than sugar . . . I mean, sugar's sweet, isn't it? And something tastes sweet or it doesn't."

"Sugar's a chemical . . .," Rhonda reasons with him, ". . . everything we eat is a chemical. And all our senses register intensity because, for instance . . .," she laughs, ". . . it's important to know whether you're being caressed or hit."

Electrons again! Light, heat, touch, sound, smell, and taste are all transformed into electric signals and sent to areas in our brains that interpret them according to type, intensity, duration, and extent. Then we can judge whether they promise something good or bad for our immediate future. The stimuli for taste buds and olfactory organs are chemicals dissolved on mucous membranes. We taste sweet, salt, sour, and bitter substances because our bodies need foods that contain sugar, salt, and vitamin C (ascorbic *acid* and therefore sour); and we need to avoid things that taste bitter, the flavor of many toxins (though we can acquire a taste for mild bitterness). Most complex flavors are combinations of tastes, or of taste and odor (the two are closely allied).

Taste buds, located mostly on different parts of the tongue, are tiny pits that contain receptors with microvilli (Latin: shaggy hairs) near nerve endings. The sizes and electronic charges of the receptors are such that only suitable ions or molecules can bond to them. When these receptors are bonded to the right molecules, they send signals to the brain. The signals from the various taste buds are all the same, and the taste we sense depends on what part of the brain receives the signal, and that depends on the position of the taste bud on the tongue.

Charley's wish to eat only natural foods is a hopeless ideal because humans haven't done that since the last time we ate a naturally barbecued animal after a grassfire on the savannah, liked the taste, and figured out how to cook our own. People did eat "more-natural" foods until the end of the nineteenth century: life expectancy then was only forty years, and natural food contributed to many early deaths and miserable lives. Besides the other hazards and diseases of life, people died of botulism, salmonellosis, and other food poisoning. Life expectancy and enjoyment were also reduced by pellagra (niacin deficiency), rickets (vitamin D deficiency), scurvy (vitamin C deficiency), beriberi (thiamine deficiency), and other miseries that could have been prevented by modern food additives.

But Charley's right that many unnecessary and some possibly harmful chemicals are being added to our food, partly because we demand them. Until the modern age, food was just food, and the main concern of its preparation was preservation by drying, salting, pickling, sugaring, and smoking, treatments that altered flavors. But meals are now an esthetic experience: vegetables must be bright colored and meat blood red. Heme, the iron-containing pigment in muscle and blood quickly turns brown in air, and even though the brown meat is fresh and perfectly safe, people won't buy it. To keep it red, butchers treat it with $NaNO_2$, which produces NO, which turns heme into bright-red stable compounds.

A few coal-tar dyes have been approved for use in American foods. Oranges, sweet potatoes, and other produce are dyed, formerly with coal-tar dyes; but natural pigments, such as beta-carotene, and their synthetic equivalents are used increasingly to impart reds, oranges, and yellows. Food colors are also derived from beets, saffron, and grape skins.

Flavor enhancers and balancers aren't needed unless you're used to them and food seems tasteless without them, in which case, the processor has to add them to have your business. Emulsifiers give us the evenness of oil dispersion we expect in peanut butter, mayonnaise, and milk products.

Over 2600 chemicals are added to foods, and more will be added as we discover more about ourselves, our tastes, and our foods. No one knows the effects of all these chemicals, alone or in combination; we're involved in a revolution of self-experiment. But, we've been experimenting on ourselves for the last several million years, and however much Charley might like us to stop the experiment, we're probably unable to go back.

DERIVATIVES OF HYDROCARBONS

Hydrocarbon derivatives are grouped into families depending on the kind of functional group(s) they contain. When present in a molecule, a particular functional group imparts certain properties to it and behaves chemically in a characteristic way. For example, many carboxylic acids, R—CO_2H (R = "rest" of molecule), have pungent, offensive odors. Most of them exhibit the typical behavior of weak acids and can be neutralized with base. Amino acids,

$$R\text{—}\underset{\displaystyle NH_3^+}{\underset{|}{CH}}\text{—}CO_2^-$$

on the other hand, contain two functional groups that react with one another to form ions. These compounds are odorless, crystalline solids that form the chemical building blocks for proteins.

HALOGENATED HYDROCARBONS

As solvents and cleaning agents, insecticides, anesthetics, refrigerants, aerosol propellants, blood extenders, fire extinguishers, and chemical intermediates, the **halogenated hydrocarbons** have many uses. They also can be poisonous, cause cancer, deplete the ozone layer, and contaminate water supplies. These are hydrocarbon molecules in which one or more hydrogen atoms have been substituted chemically by halogen atoms. The **halogens** are the Group 7A elements: F, Cl, Br, and I.

The simplest of this family are made from methane. Chloromethane (common name, methyl chloride), CH_3Cl, is used as a local anesthetic and as an industrial chemical, and bromomethane (common name, methyl bromide), CH_3Br, is a fumigant used to kill insects in stored grain. Dichloromethane (common name, methylene chloride), CH_2Cl_2, is an excellent solvent, used in paint removers and to extract caffeine from coffee to decaffeinate it. Trichloromethane (common name, chloroform), $CHCl_3$, is also an excellent solvent and was the first surgical anesthetic used. It proved to be too toxic, however, and is no longer used. Tetrachloromethane (carbon tetrachloride, or carbon tet), CCl_4, once was used as drycleaning fluid, as a fire extinguisher, as a degreaser, as a surgical anesthetic, and as an oral remedy for hookworm infestations. It causes severe kidney and liver damage, however, and these uses have been discontinued.

IUPAC naming of halogenated alkanes, alkenes, and alkynes is straightforward. The parent chain and numbering are determined in the same manner as for the hydrocarbons themselves. The halogen substituent on the parent chain is given the name fluor*o*-, chlor*o*-, brom*o*-, or iod*o*-. A number of examples are given in ■ Table 11.5.

■ **TABLE 11.5 SOME HALOGENATED HYDROCARBONS**

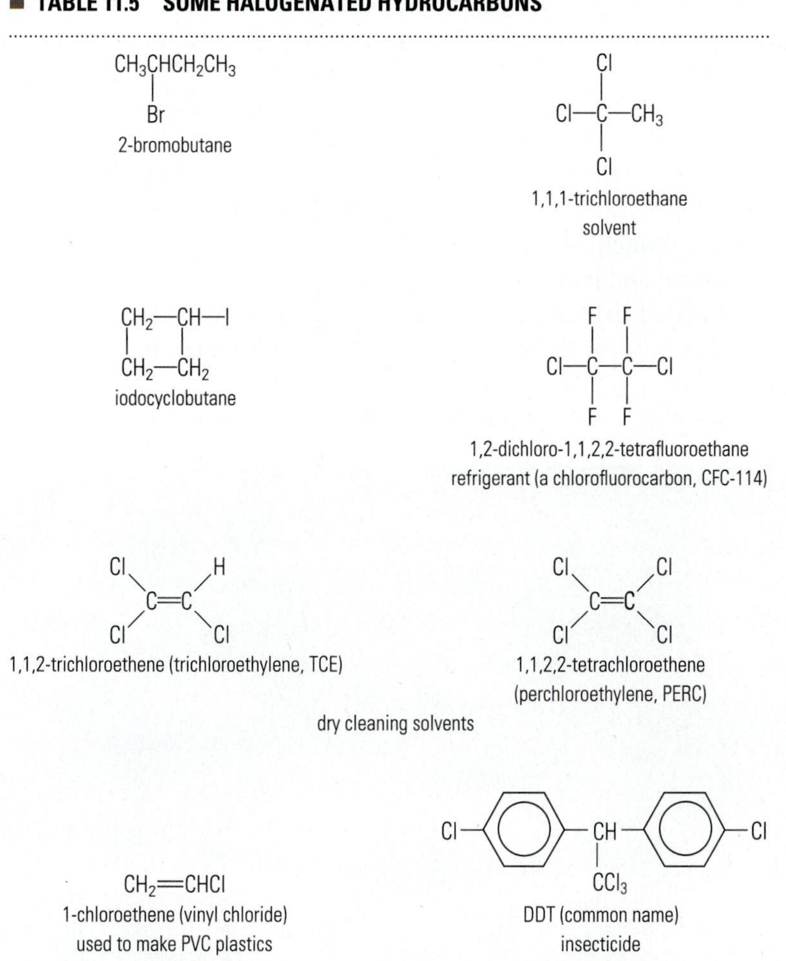

1,1,1-Trichloroethane (called 1-1-1) is one of a few chlorinated hydrocarbon solvents that are relatively safe for humans, and it is now used in place of carbon tetrachloride and similar compounds. The wide use of CFC-114 and other chlorofluorocarbons (CFCs) as refrigerants is being discontinued because they are thought to contribute to the destruction of the ozone layer in the stratosphere (Chapter 13). Unfortunately, 1-1-1 can also deplete the ozone layer, and its use will cease at the end of 1995.

The dry-cleaning solvents, 1,1,2-trichloroethene (TCE) and 1,1,2,2-tetrachloroethene (PERC) are somewhat toxic but safe to use under controlled conditions. They can be insidious environmental pollutants, however. In certain parts of this country they and other chlorinated solvents have been placed in unlined landfills or disposed of indiscriminately elsewhere. These compounds are much more dense than water, are insoluble in it, and do not adhere well to soil or rock. Thus, they tend to percolate downward through the ground until they reach a water table. Once there, they settle to the bottom and continue their downward course, contaminating more groundwater as they descend. Even water deep in the ground eventually can be polluted.

Vinyl chloride is an industrially important compound used to make vinyl plastics. It is thought to cause cancer after long-term exposure in the workplace, however, and its use is strictly regulated.

It is estimated that DDT has saved 25 million human lives worldwide by reducing insect-borne diseases (▶ Figure 11.15). However, the use of DDT as

▶ Figure 11.15
DDT is effective for eliminating insects that carry disease, such as malaria-carrying mosquitoes. Although DDT is banned in the United States, it is still used in many countries for malaria control.

11.15 Name the following halogenated hydrocarbons

a. $CH_3CH_2CH_2CHCH_3$ (with Cl on the fourth carbon)

b. $ICH_2CH_2CH_2CH_3$

c. $CH_3CH_2CH_2CH_2CHCH_2CH_2CH_3$ (with Br on the fifth carbon)

d. $CH_3CH_2CH_2Cl$

e. $CH_3CH_2CHCH_2CH_2CH_3$ (with F on the third carbon)

f. Cl

YOU TRY IT!

11.16 Write the correct structure for the following halogenated hydrocarbons.

a. 3-bromopentane
b. chlorocyclopropane
c. dichlorodifluoromethane (CFC-12)
d. 1-fluoroheptane
e. 3-iodononane

an insecticide is banned in this country. Like many halogenated compounds, DDT is insoluble in water but soluble in fats. It decomposes very slowly in the environment (it is persistent), and it tends to collect in fatty tissues of fish, birds, and mammals. It has caused severe reproductive losses in birds that eat fish, such as the bald eagle, osprey, and some gulls.

DDT is persistent in the environment.

ALCOHOLS, ETHERS, AND PHENOLS

Alcohols are probably the most important family of organic compounds. Because almost every other kind of hydrocarbon derivative (except aromatic) can be made from them, they are important raw materials for the chemical industry. Alcohols are used in perfumes, flavorings, medicines, cosmetics, antifreeze, and solvents for chemical reactions and recrystallization. Carbohydrates, such as sugars, starch, and cellulose, are alcohols that provide us with food, clothing and shelter.

Alcohols contain the hydroxyl functional group, —OH, and undergo hydrogen bonding (Chapter 7). Like water, alcohols of lower molecular mass

Alcohols are associated compounds.

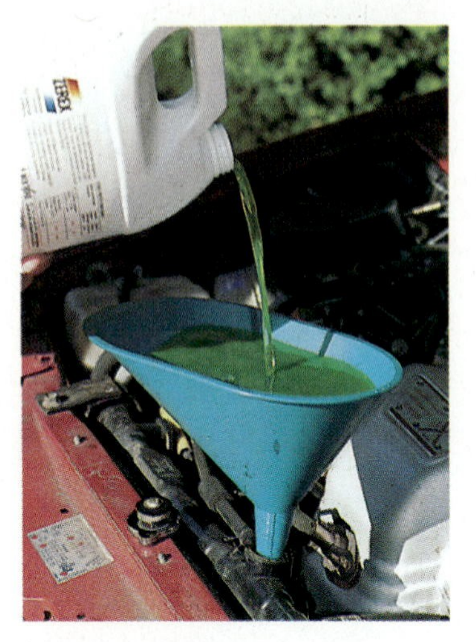

▶ Figure 11.16
Ethylene glycol is used as automobile antifreeze.

are associated liquids having high boiling points. Many are also very soluble in water. Ethylene glycol ($HO—CH_2CH_2—OH$), for example, has excellent properties for use as antifreeze because it boils at 197 °C, has a low freezing point (about –17 °C), and is completely miscible with water (▶ Figure 11.16).

Ethanol (ethyl alcohol, grain alcohol, C_2H_5OH) has been made by humans since prehistoric times, and it is the most important alcohol today. It is used as a solvent, as a motor fuel, as an important starting material for making a host of other organic compounds, and as a beverage. Historically, it was made by fermenting almost any kind of vegetable material, from sugar to cactus pulp or beets. *Fermentation* of sugar continues to provide a large amount of ethanol today:

$$\underset{\text{glucose or fructose}}{C_6H_{12}O_6} \xrightarrow{\text{yeast}} 2CH_3CH_2OH + 2CO_2 + \text{energy}$$

11.17 The carbon dioxide produced in this reaction is not wasted. What do you think are some commercial uses for it?

Ethanol is also made in large quantities by the hydration of ethylene, described earlier in this chapter. Thus, its commercial production ultimately consumes petroleum, either as the source of ethylene or as hydrocarbon fuel for machinery used to grow a crop (corn, grain, cane, etc.) that is suitable for fermentation.

Methanol (methyl alcohol, wood alcohol, CH_3OH) is made by heating wood to high temperature in the absence of air (destructive distillation), or by catalytic hydrogenation of carbon monoxide. It is used as a solvent, a raw material, and a motor fuel additive, and it is being tested as a motor fuel by itself. Sometimes it is mistakenly used as a beverage, but it is poisonous, and drinking small amounts of it (about 30 ml, or 1 oz) can cause blindness or death.

Naming simple alcohols is again an extension of what you know about hydrocarbons and halogen derivatives. Determine the parent chain as before. The presence of the hydroxyl group on the parent chain changes the ending of its name, *-ane,* to *-anol,* and the chain is numbered to give the hydroxyl group the lowest number.

TABLE 11.6 SOME COMMON ALCOHOLS

CH_3CHCH_3 (with OH on the middle carbon)

OH

2-propanol
(isopropyl alcohol)

$HOCH_2CHCH_2OH$ (with OH on the middle carbon)

OH

1,2,3-propanetriol
(glycerol)

OH

cyclopentanol

CH_3, CH_3, CH_3 (ring); $-CH{=}CH-C(CH_3){=}CH-CH{=}CH-C(CH_3){=}CH-CH_2OH$

vitamin A (retinol)

CH_3, CH_3, CH_3 $CH-(CH_2)_3-CH$ CH_3, HO

cholesterol

CH_2OH, H, O, OH, H, HO, HO, CH_2OH, H

fructose (cyclic form)
(one isomer)

Isopropyl alcohol is used as a disinfectant and as an ingredient in many perfumes and personal care products. A 70% solution of it in water, known as rubbing alcohol, has a cooling effect when allowed to evaporate from the skin and is useful for moderating high fevers. Glycerol is a nontoxic, slightly sweet syrupy liquid obtained from natural oils and fats. It is highly associated and has a great affinity for water (that is, it is **hygroscopic**). Thus, it is widely used as a moisturizer in hand lotions and cosmetics and to prevent foods from drying out. When it is treated *carefully* with a mixture of concentrated nitric and sulfuric acids, it forms glyceryl trinitrate, or nitroglycerin, a powerful explosive used in dynamite.

$$\begin{matrix} CH_2OH \\ | \\ CHOH \\ | \\ CH_2OH \end{matrix} + 3HNO_3 \xrightarrow{H_2SO_4} \begin{matrix} CH_2ONO_2 \\ | \\ CHONO_2 \\ | \\ CH_2ONO_2 \end{matrix} + 3H_2O$$

glycerol (glycerin) — glyceryl trinitrate (nitroglycerin)

Vitamin A is an alcohol used in the production of vision.

Vitamin A (retinol) is a highly unsaturated alcohol necessary for the production of vision, and a dietary deficiency of vitamin A in humans causes night blindness. Moderate doses of the vitamin can help prevent or reduce the severity of colds in some persons. (Very large doses can be toxic and should be avoided.) Cholesterol, essential to the function of almost all animal cells, is also needed by our bodies for the production of steroid hormones, such as testosterone and estrogen, the male and female sex hormones. Cholesterol is found naturally in many foods obtained from animals, but not from plants. Egg yolks, cream, red meats, and animal organs are especially rich in cholesterol. Excess amounts of it in the blood, however, can deposit on the walls of arteries, particularly of the heart, causing atherosclerosis and heart disease. Fructose, a sugar, is a polyhydroxy compound that our bodies use to produce energy. It is about twice as sweet as table sugar (sucrose). (The term *polyhydroxy* means that it contains a large number of hydroxyl groups. Glycerol is another example.)

11.18 Give the correct IUPAC name for each of the following alcohols.

a. $CH_3CH_2CH_2CH_2OH$

b. $HOCH_2CH_2CH_3$

c. $CH_3CH_2\underset{\displaystyle OH}{\underset{|}{C}}HCH_2CH_3$

d. $CH_3\underset{\displaystyle OH}{\underset{|}{C}}HCH_2CH_2CH_2CH_2CH_3$

e. OH

f. $CH_3CH_2CH_2CH_2CH_2CH_2CH_2CH_2CH_2OH$

11.19 Write the correct structure for each of the following alcohols.

a. 3-hexanol

b. cyclohexanol

c. 3-octanol

d. 2-butanol

e. 1-propanol

f. 2-pentanol

Like alcohols, **phenols** also contain the hydroxyl group. The difference between them is that, in alcohols, the hydroxyl group is bonded to a saturated (tetrahedral) carbon atom, whereas in phenols it is bonded to a carbon atom of a benzene ring, which is aromatic. Although phenols have properties in common with alcohols, they also differ in some respects. For example, they are easily oxidized (they are good reducing agents). This makes them useful as photographic developers and as antioxidants to preserve food.

Ethers are derivatives of alcohols and phenols. The ether linkage, C—O—C, contains two hydrocarbon groups bonded to an oxygen atom. An easy way to visualize ether formation is to join two alcohol molecules with the loss of water. In fact, some ethers, such as ethyl ether (diethyl ether) are made this way:

$$CH_3CH_2O\text{—}\mathbf{H} + \mathbf{HO}\text{—}CH_2CH_3 \xrightarrow{H_2SO_4} \underset{\text{ethyl ether}}{CH_3CH_2OCH_2CH_3} + \mathbf{H_2O}$$

► Figure 11.17
Ethyl ether is very volatile and highly flammable. These properties make it useful for starting very cold automobile engines.

Ethyl ether is used as an anesthetic and as a solvent, but it is very flammable and must be used with care (► Figure 11.17). Ether linkages are found in starch and cellulose (Chapter 12), in essential oils from plants (*essence:* odor or flavor), in nonionic detergents, and in foam rubber. Methyl-*tert*-butyl ether (MTBE) is used as a lead-free octane booster in gasoline.

Phenol is an important industrial chemical.

Phenol (■ Table 11.7) is an important starting material for making pharmaceuticals and plastics. It is used by itself in numerous deodorant and disinfectant preparations and as an antiseptic. It acts as a mild anesthetic and is used, for example, to relieve sore throats. However, phenol is poisonous and can burn the skin, so other antiseptics and disinfectants, such as hexylresorcinol, have replaced it in some applications.

11.20 The polyethoxylate molecule shown in Table 11.7 acts as a nonionic detergent. Identify its hydrophilic and hydrophobic portions and show how it could form a micelle. What kinds of attractive forces would cause water to solvate the hydrophilic end of the molecule?

Eugenol (oil of cloves) and isoeugenol (oil of nutmeg) are isomers.

Note that BHA, eugenol, and isoeugenol are ethers as well as phenols. Eugenol and isoeugenol differ in structure only in the position of the carbon–carbon double bond on the side chain, yet the flavor and fragrance of cloves are quite different from those of nutmeg. Vanillin (► Figure 11.18) has three functional groups.

► Figure 11.18
Vanilla plants are members of the orchid family. The flavoring in vanilla extract is vanillin. Note that vanillin is an ether, an aldehyde, and a phenol:

H, C, O, OCH_3, OH

vanillin

■ TABLE 11.7 SOME TYPICAL PHENOLS AND ETHERS

OH

phenol
(carbolic acid)

CH_2OH

benzyl alcohol
(an alcohol, not a phenol)

CH_3
CH_3OCCH_3
CH_3

methyl-*tert*-butyl ether
MTBE, a motor fuel additive

OH, CH_3, $C-CH_3$, CH_3, OCH_3

butylated hydroxyanisole (one isomer)
BHA, an antioxidant

OH, OCH_3, $CH_2CH{=}CH_2$

eugenol
oil of cloves

OH, OCH_3, $CH{=}CH-CH_3$

isoeugenol
oil of nutmeg

OH, CH_3, CH, CH_3, CH_3

thymol
oil of thyme and mint

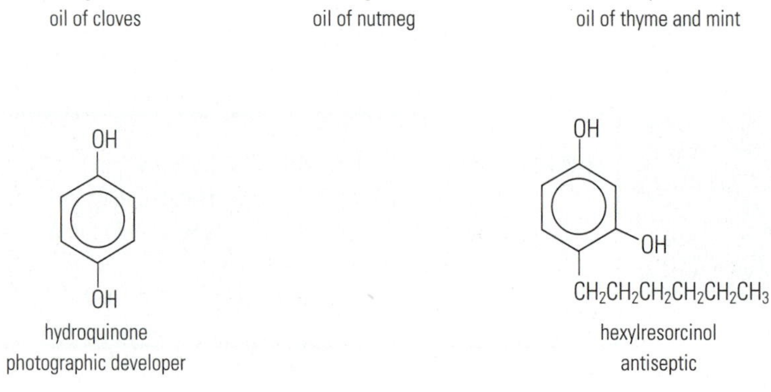

hydroquinone
photographic developer

hexylresorcinol
antiseptic

$CH_3(CH_2)_{10}CH_2O(CH_2CH_2O)_8CH_2CH_2OH$

a typical polyethoxylate
a nonionic detergent

PHOTOGRAPHIC DEVELOPERS

Phenolic compounds, such as BHA and BHT, are used as food preservatives because they act as antioxidants. That is, they react with atmospheric oxygen more readily than many foods do and thereby prevent them from spoiling in the air. Like BHA and BHT, many other phenols are easily oxidized: that is, they are mild reducing agents. This reducing property of a phenol called hydroquinone is exploited to develop photographic images.

Photographic films are composed of a polyester (plastic) film backing coated with an emulsion, a layer of gelatin with silver bromide crystals suspended in it. When photons of light strike a crystal of silver bromide, a very small number of the silver ions are reduced to free silver, and the bromide ions are oxidized to bromine atoms:

$$AgBr \xrightarrow{\text{light}} Ag + Br$$

The crystals of silver bromide that contain four or more free silver atoms (formed in this manner) constitute a so-called latent image. These silver atoms serve as a catalyst that causes *all* of the silver ions in the crystal to be reduced to silver metal when the crystals are treated with a mild reducing agent called a developer. Hydroquinone, a common developer, reduces the silver ion in the crystal as follows [note that the hydroquinone loses hydrogen (is oxidized) to form quinone]:

$$2Ag^+ \text{ (in crystals containing Ag)} + \underset{\text{hydroquinone}}{HO{-}C_6H_4{-}OH} \longrightarrow \underset{\substack{\text{photographic image}\\ \text{(negative)}}}{2Ag} + \underset{\text{quinone}}{O{=}C_6H_4{=}O} + 2H^+$$

Crystals that have not been sufficiently exposed to light react with the developer at a much slower rate. Thus, the developer is selective: it reduces exposed silver crystals

PHOTOGRAPHIC DEVELOPERS (Continued)

(grains) to produce a dark silver image but does not reduce the unexposed crystals. In time, however, the hydroquinone also will reduce the silver ion in the unexposed crystals, turning the entire film black with free silver. To stop the reaction before this happens, the film is placed in a solution of acetic acid, appropriately named a stop bath.

The unexposed silver bromide that remains in the emulsion also will eventually form free silver and turn black on exposure to light. To remove the silver bromide, the

(a) Object

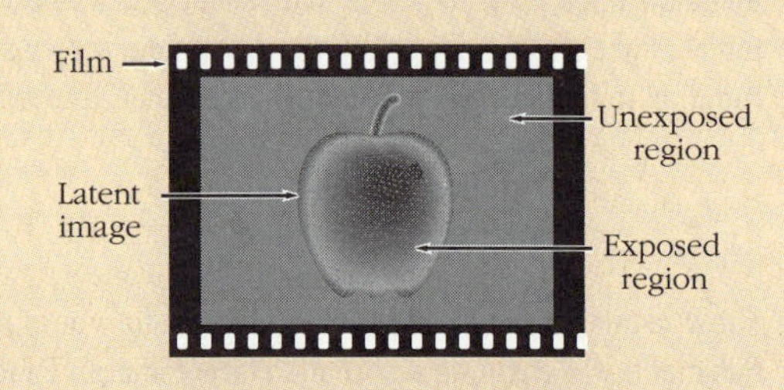

(b) Film After Exposure to Light

(c) The Negative After Being Fixed

(d) The Positive Black and White Print

Figure 1
The Photographic Process
(a) The object of the photograph. (b) In the exposed region a latent image forms when a small proportion of the silver ions is reduced to silver atoms in each grain of AgBr. No silver atoms form in the unexposed region. (c) The film is placed in the developer. Silver metal forms in the region that was exposed, producing the negative. The negative is fixed by removing excess AgBr from the unexposed region. The exposed region is dark, and the unexposed region is transparent. (d) Light is passed through the negative onto light-sensitive paper to produce the positive print. The exposed region is lightly colored and the unexposed region is dark.

film is immersed in a solution of sodium thiosulfate ($Na_2S_2O_3$), called *hypo* or *fixer,* which dissolves the AgBr as follows:

$$AgBr(s) + 2S_2O_3^{2-} \longrightarrow Ag(S_2O_3)_2^{3-} + Br^-$$

The film is then washed with water and dried. Where light has exposed the silver crystals in the film, fine, black particles of silver metal remain. Unexposed areas are transparent. This is called a negative, because light areas in the picture are black and dark areas are light.

To produce a picture that matches the original scene, light is passed through the negative onto a piece of photosensitive paper coated with a similar emulsion of gelatin and silver bromide. The exposed paper is developed, stopped, and fixed like the film was. In this manner, light and dark areas on the negative are reversed to produce a positive print.

Besides hydroquinone, several other phenolic compounds are used as photographic developers. Note that they have similar structural features: namely, at least two hydroxyl groups or one hydroxyl group and an amino group (—NH— or —NH_2) attached to an aromatic ring.

gallic acid

metol
(*para*-methylaminophenol)

amidol
(2, 4-diaminophenol)

BHA
(butylated hydroxyanisole,
a mixture of two isomers)

BHT
(butylated hydroxytoluene)

antioxidants used as food preservatives

ALDEHYDES AND KETONES

Aldehydes and ketones are known as **carbonyl compounds** because they all have the **carbonyl group,**

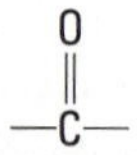

$$-\overset{\overset{\displaystyle O}{\|}}{C}-$$

as their functional group. These compounds comprise sugars, flavors and fragrances, hormones, solvents, pharmaceuticals, and many other substances. Many are intermediates in the manufacture of other useful organic chemicals, and several are used to make plastics and explosives.

The carbonyl carbon atom of **aldehydes** is bonded to at least one hydrogen atom, whereas that of **ketones** is bonded to two carbon atoms. The simplest aldehyde, methanal (formaldehyde), has two hydrogen atoms bonded to its carbonyl carbon atom.

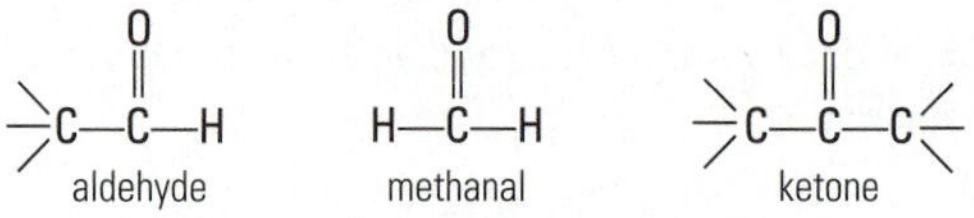

$$\underset{\text{aldehyde}}{>C-\overset{\overset{\displaystyle O}{\|}}{C}-H} \qquad \underset{\text{methanal (formaldehyde)}}{H-\overset{\overset{\displaystyle O}{\|}}{C}-H} \qquad \underset{\text{ketone}}{>C-\overset{\overset{\displaystyle O}{\|}}{C}-C<}$$

One common way of making aldehydes and ketones is by oxidation of the corresponding alcohol. For example, formaldehyde can be prepared by heating methanol in air with a copper catalyst.

$$2H_3COH + O_2 \xrightarrow{Cu} \underset{\text{formaldehyde}}{2H_2C{=}O} + 2H_2O$$

2-Propanone (acetone) is made from 2-propanol (isopropyl alcohol) in a similar manner,

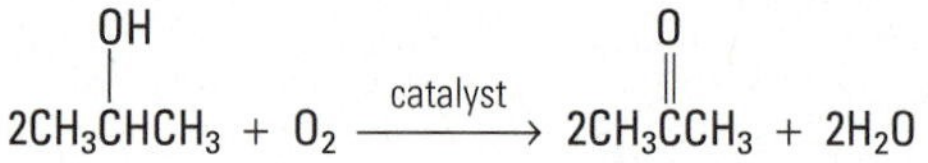

$$\underset{\text{2-propanol}}{2CH_3\overset{\overset{\displaystyle OH}{|}}{C}HCH_3} + O_2 \xrightarrow{\text{catalyst}} \underset{\text{acetone}}{2CH_3\overset{\overset{\displaystyle O}{\|}}{C}CH_3} + 2H_2O$$

Formaldehyde, a gas, is used in making various resins (a kind of plastic), including Bakelite, Melmac, Formica, and urea-formaldehyde resins and foams. (See Chapter 12.) It has a choking, biting odor, is very irritating to the eyes and nose, and is suspected of causing cancer. Because traces of it are present in the resins used to bind building materials, such as plywood and particle board, and to make some kinds of insulation, it can become an indoor air pollutant. A 40% solution of formaldehyde in water is used as a preservative of biological specimens, as an embalming fluid, and as a disinfectant.

Acetone and 2-butanone (methylethylketone, MEK) are excellent solvents for a number of organic compounds and are used, for example, to dissolve plastics, fats, adhesives, and varnishes. They are also present in paint and nail polish removers (▶ Figure 11.19).

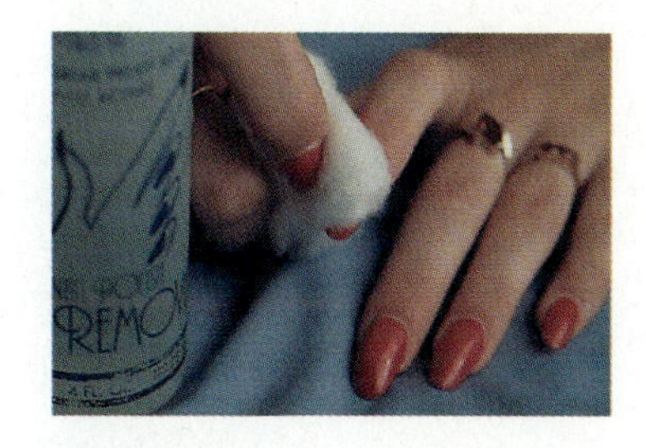

▶ Figure 11.19
Acetone is an excellent solvent for many organic compounds. It dissolves fingernail polish with ease.

Again, naming simple aldehydes and ketones builds on existing rules. Because the aldehyde group must reside at the end of a chain, its location is always carbon 1. This number is not written; it is simply understood. The ending for aldehydes with saturated hydrocarbon chains is *-anal.* In ketones, the carbonyl group is given as low a number as possible, and the ending is *-anone.* Many aldehydes and ketones have common names, and these are often used in place of their systematic names.

■ **TABLE 11.8 SOME COMMON ALDEHYDES AND KETONES**

$CH_3CH_2CH_2C(=O)—H$	butanal (butyraldehyde)
$CH_3C(=O)CH_3$	2-propanone (acetone)
$CH_3C(=O)CH_2CH_3$	2-butanone (methylethylketone, MEK)
$CH_3(CH_2)_4CH{=}CHCH_2C(=O)—H$	3-nonenal odor of freshly ground coffee
$CH_3CH_2C(=O)CH_2CH_2CH(CH_3)CH_3$	6-methyl-3-heptanone
cyclohexane ring with C=O	cyclohexanone
benzene ring with C(=O)H	benzaldehyde
benzene ring—$CH{=}CH—C(=O)—H$	cinnamaldehyde oil of cinnamon
steroid ring system with O=, CH_3, CH_3, CH_3—C=O	progesterone female sex hormone
ring with CH_2OH, H, OH, H, HO, OH, H, H, OH, H—C=O	glucose (one isomer) a sugar

11.21 Give the IUPAC names for each of the following aldehydes and ketones.

a. CH_3CH_2CHO

b. $CH_3CH_2CH_2CH_2CH_2CHO$

c. $CH_3CH_2\overset{\overset{O}{\|}}{C}CH_2CH_3$

d. $CH_3CH_2CH_2CH_2CH_2\overset{\overset{O}{\|}}{C}CH_3$

11.22 Write the correct structure for each of the aldehydes and ketones named here:

a. pentanal
b. nonanal
c. 3-octanone
d. cyclopentanone

Cyclohexanone is used as a solvent for many plastics and rubbers and is a component in glue for PVC plastic pipe. Sometimes, traces of it are present in carpet backing, and it can contribute to indoor air pollution. Benzaldehyde has an almond-like odor. It is an aromatic aldehyde, with its carbonyl group attached to an aromatic ring. The small amount of 3-nonenal present in freshly ground coffee has a pleasant aroma, but in higher concentrations its odor is unbearable. Aldehydes are easily oxidized to carboxylic acids (see next section), and the aldehyde group in glucose can be oxidized by Cu^{2+} and Ag^+ ions. In the process, glucose reduces the metal ion and thus is called a reducing sugar. An interesting application of this reaction is in the silvering of mirrors, called the Tollens' reaction (▶ Figure 11.20):

$$\underset{\text{glucose}}{C_5H_{11}O_5CHO} + \underset{\text{"ammoniacal silver"}}{2Ag(NH_3)_2^+} + 3OH^- \longrightarrow \underset{\text{gluconate ion}}{C_5H_{11}O_5CO_2^-} + \underset{\text{silver mirror}}{2Ag} + 4NH_3 + 2H_2O$$

CARBOXYLIC ACIDS

Some of them stink. Most taste sour. They are important components of soaps, preservatives, sunscreens, flavors, body odor, vitamins, and bee stings, and they serve as starting materials for synthetic fibers. (See polymers, Chapter 12.)

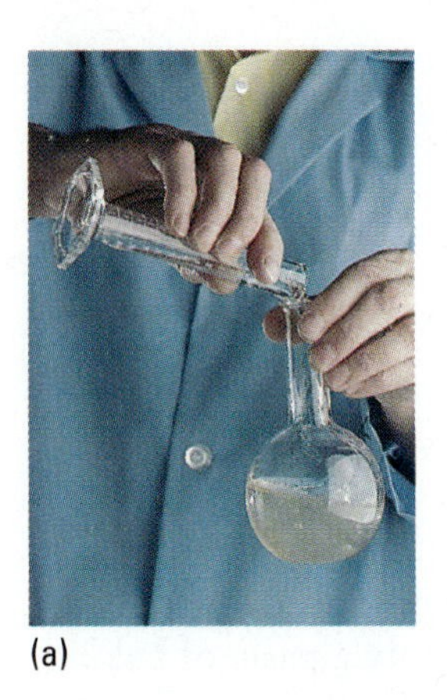
(a)

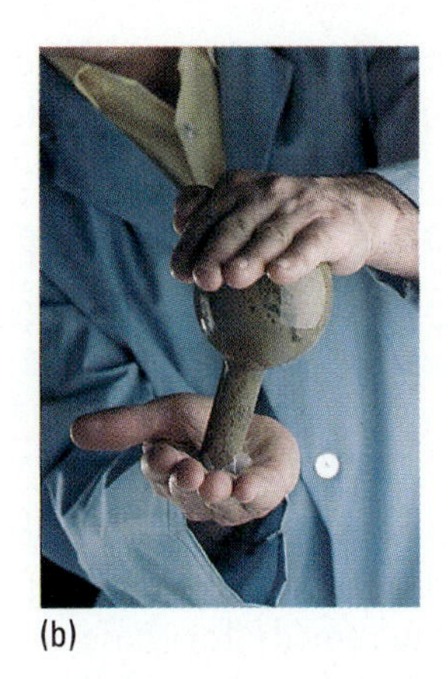
(b)

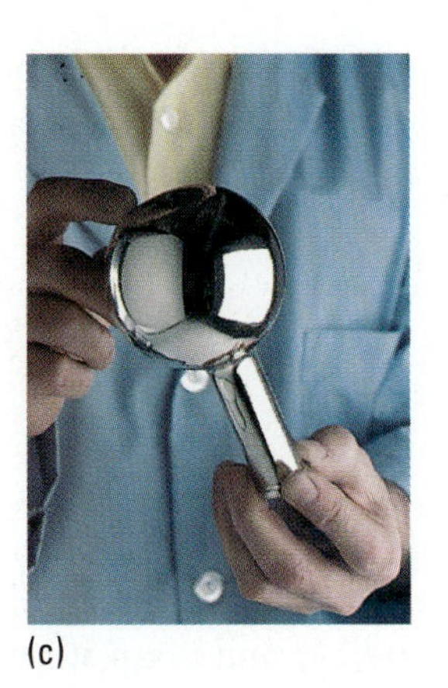
(c)

▶ Figure 11.20
Making a Silver Mirror. *(a) An aldehyde is added to a flask containing a solution of silver nitrate and ammonia. (b) The opening of the flask is covered with plastic film, and the contents of the flask are mixed. (c) The silver ion in solution is reduced to free silver, which deposits on the walls of the flask as a mirror.*

Carboxylic acids contain the **carboxyl group,**

$$-\overset{\overset{\displaystyle O}{\|}}{C}-O-H$$

Many are weak acids because this functional group can ionize slightly in water solution to furnish H^+. Ethanoic acid (acetic acid, also written as $HC_2H_3O_2$), found in vinegar (▶ Figure 11.21), ionizes to produce H^+ and acetate ion as follows,

Carboxylic acids are weak acids.

$$\underset{\text{acetic acid}}{CH_3-\overset{\overset{\displaystyle O}{\|}}{C}-O-H} \rightleftharpoons \underset{\text{acetate ion}}{CH_3-\overset{\overset{\displaystyle O}{\|}}{C}-O^-} + H^+$$

Most carboxylic acids can be neutralized with bases to form their corresponding salts,

$$CH_3-\overset{\overset{\displaystyle O}{\|}}{C}-O^- + H^+ + Na^+ + OH^- \longrightarrow \underset{\substack{\text{a solution of the salt of the acid} \\ \text{(sodium acetate)}}}{CH_3-\overset{\overset{\displaystyle O}{\|}}{C}-O^- + Na^+ + H_2O}$$

Carboxylic acids typically are made by oxidation of the corresponding alcohol or aldehyde. Acetic acid in vinegar is produced by oxidation of the ethanol in fermented fruit juices by the action of *Acetobacter* bacteria,

$$CH_3CH_2OH + O_2 \xrightarrow{\textit{Acetobacter}} CH_3\overset{\overset{\displaystyle O}{\|}}{C}OH + H_2O$$

▶ Figure 11.21
The active component in vinegar is acetic acid, a typical weak acid.

Pure acetic acid is prepared by oxidation of various hydrocarbons, or of ethanal, with oxygen in the presence of a catalyst, for example,

$$2CH_3\overset{\overset{\displaystyle O}{\|}}{C}{-}H + O_2 \xrightarrow{\text{catalyst}} 2CH_3\overset{\overset{\displaystyle O}{\|}}{C}OH$$

ethanal — ethanoic acid (acetic acid)

The name for formic acid is derived from the latin: *formica,* ant.

Methanoic acid (formic acid), the simplest carboxylic acid, was originally obtained from ants. It is partly responsible for the pain produced by ant bites and bee and wasp stings. One remedy for these is to apply a paste of baking soda to the affected area to neutralize the acid. This is only partially effective, though, because other poisons are present in the bites and stings as well.

Carboxylic acids are named by placing the ending, *-anoic acid,* on the name of the parent chain. As for aldehydes, the carboxyl group must be numbered 1, and the number is not written in the name. Not surprisingly, the common names of many of the carboxylic acids are used more frequently than their systematic names.

Butanoic acid (■ Table 11.9), commonly called butyric acid, is a foul-smelling liquid formed when butter turns rancid. It is also a contributor to body odor. Stearic acid, a fatty acid, is obtained from animal fats, particularly beef tallow (see glycerides later in this chapter). Its sodium and potassium salts are commonly used soaps:

$$CH_3(CH_2)_{16}\overset{\overset{\displaystyle O}{\|}}{C}O^-Na^+$$

sodium stearate, a soap

Linoleic acid, another fatty acid, is present as glycerides (see later) in oils extracted from plants, such as canola oil and soybean oil. It is the fatty acid primarily responsible for polyunsaturation in fats and is an *essential fatty acid:* that is, the body cannot synthesize it and it must be taken in the diet.

Sodium benzoate, a salt of benzoic acid, is used as a preservative in foods, as is calcium propionate, $(CH_3CH_2CO_2^-)_2Ca^{2+}$. *para*-Aminobenzoic acid (PABA) and its derivatives absorb harmful ultraviolet light and are used in sunscreens and sun-blocking preparations. Oxalic acid, a dicarboxylic acid, is found in spinach, tomatoes, and rhubarb stalks. It is poisonous, but eating the small amounts of it found in these foods is not harmful. However, rhubarb leaves contain enough of it to be dangerous, particularly to children. It is used in tanning leather, as a bleach, and as a rust and ink remover.

TABLE 11.9 SOME COMMON CARBOXYLIC ACIDS

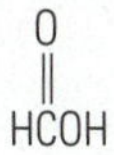

$HCOOH$ (HCOH with C=O)	methanoic acid (formic acid)
$CH_3CH_2CH_2COOH$	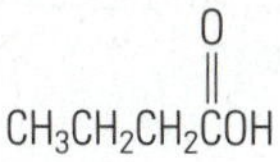 butanoic acid (butyric acid)
$CH_3CH(CH_3)CH_2CH_2CH_2CH_2COOH$	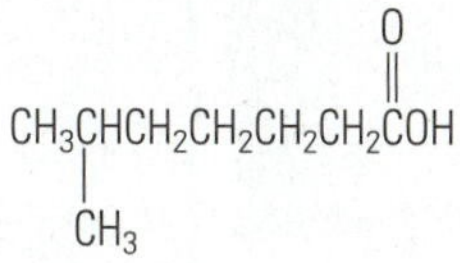 6-methylheptanoic acid
$CH_3(CH_2)_{16}COOH$	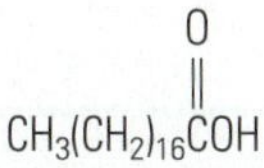 octadecanoic acid (stearic acid)
$CH_3(CH_2)_4CH{=}CHCH_2CH{=}CH(CH_2)_7COOH$	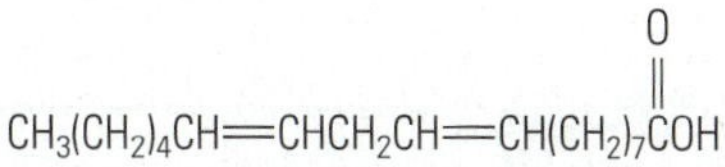 9,12-octadecadienoic acid (linoleic acid)
C_6H_5COOH	benzoic acid
$C_6H_5COO^- Na^+$	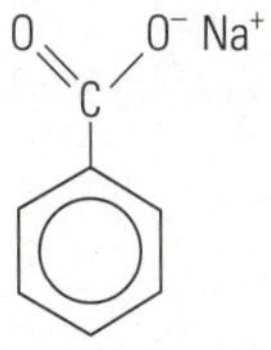 sodium benzoate preservative
$H_2N{-}C_6H_4{-}COOH$	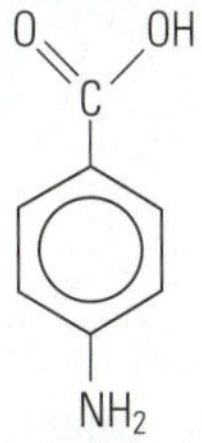 *para*-aminobenzoic acid (PABA)
$HOOC{-}COOH$	oxalic acid
ring structure with O, HO, OH, $CHCH_2OH$ / OH	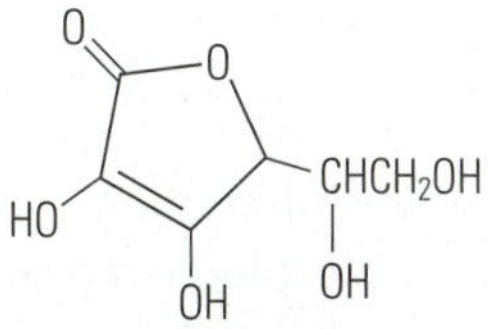 ascorbic acid (vitamin C)
$CH_3CH(OH)C(=O){-}OH$	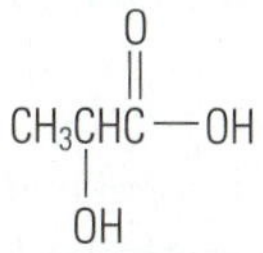 lactic acid

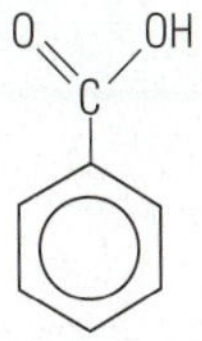

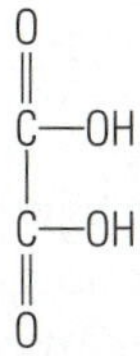

Ascorbic acid is an organic acid that is not a carboxylic acid. Not only is it a vitamin essential for human health, but also it is an antioxidant, used for example on freshly cut fruits such as apples and bananas to keep them from turning brown in the air. The claim that taking large doses of vitamin C prevents or moderates colds has not yet been firmly established. Some people apparently benefit from taking it whereas others do not. Lactic acid, a

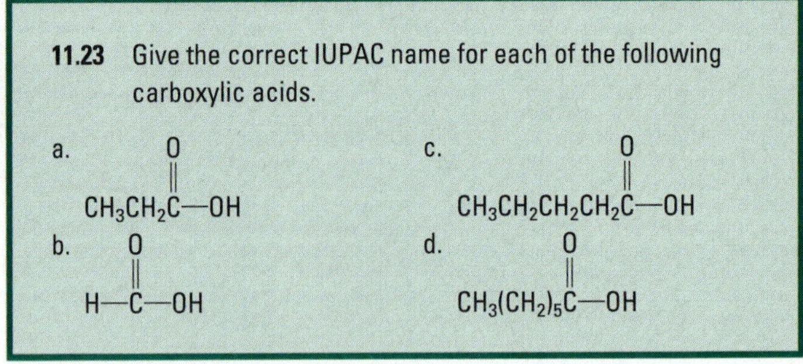

11.23 Give the correct IUPAC name for each of the following carboxylic acids.

a. $CH_3CH_2\overset{\overset{O}{\|}}{C}—OH$

b. $H—\overset{\overset{O}{\|}}{C}—OH$

c. $CH_3CH_2CH_2CH_2\overset{\overset{O}{\|}}{C}—OH$

d. $CH_3(CH_2)_5\overset{\overset{O}{\|}}{C}—OH$

11.24 Write the correct structure for each of the following carboxylic acids.

a. butanoic acid
b. decanoic acid
c. ethanoic acid
d. octanoic acid

Strenuous exercise produces lactic acid in muscles, which makes them sore.

hydroxyacid, is responsible for the tart taste of yogurt, sour cream and milk, sourdough, and sauerkraut. It is also generated by muscular activity in the body. Usually, the body decomposes it about as fast as it forms, but vigorous exercise can cause it to build up in the muscles, making them sore.

AMINES

Amines can be viewed as derivatives of ammonia, in which one or more hydrogen atoms on the ammonia molecule is replaced by an organic group. Thus, methylamine is an ammonia molecule in which a hydrogen atom has been replaced by a methyl group. Similarly, in diethylamine two hydrogen atoms on the nitrogen atom have been replaced with ethyl groups.

$\mathbf{H}—NH_2$	$\mathbf{CH_3}—NH_2$	$\mathbf{CH_3CH_2}—NH—\mathbf{CH_2CH_3}$
ammonia	methylamine	diethylamine

Amines are weak bases.

Like ammonia, most amines are weak bases, and those of smaller molecular size hydrolyze in water solution to produce substituted ammonium ions and hydroxide ions:

$$\underset{\text{methylamine}}{CH_3NH_2} + H_2O \rightleftharpoons \underset{\text{methylammonium ion}}{CH_3NH_3^+} + OH^-$$

Most amines are associated compounds, and many are liquids that have characteristically fishy odors (▸ Figure 11.22).

Naming amines is different: each separate organic group bound to the nitrogen atom is named as a substituent, and the parent is *-amine.* If one organic group bound to nitrogen is a benzene ring, then the amine is an aniline.

Amines are named differently from other organic compounds.

Aniline, in which the amino group (—NH_2) is bonded to a benzene ring, is an aromatic amine. It is an important starting material for a host of aromatic industrial compounds, including dyes and sulfa drugs. Many aromatic amines are toxic and are readily absorbed through the skin. 1-Naphthylamine and 2-naphthylamine, aromatic amines that are isomers, are used to make dyes and food colorings. 2-Naphthylamine causes cancer in humans, but 1-naphthylamine does not.

Nicotine is a poisonous, narcotic drug obtained from tobacco. It is sometimes used as an insecticide. Amphetamine is a stimulant that has been used, legitimately and otherwise, to reduce fatigue and prevent sleep, to suppress appetite, to produce euphoria, and to treat narcolepsy. Its use can be addictive. Interestingly, amphetamines are used in the management of hyperactivity in children, because they can have a calming effect on them.

1,6-Diaminohexane is used in the manufacture of Nylon 66, one of the more common types of Nylon in use today. The dioctadecyldimethylammonium ion is a cation made up of a positively charged nitrogen atom bonded to four carbon atoms. The ion, called a *quaternary ammonium ion,* is analogous to an ammonium ion, NH_4^+, with alkyl groups substituted for hydrogen atoms. The example shown, as the chloride, is used as a fabric softener (see ■ Table 11.10). Other quaternary ammonium salts are used as cationic

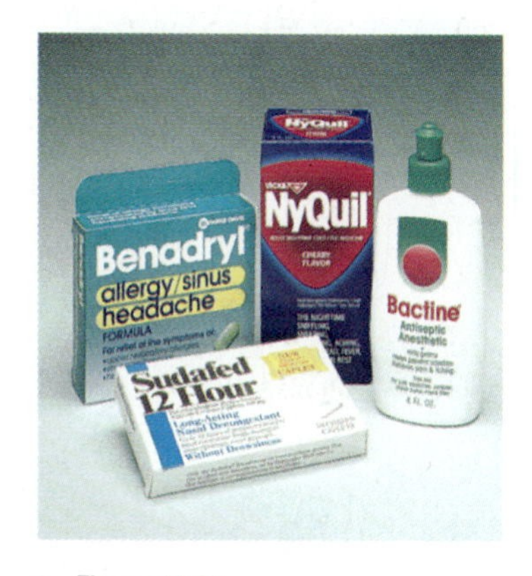

▸ Figure 11.22

Each of these over-the-counter medications contains the salt of an amine as the active ingredient.

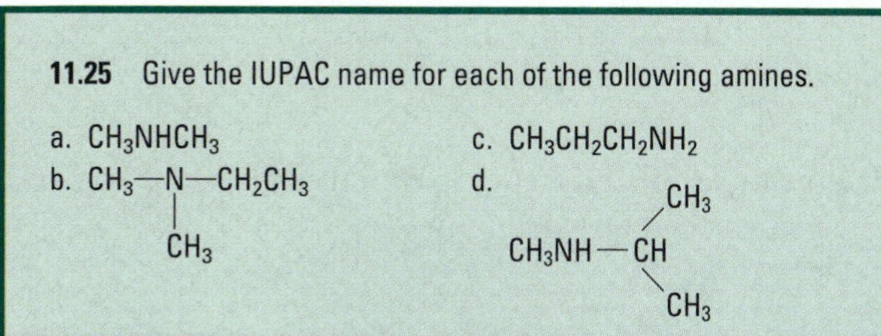

11.25 Give the IUPAC name for each of the following amines.

a. CH_3NHCH_3

b. $CH_3{-}N{-}CH_2CH_3$ (with CH_3 on N)

c. $CH_3CH_2CH_2NH_2$

d. $CH_3NH{-}CH(CH_3)_2$

11.26 Write the structure for each of the following amines.

a. isopropylamine

b. methylethylpropylamine

c. diethylamine

d. aniline

TABLE 11.10 SOME REPRESENTATIVE AMINES

$CH_3CH_2NH_2$ ethylamine	$CH_3CH(CH_3)N(H)CH_3$ methylisopropylamine	$CH_3N(CH_3)CH_3$ trimethylamine
NH_2 aniline	N, N, CH_3 nicotine	C_6H_5—CH_2—$CH(NH_2)$—CH_3 1-phenyl-2-aminopropane (amphetamine, benzedrine)
$H_2NCH_2(CH_2)_4CH_2NH_2$ 1,6-diaminohexane (1,6-hexamethylene-diamine)	NH_2 1-naphthylamine (α-naphthylamine)	NH_2 2-naphthylamine (β-naphthylamine)
$(CH_3(CH_2)_{16}CH_2)_2N^+(CH_3)_2\ Cl^-$ dioctadecyldimethylammonium chloride		

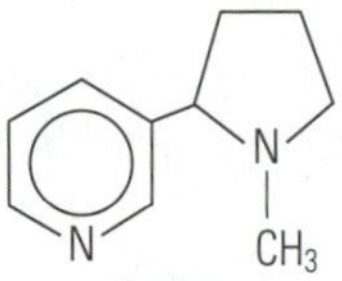

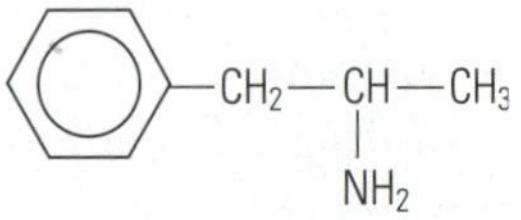

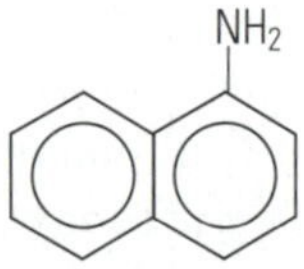

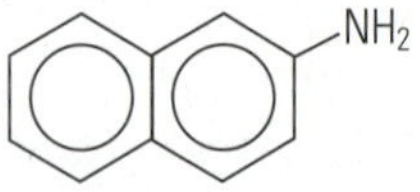

detergents, as disinfectants in mouthwashes and bathroom cleaners, and as catalysts in chemical reactions.

ESTERS AND AMIDES

Esters make up fats and oils, flavors, and fragrances. Amides form the basis for proteins such as wool, silk, and muscle. Both kinds of compounds are used to make synthetic fibers and pain relievers. (See Chapter 12 for fibers.)

Both are derivatives of carboxylic acids, formed by a process called **condensation.** In condensation reactions, two molecules are joined to make a larger molecule, and a small molecule, such as water, is split out, or *lost.* For example, acetic acid and ethanol react when heated with a catalytic amount of a strong acid, such as H_2SO_4, to produce the **ester,** ethyl acetate, and water.

$$\underset{\substack{\text{ethanoic acid}\\\text{(acetic acid)}}}{CH_3\overset{O}{\overset{\|}{C}}\text{—}OH} + \underset{\substack{\text{ethanol}\\\text{(ethyl alcohol)}}}{H\text{—}OCH_2CH_3} \xrightarrow{H_2SO_4} \underset{\substack{\text{ethyl ethanoate}\\\text{(ethyl acetate)}\\\text{an ester}}}{CH_3\overset{\mathbf{O}}{\overset{\|}{\mathbf{C}}}\text{—}\mathbf{O}CH_2CH_3} + H_2O$$

(arrow label: new bond)

The ester functional group is shown in bold type. Esters typically have pleasant odors (▶ Figure 11.23). Many foul-smelling carboxylic acids are converted to agreeable fragrances and flavors by use of this condensation reaction; for example, rancid butyric acid and methanol react to make methyl butyrate, found in apples:

$$\underset{\substack{\text{butanoic acid}\\\text{(butyric acid)}}}{CH_3CH_2CH_2\overset{O}{\overset{\|}{C}}\text{—}OH} + \underset{\substack{\text{methanol}\\\text{(methyl alcohol)}}}{H\text{—}OCH_3} \xrightarrow{H_2SO_4} \underset{\substack{\text{methyl butanoate}\\\text{(methyl butyrate)}\\\text{an ester found in apples}}}{CH_3CH_2CH_2\overset{\mathbf{O}}{\overset{\|}{\mathbf{C}}}\text{—}\mathbf{O}CH_3} + H_2O$$

▶ Figure 11.23
Isopentyl acetate is an ester that gives bananas their flavor and fragrance.

$$\underset{\text{isopentyl acetate}}{(CH_3)_2CHCH_2CH_2O\text{—}\overset{O}{\overset{\|}{C}}\text{—}CH_3}$$

Amides are formed in a similar manner from ammonia or amines and carboxylic acids. Because amines are bases, an acid–base reaction first produces a salt, which must then be heated to drive off water to yield the amide (group shown in bold).

$$\underset{\substack{\text{ethanoic acid}\\\text{(acetic acid)}}}{CH_3\overset{O}{\overset{\|}{C}}\text{—}OH} + NH_3 \longrightarrow \underset{\substack{\text{ammonium ethanoate}\\\text{(ammonium acetate)}\\\text{a salt}}}{CH_3\overset{O}{\overset{\|}{C}}\text{—}O^-NH_4^+} \xrightarrow{\text{heat}} \underset{\substack{\text{ethanamide}\\\text{(acetamide)}\\\text{an amide}}}{CH_3\overset{\mathbf{O}}{\overset{\|}{\mathbf{C}}}\text{—}\mathbf{N}H_2} + H_2O$$

More complex amides can be made by substituting organic groups for one or both of the hydrogen atoms on the nitrogen atom of the amide group. The pain reliever, acetaminophen (Tylenol), is an example. Note that it is also a phenol.

$$\underset{\text{acetaminophen}}{HO\text{—}C_6H_4\text{—}NH\text{—}\overset{O}{\overset{\|}{C}}\text{—}CH_3}$$

We shall not emphasize the naming of esters and amides in this text, but you may want to be able to identify them in your everyday experience. Esters are named like metal salts of carboxylic acids, such as sodium benzoate, a food preservative. That is, the cation (Na^+) is named first and the anion, in this case the benzoate ion, last.

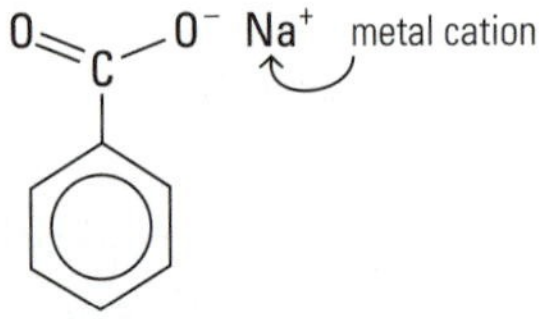

metal cation

alcohol portion of ester

sodium benzoate

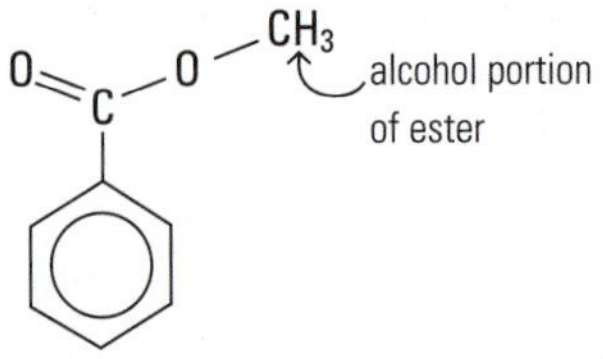

methyl benzoate

Esters are named the same way, except that the alcohol portion of the ester is named first, using its substituent name: methyl, ethyl, isopropyl, and so on. Amides are usually simply called amides. Some examples of useful esters and amides follow in ■ Table 11.11.

Methyl salicylate is used in small amounts to give a wintergreen flavor to confections and medicines. An ingredient in analgesic balm, it soothes sore muscles by producing a warming sensation when rubbed on the skin. Acetyl salicylic acid, or aspirin, is still the most widely used remedy for aches, pains, and fever. Small doses of the drug taken regularly are believed to reduce the likelihood of heart attacks. Note that aspirin is an ester of acetic acid, and that the alcohol portion of the molecule is a phenol. Methyl salycilate is an ester of salicylic acid,

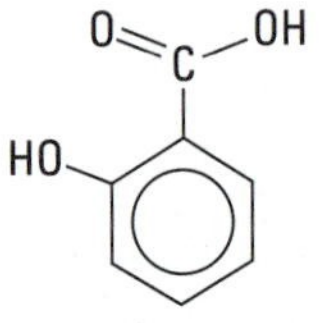

salicylic acid
(from willow bark, genus *salix*)

Meperidine, or Demerol, is a commonly used synthetic pain reliever. Like many narcotics, it is habit forming and its use is strictly controlled. It is both an ester and an amine.

Cocamide DEA is an amide used as a nonionic detergent in dishwashing liquids and shampoos. Its hydrophilic end contains two hydroxyl groups and the very polar amide functional group. Lysergic acid diethylamide (LSD) is a potent, unpredictable, dangerous hallucinogen, which when taken in *micro-gram* quantities causes altered mental states ranging from euphoria to crying episodes. It dramatically affects visual perception and alters the perceived flow of time. Besides producing "bad trips," use of the drug can lead to flash-

TABLE 11.11 SOME USEFUL ESTERS AND AMIDES

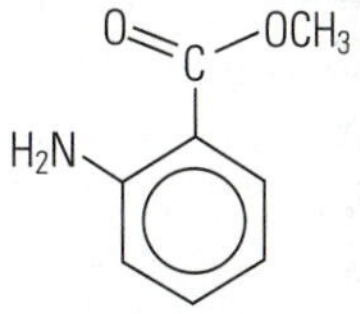

methyl anthranilate artificial grape flavor	$CH_3(CH_2)_{12}C(=O){-}N(CH_2CH_2OH)_2$ cocamide DEA nonionic detergent
methyl salicylate oil of wintergreen	acetylsalicylic acid (aspirin)
meperidine (Demerol)	lysergic acid diethylamide (LSD)
$H_2N{-}CH(CH_2COOH){-}C(=O){-}NH{-}CH(CH_2C_6H_5){-}C(=O){-}OCH_3$ aspartame	$CH_2OC(=O)(CH_2)_{16}CH_3$ $CHOC(=O)(CH_2)_{16}CH_3$ $CH_2OC(=O)(CH_2)_{16}CH_3$ glyceryl tristearate

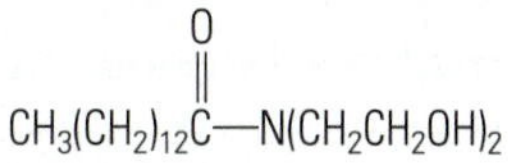

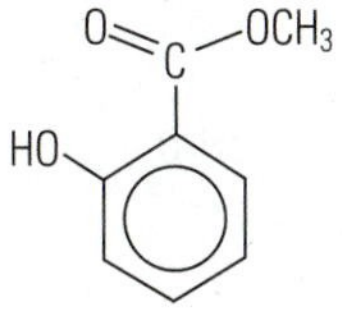

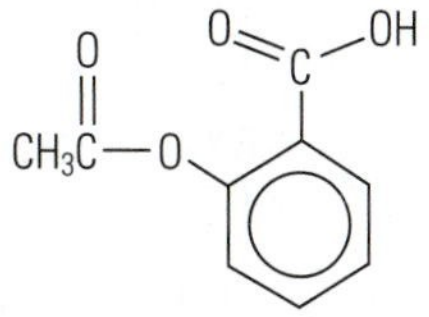

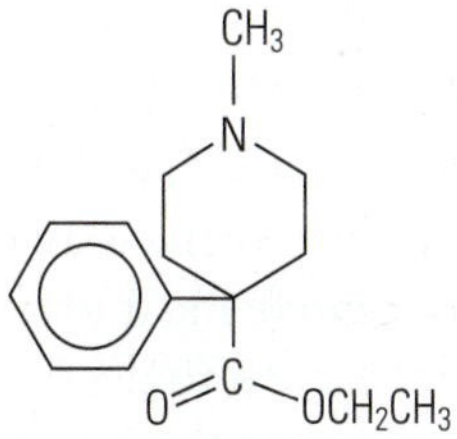

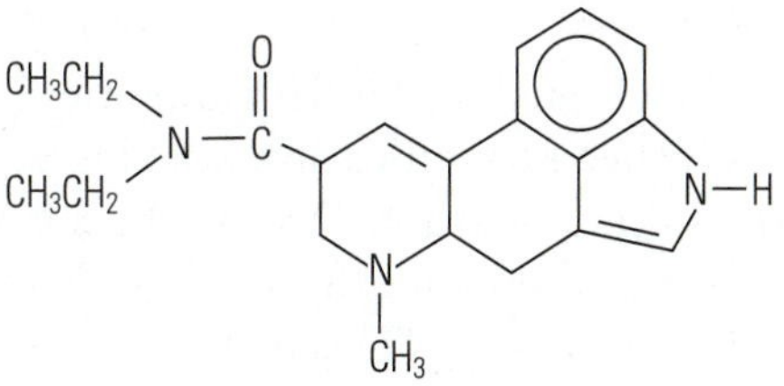

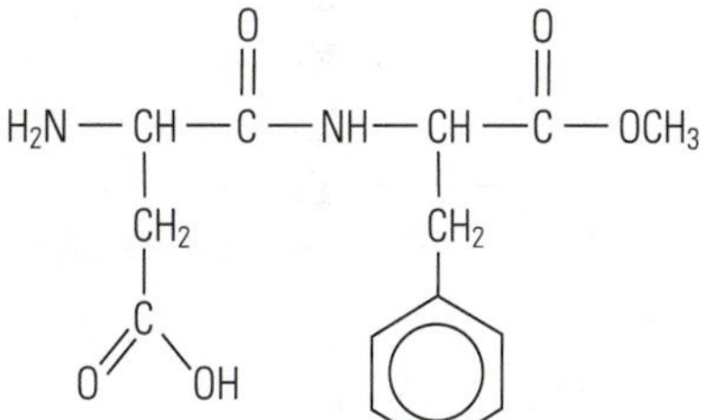

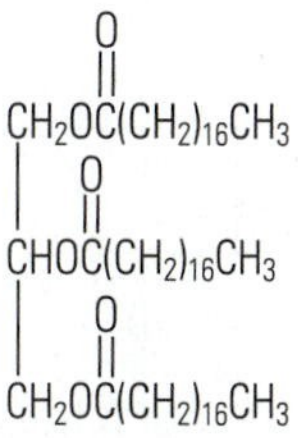

backs, panic attacks, and prolonged psychoses. The artificial sweetener aspartame is better for you, although it is not safe for people who suffer from phenylketonuria. About 160 times sweeter than table sugar, it contains four functional groups: an amide, an ester, an amino group, and a carboxyl group.

Grandma's lye soap (good for pots and kettles, dirty dishes, hands and faces, and everything around the place) was made from animal fat and lye (or

Triglycerides are esters of fatty acids and glycerol.

sometimes leachings from wood ashes). Animal fats, particularly from beef, venison, bear, and pork, are made up of glyceryl tristearate and other **triglycerides,** which are esters of long-chain fatty acids and glycerol. This reaction is carried out (without the sulfuric acid) in the cells of plants and animals, and of us.

$$\begin{array}{l} CH_2OH \\ | \\ CHOH \\ | \\ CH_2OH \end{array} + 3CH_3(CH_2)_{16}CO_2H \xrightarrow{H_2SO_4} \begin{array}{l} CH_2OC(=O)(CH_2)_{16}CH_3 \\ | \\ CHOC(=O)(CH_2)_{16}CH_3 \\ | \\ CH_2OC(=O)(CH_2)_{16}CH_3 \end{array} + 3H_2O$$

glycerol (glycerin) — stearic acid (a fatty acid) — glyceryl tristearate (a triglyceride) (a fat)

(Labels on the product: glycerol portion; fatty acid portions)

Grandma heated the fat with a solution of lye to cleave it (break it apart) in a process called **saponification** (making soap). She then added salt (lots of it) to the resulting mixture, causing the soap to float to the surface as curds, where she could skim it off, rinse it with salt water to remove most of the lye, and press it into cakes. (This process, called *salting out,* is used to decrease the solubility of organic compounds in water.) If she didn't rinse the NaOH out well enough, using the soap could be pretty rough on the skin.

$$\begin{array}{l} CH_2OC(=O)(CH_2)_{16}CH_3 \\ | \\ CHOC(=O)(CH_2)_{16}CH_3 \\ | \\ CH_2OC(=O)(CH_2)_{16}CH_3 \end{array} + 3Na^+ + 3OH^- \xrightarrow{heat} \begin{array}{l} CH_2OH \\ | \\ CHOH \\ | \\ CH_2OH \end{array} + 3CH_3(CH_2)_{16}C(=O)O^- + 3Na^+$$

glyceryl tristearate (a triglyceride) — glycerol — solution of sodium stearate (a soap)

► Figure 11.24
We use a large variety of soaps and detergents to keep ourselves clean.

Soap is still made this way today—with a few refinements of course (► Figure 11.24). Usually, Grandma didn't bother to save the glycerol in the remaining solution. However, it is an important industrial chemical (see alcohols), and it is usually recovered on a commercial scale by fractionation.

DID YOU LEARN THIS?

Fill in the blanks with the most appropriate word or phrase from the following list. You may use any word or phrase more than once.

aspirin	ethyl ether
LSD	saturated
PERC	unsaturated
lactic acid	aromatic
fermentation	isomers
catalyst	substituent
fractionation	planar
hydrogenation	twisted
catenation	zigzag

a. _______ Used as an anesthetic
b. _______ A process used to make ethyl alcohol
c. _______ Describes the geometry of the benzene ring
d. _______ Found in sour milk and sore muscles
e. _______ A means of separating the components in petroleum
f. _______ The ability of atoms to repeatedly bond to one another
g. _______ An ester
h. _______ The methyl group on methylcyclohexane
i. _______ Describes alkenes and alkynes
j. _______ An amide
k. _______ The relationship of pentane and neopentane
l. _______ Describes the chemical behavior of benzene and its derivatives
m. _______ A chlorinated hydrocarbon
n. _______ Changes the speed of a chemical reaction but is not itself consumed in the process

EXERCISES

1. What was the theory of vitalism? What experimental evidence supported it? How was the theory of vitalism refuted?
2. What distinguishes organic compounds and organic chemistry from inorganic compounds and inorganic chemistry? List some examples in which there might be some overlap between the two fields of chemistry.

3. How is the bonding in carbon unique? What kinds of bonding possibilities does it offer to provide the huge number of organic compounds known today?
4. What are the two main classes of organic compounds? How do they differ?
5. What is a functional group?
6. What is the difference between a saturated hydrocarbon and an unsaturated hydrocarbon? How do they differ in reactivity?
7. What are aromatic compounds? What structural feature makes compounds aromatic? What feature of the molecular orbitals in benzene makes the double bonds so stable and unreactive? Compare the kinds of reactions that aromatic compounds typically undergo with those of alkenes and alkynes.
8. Compare the physical properties of saturated fats with those of unsaturated fats.
9. Without looking up your answer, tell which of the two compounds, octane or iso-octane (2,2,4-trimethylpentane) would have the higher boiling point. Justify your answer in terms of intermolecular forces.
10. Which would you expect to have the higher boiling point?

    a. ethane or hexane
    b. $CH_3(CH_2)_9CH_3$ or $CH_3(CH_2)_{13}CH_3$
    c. 1-butanol or pentane
    d. 2-methylhexane or 2-hexanone

11. What chemical reaction provides most of the energy used in this country today?
12. Name two major sources of hydrocarbons. Which provides mostly alkanes, alkenes, and cycloalkanes? Which provides mostly aromatic hydrocarbons?
13. What property of hydrocarbons is exploited in a fractionating tower? What kinds of hydrocarbons are obtained from the top of the tower? From the middle of the tower? From the bottom?
14. What is catalytic cracking? Write an equation using octadecane, $C_{18}H_{38}$, to show how the process works.
15. List two major uses of ethylene. What is acetylene used for?
16. What happens when a hydrocarbon is burned in an insufficient quantity of oxygen? Why is this dangerous to do in an enclosed space?
17. List at least four uses of halogenated hydrocarbons. What health or environmental problems can be caused by the use of halogenated

hydrocarbons? Do you think we could live very well without using at least some halogenated hydrocarbons?

18. Why are alcohols the most important class of hydrocarbon derivatives?
19. Why are polyhydroxy alcohols, such as ethylene glycol, glycerol, and propylene glycol so soluble in water? How does their structure account for their high boiling points?
20. There are two ways that ethanol is made commercially. Write equations to illustrate each of them.
21. What is the difference between an alcohol and a phenol? Identify the following compounds as alcohols, phenols, or ethers. Some may fit more than one class.

a. 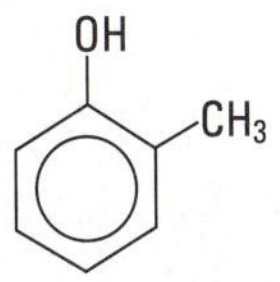

e. 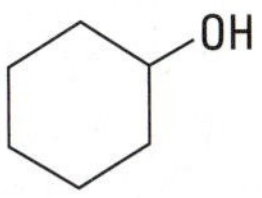

b. 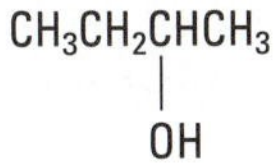

f. 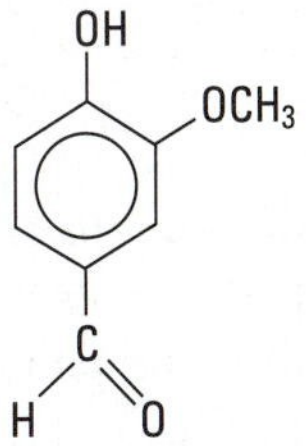

c. $CH_3OCH_2CH_2OCH_3$

g. 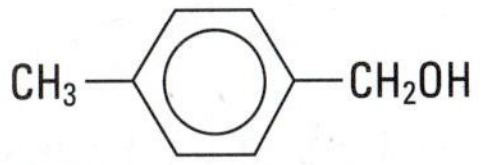

d. 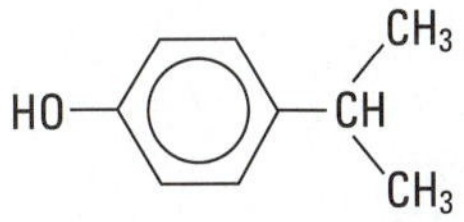

h. 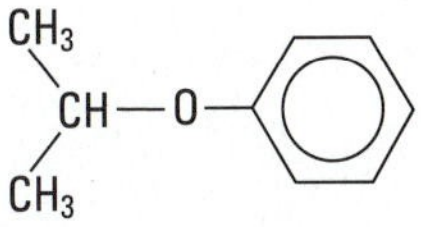

22. What structural feature do aldehydes and ketones have in common? Write down how they differ.
23. In what products is formaldehyde used in a way that could make it an indoor pollutant?
24. Write down the Tollens' reaction. Which chemical species is oxidized? Which is reduced? Which is the oxidizing agent? The reducing agent?

25. Solid baking soda ($NaHCO_3$) is often placed in refrigerators to deodorize and "sweeten" them. List several compounds you might expect to be present in refrigerators that would produce a sour smell. Write an equation to show how baking soda would react with one of these compounds to remove its odor. (Hint: see neutralization, Chapter 8.)
26. A solution of dimethylamine in water feels slippery when rubbed between the fingers. What is causing the slipperyness? Would the solution turn red litmus blue, blue litmus red, or neither? Write an equation to show the reaction, if any, of dimethylamine with water.
27. The label on a commercial toilet bowl cleaner lists tetradecyldimethylbenzylammonium chloride as an ingredient that serves as a disinfectant and a detergent. Its structure is,

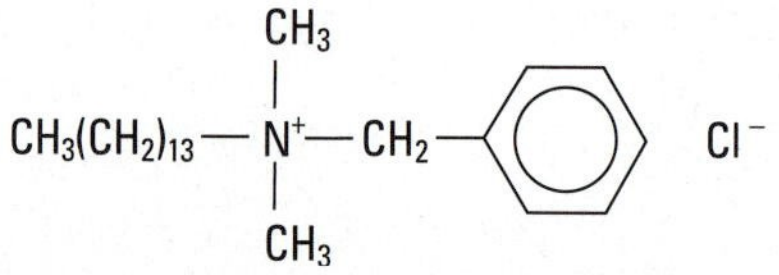

$$CH_3(CH_2)_{13}-\overset{\displaystyle CH_3}{\underset{\displaystyle CH_3}{\overset{|}{\underset{|}{N^+}}}}-CH_2-C_6H_5 \quad Cl^-$$

tetradecyldimethylbenzylammonium chloride

Make a drawing to show how this compound could form a micelle.
28. Write equations to show how esters could be formed by reaction of the following pairs of carboxylic acids and alcohols in the presence of a sulfuric acid catalyst.

a. $CH_3\overset{\overset{O}{\|}}{C}OH + CH_3(CH_2)_2CH_2OH \xrightarrow{H_2SO_4}$

b. $C_6H_5\overset{\overset{O}{\|}}{C}OH + CH_3OH \xrightarrow{H_2SO_4}$

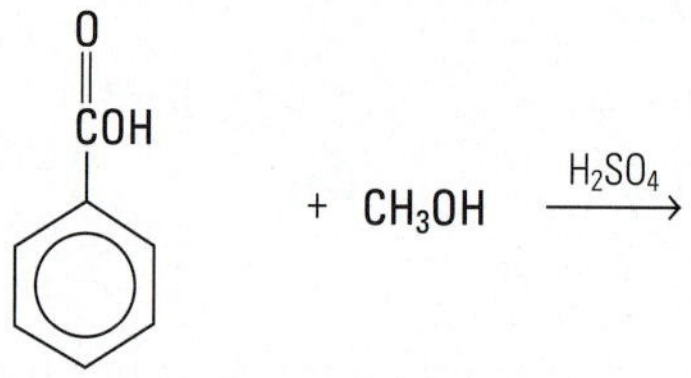

c. $CH_3(CH_2)_2\overset{\overset{O}{\|}}{C}OH + CH_3CH_2OH \xrightarrow{H_2SO_4}$

29. Glyceryl tripalmitate, the structure of which follows, is a triglyceride found in palm oil. Is it a saturated fat, a monounsaturated fat, or a polyunsaturated fat? Write the equation for its saponification

with hot sodium hydroxide solution. What is the formula or palmitic acid?

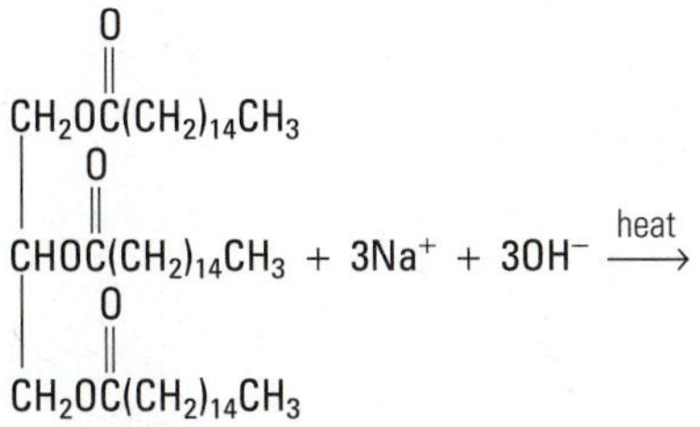

$$\begin{array}{l} \overset{\displaystyle O}{\|} \\ CH_2OC(CH_2)_{14}CH_3 \\ |\overset{\displaystyle O}{\|} \\ CHOC(CH_2)_{14}CH_3 + 3Na^+ + 3OH^- \xrightarrow{\text{heat}} \\ |\overset{\displaystyle O}{\|} \\ CH_2OC(CH_2)_{14}CH_3 \end{array}$$

glyceryl tripalmitate
(a triglyceride)

30. Glyceryl trilinolenate, the structure of which is shown below, is a triglyceride which makes up about 80% of tung oil. Is it a saturated fat, a monounsaturated fat, or a polyunsaturated fat? Write the equation for its saponification with a hot sodium hydroxide solution. What is the formula for linolenic acid?

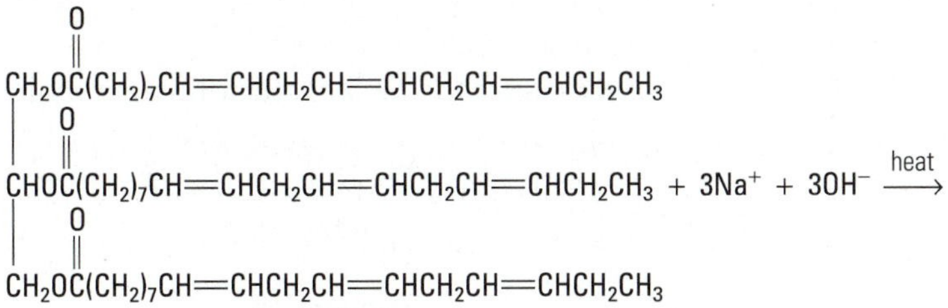

$$\begin{array}{l} \overset{\displaystyle O}{\|} \\ CH_2OC(CH_2)_7CH{=}CHCH_2CH{=}CHCH_2CH{=}CHCH_2CH_3 \\ |\overset{\displaystyle O}{\|} \\ CHOC(CH_2)_7CH{=}CHCH_2CH{=}CHCH_2CH{=}CHCH_2CH_3 + 3Na^+ + 3OH^- \xrightarrow{\text{heat}} \\ |\overset{\displaystyle O}{\|} \\ CH_2OC(CH_2)_7CH{=}CHCH_2CH{=}CHCH_2CH{=}CHCH_2CH_3 \end{array}$$

glyceryl trilinoleate
(a triglyceride)

31. Identify all of the functional groups in each of the following molecules.

a. 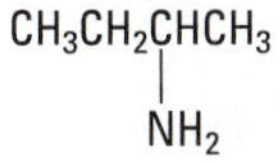 $CH_3CH_2\underset{\displaystyle NH_2}{\underset{|}{C}}HCH_3$

b. $CH_3(CH_2)_8\overset{\displaystyle O}{\overset{\|}{C}}{-}H$

c. (cyclopentane ring)—OH

g. $CH_3C{\equiv}CH$

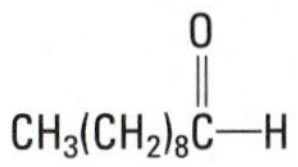

h. 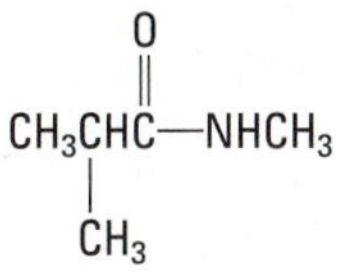 $CH_3\underset{\displaystyle CH_3}{\underset{|}{C}}H\overset{\displaystyle O}{\overset{\|}{C}}{-}NHCH_3$

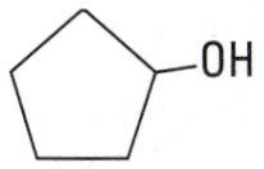

i. 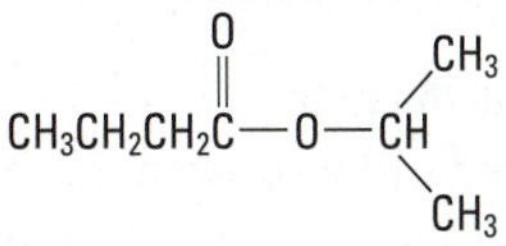 $CH_3CH_2CH_2\overset{\displaystyle O}{\overset{\|}{C}}{-}O{-}CH(CH_3)_2$

d. $HO-C_6H_4-NH_2$ (OH and NH_2 para on benzene ring)

e. $CH_3CH(OH)C(=O)-H$

f. cyclobutanone (square ring with =O)

j. $C_6H_5-CH{=}CH-C(=O)-OCH_3$

k. $CH_2OC(=O)(CH_2)_{10}CH_3$ / $CHOC(=O)(CH_2)_{10}CH_3$ / $CH_2OC(=O)(CH_2)_{10}CH_3$

l. $CH_3CH_2CH_2CH_2OCH_2CH_3$

32. Give the *common names* for each of the following molecules.

a. $CH_3C(=O)CH_3$

b. $H_2C{=}CH_2$

c. $CHCl_3$

d. $CH_3C(=O)-OH$

e. CH_3CH_2OH

f. CH_3OH

33. Write the structures of the molecules whose common names are given.

a. isopropyl alcohol
b. acetylene
c. methylethylketone
d. formic acid
e. formaldehyde
f. ethyl acetate
g. ethylene glycol
h. isobutane

THINKING IT THROUGH

Reindeer in the Cold

Lipids are fatlike substances that are found in all living cells and are essential to their function. In reindeer, the lipids in cells near the hooves are more unsaturated than those in cells nearer the body. How do you suppose this helps the reindeer survive?

Fat in the Fire

Acrolein is an aldehyde formed by the pyrolysis (decomposition by heating) of fats. It is a lachrymator (causes tears) and is the substance that stings the eyes and nose when fat is spilled on a hot burner or spatters in a broiler. The IUPAC name of acrolein is prop*enal.* Write its structure. What part of a fat molecule do you suppose it comes from? (Hint: what molecules go together to make up a fat?) Water is also a product of the pyrolysis. Try to write an equation to show how the acrolein might be formed.

Keeping Warm Safely

Why is it hazardous to use a charcoal grill for cooking or heating in an enclosed space, such as a tent or a cabin? Why is it not dangerous to burn charcoal or wood in a properly designed fireplace or stove?

Formulating Drugs

Many amine drugs are formulated to be used as their hydrochloride or hydrobromide salts. Examples are pseudoephedrine hydrochloride (a nasal decongestant), loperamide hydrochloride (an antidiarrheal), and dextromethorphan hydrobromide (a cough suppressant). What acids would be used to make these salts? Assuming that an amine drug had the generic structure $R—NH_2$, where R— is a large organic group, write an equation to illustrate the formation of its salt with either of the two acids. Why do you suppose it is often desirable to administer amine drugs in the form of their salts instead of the free amines?

ANSWERS TO "DID YOU LEARN THIS?"

a. ethyl ether
b. fermentation
c. planar
d. lactic acid
e. fractionation
f. catenation
g. aspirin
h. substituent
i. unsaturated
j. LSD
k. isomers
l. aromatic
m. PERC
n. catalyst

AN ORGANIC CHEMIST IN THE PHARMACEUTICAL INDUSTRY

Todd A. Blumenkopf

Senior Research Scientist, Burroughs Wellcome Co. North Carolina

B. S. Chemistry, University of California, Los Angeles

Ph.D. Chemistry, University of California, Berkeley

I work as an organic chemist for a pharmaceutical company, synthesizing molecules that are tested to determine whether they can be used as drugs to fight or prevent disease. The biological properties of a chemical compound are determined by its structure and can be refined by varying the structures of the molecules synthesized. The chemist, therefore, is positioned at the very beginning of the drug development process, trying to unravel the relationship between chemical structure and biological activity, and to discover promising new compounds for study. The branch of organic chemistry that deals with the discovery of therapeutic chemical agents is called medicinal chemistry.

A new molecule can be synthesized by a series of chemical reactions and, as a synthesis chemist, I attempt to predict which of a variety of possible routes is likely to be successful. I may rely on my understanding of reactions known in the chemical literature, or a new reaction may have to be invented. Every day on the job is different because almost every reaction I run is different from the one I ran before. Sometimes, the selected route to a compound fails because of unexpected reactions; indeed, observation of unexpected results often leads to the development of useful, new reactions. I must consider what I learned about chemical bonding and stoichiometry (quantitative relationships that govern chemical composition and chemical change) as I investigate new reactions. I must also prove that I have actually synthesized my target molecules. This is done by examining the physical properties of molecules, often by observing their behavior when exposed to light or to magnetic fields. The spectra of molecules obtained in this manner provide clues to the structure of the compound. I then determine the structure of the molecule consistent with all the data.

One of the most exciting aspects of medicinal chemistry is the opportunity to interact with scientists from a variety of backgrounds. We work as a team to decide what properties a drug must have. We also consider biological properties to avoid, properties that might generate unacceptable or even toxic side effects. The drug discovery process requires frequent collaboration between chemists, biochemists, biologists, and physicians. Compounds synthesized by chemists are evaluated by collaborating biologists in a series of tests. Initial tests, often run on hundreds or even thousands of chemical compounds, provide data that are used to select the most promising compounds to undergo further testing. The chemist considers these initial biological data in selecting which new compounds to make next.

The real challenge for medicinal chemists is deciding what compounds to make. For example, to treat a bacterial or viral infection, a chemist may design a drug that selectively interferes with the biochemical processes of the invading organism. We may get ideas by examining the differences in metabolic pathways between a human and the organism and choose to synthesize a molecule that mimics an essential chemical component of the organism. Scientists may choose to exploit differences in properties of specific enzymes between the human host and a parasite. We can also use a variety of computer-based methods to examine the structures of enzymes determined by X-ray crystallography. Many approaches can be used to examine a problem, and the methods used vary depending on the disease or condition of interest. I am now working on a project whose outcome (hopefully!) will be a new drug to treat cancer.

Developing drugs is a very expensive and time-consuming process. More than $100 million can be spent to develop a single new drug, with ten years or more elapsing from the time a research program is started to the time the new drug reaches the marketplace. Thus, a successful medicinal chemist must be curious, creative, motivated, and patient!

I have been paraplegic since birth as a result of a congenital spinal cord injury known as spina bifida. I use a wheelchair. I have found that many educators and employers unacquainted with people with disabilities perceive nonexistent limitations in my disability. Some people have difficulty recognizing that I accomplish my job safely and effectively, although I may accomplish it by slightly different means than my nondisabled colleagues. Although my laboratory was modified somewhat to provide better access to my workspace (for example, the surface of my fume hood was lowered), the only difference between most of my work and that of my colleagues is that I perform my experiments sitting down.

Occasionally, I have had to challenge unnecessary restrictions imposed by people unfamiliar with the capabilities of persons with disabilities. Yet, I find this need to challenge to be consistent with successful research itself, because some of our most significant discoveries have been made by people who challenged conventional thinking. Like their able-bodied peers, people with disabilities must take positive action on their own behalf to obtain opportunities for a good education and for meaningful careers. I have found that even the most skeptical persons eventually accept my ability to work independently in the chemistry lab.

Chemistry is a rapidly expanding discipline, capable of accommodating people with a wide assortment of interests and abilities. It can also accommodate people with a vast range of disabilities.

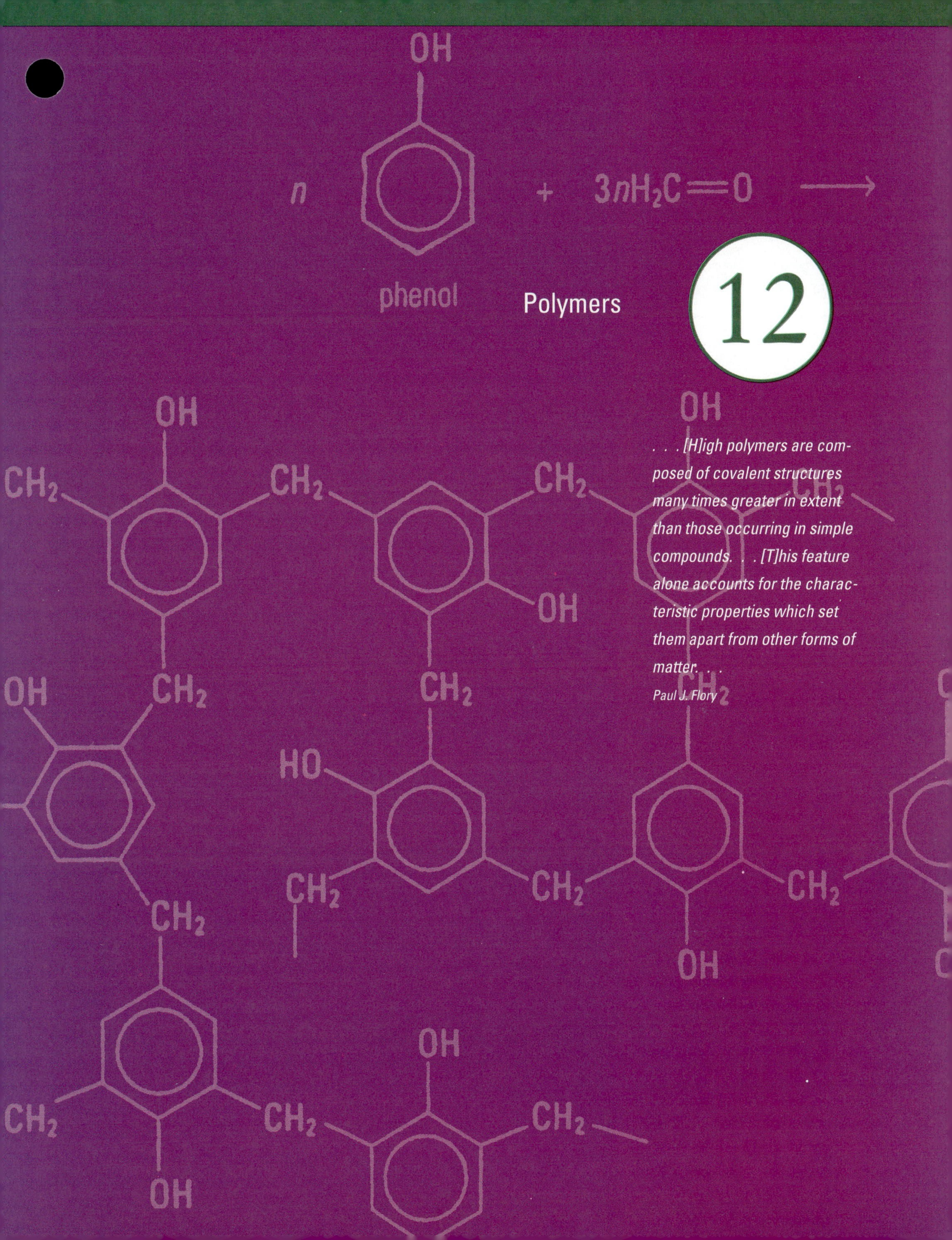

12 Polymers

. . . [H]igh polymers are composed of covalent structures many times greater in extent than those occurring in simple compounds. . . [T]his feature alone accounts for the characteristic properties which set them apart from other forms of matter. . .

Paul J. Flory

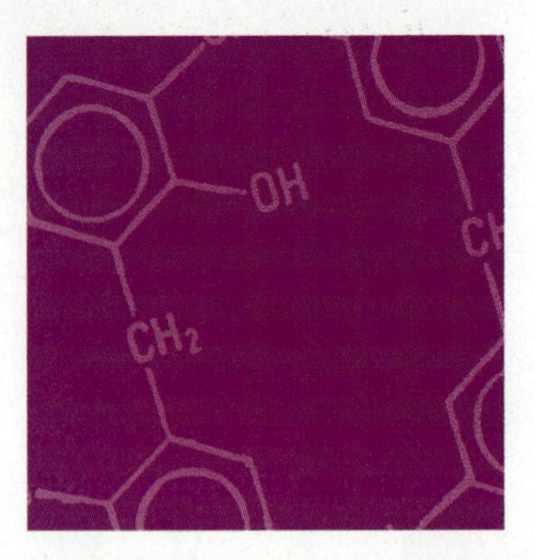

A major portion of the chemical industry is devoted to the chemistry of polymers, and more chemists are engaged in polymer chemistry than in any other field of chemistry. To understand why this is so, you need only consider some of the things you use every day: shoes, tires, pens, paints, foam cups, food, clothing, compact discs, rope, toothbrushes, plastic bags and wrap, glue, dental floss, cushions, seat belts, backpacks, eyeglasses, carpets, water pipes, and beverage bottles. All of these items, and thousands more, are made of polymers, materials composed of very large molecules. Polymers can be elastic, tough, slippery, sticky, lightweight, strong, waterproof, water soluble, moldable, foamable, gelatinous, insulating, oil resistant, transparent, and edible. Often, properties can be combined. For example, polycarbonate polymers such as Lexan are lightweight, highly impact resistant, and transparent, making them well suited for eyeglass lenses, safety helmets, telephones, and bullet proof windows.

WHAT POLYMERS ARE

Macromolecules are very large molecules.

Polymers are often called plastics, but plastics comprise only a portion of the compounds classified as polymers. **Polymers** are composed of very large molecules, called **macromolecules,** that contain as many as 10^4 to 10^6 atoms. They are made up by the repetitious combination of thousands of small molecules called **monomers.** The word, *polymer* means *of many parts* in Greek, and the monomer is the small molecule that forms the part, or the repeating unit (▸ Figure 12.1). (In a string of sausages, one sausage is the repeating unit.)

CLASSES OF POLYMERS

There are four classes of polymers: **thermoplastics, fibers, elastomers,** and **thermosetting resins.**

There are four classes of polymers based roughly upon their physical state. A *plastic* (or more properly, **thermoplastic**) is a polymer that is a solid at room temperature and can be shaped by heating and molding. These polymers,

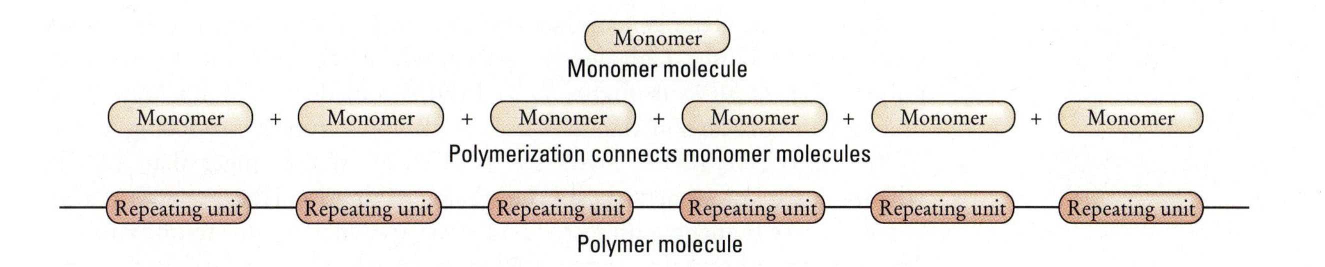

▶ Figure 12.1
Monomers, Polymers, and Repeating Units

such as polyethylene and Plexiglas, can be heated and molded under pressure into many kinds of objects and can be formed into tubes, sheets, and films. **Fibers** are fine threads formed by squeezing molten or dissolved polymer through a *spinneret,* a die with tiny holes in it. Once the thread is formed, it is stretched, or drawn, in the direction of the fiber, which gives it considerable strength. Nylon, Dacron, Orlon, and polypropylene are polymers that can be drawn into fibers. **Elastomers** are polymers that can be stretched considerably and then return to their original shapes. Natural rubber, polyisobutylene, and styrene–butadiene rubber (SBR) used in automobile tires are common examples. Finally, **thermosetting resins** (sometimes simply called resins) are polymers that solidify into a rigid mass when heated and molded; and they cannot be remelted. These kinds of polymers are also used as adhesives. Examples are Melamine, Bakelite, and epoxy glues.

ADDITION POLYMERS

Polymers also are classified according to the manner in which they are chemically produced. **Addition polymers** (also called *chain-growth polymers*) are formed by successive addition of monomer molecules to one another to form a long polymer chain. For example, polyethylene, the simplest addition polymer, is made by treating ethylene, a gas, with peroxides or special catalysts (▶ Figure 12.2).

(a)

$$CH_2 \overset{1}{=} CH_2 + CH_2 \overset{2}{=} CH_2 + \mathbf{CH_2 \overset{3}{=} CH_2} + CH_2 \overset{4}{=} CH_2 + CH_2 \overset{5}{=} CH_2 \longrightarrow$$

large number of ethylene monomer molecules

$$-CH_2 \overset{1}{-} CH_2 - CH_2 \overset{2}{-} CH_2 - \mathbf{CH_2 \overset{3}{-} CH_2} - CH_2 \overset{4}{-} CH_2 - CH_2 \overset{5}{-} CH_2 -$$

polyethylene polymer containing a large number of repeating units

(b)

$$n CH_2{=}CH_2 \longrightarrow +\!\!\left(CH_2 - CH_2 \right)\!\!+_n$$

n is very large, up to 3500

▶ Figure 12.2
Formation of an Addition Polymer: Polyethylene from ethylene. *(a) Molecules of ethylene monomer are connected end-to-end to form a polymer with a long chain of repeating units. The monomer and the repeating unit are shown in bold. (b) The convention for writing a polymerization reaction and the structure of a polymer is to show the repeating unit in parentheses, followed by subscript* n *(a large number).*

More polyethylene is manufactured than any other polymer. It is usually made in two kinds: high-density polyethylene (HDPE) and low-density polyethylene (LDPE) (▶ Figure 12.3). HDPE is made up of long, *linear* (unbranched) hydrocarbon chains that pack closely together, producing a relatively orderly, crystalline structure. It is stiffer and stronger than LDPE, which consists of somewhat shorter chains having branches, or *side chains.* Some of these branches connect, or *crosslink,* two main chains to one another. This prevents the chains from packing as closely together and produces a lower density solid, which is less rigid, more flexible, somewhat waxy, and transparent.

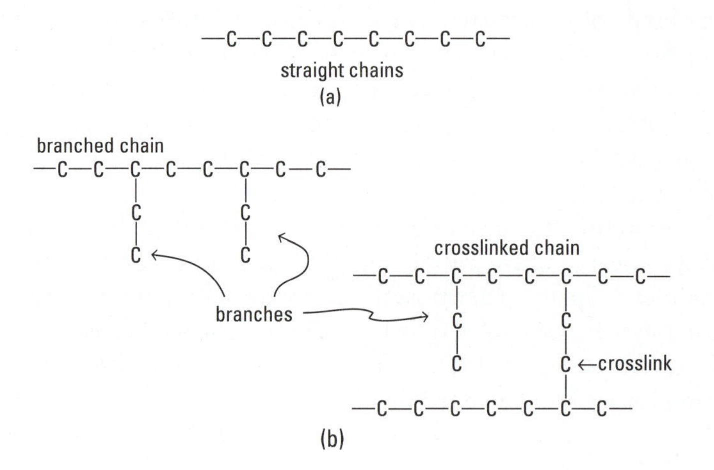

▶ Figure 12.3
Polyethylene Structures. *(a) High-density polyethylene (HDPE) has long, straight chains. (b) Low-density polyethylene (LDPE) has branched and crosslinked chains.*

▶ Figure 12.4
Many different items are made of polyethylene, from plastic containers and waste bags to laboratory ware.

Both HDPE and LDPE are thermoplastics. HDPE is used in molded items such as bottle caps, toys, rope, and electronics cabinets. Most large milk containers and those for motor oils, bleaches, detergents, and other liquid products are made of HDPE. The greater flexibility of LDPE makes it suitable for use in blister packaging, plastic bags, food storage containers, plastic wrap, waterbottles, and electrical wire insulation (▶ Figure 12.4).

SUBSTITUTED POLYETHYLENES

Many useful polymers have substituted polyethylene chains.

Other common addition polymers share the backbone of the polyethylene chain, except that one or more of the hydrogen atoms bound to each repeating unit has been replaced with a substituent group. For example, polypropylene, used to manufacture rope, indoor–outdoor carpeting, and upholstery fabrics, is made from propylene, a monomer in which a hydrogen atom on ethylene has been replaced with a methyl group:

$$n\mathrm{CH_2{=}CH(CH_3)} \longrightarrow \mathrm{+\!\!\!-CH_2{-}CH(CH_3)\!-\!\!\!+_{\,n}}$$

propylene → polypropylene

Other monomers and polymers that share this feature are shown in ▸ Figure 12.5 and ■ Table 12.1.

Two interesting, though less well known, relatives of polyethylene form resins. Polyvinylpyrrolidone, or PVP, is a holding resin used in hair sprays and mousses. In hair sprays it is simply dissolved in a suitable solvent. When

■ **TABLE 12.1 POLYMERS RELATED TO POLYETHYLENE: THEIR PROPERTIES AND USES**

Name	Formula: Monomer	Formula: Polymer	Recycling symbol	Properties* and uses
Polyethylene	$H_2C{=}CH_2$	$-[CH_2-CH_2]_n-$	2 HDPE, 4 LDPE	Unreactive, flexible, impermeable to water vapor. Packaging films, containers, toys, housewares
Polypropylene	$CH_3CH{=}CH_2$	$-[CH(CH_3)-CH_2]_n-$	5 PP	Lowest density of any plastic. Indoor-outdoor carpeting, upholstery, pipes, bottles
Polyisobutylene	$CH_2{=}C(CH_3)_2$	$-[CH_2-C(CH_3)_2]_n-$	7 OTHER	Elastomer. Inner tubes, truck and bicycle tires
Poly(vinyl chloride) (PVC)	$H_2C{=}CHCl$	$-[CH_2-CH(Cl)]_n-$	3 V	Self-extinguishing to fire. Pipe, siding, floor tile, raincoats, shower curtains, imitation leather upholstery, garden hoses
Polytetrafluoroethylene (Teflon)	$F_2C{=}CF_2$	$-[CF_2-CF_2]_n-$	7 OTHER	Very unreactive, nonstick, relatively high softening point. Liners for pots and pans, greaseless bearings, artificial joints, heart valves, plumbers' tape, fabrics
Polystyrene	$C_6H_5-CH{=}CH_2$	$-[CH(C_6H_5)-CH_2]_n-$	6 PS	Housings for large household appliances such as refrigerators, auto instruments and panels, clear cups and food containers, and foam cups and packing
Polyacrylonitrile (Orlon)	$H_2C{=}CHCN$	$-[CH_2-CH(CN)]_n-$	7 OTHER	Carpets and knitwear
Poly(methyl methacrylate) (Lucite, Plexiglas)	$H_2C{=}C(CH_3)COOCH_3$	$-[CH_2-C(CH_3)(COOCH_3)]_n-$	7 OTHER	Substitute for glass, airplane windows, contact lenses, fiber optics, paint
Poly(vinyl acetate)	$H_2C{=}CH-O-C(=O)-CH_3$	$-[CH_2-CH(O-C(=O)-CH_3)]_n-$	7 OTHER	Adhesives, paint, chewing gum, safety glass

*Those that are outstanding or unusual.

Source: Umland, General Chemistry, *© 1993 by West Publishing Company. Adapted with permission.*

▶ Figure 12.5
Automobile Safety Glass. *A layer of poly(vinyl acetate) between two sheets of tempered glass prevents sharp fragments of glass from flying about in a collision. Poly(vinyl acetate) is a substituted polyethylene.*

sprayed on the hair, evaporation of the solvent leaves the sticky resin to hold the hair in place. A mousse is a foam preparation containing resins such as PVP. PVP is also used medically as a blood plasma substitute.

$$-CH_2-\underset{\text{pyrrolidone}}{CH}-CH_2-\underset{\text{pyrrolidone}}{CH}-CH_2-\underset{\text{pyrrolidone}}{CH}-CH_2-\underset{\text{pyrrolidone}}{CH}-CH_2-\underset{\text{pyrrolidone}}{CH}-$$

polyvinylpyrrolidone (PVP)

Super Glue is remarkable stuff. One drop of it can hold a 2000-lb weight. It is simply the pure monomer, methyl α-cyanoacrylate in a tube, which polymerizes rapidly in the presence of environmental *traces* of basic substances, even water, to form a very strong resin. You may have experienced inadvertently gluing your fingers together with it. Your skin is an excellent source of bases which initiate this polymerization.

$$n\,CH_2{=}C(CN)(C(=O)OCH_3) \xrightarrow{\text{a trace of base}} \left[CH_2-C(CN)(C(=O)OCH_3) \right]_n$$

methyl α-acyanoacrylate → Super Glue

THE POLYBUTADIENES

Another group of addition polymers share a different backbone; that of a polybutadiene chain.

$$n\,CH_2{=}CH-CH{=}CH_2 \longrightarrow \left[CH_2-CH{=}CH-CH_2 \right]_n$$

1,3-butadiene → polybutadiene

Many of the polybutadienes are elastomers.

Many of these polymers are elastomers that are used in rubber goods; tires, inner tubes, gaskets, hoses, automobile bumpers, and shoe soles (■ Table 12.2). Note that polybutadiene chains contain double bonds. This structural feature makes the polybutadienes elastic, but it also contributes to their tendency to decompose in the environment; for example, tires harden and crack in polluted air.

Polybutadiene is used to make tires, rubber belts, and hoses. As with polyethylene, replacing a hydrogen atom on the chain of polybutadiene with a

12.1 Identify the repeating unit in each of the polymers given. Write the structure of the monomer that might be used to make each polymer.

a. $-CH_2-CH(NO_2)-CH_2-CH(NO_2)-CH_2-CH(NO_2)-CH_2-CH(NO_2)-CH_2-CH(NO_2)-$

b. $-CH_2-CCl_2-CH_2-CCl_2-CH_2-CCl_2-CH_2-CCl_2-CH_2-CCl_2-$

c. $-CFCl-CF_2-CFCl-CF_2-CFCl-CF_2-CFCl-CF_2-CFCl-CF_2-$ (Kel-F)

d. $-CH_2-CH(O-H)-CH_2-CH(O-H)-CH_2-CH(O-H)-CH_2-CH(O-H)-CH_2-CH(O-H)-CH_2-CH(O-H)-$

substituent group changes the properties of the polymer. Natural rubber, which is also used in tires and many other rubber goods, is polyisoprene. In isoprene, the monomer unit, a hydrogen atom on butadiene has been replaced by a methyl group.

$$n\,CH_2{=}CH{-}C(CH_3){=}CH_2 \longrightarrow \left(CH_2{-}CH{=}C(CH_3){-}CH_2 \right)_n$$

isoprene — polyisoprene (natural rubber)

Polyisoprene is now made synthetically on a commercial scale.

■ **TABLE 12.2 SOME COMMON BUTADIENE ELASTOMERS**

Name	Formula: Monomer	Formula: Polymer	Properties* and Uses
Polybutadiene	$H_2C{=}CH{-}CH{=}CH_2$	$\left(H_2C{-}CH{=}CH{-}CH_2 \right)_n$	Tires, belts, hoses, metal can coatings
Polychloroprene (Neoprene)	$H_2C{=}CH{-}C(Cl){=}CH_2$	$\left(H_2C{-}CH{=}C(Cl){-}CH_2 \right)_n$	Oil, solvent, and abrasion resistant. Used in severe service applications such as automotive hoses and belts, gasoline hoses, wire and cable covering, conveyor belts. Not good for tires.
Polyisoprene	$H_2C{=}CH{-}C(CH_3){=}CH_2$	$\left(H_2C{-}CH{=}C(CH_3){-}CH_2 \right)_n$	Tires, rubber foot gear, caulks and sealants, adhesives

*Those that are outstanding or unusual.

12.2 Isoprene not only forms the monomer unit in natural rubber, but it also is a repeating unit in many smaller molecules of natural origin (natural products). The idea that the isoprene skeleton

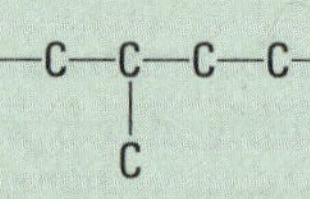

$$\begin{array}{c} -C-\underset{\displaystyle C}{\underset{|}{C}}-C-C- \end{array}$$

is a repeating unit that forms these molecules is called the isoprene rule. (It does not matter how the skeletons are joined or where double bonds and hydrogen atoms are.) The structure of myrcene is an example:

$$\underbrace{CH_3-\overset{\displaystyle CH_3}{\overset{|}{C}}=CH-CH_2}_{\text{isoprene unit}}-\underbrace{CH_2-\overset{\displaystyle CH_2}{\overset{\|}{C}}-CH=CH_2}_{\text{isoprene unit}}$$

myrcene (oil of bay)

Find the isoprene units in the structures of the following natural products and circle them.

a.

O, CH_3, CH_3, C, CH_2

carvone (spearmint oil)

b.

CH_3, CH_3, CH_3, CH_2

caryophyllene (cloves)

c.

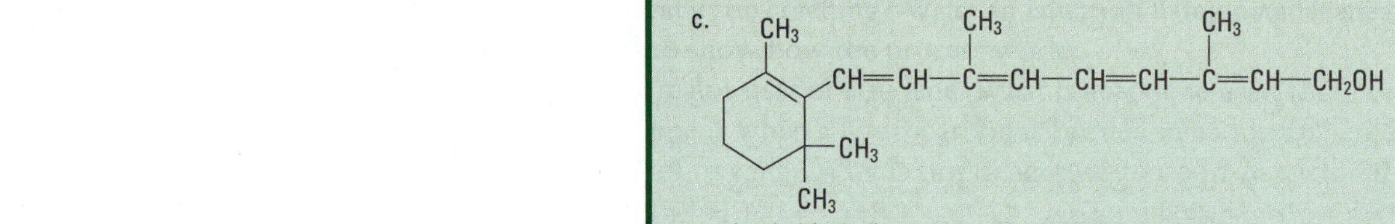

CH_3, $CH=CH-\overset{\displaystyle CH_3}{\overset{|}{C}}=CH-CH=CH-\overset{\displaystyle CH_3}{\overset{|}{C}}=CH-CH_2OH$, CH_3, CH_3

vitamin A (retinol)

Substituting a chlorine atom for the methyl group on the backbone of polyisoprene produces a rubber that is resistant to gasoline, oil, organic solvents, abrasion, and environmental exposure. It is called polychloroprene, or Neoprene.

$$nCH_2{=}CH{-}\underset{\displaystyle Cl}{\underset{|}{C}}{=}CH_2 \longrightarrow {+}CH_2{-}CH{=}\underset{\displaystyle Cl}{\underset{|}{C}}H{-}CH_2{+}_n$$

chloroprene — polychloroprene (Neoprene)

All of the polymers we have encountered so far have been **homopolymers.** That is, they are made up of just one kind of monomer unit. Often, however, desirable properties can be imparted to an addition polymer by incorporating a second (and sometimes a third) monomer unit into its chain. This kind of polymer is called a **copolymer.** A good example is SBR (styrene–butadiene rubber), made by polymerizing 1,3-butadiene mixed with styrene so that the ratio of the monomers is about 3:1. This produces a copolymer with the following approximate structure. Inserted in the butadiene backbone, at about every fourth repeating unit, is a styrene unit. The pattern of this insertion is not entirely regular.

Copolymers contain more than one kind of monomer unit.

$$-CH_2CH{=}CHCH_2-CH_2CH{=}CHCH_2-CH_2\underset{\displaystyle C_6H_5}{\underset{|}{C}H}-CH_2CH{=}CHCH_2-$$

butadiene unit — butadiene unit — styrene unit — butadiene unit

styrene–butadiene rubber (SBR)

SBR has properties similar to natural rubber and has largely replaced the latter in the manufacture of automobile and truck tires (► Figure 12.6).

VULCANIZATION

Natural rubber itself is not a very useful substance. When hot, it becomes sticky and gummy. When cold, it turns hard and brittle. Exposed to the air over time, it decomposes into gummy flakes and dust. Although many inventors had tried to make it into useful products such as waterproof clothing, boots, and inflatable pillows and life preservers, about all it really was good for prior to 1839 was rubbing out pencil marks on paper. Hence its English name: *rubber.*

► Figure 12.6
Most tires are made of SBR rubber, a copolymer.

12.3 Write the structure of the monomers used to make each of the following addition copolymers.

a.

$$-CH_2CH-CH_2-\underset{Cl}{\overset{Cl}{C}}-CH_2CH-CH_2-\underset{Cl}{\overset{Cl}{C}}-$$

(with Cl on the first CH and the second CH)

Saran

b.

$$-CH_2CH{=}\overset{CH_3}{C}-CH_2-CH_2-\underset{CH_3}{\overset{CH_3}{C}}-CH_2CH{=}\overset{CH_3}{C}-CH_2-CH_2-\underset{CH_3}{\overset{CH_3}{C}}-$$

butyl rubber

Charles Goodyear had been trying for about five years to make rubber useful when, in the winter of 1839, he accidentally dropped a sample of natural rubber, coated with sulfur and white lead, on a hot stove. The sample charred but did not melt, and it remained elastic when he nailed it outside in cold weather. Immediately realizing that heating the rubber with the sulfur had caused the desirable changes, he set about to understand the process that today is called **vulcanization.** (Goodyear did not invent this term: one of his competitors, Thomas Hancock, did.)

How does heating with sulfur change natural rubber? It causes bridges of sulfur atoms to form between the polyisoprene chains (► Figure 12.7). This

$$-CH_2-CH{=}\overset{CH_3}{C}-CH_2-CH_2-CH{=}\overset{CH_3}{C}-CH_2-CH_2-CH{=}\overset{CH_3}{C}-CH_2-$$

$$-CH_2-CH{=}\underset{CH_3}{C}-CH_2-CH_2-CH{=}\underset{CH_3}{C}-CH_2-CH_2-CH{=}\underset{CH_3}{C}-CH_2- \quad + \quad S \xrightarrow{heat}$$

natural rubber

$$-CH_2-CH{=}\overset{CH_3}{C}-CH-CH_2-CH{=}\overset{CH_3}{C}-CH_2-CH_2-CH{=}\overset{CH_3}{C}-CH-$$

$$S_x \qquad\qquad S_x \qquad + \quad H_2S$$

$$-CH_2-CH{=}\underset{CH_3}{C}-CH-CH_2-CH{=}\underset{CH_3}{C}-CH_2-CH-CH{=}\underset{CH_3}{C}-CH_2-$$

vulcanized rubber

► Figure 12.7
Vulcanization of Natural Rubber. *Heating natural rubber with sulfur causes sulfur bridges, called crosslinks, to form between the polyisoprene chains. The S_x in this equation means that the number of sulfur atoms in a crosslink is variable. (Note that the equation is not balanced.) Today, natural rubber is often crosslinked by irradating it with high-energy electromagnetic radiation, such as X-rays.*

bridging, called **crosslinking,** forms a more rigid structure, locking the chains together and making the rubber harder, stronger, less tacky, and more elastic.

The polymer chains in natural rubber (and other elastomers) have an irregular shape and do not pack well in the solid. Also, the intermolecular forces between the chains are weak. When the rubber is stretched, the chains are forced to line up with one another. However, intermolecular forces are not sufficient to hold them in alignment when the stretching force is removed, so the chains revert to their original, more disorderly arrangement, which is favored by entropy (Chapter 10). During stretching, however, the chains can slide past one another, and so natural rubber has little strength.

Vulcanization enhances the elastic properties and strength of natural rubber (► Figure 12.8). In unstretched rubber, the chains are folded and coiled. Stretching causes the chains to straighten. However, the sulfur crosslinks, which were put in place when the chains were folded, tend hold them in that alignment, resisting the force to straighten them. Thus, when the stretching force is removed, the crosslinks help pull the chains back to their original shapes. The more sulfur crosslinks, the stiffer the rubber is. Rubber formulated with less than 5% sulfur by weight is quite elastic, whereas the sulfur content of very hard rubber, like that used to make combs and bowling balls, is 25–50% sulfur.

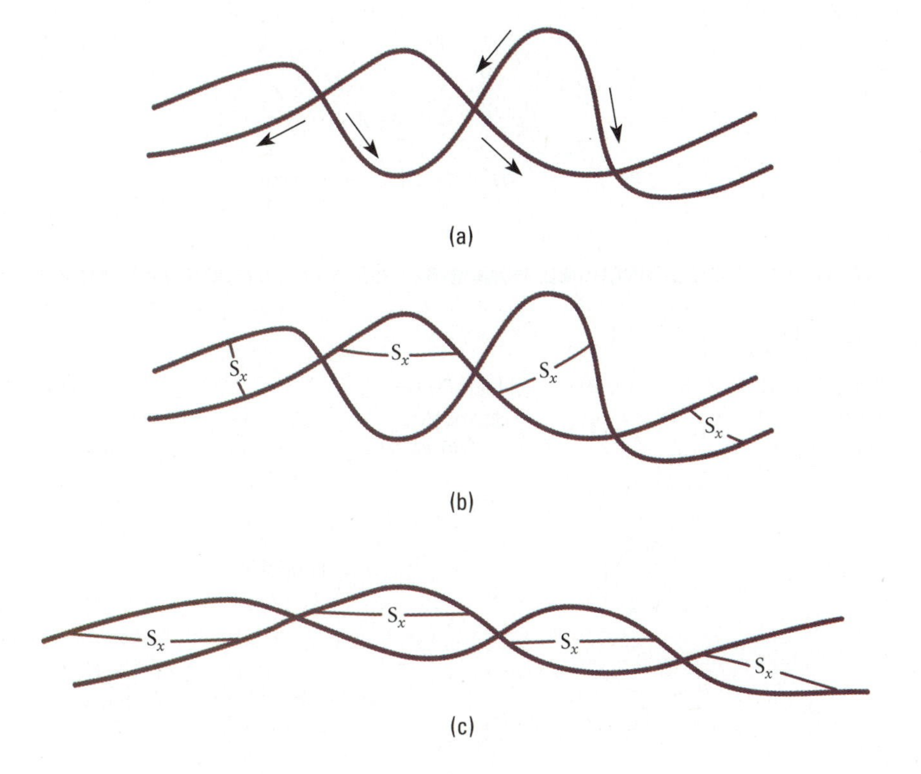

► Figure 12.8

Vulcanization strengthens natural rubber and makes it more elastic. (a) Natural rubber has little strength because the polyisoprene chains can slide past each other when a stretching force is applied. (b) In vulcanization, sulfur crosslinks are put in place when the chains are tangled and folded. (c) When this rubber is stretched, the crosslinks hold the chains in alignment, and help to pull them back into their original shapes when the stretching force is released.

CONDENSATION POLYMERS

A monomer unit must be **bifunctional** to form a **condensation polymer**.

Condensation polymers, also called *step-reaction polymers,* constitute the second chemical class of polymers. Most of these are formed by condensation reactions, similar to those we encountered in forming esters and amides in Chapter 11. Recall that in condensation reactions a small molecule, such as water, is split out when two larger molecules containing different functional groups are joined. To form a condensation polymer, a monomer unit must be **bifunctional.** That is, it must contain two functional groups (the same or different) (■ Table 12.3). This is analogous to forming a human chain: every person in it must use two hands.

The formation of polyethylene terephthalate, or PET, a common polyester, illustrates how a condensation polymer is formed. First, a molecule of terephthalic acid is condensed with a molecule of ethylene glycol to form the ester (with the loss of a molecule of water).

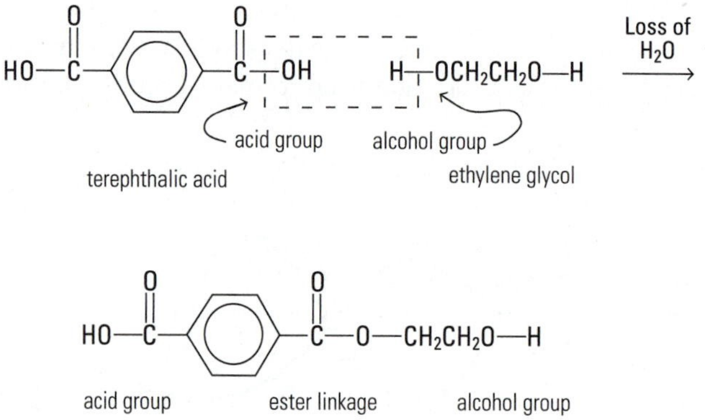

$$\mathrm{HO{-}\overset{\displaystyle O}{\overset{\|}{C}}{-}C_6H_4{-}\overset{\displaystyle O}{\overset{\|}{C}}{-}OH \quad + \quad H{-}OCH_2CH_2O{-}H} \xrightarrow{\text{Loss of } H_2O}$$

acid group — alcohol group

terephthalic acid — ethylene glycol

$$\mathrm{HO{-}\overset{\displaystyle O}{\overset{\|}{C}}{-}C_6H_4{-}\overset{\displaystyle O}{\overset{\|}{C}}{-}O{-}CH_2CH_2O{-}H}$$

acid group — ester linkage — alcohol group

■ TABLE 12.3 SOME BIFUNCTIONAL MONOMERS USED IN CONDENSATION POLYMERS

Structure	Name	Type
$\mathrm{HO\overset{O}{\overset{\|}{C}}(CH_2)_4\overset{O}{\overset{\|}{C}}OH}$	adipic acid	a dicarboxylic acid
$\mathrm{HOCH_2CH_2OH}$	ethylene glycol	a dialcohol
$\mathrm{H_2N(CH_2)_6NH_2}$	1,6-diaminohexane	a diamine
$\mathrm{HO{-}C({=}O){-}C_6H_4{-}C({=}O){-}OH}$	terephthalic acid	a dicarboxylic acid
$\mathrm{H_2N(CH_2)_5\overset{O}{\overset{\|}{C}}OH}$	6-aminohexanoic acid (aminocaproic acid)	an amino acid

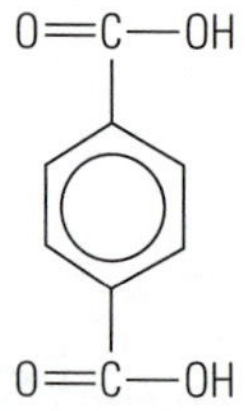

This ester, called a *dimer*, has an unreacted acid group and alcohol group, both of which are free to react with another molecule of alcohol and acid, respectively, to form a four-unit molecule called a *tetramer.*

$$H-OCH_2CH_2O-\overset{O}{\overset{\|}{C}}-C_6H_4-\overset{O}{\overset{\|}{C}}-O-CH_2CH_2-O-\overset{O}{\overset{\|}{C}}-C_6H_4-\overset{O}{\overset{\|}{C}}-OH$$

alcohol group — ester linkages — acid group

Because it also has free alcohol and acid functional groups, this tetramer can react further with more alcohol and acid, and again stepwise many times, to form a very long chain of alcohol and acid units, joined by ester linkages.

$$\left(\overset{O}{\overset{\|}{C}}-C_6H_4-\overset{O}{\overset{\|}{C}}-O-CH_2CH_2-O\right)_n$$

polyethylene terephthalate, PET
a polyester

PET fibers are marketed under the trade name Dacron and are used primarily in clothing, tire cord, automotive upholstery, large soft-drink bottles, and stuffing for pillows and sleeping bags. It can also be made into a film, called Mylar. In this form it is used in magnetic recording tape, zippers, and in bags for boil-in foods (▶ Figure 12.9).

Polyamides, such as Nylon-66, can be formed in a similar manner. In this case adipic acid is condensed with the diamine, 1,6-hexanediamine.

$$nHO-\overset{O}{\overset{\|}{C}}(CH_2)_4\overset{O}{\overset{\|}{C}}-OH + nHN(H)(CH_2)_6NH_2 \xrightarrow{\text{Loss of } H_2O}$$

adipic acid — 1,6-hexanediamine

$$-NH-\overset{O}{\overset{\|}{C}}(CH_2)_4\overset{O}{\overset{\|}{C}}-NH(CH_2)_6NH-\overset{O}{\overset{\|}{C}}(CH_2)_4\overset{O}{\overset{\|}{C}}-NH(CH_2)_6NH-\overset{O}{\overset{\|}{C}}(CH_2)_4\overset{O}{\overset{\|}{C}}-NH(CH_2)_6-$$

amide linkages — amide linkages

a section of the polyamide chain in Nylon 66

or,

$$\left((CH_2)_6NH-\overset{O}{\overset{\|}{C}}(CH_2)_4\overset{O}{\overset{\|}{C}}-NH\right)_n$$

▶ Figure 12.9
The Gossamer Albatross was the first human-powered aircraft to fly across the English Channel. Its wings and pilot's compartment were covered with Mylar, a polyester developed by E. I. DuPont DeNemours Co.

CATALYTIC CONVERTERS

This morning's thoughts of food, and reading the carbon-chemistry chapter this afternoon, have brought up the question of reality again. Foods are chemicals, but when we eat, we're not just dumping reactants into test-tube stomachs. Something else is happening.

Rhonda had glanced at your book while you were reading: "Organic chemistry. Now you're getting somewhere!" She's enthusiastic about biochemistry.

"But Rhonda , all these reactions seem to take a lot of heat, and I don't see how anything like them can take place in our bodies."

"Well, I can't explain all the details, but I can give you an idea of how it all works."

To react, the right molecules or atoms have to collide in the right orientation with a certain amount of energy (activation energy). The way we cause many reactions to proceed in the laboratory is to heat the reactants to high temperatures. That causes huge numbers of collisions and some of them are certain to be right. Of course, biology can't do that. Most biological reactions are catalyzed by proteins called enzymes. Catalysts are substances that either lower the activation energy of a reaction or hold the reactants in an optimal orientation for reaction, or both.

An example of catalyzed reactions takes place in catalytic converters. In a converter, molecules of NO and CO, noxious gases in automobile exhaust, are adsorbed by (stick to) the surfaces of beads of platinum. Some of the NO molecules dissociate into atoms of nitrogen and oxygen. The attraction between pairs of absorbed nitrogen atoms, and between nitrogen atoms and the nitrogen in undissociated NO, is stronger than that between the nitrogen and oxygen. Thus, harmless N_2 forms and, having no attraction to the platinum, leaves its surface. The leftover oxygen is absorbed by the platinum, where it oxidizes the CO into harmless CO_2, which also leaves. The platinum ends up alone and unaffected, as a catalyst should.

Proteins are complex amino acid polymers made up of hundreds to millions of atoms. They're involved in every aspect of life, the structure of fingernails and hair, hormones, muscle action, and the transport of nerve signals and oxygen (hemoglobin is a protein). Enzymes are proteins that catalyze most of the chemical reactions that constitute living. There are thousands of types. Each catalyzes one particular reaction or a particular part of a reaction: for instance, hydrolyzing or cleaving a carbohydrate, fat, or protein polymer. The electron arrangements of enzymes cause them to attract and hold particular molecules in optimum orientations for reaction. Just as in catalytic converters, molecules that are lightly bonded to an enzyme dissociate or link up with other molecules. The products of such reactions become unattractive to the enzyme and are let go, leaving the enzyme ready for another reaction.

Hundreds of such reactions take place simultaneously in every cell of your body (except red blood cells, which are specialized for transporting oxygen and carbon dioxide); and a molecule of a certain enzyme has been observed to catalyze more than 3×10^6 reactions every minute. Yet the reactions must be rigidly controlled: too much or too little of a product and a whole system malfunctions. The amount of product catalyzed by an enzyme or a set of enzymes may be governed by another enzyme or by the amount of available reactants produced by another enzyme, which may be controlled by yet another enzyme.

Replication of the famous DNA (deoxyribonucleic acid) is catalyzed by cooperation among many kinds of enzymes. The DNA molecule, which contains our genes, is made of two strands of polymers connected in parallel. When a cell divides, the DNA must be duplicated for the new cell. DNA is stored in cells as a space-saving double helix (like a rubber band wound until it coils and then wound until it coils again). In replication, the double helix is uncoiled by enzymes, and the two strands are unzipped by other enzymes that break the hydrogen bonds holding them together. Enzymes copy both strands, accepting the appropriate monomer building blocks from other enzymes. As soon as parts of the old strands are copied, they're rezipped by enzymes that reestablish the hydrogen bonds that hold them together. Enzymes check the new copies for errors, correct the errors, and then zip them together. After that, other sets of enzymes coil both molecules of DNA into their helices and double helices. All of these complexities are caused by ordinary chemical reactions such as acid–base, hydrolysis, and condensation, each one catalyzed by a particular enzyme made from the blueprints in the genes in DNA.

"You mentioned hemoglobin. I've always wondered how it can pick up oxygen at one moment and leave it off the next and pick up CO_2. Is that done by enzymes?"

"Partly. Several reactions are involved, but one of them is very simple: oxygen pressure is greater in the lungs than in the tissues, and CO_2 pressure is greater than oxygen pressure in the tissues. That's obvious, if you think about it. So what is bound to happen? Remember Henry's law about the solubility of gases?"

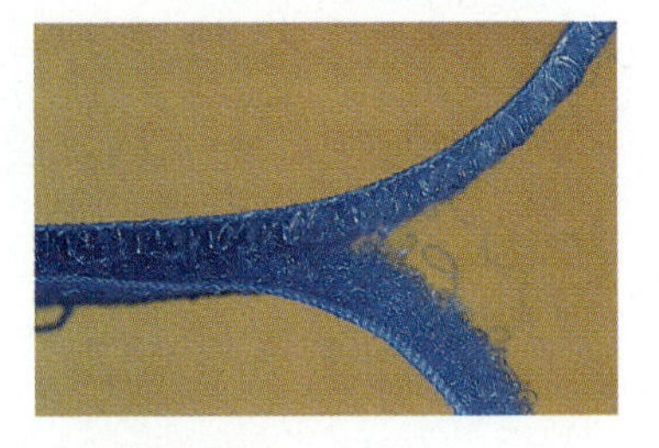

▶ Figure 12.10
This convenient hook-and-loop fastener is made of Nylon.

The 66 in the name means that each monomer unit has six carbon atoms. Nylon-6, made from 6-aminohexanoic acid, has the following structure,

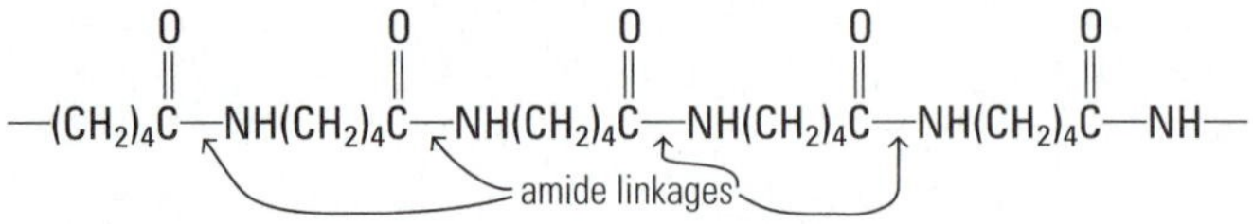

$$-(CH_2)_4\overset{O}{\overset{\|}{C}}-NH(CH_2)_4\overset{O}{\overset{\|}{C}}-NH(CH_2)_4\overset{O}{\overset{\|}{C}}-NH(CH_2)_4\overset{O}{\overset{\|}{C}}-NH(CH_2)_4\overset{O}{\overset{\|}{C}}-NH-$$

amide linkages

or,

$$+(CH_2)_5\overset{O}{\overset{\|}{C}}-NH+_n$$

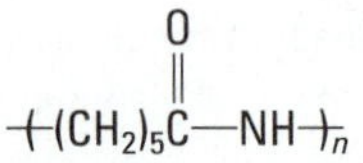

Nylon-6 repeating unit

The Nylons can be drawn into very strong fibers, which are used to make tire cord, hosiery, rope, fishing line, parachutes, surgical sutures, carpets, and clothing. They are thermoplastic and also can be molded into gears, valves, fasteners, door latches, and Velcro (▶ Figure 12.10).

The Nylons have chemical structures similar to those of silk and wool.

The Nylons mimic the chemical structure of the natural fibers silk and wool, which are also polyamides. Instead of having many carbon atoms between amide linkages, however, the polyamide chains of silk and wool have only one. ▶ Figure 12.11 illustrates these similarities and differences. Another natural fiber is shown in ▶ Figure 12.12.

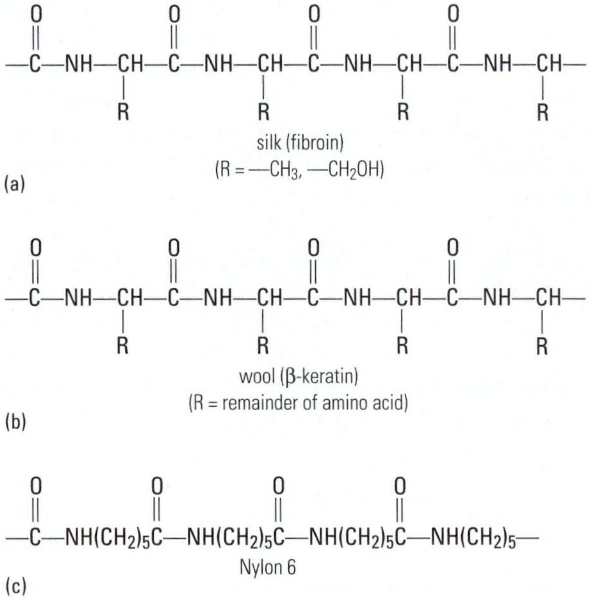

$$-\overset{O}{\overset{\|}{C}}-NH-\underset{R}{\underset{|}{CH}}-\overset{O}{\overset{\|}{C}}-NH-\underset{R}{\underset{|}{CH}}-\overset{O}{\overset{\|}{C}}-NH-\underset{R}{\underset{|}{CH}}-\overset{O}{\overset{\|}{C}}-NH-\underset{R}{\underset{|}{CH}}-$$

(a) silk (fibroin)
($R = -CH_3, -CH_2OH$)

$$-\overset{O}{\overset{\|}{C}}-NH-\underset{R}{\underset{|}{CH}}-\overset{O}{\overset{\|}{C}}-NH-\underset{R}{\underset{|}{CH}}-\overset{O}{\overset{\|}{C}}-NH-\underset{R}{\underset{|}{CH}}-\overset{O}{\overset{\|}{C}}-NH-\underset{R}{\underset{|}{CH}}-$$

(b) wool (β-keratin)
(R = remainder of amino acid)

$$-\overset{O}{\overset{\|}{C}}-NH(CH_2)_5\overset{O}{\overset{\|}{C}}-NH(CH_2)_5\overset{O}{\overset{\|}{C}}-NH(CH_2)_5\overset{O}{\overset{\|}{C}}-NH(CH_2)_5-$$

(c) Nylon 6

▶ Figure 12.11
Chemical Similarity of Silk, Wool, and Nylon-6. *All are polyamides, but the natural polymers silk (a) and wool (b) have only one carbon atom between amide linkages whereas Nylon-6 (c) has many carbon atoms. Silk and wool differ in the nature of the substituents attached to their polyamide chains.*

TABLE 12.4 SOME COMMON CONDENSATION POLYMERS AND THEIR USES

Name	Formula: Monomers	Formula: Polymer	Recycling Symbol	Properties* and Uses
Polyamide (Nylon 66)	$HOOC(CH_2)_4COOH$ and $H_2N(CH_2)_6NH_2$	$-\!\!\!+ C(=O)(CH_2)_4C(=O)NH(CH_2)_6NH +\!\!\!-_n$	7 OTHER	Strong, wear-resistant. Hosiery, tire cord, door latches, gears, rope, fishing line, parachutes, surgical sutures, carpets, clothes, fasteners, Velcro
Polyester or polyethylene terephthalate (Dacron fiber and Mylar film)	$CH_3OOC-C_6H_4-COOCH_3$ and $HOCH_2CH_2OH$	$-\!\!\!+ OCH_2CH_2OC(=O)-C_6H_4-C(=O) +\!\!\!-_n$	1 PETE	Large soft-drink bottles, clothing, tirecord, firehose, stuffing for pillows, sleeping bags, comforters, magnetic recording tape, zippers, bags for boiling foods, upholstery
Polycarbonates (Lexan)	$C_6H_5-O-C(=O)-O-C_6H_5$ and $HO-C_6H_4-C(CH_3)_2-C_6H_4-OH$	$-\!\!\!+ O-C_6H_4-C(CH_3)_2-C_6H_4-O-C(=O) +\!\!\!-_n$	7 OTHER	Highly impact resistant, clear, can be autoclaved. Safety helmets, housings for machinery, eyeglass lenses, bullet proof windows, telephones, sterile labware
Polyurethanes	$O{=}C{=}N-C_6H_3(CH_3)-N{=}C{=}O$ and HO(short polymer)OH	$-\!\!\!+ C(=O)-NH-C_6H_3(CH_3)-NH-C(=O)-O(\text{short polymer})O +\!\!\!-_n$	7 OTHER	Stretchable Spandex and Lycra fibers, other foams and foam rubbers, skin substitutes for burn victims, paints, synthetic leather, rubber, adhesives
Polyoxymethylene (Delrin)	$H_2C{=}O$	$-\!\!\!+ CH_2O +\!\!\!-_n$	7 OTHER	Thermoplastic, strong. Replacement for metal in automotive parts, plumbing, power tool housings
Polyoxyethylene (polyethers)	CH_2-CH_2 (ring closed by O)	$RO -\!\!\!+ CH_2CH_2O +\!\!\!-_n$ (n = 5–15)	7 OTHER	Water soluble waxes, low-sudsing household detergents, manufacture of water-base paints, textile processing, papermaking, monomers for polyurethanes

*Those that are outstanding or unusual.

▶ Figure 12.12
Spiderwebs are constructed of natural polyamide fibers that are very strong.

12.4 Identify the repeating unit in each of the following condensation polymers. Name each kind (polyester, polyamide, etc.).

a. $-NHCH_2CH_2CH_2\overset{O}{\overset{\|}{C}}NHCH_2CH_2CH_2\overset{O}{\overset{\|}{C}}NHCH_2CH_2CH_2\overset{O}{\overset{\|}{C}}NHCH_2CH_2CH_2\overset{O}{\overset{\|}{C}}-$

b. $-\overset{O}{\overset{\|}{C}}CH_2\overset{O}{\overset{\|}{C}}OCH_2CH_2O\overset{O}{\overset{\|}{C}}CH_2\overset{O}{\overset{\|}{C}}OCH_2CH_2O\overset{O}{\overset{\|}{C}}CH_2\overset{O}{\overset{\|}{C}}OCH_2CH_2O-$

c. $-CH_2CH_2CH_2OCH_2CH_2CH_2OCH_2CH_2CH_2OCH_2CH_2CH_2O-$

RESINS: SPACE-NETWORK POLYMERS

A **space-network polymer** is a huge three-dimensional molecule.

To make a space-network polymer, one monomer must be trifunctional to cause crosslinking.

The polyesters and polyamides we have seen so far have been formed from monomers that have only two functional groups. This produces a long polymer chain with no branches or crosslinks. A monomer with three or more reaction sites, however, can form a polymer in which the chains extend in all directions, connecting the monomer units in a crosslinked, three-dimensional network. What is formed is essentially one huge, rigid molecule called a **space-network polymer.** Space-network polymers form the basis of many thermosetting resins, such as those made of formaldehyde with phenol, urea, or melamine, as well as alkyd resins and epoxy glues.

Alkyd resins, used in tough, durable baked-on finishes for large household appliances such as stoves and washing machines are good examples of space-network polymers. These three-dimensional polyesters are made by heating terephthalic acid with glycerol (with the loss of water). The glycerol molecule is trifunctional and causes extensive crosslinking of the polyester chains.

$$3nHO-\overset{O}{\overset{\|}{C}}-C_6H_4-\overset{O}{\overset{\|}{C}}-OH + 2nHOCH_2\underset{\underset{OH}{|}}{CH}CH_2OH \longrightarrow$$

terephthalic acid (T)   glycerol (G)

```
— T—G—T—G—T—     T—
     |     |      |
     T     T—G—T—G
     |        |   |              +  6nH2O
 —T—G—T—G—T       T—G—
           |         |
          —T
```

$+ 6nH_2O$

an alkyd resin (Glyptal)
a polyester

The oldest thermosetting polymer is Bakelite, a phenol-formaldehyde resin introduced in 1909. It is formed by heating phenol with formaldehyde in the presence of acid or base.

$$n\,C_6H_5OH + 3n\,H_2C{=}O \longrightarrow \text{(a phenol-formaldehyde resin)} + 3n\,H_2O$$

phenol

a phenol-formaldehyde resin (Bakelite)

Phenol-formaldehyde resins (phenolics) are used in electrical insulators, knobs, and distributor caps, but their major use is in adhesives for plywood and particleboard (► Figure 12.13).

Another resin used primarily for adhesives and molded products is formed by condensation of formaldehyde with urea. Like phenolics, urea-formaldehyde resins contain traces of formaldehyde, which is toxic. Commonly used to make interior building materials, these resins can release small amounts of the gas, contributing to indoor air pollution and "sick-building" syndrome.

► Figure 12.13
Phenol-formaldehyde and urea-formaldehyde resins are used as adhesives in particleboard, plywood, and other interior building materials.

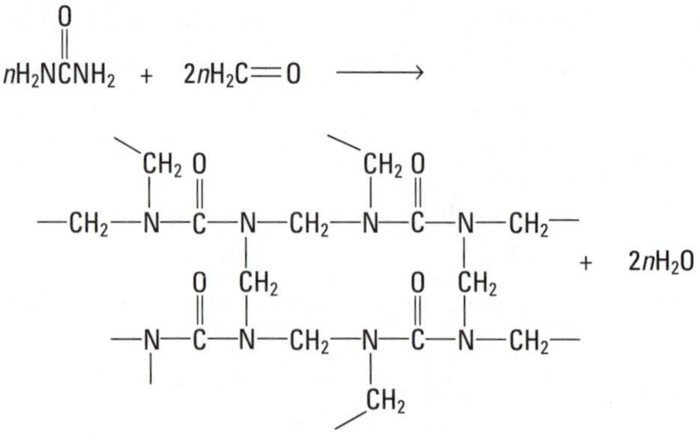

$$n\text{H}_2\text{N}\overset{\text{O}}{\overset{\|}{\text{C}}}\text{NH}_2 + 2n\text{H}_2\text{C}{=}\text{O} \longrightarrow$$

A urea-formaldehyde resin
(equation not balanced)

$$+ \; 2n\text{H}_2\text{O}$$

The condensation of formaldehyde with melamine produces a polymer used in dishes (Melmac) and countertops (Formica). These materials are very durable and wear resistant.

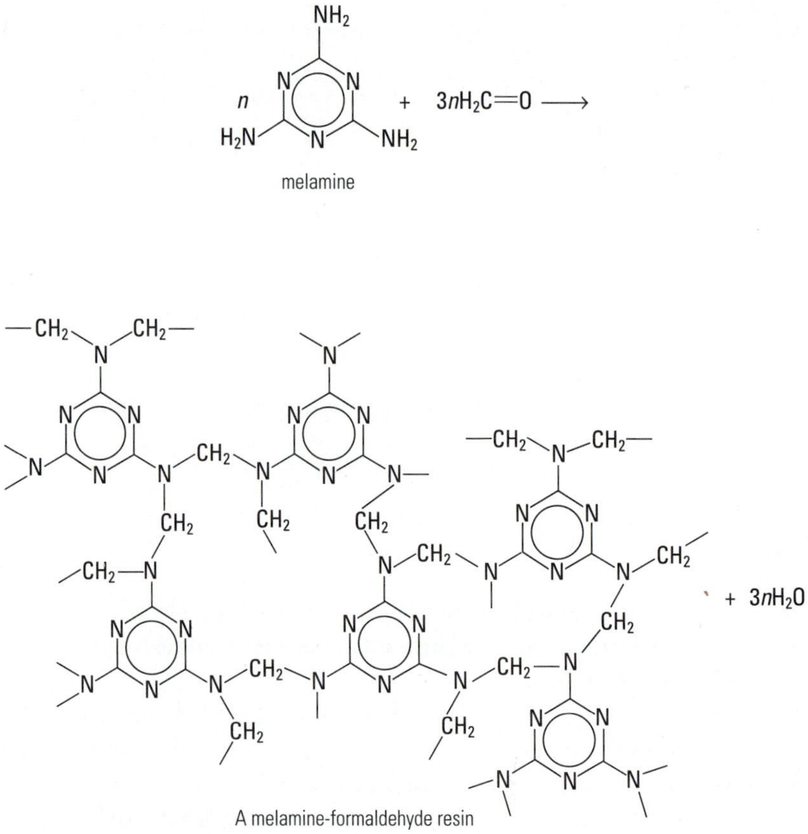

$$n\ \text{melamine} + 3n\text{H}_2\text{C}{=}\text{O} \longrightarrow$$

melamine

A melamine-formaldehyde resin
(equation not balanced)

$$+ \; 3n\text{H}_2\text{O}$$

FIBERS

Fibers are threadlike materials that have considerable strength when pulled from the ends. (This is called tensile strength.) Some synthetic fibers, like the aramid polyamide fibers, are so strong that on a weight-for-weight basis they are stronger than steel. The polymer chains in fibers are mostly linear, as are those in elastomers, but with one important difference. In elastomers, the chains are looped and coiled, and the intermolecular forces between them are weak. In fibers, the intermolecular forces between the chains are much greater, causing the chains to line up alongside one another in a highly ordered arrangement. These strong intermolecular forces prevent the chains from slipping past one another when the fiber is pulled. Also, the structure of the repeating units lets the chains nestle closely together. Most fibers are drawn (stretched) along their length after they are spun to cause the polymer chains to align as closely with one another as possible (► Figure 12.14). The high degree of order combined with the strong attraction of the chains for one another gives a fiber its strength. (Recall that elastomers can be made stronger by crosslinking to prevent the chains from sliding over one another.)

► Figure 12.14
A Spinneret. *Synthetic fibers are made by forcing molten or dissolved polymer through small holes in a spinneret, made of gold-platinum alloy, stainless steel, or titanium. A spinneret 10 cm in diameter can contain as many as 15,000 to 50,000 holes as small as 0.05 mm in diameter.*

The polymer chains of most fibers are held in alignment with one another either by hydrogen bonds or by strong dipole–dipole attractions. Chains in the polyamides (Nylons) and polyurethanes (Spandex) are hydrogen bonded to one other, whereas those in polyesters (Dacron) and polyacrylonitrile (Orlon, Acrilan) are held by strong dipole–dipole attractions. ► Figure 12.15 illustrates the two kinds.

Most fibers are drawn to give them strength.

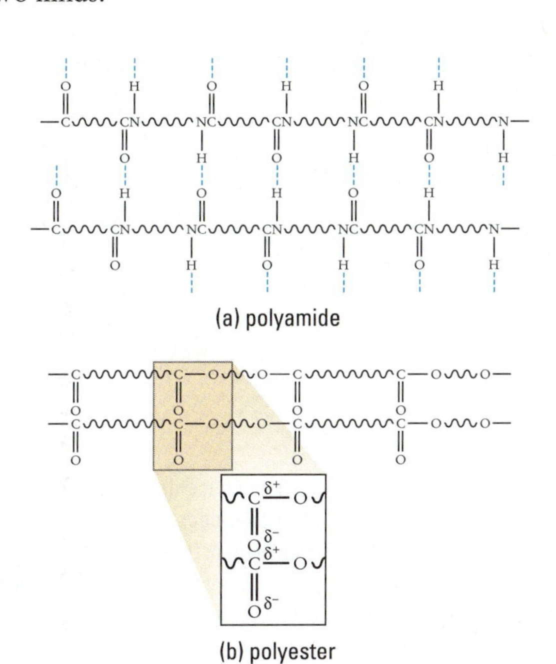

► Figure 12.15
Intermolecular Forces in Fibers. *(a) Polyamide chains are held in alignment by strong hydrogen bonds. (b) Polyester chains are held by dipole–dipole attractions (inset).*

MAKING A PERMANENT WAVE

Hair is a fiber with a structure much like wool (Figure. 12.11). and the polyamide chains in it are held in alignment with one another by three kinds of forces (Figure 1). One of these attractive forces is due to hydrogen bonding between the chains (illustrated in Figure 12.15 for Nylon). A second attractive force results from disulfide links, which crosslink the chains much like the sulfur bridges do in vulcanized rubber. The third attraction is from salt bridges, in which ammonio groups, —NH_3^+, on one chain and carboxylate groups —CO_2^-, on another chain are held together by electrostatic attraction. The combination of these forces is sufficiently strong to hold the dry hair fiber in a fairly well-defined shape: curly, not-so-curly, or straight.

hydrogen bonding | disulfide links | salt bridges

Figure 1

Wetting the hair fiber causes it to swell and become stretchable. Water molecules intrude between the hydrogen bonds and salt bridges that hold the polymer chains together. This separates the chains and allows them to rearrange when the hair fiber is stretched around a roller, "scrunched," or straightened. Drying the hair forces the water molecules from between the chains, and the hydrogen bonds and salt bridges reform, giving the hair fiber its new shape (Figure 2).

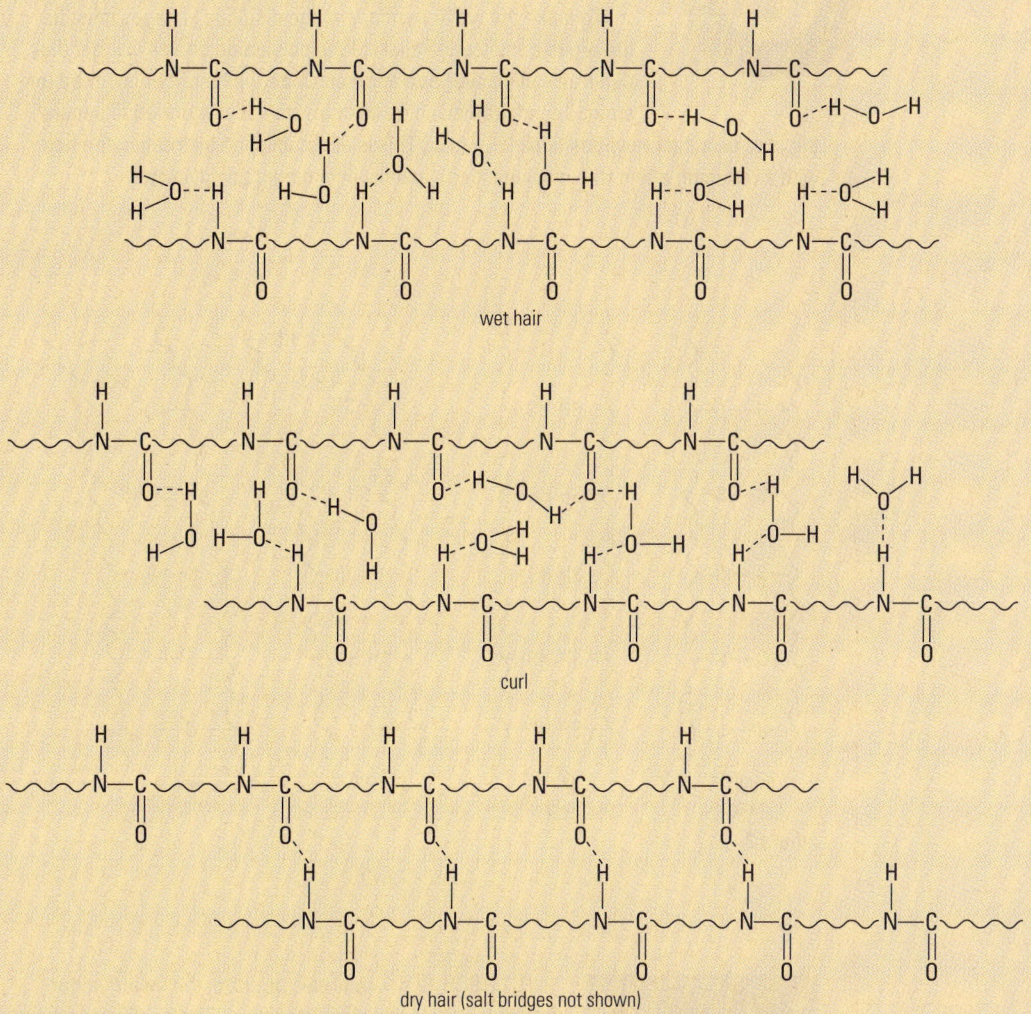

wet hair

curl

dry hair (salt bridges not shown)

Figure 2

This *wet wave* holds for a while, until the hair is washed or the humidity gets high, whereupon the hair reverts to its original shape. It does so because the disulfide linkages that crosslink the polyamide chains have not been broken in the wet wave process. These linkages behave just like the sulfur bridges in vulcanized rubber do; they pull the separated chains back into their original alignment.

MAKING A PERMANENT WAVE (Continued)

To make the wave *permanent,* the disulfide linkages must be broken to completely separate the polyamide chains, so that they can freely slide over one another. Once the hair fiber is curled (or straightened), the disulfide linkages must be reformed to secure the chains in their new positions (Figure 3). This is accomplished by treating the hair with a reducing agent (the waving lotion), usually ammonium thioglycolate

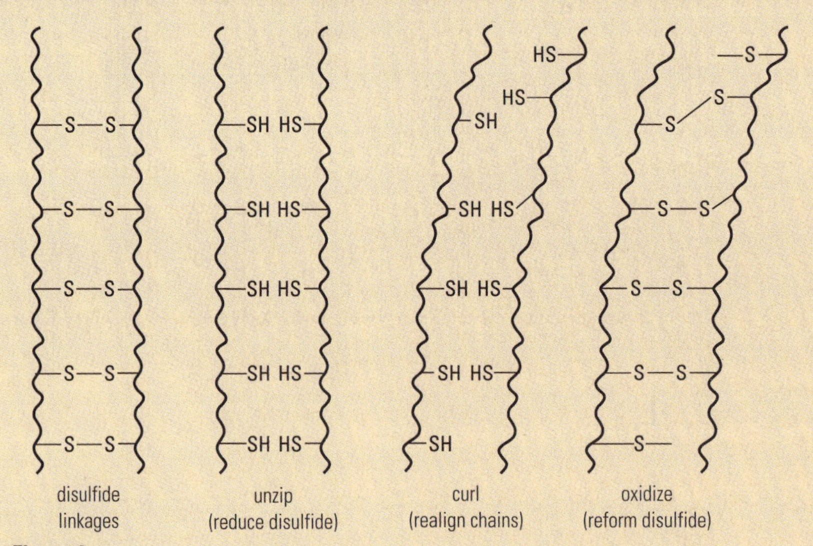

Figure 3

($NH_4^+ HSCH_2CO_2^-$), to break, or unzip, the sulfur–sulfur bonds that hold the chains together. The hair fiber is then set, and the sulfur–sulfur bonds are reformed using an oxidizing agent (the neutralizer) such as hydrogen peroxide (H_2O_2), potassium bromate ($KBrO_3$), or sodium perborate ($NaBO_2 \cdot H_2O_2 \cdot 3H_2O$). The newly formed covalent bonds do not break in the presence of water, so the hair fiber retains its new shape (Figure 4).

Figure 4
A permanent wave actually reforms the hair.

14. Nylon has been called synthetic wool. How is nylon chemically similar to wool and silk? How does it differ?
15. How does the structure of the polymers that make good fibers differ from the structure of polymers that make good elastomers? Would HDPE make a good fiber? Why or why not?
16. How does drawing a fiber alter its structure to increase its strength?
17. Identify the repeating unit(s) in each of the following condensation polymers. What kind is each (polyamide, etc.)?

a. $-CH_2-C_6H_{10}-CH_2OC(=O)-C_6H_4-C(=O)OCH_2-C_6H_{10}-CH_2OC(=O)-C_6H_4-C(=O)O-$

Kodel

b. $-O-C_6H_4-C(=O)O-C_6H_4-C(=O)O-C_6H_4-C(=O)O-C_6H_4-C(=O)O-C_6H_4-C(=O)-$

A-Tell

c. $-NHCH(CH_3)CH_2C(=O)NHCH(CH_3)CH_2C(=O)NHCH(CH_3)CH_2C(=O)NHCH(CH_3)CH_2C(=O)-$

d. $-C(=O)NH(CH_2)_6NHC(=O)OCH_2CH_2OC(=O)NH(CH_2)_6NHC(=O)OCH_2CH_2O-$

e. $-O-Si(C_6H_5)_2-O-Si(C_6H_5)_2-O-Si(C_6H_5)_2-O-Si(C_6H_5)_2-O-Si(C_6H_5)_2-O-Si(C_6H_5)_2-$

18. Explain why thermosetting resins are usually not plastic.
19. List the three basic components in a paint. How do oil-based paints differ from water-based points in terms of

a. solvent? b. kind of binder?

20. Write the chemical reaction that takes place when drying-oils dry. How does this reaction differ chemically from the drying of the binders in water-based paints?
21. What is the difference in chemical structure between starch and cellulose? How are they similar? How is each produced in nature? Which one is easily metabolized by the human body?
22. Write an equation to show how an alcohol is converted to a xanthate. What happens when the xanthate is treated with acid? Write an equation for this process. How does conversion of cotton to viscose, followed by its regeneration in acid, produce a better fiber?
23. Why has cellulose acetate replaced cellulose nitrate for use in photographic film?
24. What makes a cellulose nitrate fire so difficult to extinguish?
25. What is the chemical similarity between silicones and the silicas, such as quartz? How is the structure of silicones different from that of quartz? What useful properties do silicones have? What are they used for?
26. List three reasons that the manufacture of many polymers is highly energy intensive. Why does burning petroleum and natural gas for fuel compromise the availability of polymers in the future?

THINKING IT THROUGH

Tinting Paints in the Store

Many water- and oil-based architectural paints are supplied to stores as the *tint base,* essentially a plain white paint. When the paint is sold, pigments, or tints, are mixed with this base to produce the customer's choice of color. To simplify stocking and mixing of colors, these pigments are often suspended in propylene glycol to make them compatible with both water- and oil-based paints. What properties does propylene glycol have that would permit its use with both kinds of solvents? The structure of propylene glycol is $CH_3CH(OH)CH_2OH$.

Silly Putty

Silly Putty is a silicone polymer that can be molded like putty. If permitted to stand, it flows slowly into a puddle, like a thick liquid. Yet it bounces when thrown against a hard object, such as a wall or floor. What general kind of chemical structure do you think it could have that would give it these properties?

Biodegradable Plastics

Although natural polymers such as wool, silk, and cellulose are biodegradable (are decomposed in the natural environment), many synthetic polymers such as polyethylene, polypropylene, polystyrene, and Teflon are not. Instead, they remain essentially unchanged in the environment for long periods of time. What structural differences between the two kinds of polymers could account for this behavior? In what way could you change the structure of, say, polyethylene to make it biodegradable? How would this structural change affect the properties of the polymer, if at all?

Rubber Bands and the Second Law

Stretch a rubber band across your forehead or upper lip. Then, allow it to relax quickly and shorten. Try this again, and notice whether the rubber band changes temperature. Use the concepts of entropy and work to propose an explanation for what you observe.

ANSWERS TO "DID YOU LEARN THIS?"

a. vulcanization
b. cellulose
c. pigment
d. spinneret
e. drying oil
f. sulfur
g. copolymer
h. binder
i. petrochemicals
j. space-network
k. crosslinking
l. thermoplastic
m. macromolecule
n. rayon(s)

SOIL MICROBIOLOGY AND CHEMISTRY: A "NATURAL" ROMANCE

Diann Jordan

Soil Microbiologist

Assistant Professor, University of Missouri, Columbia

B. S. Biology/Chemistry, Tuskegee University

M. S. Plant and Soil Science, Alabama A&M University

Ph.D. Soil Science, Michigan State University

Although I did well in my high school chemistry class, I can't say it was my first love in college. The romance began much later. I had always liked *biology* better than chemistry. As I progressed through graduate school, I began to realize how important my background in chemistry was and how beneficial chemistry would be in my chosen career. Understanding chemistry is important in even the most routine lab jobs. For instance, a simple solution of alcohol and water is routinely used as a disinfectant in a typical microbiology laboratory. I use my knowledge of chemistry every day in the laboratory.

I took my first college chemistry course during a summer program for pre-med majors at Fisk University. I did well. The summer experience exposed students to general college chemistry and cell biology, offered a seminar course, and provided a chance to view medical researchers and doctors at a neighboring medical school hospital. Learning science during the summer was exciting. The program gave us a headstart on general chemistry in the fall and thus made general chemistry a lot easier. I especially enjoyed the laboratory exercises during that fall term.

However, the romance was not quite the same when I met organic chemistry. It was quite an ordeal but I survived. As I reflect back on the experience, it really wasn't as bad as it seemed at the time. I did a good job in the laboratory section but had difficulty integrating this information with the lecture.

Although I spent my childhood years on a small farm with my grandparents, I never really considered a career in the agricultural sciences until years later. After graduating from college, I worked for a year as a microbiology lab technician at a university. That experience taught me that I had good ideas, but if I wanted to put them into action I needed to further my education. During this time, I met a plant pathologist who was studying the effects of soilborne pathogens on plants. In addition to talking about his research, he began telling me about all the opportunities in the plant and soil sciences. After several discussions, I became even more convinced that quitting my job and going back to graduate school was the right decision. The next fall, I began graduate studies in soil science with an emphasis in microbiology.

Soil microbiologists study organisms that live in the soil, their metabolic activities, the role they play in the energy flow and the cycling of nutrients for plant production, and more recently the impact such organisms have on environmental problems (microbial degradation of pesticides or xenobiotics that may contaminate our drinking water supply). I have been particularly interested in the role of soil organisms in nutrient cycling of elements like nitrogen (N) and carbon (C), which are essential macro-nutrients for proper plant growth and development.

During my graduate studies, I became involved in a study on how farmers, particularly those in the tropics, could use leguminous plants in their cropping systems as a more economical source of N than fertilizer N (commercially produced N). Inorganic N fertilizer is very expensive for small farmers in the tropics. Therefore, finding alternative means of producing crops without inorganic N fertilizer is important for their economic survival, and the success of those crops is an important part of the farmers' livelihood. Legumes, in combination with bacteria *(Rhizobium sp.)* coexisting in a nodular structure on their root systems, are able to fix atmospheric N_2 to usable forms for plant growth and production. It is a symbiotic relationship in which both the plant and the bacterium benefit from the products that are being synthesized. The enzyme, nitrogenase, reduces N_2 molecules using reductants that originate from carbohydrates that are produced by the plant and are metabolized through the electron transport system. The ammonia, NH_3, produced is used by the plant for protein (amino acid) synthesis:

$$N{\equiv}N + 6H^+ \xrightarrow{\text{nitrogenase}} 2NH_3$$

This process can also occur chemically to a small extent, either by lightning discharges in the atmosphere or during the production of nitrogen fertilizers, which requires massive inputs of energy (petroleum).

Many legume plants grown during the cooler months are completely killed by herbicides shortly before the summer food crop is planted. The dead plant materials (commonly called plant residues) are left on the soil surface or are incorporated into the soil where they are decomposed by microbes. These plant residues may contain high amounts of organic N and will release this N to the next food crop as they decompose. Thus, legume residues are a source of fertilizer. As I began to understand the microbial and biochemical processes involved in the decomposition process, it suddenly occurred to me how neatly my organic chemistry studies fit into a very practical problem I was studying. In my organic chemistry course, we studied the processes in which organic compounds are broken down into inorganic products and the reactions involved in these processes. What I was studying was actually the same scenario, only in a more applied situation.

Corn needs potassium to grow.

Although I use many of the basic principles I learned in general chemistry every day in my lab, many aspects of chemistry are also important in understanding the broader soil/plant system. Now as a new college professor, I will encourage my students to study chemistry and take courses beyond those required in the basic curriculum. Even if they don't understand the immediate application, I know they will in time. Microbiologists really serve themselves well when they have a strong chemistry background. From making chemical solutions of nutrient media used in culturing microorganisms to understanding the detailed biochemical pathways of microorganisms, the two subjects are intertwined in a basic and complex way. Even though biology was and still is my first love, I also have an ongoing courtship with chemistry. Soil microbiology and chemistry—it's a natural romance.

13

Chemistry and the Environment: An Overview

What is man without the beasts?
If all the beasts were gone,
Men would die from a great
loneliness of spirit.
For whatever happens to the
beasts,
Soon happens to man.
All things are connected.

Attributed to Chief Seattle,
Suquamish Indian Tribe

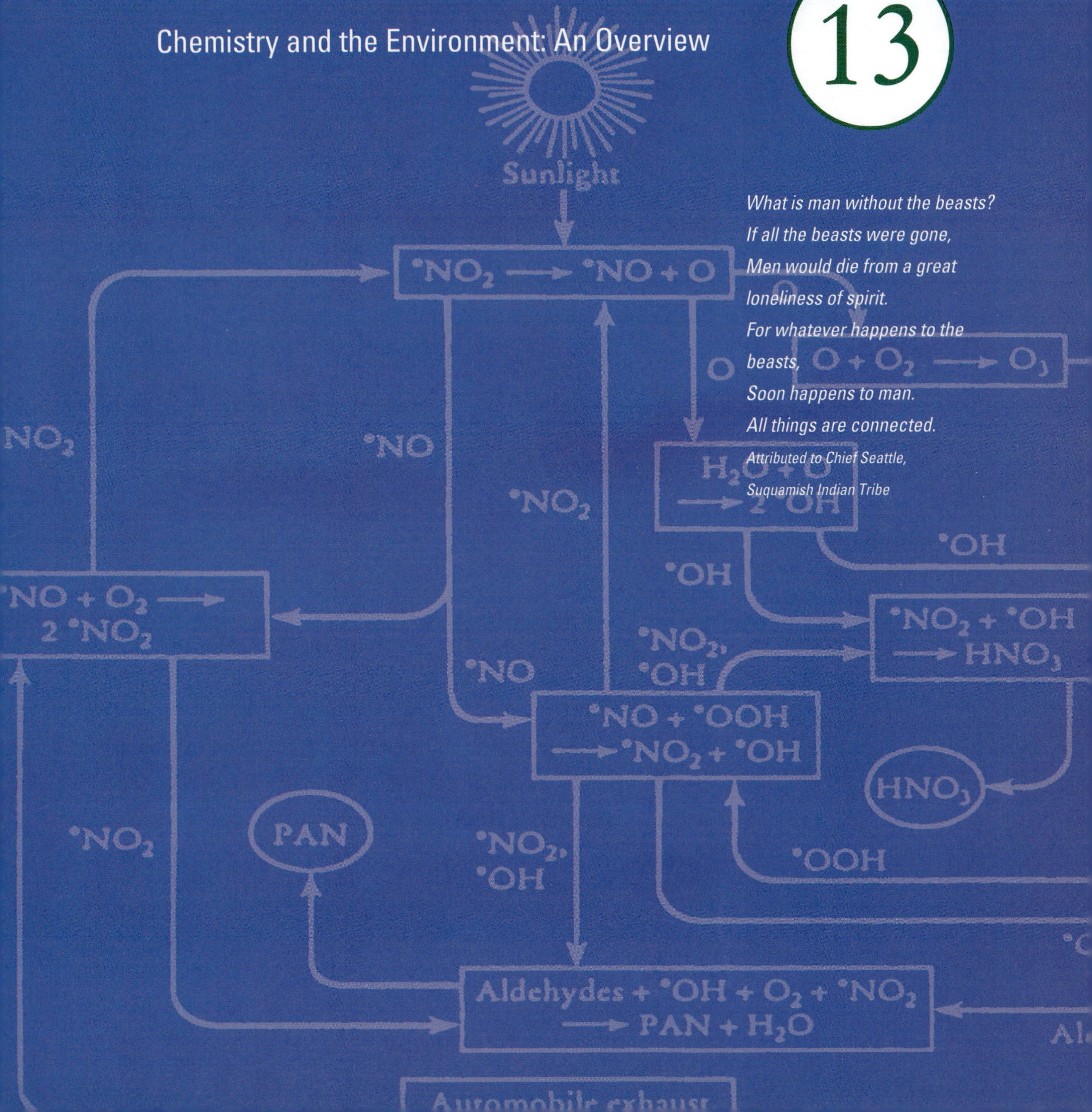

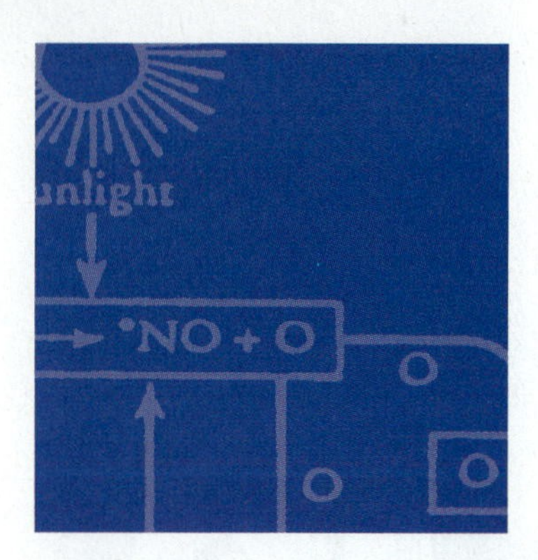

Human activity disturbs the environment and changes it. It has always been this way, and it always will be as long as humans exist. It cannot be avoided: we must change our environment to live.

Early humans gathered plants, hunted animals, and caught fish for food, and found shelter wherever they could. Later, they lived in caves or in dwellings of stone or clay, of animal skins or woven grasses, or of wood from nearby forests. They used fire to keep warm, cook their food, and forge metals. Importantly, they learned to farm, which freed many from subsistence living.

Serious environmental damage from human activity is not just a modern phenomenon. Many early cultures also suffered the consequences of overpopulation, overuse of natural resources, pollution, and destruction of their environments. For example, Plato (427–347 B.C.) wrote that the once-forested mountains and fertile plains of Greece had been "denuded" of vegetation. Their soils had washed away, leaving little land upon which to grow food, and there were no trees left to use for roofing timbers. He knew the importance of the mountain forests, which had once absorbed the rainfall and delivered it gradually to the lands below, and observed that the water was lost to the sea over now barren ground. The land in Greece is in very poor condition today.

Although most early cultures damaged or destroyed their environments without really knowing why, like Plato some recognized environmental relationships. The Roman Virgil noted that using fertilizers would restore infertile soil but lamented that pests and weeds were beyond human control. Han Fei-Tzu in China (ca. 500 B.C.) wrote that "The life of a nation depends upon having enough food to eat, not upon the number of people." Today we have at our disposal a large body of scientific knowledge, and theories and laws that work, which we can use to describe, understand, and care for our environment. Even so, the decisions we make will not be easy, because environmental systems are complex, interrelated, and sometimes not well understood. How will we make these decisions?

RISKS VERSUS BENEFITS

Human activity changes the environment. Today large numbers of us occupy almost every corner of the globe. Our technology is powerful and its use is widespread. Thus, our power to change the environment—for the good or otherwise—is very great. We would, of course, like to have all the *benefits* (the positive results of technology) and none of the *risks* (the negative effects of that technology), but this is not possible. There is no such thing as a risk-free human endeavor. Instead, we must weigh the anticipated risks of any activity against its potential benefits to determine whether it is worthwhile. This is called **risk–benefit analysis.**

Risk–benefit analyses can help us make environmental decisions.

Deciding what the risks and benefits are, and what is worthwhile, depends on a number of factors. For example, how great will the benefits of an activity be and how severe will be its risks? If the anticipated benefits are high and the risks low, an activity may be worthwhile; but if its benefits are low and its risks are high, we may not want to pursue it. Aspirin is a useful drug for relieving aches and pains, reducing the likelihood of heart attacks and strokes, and reducing fevers. It is not totally safe, however, and can cause allergic reactions or mild stomach irritation in some people, and it is sufficiently toxic that an overdose can cause poisoning, particularly in children. It can also contribute to the onset of a potentially fatal disease, Reye's syndrome, in children with flu or chicken pox. The risks of using aspirin can for the most part be avoided by intelligent use, and they are very small compared with its benefits. It has been the most widely used pain reliever for almost a century (since 1899). On the other hand, the use of anabolic (muscle-building) steroids by athletes to increase strength and endurance in competition has risks that far outweigh this benefit. In men the use of these steroids can cause changes in sexual desire, atrophy (shrinking) of the testicles, enlargement of the breasts, baldness, and decreased sperm production. In women they can cause male-pattern baldness, changes in the menstrual cycle, deepening of the voice, and growth of facial hair. Worse, anabolic steroids are toxic to the liver, and there is some evidence that they can cause liver cancer. Thus, the small "edge" in competition gained by using steroids is far offset by the deleterious effects they can cause.

EASTER ISLAND

In about 400 A.D., fifty Polynesians arrived by canoe at Easter Island in the southern Pacific Ocean. This island has an area of only 64 square miles, and at that time it had no land mammals, only about 30 plant species, and scarce fresh water. Forested with giant palm trees, it was completely isolated from the outside world. Yet in this closed system the Polynesians thrived, gradually clearing the forest to plant crops they had brought with them in their canoes, and catching the abundant fish offshore. With their prosperity they could afford to carve, over centuries, more than 800 of the now-famous giant stone statues, called *moai,* which stare outward from the island's shores (Figure 1). It was an impressive feat. The *moai,* which stand as high as 40 feet and weigh as much as 50 tons, were hewn from soft volcanic rock (called tuff) in an inland quarry. Once cut, they were hauled over land to their places all around the island by hundreds of people using wooden rollers and sledges. Some statues even were lowered over cliffs with ropes woven from tree bark.

But the Polynesians eventually overused and destroyed their environment. Their numbers grew and eventually reached about 7,000, too many people for the island to support. Rival clans divided the island into territories and competed bitterly with one another, not only for available resources, but in carving bigger and better *moai.* The forest disappeared, and it became harder to grow food. By about the year 1500 the clans had fallen into open warfare and, because food was scarce, desperate cannibalism. Today, Easter Island is barren of trees and agriculture is difficult. About 2,800 people live on the island, its economy supported largely by tourism.

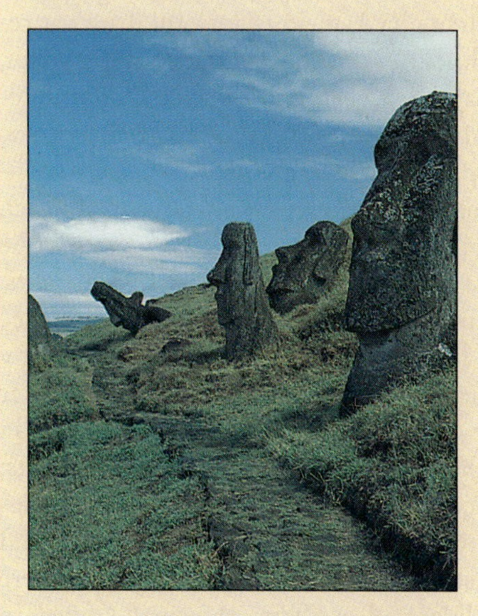

Figure 1
MOAI

Other considerations relate to who will benefit from an activity and who will take the risks. Will those who benefit also take the risks, or will someone else? Will a large number of people benefit (or be at risk) or a small number? Making decisions in these cases may not be straightforward. A company applied for a state permit to build a medical waste incinerator on the Snake River near American Falls, Idaho. Large quantities of hazardous medical waste, from hospitals and other medical facilities serving populations along the U.S. west coast, were to be shipped there and burned at high temperatures. This would convert the waste into harmless gases and ash, and there would be no pollution. Residents were urged to support the project because it would provide jobs and thereby help the local economy. However, a group of citizens who had some understanding of science examined the project closely and found that these claims were not true. The incinerator would indeed cause unacceptable air pollution, particularly by emitting mercury and other heavy metals. The nearby Snake River is winter host to a small population of bald eagles and is a popular water recreational area, which provides the town with some of its income. In addition, the safeguards for handling and destroying the dangerous biological material were found to be inadequate, and so were those for disposal of the ash, which also would contain heavy metals. Last, it was discovered that the company could not have obtained permits to build its incinerator in the states from which much of the waste was to come. What was of benefit to millions of people on the west coast and to the company's investors was a risk the residents of American Falls did not want to take. The State of Idaho agreed that the project was not worthwhile and denied the permit.

Finally, people often take a present benefit and pass the risk on to future generations. An excellent example is our gluttonous consumption of petroleum (Chapter 10). What will be the source of important petrochemicals in the year 2100? Similarly, the generation of electrical power with nuclear fuels produces radioactive wastes, some of which have lifetimes of thousands of years. Improper treatment and containment of these materials could present dangers far into the future. Many of the environmental problems that we face today are of this kind: the possibility of human-caused global warming;

13.1 Does a high-risk/low-benefit situation mean that a certain course of action should never be taken? Under what circumstances might it be justified to undertake a high-risk activity even though the anticipated benefits might be small? Give an example.

continued destruction of thousands of species of plants and animals; deforestation; irreversible erosion of farmable land; pollution of air, water, and land with toxic substances; and the waste of natural resources.

If we believe that future generations have a right to live in a decent environment, we must not repeat these kinds of mistakes. We also must try to rectify those already made, if it makes sense in terms of the resources committed to do so. The decision of what is of benefit to us is a social one, not one of science, and so is the determination of acceptable risk. However, scientific knowledge *can* help us immensely in evaluating these risks, and often can provide us with options for intelligent action.

In the following sections, some representative environmental concerns are presented in overview to show how science helps us understand their nature. Because it is the author's belief that a problem cannot be solved until it is understood (and because space in a book is limited), emphasis has been placed on the problems themselves rather than on ways of solving them. (There *are* ways to address them.) What follows is intended to provide points of departure for class discussions, for thinking in perspective about the environmental problems we face, and for evaluating proposed mitigation, remediation, and prevention strategies in a critical and well-informed manner.

POLLUTANTS

"The dose makes the poison."

It has been said that "the dose makes the poison." Or, *all* substances are poisonous if taken in large enough quantity and *none* are if their concentration is low enough. Most people do not consider table salt (NaCl) to be dangerous, yet ingesting even moderate amounts of it can be life-threatening to persons with high blood pressure or heart disease. A large number of living organisms thrive in the oceans, which are slightly salty, but few can survive in the highly saline waters of the Dead Sea, Mono Lake, or the Great Salt Lake.

So it is with pollutants. A **pollutant** is a substance, placed in the environment by human activity in greater than natural amounts, that has a detrimental effect on that environment or on something of value within it. Phosphates, which are essential nutrients for plants, are found dissolved in low concentrations in almost all natural bodies of water and are present in most soils. Introduced in large quantities into watercourses by runoff from overfertilized

13.2 Salt-cured foods, such as salted meat and fish, are resistant to spoilage. How does salt work to make it a good preservative?

agricultural lands, or from detergents in domestic wastewater, they become pollutants that stimulate excessive growth of algae, choking out other forms of aquatic life. **Contaminants** are somewhat different. Although they change the normal makeup of the environment, they do not appear to harm it like pollutants do.

AIR, EARTH, WATER, AND LIFE

Our physical environment consists of four basic parts: the atmosphere, the lithosphere, the hydrosphere, and the biosphere. The blanket of gases that surrounds the earth is the **atmosphere.** It is our gaseous environment. The atmosphere protects life on earth from the harsh conditions of outer space and provides carbon dioxide (used by plants for photosynthesis), oxygen (used by animals and plants for respiration), and nitrogen (converted by bacteria to nutrients for plants via *nitrogen fixation*). Air also serves as a medium for the transport of water, as rain and snow, from the oceans to the land.

The **lithosphere** is the thin outer shell of the solid earth, a layer of rock about 100 km thick that is the solid environment we live on. It includes the earth's crust (5–40 km thick) and the upper part of the mantle, and it is made up of various minerals and soils.

The lithosphere is also the "container" for the **hydrosphere,** our liquid environment. This is the system of waters that covers the earth, including lakes and oceans, rivers and groundwater, rain, snow, glaciers, and polar ice caps.

The **biosphere** is our living environment, the part of our environment that supports life (► Figure 13.1) It exists where air (the atmosphere), land (the lithosphere) and water (the hydrosphere) meet one another, and it reaches from the top of the highest mountains to the deepest oceans. (At these extremes, however, life is sparse because conditions are severe and opportunities for survival are poor.) With the exception of sunlight as an external source of energy, the biosphere is a **closed system.** That is, no materials can be brought into it from the outside and *none can be taken out.* This means that everything necessary for life must be recycled, over and over. It also means that when pollutants are introduced into the biosphere they become a part of it. We cannot throw our garbage away.

The **biosphere** is a **closed system.**

These classifications are made for convenience in talking about our environment, but in reality they are closely interrelated. For example, we have seen (Chapter 8) that the oceans, which are part of the hydrosphere, assist in removing carbon dioxide, a suspected greenhouse gas, from the atmosphere. Marine animals, part of the biosphere, use dissolved carbonates to form their shells, which become part of the lithosphere. Green vegetation, another part of the biosphere, also removes carbon dioxide from the atmosphere and replaces it with oxygen during photosynthesis (Chapter 9).

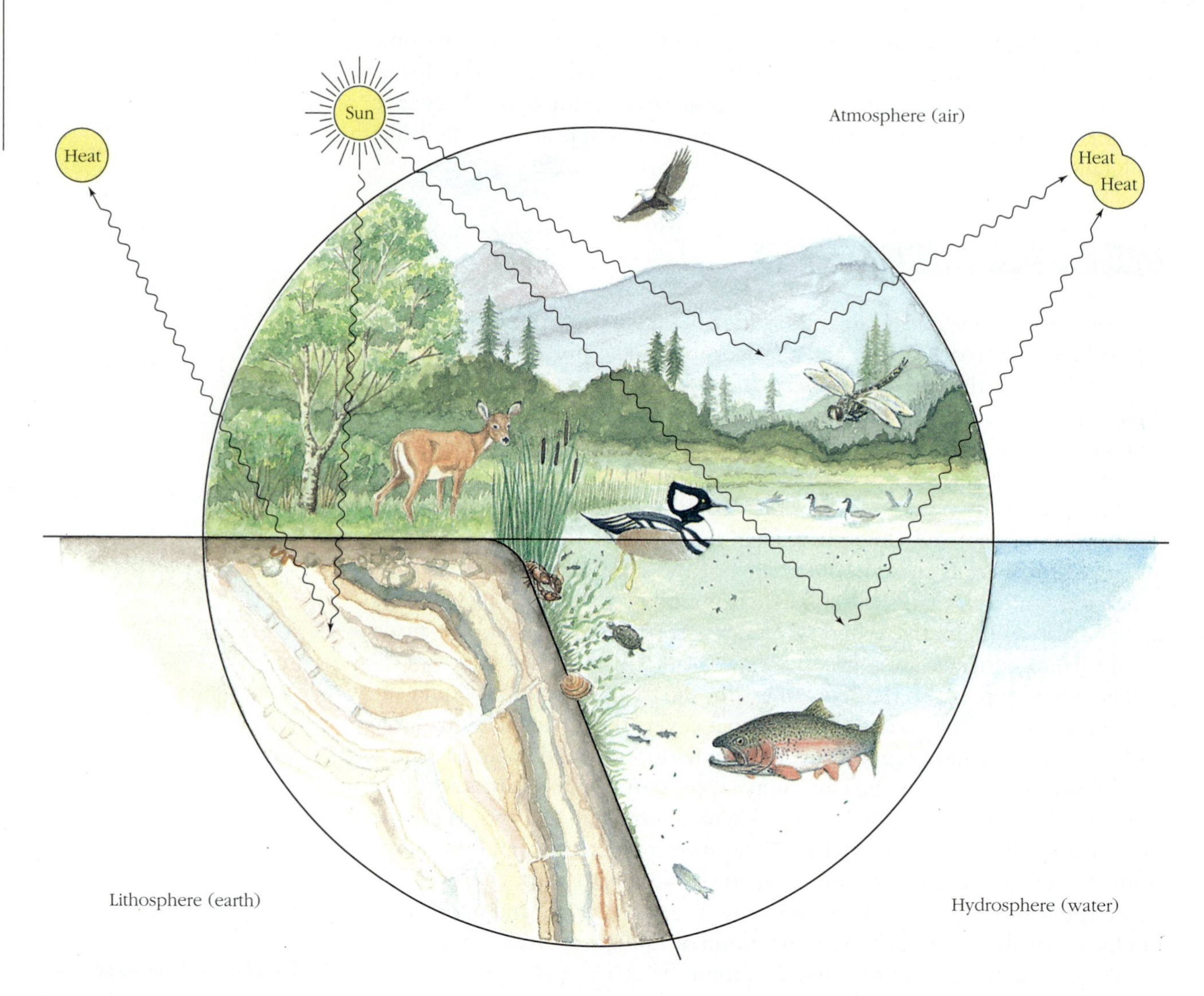

► Figure 13.1
The Biosphere. *Life exists at the interface of the lithosphere, the atmosphere, and the hydrosphere. It is essentially a closed system. Nothing can enter except sunlight, and nothing can leave except heat.*

For hundreds of millions of years, life on Earth has adapted to changes in its environment caused by natural events such as floods, droughts, ice ages, forest and grassland fires sparked by lightning, volcanic eruptions—possibly even impact by asteroids—and has continued to evolve. Today, at the end of this progression so far, large numbers of humans are creating far greater environmental disruption than was caused by any of these natural disasters of the past. We are altering the atmosphere, the lithosphere, and the hydrosphere at an unprecedented rate, primarily by dumping enormous amounts of chemicals into them. In this section we shall examine the chemistry of some well-known and representative examples of chemical pollution in each of these realms, and conclude with a short discussion of the implications they have for the biosphere.

THE ATMOSPHERE

Burning fossil fuels is the primary source of atmospheric pollution.

The primary source of atmospheric pollution by far is our combustion of large quantities of fossil fuels: burning coal, fuel oil, and natural gas to produce electricity, drive industrial processes and to heat homes, and using gasoline and diesel fuel to power transportation and agricultural machinery. Combustion of hydrocarbons produces primarily carbon dioxide and water as gaseous products (Chapters 9 and 11). However, petroleum and coal contain small amounts of impurities, the most important of which are sulfur compounds. These produce sulfur dioxide, SO_2, when the fuel is burned (▶ Figure 13.2). Sulfur compounds are usually removed when petroleum is refined into transportation fuels and heating oil, but it is expensive in terms of money and energy to remove these compounds from coal. Another option is to remove the sulfur dioxide from the combustion gases of coal- and oil-fired processes, as in electric power plants. The technology exists to do this, but it also costs money and energy. Thus, a considerable amount of sulfur dioxide is allowed to escape into the atmosphere; it is a major "acid precursor," which forms acid rain, discussed in Chapter 8.

▶ Figure 13.2
Sulfur Awaiting Shipment on a Dock at Vancouver, British Columbia. *The sulfur, a contaminant in natural gas produced in Alberta, is removed so as to provide gas that will not produce sulfur oxides when it is burned. Sale of the sulfur as a raw material makes the process economically feasible.*

Photochemical Oxidative Smog

Transportation fuels contain only small amounts of sulfur compounds and therefore do not produce much SO_2 when they are burned in gasoline, diesel, and jet engines. However, combustion in these engines takes place at very high temperatures. This causes the *nitrogen* in the air used for combustion to be oxidized to nitric oxide (nitrogen monoxide).

$$\underset{\text{nitrogen}}{N_2} + \underset{\text{oxygen}}{O_2} \longrightarrow \underset{\text{nitric oxide}}{2NO}$$

In the atmosphere, nitric oxide reacts further with oxygen to produce nitrogen dioxide, NO_2, a red-brown gas that is toxic and irritating to eyes and lungs,

$$\underset{\text{nitric oxide}}{2NO} + \underset{\text{oxygen}}{O_2} \longrightarrow \underset{\text{nitrogen dioxide}}{2NO_2}$$

Nitric oxide and nitrogen dioxide are molecules that contain an *odd* number of valence electrons, eleven and seventeen, respectively:

$$\underset{\substack{\text{nitric oxide} \\ \text{11 valence electrons}}}{:\dot{N}::\ddot{O}} \qquad \underset{\substack{\text{nitrogen dioxide} \\ \text{17 valence electrons}}}{:\ddot{O}:\dot{N}::\ddot{O}}$$

Free radicals have unpaired electrons.

Note that in each molecule the nitrogen atom bears an *unpaired electron*, represented by the single dot. Chemical species (molecules, fragments of molecules, and atoms) that contain unpaired electrons are called **free radicals,** and most of them are quite reactive chemically. When a free radical reacts with a molecule having an even number of valence electrons, it usually produces a new free radical and some kind of molecular species with an even number of valence electrons. The new free radical can react similarly to yield another free radical, and so on, in a continuing process. This is called a **radical chain reaction.** Once *initiated* (started) it can repeat itself over and over many times before two free radicals combine and the chain is broken, or terminated.

Nitrogen oxides contribute to **photochemical oxidative smog.**

In the presence of sunlight $\cdot NO_2$ also reacts with oxygen and other air pollutants to form what is known as **photochemical oxidative smog,** or simply photochemical smog, the yellow-brown haze that blankets many large modern cities during daytime (► Figure 13.3). The color of this haze is due primarily to $\cdot NO_2$. Photochemical smog is a soup of many compounds formed by simultaneous free-radical reactions that are initiated by sunlight. Most of them do not proceed to any extent in the dark. This model explains why

13.3 Indicate which of the following chemical species are free radicals.

a. $:\ddot{\underset{..}{Cl}}:\ddot{\underset{..}{Cl}}:$

b. $H:\ddot{\underset{..}{S}}\cdot$

c. $:\ddot{\underset{..}{Br}}\cdot$

d. $:\ddot{\underset{..}{Cl}}:^-$

e. $\cdot\ddot{\underset{..}{O}}:\ddot{\underset{..}{O}}:H$

f. $:N:::N:$

g. $:\ddot{\underset{..}{O}}:H^-$

h. $H:\ddot{\underset{..}{O}}:\ddot{N}:\ddot{\underset{..}{O}}:$ (with $:O:$ bonded to N)

i. $H:\ddot{\underset{..}{C}}\cdot$ (with H above and below C)

j. $H\cdot$

smog forms during daylight hours, usually beginning about midmorning, and dissipates at night.

One of the most important photochemical reactions is the formation of *ozone*, O_3, a pungent gas that is a strong oxidant and a respiratory irritant. Not only can it be a threat to health, but it also causes economic loss by attacking rubber goods, causing them to harden and crack, and by damaging crops (► Figure 13.4, p. 482). The odor of ozone is familiar: it is produced by sparking electrical devices and by lightning. Formation of ozone in smog is

► **Figure 13.3**
Photochemical Oxidative Smog Over Mexico City. *The yellow-brown color of the smog is due. primarily to the presence of* $^{\bullet}NO_2$.

IONIZING RADIATION

Some Risks and Benefits

The spontaneous decay of radioactive atoms produces three kinds of energetic radiation: α-particles, β-particles, and γ-rays (alpha particles, beta particles, and gamma rays). Alpha particles, being helium ions ($^4_2He^{2+}$), have a high charge and mass (for particles of this kind) and move relatively slowly (about one-tenth the velocity of light), so they interact strongly with matter in their path. As a result, they have a low penetrating power (Figure 1a) and are easily stopped by a few sheets of paper, a sheet of aluminum foil, clothing, or the outer layer of skin.

Beta particles are high-speed particles that are produced by radioactive decay and that are identical to electrons. Depending on their source, they have a wide range of energies. Because beta particles have a negative charge, they also interact strongly with matter when they strike it; but being lighter, faster (about the velocity of light), and of unit charge they usually penetrate more deeply than do alpha particles (Figure 1b). High-energy beta particles can penetrate clothing and 1–2 millimeters of skin, but are stopped by a 4-mm sheet of plastic or a thin piece of metal.

Gamma rays consist of high-energy electromagnetic radiation similar to X-rays, but gamma rays have even shorter wavelengths and hence more energy. Because they have no charge or mass, they can penetrate very deeply into matter before they have given up all of their energy. For example, gamma rays can pass through the human body, several centimeters of aluminum, or even a thick concrete wall. On the average, a 3-cm sheet of lead will stop 90% of the gamma rays in a beam (Figure 1c); a much thicker sheet of lead is needed to stop them all. Gamma rays are particularly dangerous because they can pass through a large amount of matter while carrying a considerable amount of energy. Although they do not interact with matter as readily as alpha

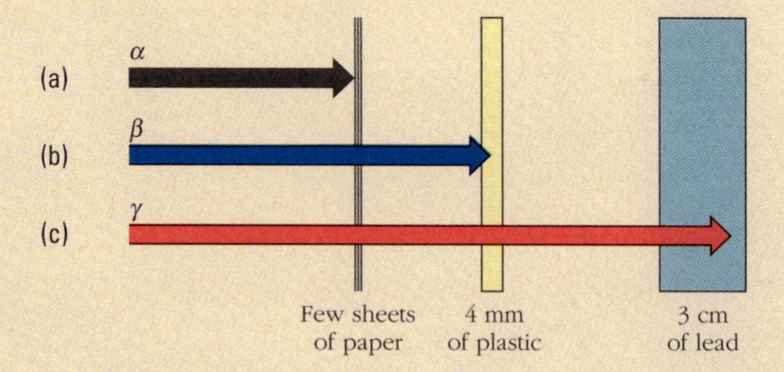

Figure 1
Penetrating Ability of Alpha, Beta, and Gamma Radiation.
(a) Alpha radiation penetrates least and is absorbed by several sheets of paper or clothing. (b) High-energy beta radiation penetrates the paper but is stopped by a sheet of plastic or a thin sheet of metal. (c) Gamma radiation is the most penetrating. A 3-cm sheet of lead absorbs about 90% of the gamma particles in a beam.

and beta particles do, they leave long trails of ions and fragments of molecules behind them.

Because alpha particles, beta particles, and gamma rays interact with matter to form ions, they are classified as ionizing radiation. Radiation having longer wavelengths (infrared, microwave, visible, and ultraviolet) usually does not form ions when it interacts with matter.

Alpha particles are actually quite energetic (more so than beta particles); but because they interact strongly with matter, they rapidly expend their energy at the surface and form ions. For this reason, alpha particles are not harmful to the outside of the body because clothing or an outer dead layer of skin stops them. When introduced into the body, however, by eating or breathing or through a wound, they form ions that come in direct contact with cells, causing considerable damage. For example, radon, found in homes in many parts of the United States, is a colorless, odorless, radioactive gas that emits alpha particles when it decays. Breathing air containing radon thus deposits alpha particles directly in sensitive lung tissue. In nonsmokers, radon is believed to be the major cause of lung cancer; in smokers, it multiplies the risk of lung cancer by a factor of ten.

Because beta particles penetrate clothing and skin better than do alpha particles, they can cause severe burns to tissues on the outside of the body. Inside the body, however, they do less damage than alpha particles. Exposure to beta particles can also cause cataracts and skin cancer.

Gamma rays typically produce ions from atoms and molecules by knocking electrons from them. For example, a molecule, M, can interact with a gamma ray to form an ion, $\cdot M^+$, and an electron,

$$M \xrightarrow{\text{gamma ray}} \cdot M^+ + e^-$$

Because molecules usually have an even number of valence electrons, removing an electron in this manner produces an ion that is also a free radical; that is, it is a chemical species having an odd number of electrons. This radical cation ($\cdot M^+$) is not only an oxidizing agent that can cause undesirable redox reactions to occur in the body, but also it can decompose on its own, destroying the molecule M. Ionizing radiation can also cause the rupture of covalent bonds in molecules, leading to the degradation of their structures. In any case, molecules necessary for proper function of living cells are destroyed, and the cells are damaged or killed. Thus, exposure to high levels of ionizing radiation can lead to radiation sickness or death. Lesser amounts of exposure can cause cancer and genetic defects.

Although ionizing radiation can be a threat to health, it can also be very useful. The penetrating ability of gamma rays (like X-rays) makes them useful for medical diagnosis, as well as for radiation therapy, whereby gamma rays are focused on a tumor or cancer inside the body, destroying it but leaving the surrounding tissue undamaged. Gamma rays were used to eliminate the screwworm (whose larvae feed on the live flesh of domestic and wild animals) in the southern United States in 1966. A large

IONIZING RADIATION (Continued)

number of male screwworms was sterilized by radiation and released to breed with fertile females in the wild. Most of the females mated with the sterile males, which were in large proportion, and thus they produced no offspring. Because ionizing radiation kills bacteria, molds, yeasts, and insects, it is also used to sterilize medical supplies and other health care products, to treat sewage, to kill insects in harvested grain, and to preserve various kinds of food.

Gamma radiation from the radioactive decay of cobalt-60 or cesium-137 is used to irradiate various kinds of foods to prevent their spoilage (Figure 2). Irradiation can be used to inhibit the sprouting of onions and potatoes; to kill insects on fruits, vegetables, and grains; to retard microbial spoilage of many fresh foods by reducing the levels of bacteria, molds, and yeasts; to sterilize food packaging materials; and to sterilize a wide variety of cooked or prepared meat, fish, and poultry products to make them keep longer.

Although foods irradiated with gamma rays *are not made radioactive,* their chemical composition is modified somewhat. (Other methods of preservation also change the chemical composition of foods, such as canning and smoking.) The compounds formed in foods by the irradiation process are called radiolytic products, and although about 90% of these compounds are also found in foods that have not been irradiated, about 10% of them appear to be new. These compounds, called unique radiolytic products (URP), may present no hazard, but research to determine their safety continues.

The Food and Drug Administration (FDA) has approved food irradiation on a low-dose basis to replace postharvest chemical fumigation of some foods, to kill *Salmo-*

initiated by **photodissociation** (dissociation caused by light) of nitrogen dioxide, which produces nitric oxide and a highly reactive oxygen atom.

$$\underset{\text{nitrogen dioxide}}{\cdot NO_2} \xrightarrow[\text{photodissociation}]{\text{sunlight}} \underset{\text{nitric oxide}}{\cdot NO} + \underset{\text{oxygen atom}}{O}$$

The oxygen atom then combines with an oxygen molecule in the atmosphere, forming ozone:

$$\underset{\text{oxygen atom}}{O} + \underset{\text{oxygen molecule}}{O_2} \longrightarrow \underset{\text{ozone}}{O_3}$$

▶ Figure 13.4
A Tire Cracked by Atmospheric Ozone. *The ozone formed in photochemical smog attacks items made of rubber, causing them to deteriorate.*

nella bacteria in poultry, and to kill the parasitic worms *Trichinella* in fresh pork. It also permits high-dose treatment of teas, dried spices, and other seasonings to rid them of microorganisms and insects. Labels on foods that have been treated with radiation must carry a special international symbol and a statement that they have been irradiated. However, labels on commercially prepared foods that contain irradiated ingredients, such as spices, need not be so marked.

Whether irradiation of foods will gain widespread use is still uncertain. Aside from concerns about the formation of unique radiolytic products, many people have strong

Figure 2
These mushrooms were gathered at the same time. The mushrooms on the left were preserved by radiation.

negative feelings about radiation in general. Also, irradiation technology requires the use of radioisotopes, the use of which must be carefully monitored, not only to provide safety for workers but also to prevent escape of radioactive material into the environment during transport and use, and later during storage as waste. Some people argue that there are other safe, cost-effective ways to keep foods from spoiling, and that the risks of irradiating food outweigh the benefits it provides.

Oxygen atoms also can react with water in the atmosphere to form hydroxyl free radicals, $\cdot OH$:

$$H_2O + O \longrightarrow \underset{\text{hydroxyl free radicals}}{2\ \cdot OH}$$

These hydroxyl radicals can react with hydrocarbons in the atmosphere. Besides $\cdot NO$, engine exhaust also contains carbon monoxide and unburned hydrocarbons, emitted as a result of incomplete combustion. Hydrocarbons also enter the atmosphere from filling gas tanks, from using solvents such as “mineral spirits” or “naphtha,” from petroleum refining, from aerosol sprays and oil-based paints, from vegetation, and from decomposition of organic matter in the absence of air (anaerobic decomposition).

Hydroxyl free radicals react with these hydrocarbons and oxygen in the air, forming more ozone, hydroperoxide radicals, and aldehydes and ketones:

$$\text{Hydrocarbons} + {}^{\bullet}OH + O_2 \longrightarrow O_3 + \underbrace{{}^{\bullet}OOH}_{\text{hydroperoxide radical}} + H_2O + \text{aldehydes and ketones}$$

(equation is unbalanced)

The hydroperoxide radicals subsequently react with nitric oxide to produce more nitrogen dioxide and hydroxyl radicals, in what is called a *chain-propagation step* that starts the cycle over again,

$${}^{\bullet}NO + {}^{\bullet}OOH \longrightarrow {}^{\bullet}NO_2 + {}^{\bullet}OH$$

Hydrocarbons contribute to smog formation, producing aldehydes, ketones, and **PAN**.

Many of the aldehydes and ketones formed are irritants to eyes and mucous membranes. The aldehydes also react further with hydroxyl radicals, oxygen, and $^{\bullet}NO_2$ to form **peroxyacylnitrates (PAN),** which, like ozone, are strong oxidizing agents that are irritating to the respiratory tract. The overall reaction by which they are formed is

$$\underset{\text{an aldehyde}}{RCH(=O)} + {}^{\bullet}OH + O_2 + {}^{\bullet}NO_2 \longrightarrow \underset{\text{a peroxyacylnitrate (PAN)}}{RC(=O)\text{—}O\text{—}O\text{—}NO_2} + H_2O$$

Peroxyacylnitrates are sufficiently stable to migrate from smoggy areas into unpolluted regions before they decompose into $^{\bullet}NO_2$ and other free-radical products. Thus, they serve to export smog damage some distance from its source.

Like SO_2, nitrogen dioxide is an acid precursor and reacts in a number of ways in the atmosphere to produce nitric acid, HNO_3, a component of acid rain. One pathway to nitric acid is the reaction of nitrogen dioxide with hydroxyl radicals,

$${}^{\bullet}NO_2 + {}^{\bullet}OH \longrightarrow HNO_3$$

Note that when two free radicals combine in this manner they form a molecule in which all the electrons are paired. This process, called a *termination step,* removes free radicals from the atmosphere and serves to stop other free-radical chain processes that are in progress. The formation of PAN is also a reaction of this kind, except that it is reversible. ▶ Figure 13.5 illustrates the connections among the primary photochemical reactions in the atmosphere that cause photochemical smog.

Limiting the formation of photochemical smog in the atmosphere requires reducing the quantities of precursor molecules that form it, primarily $^{\bullet}NO$,

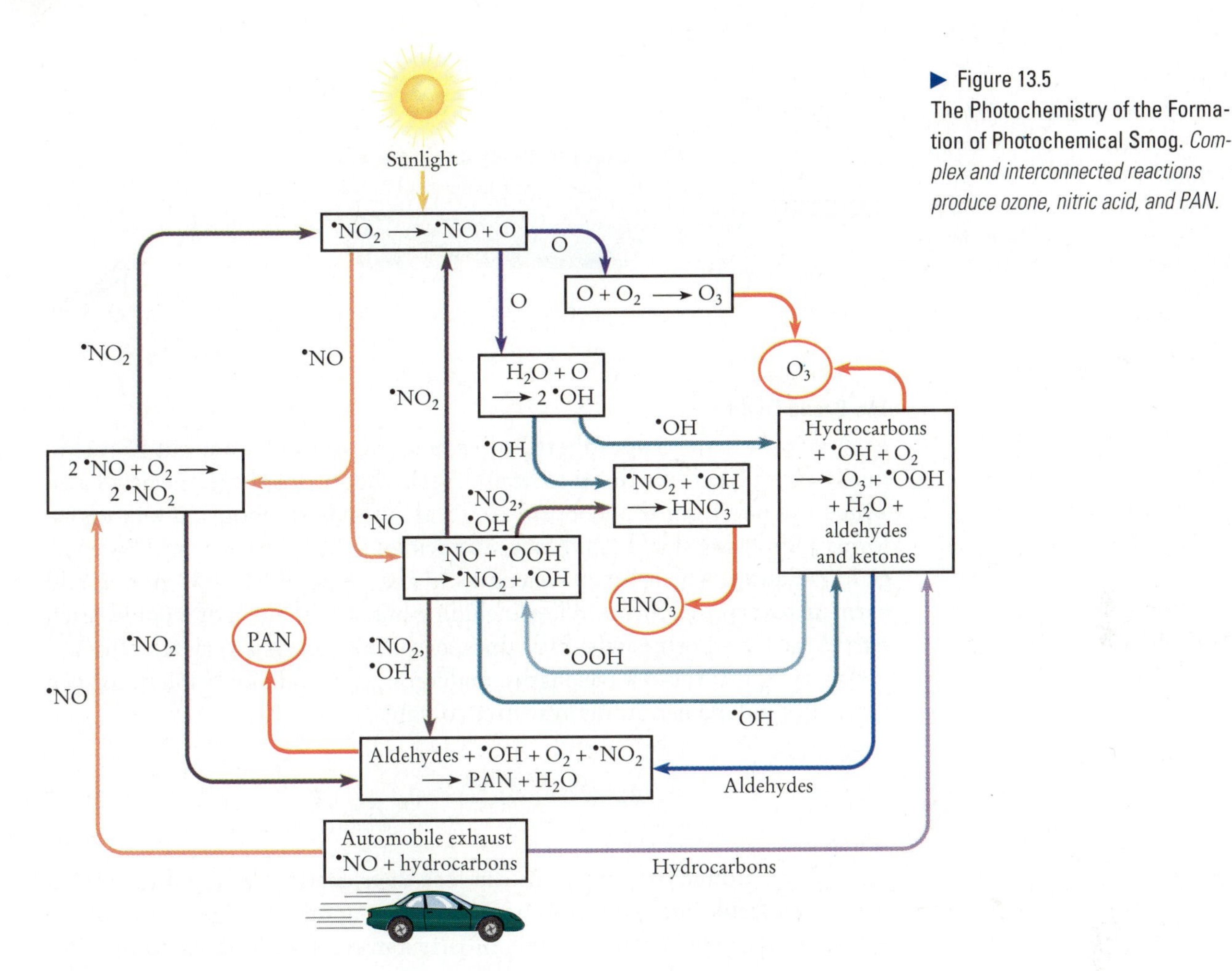

▶ Figure 13.5
The Photochemistry of the Formation of Photochemical Smog. *Complex and interconnected reactions produce ozone, nitric acid, and PAN.*

CO, and hydrocarbons. Automobile emissions are the primary source of these compounds, and efforts to control them began in the early 1970s. Strategies that have met with some success include (a) increasing fuel efficiency with improved engines and aerodynamic design (thus reducing emissions per mile driven), (b) using catalytic converters to reduce emissions of hydrocarbons, •NO, and CO in exhaust (▶ Figure 13.6), (c) limiting hydrocarbon losses with closed fuel systems, and (d) reformulating fuels with additives such as methanol or MTBE to reduce the amount of CO formed in exhaust. Although the number of automobiles in use today is much greater than in 1975, average amounts of NO_x (nitrogen oxides) in urban air are about the same. (About 40% of the NO_x is introduced by vehicles and the rest by electric power generating plants.) Carbon monoxide and ozone levels have decreased by about 40% and 22%, respectively. These are good signs, but more needs to be done.

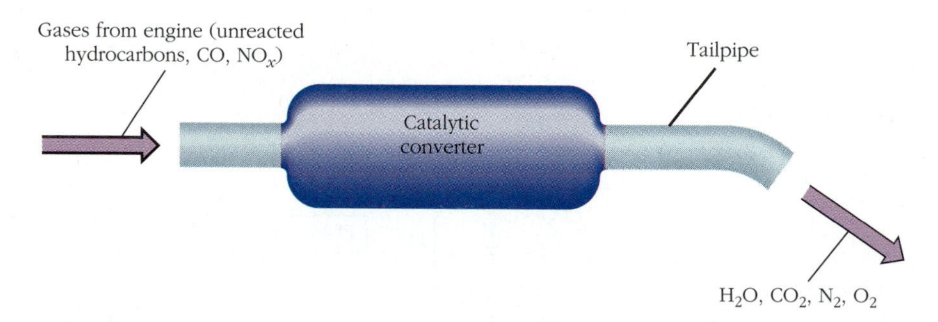

► Figure 13.6
Catalytic Converter. *Unburned hydrocarbons, CO, and NO_x react in the converter to form water, CO_2, N_2, and O_2. The hydrocarbons form CO_2 and H_2O, CO is oxidized to CO_2, and the nitrogen oxides (NO_x) are reduced to N_2 and O_2.*

The Ozone Layer

In the stratosphere, ozone absorbs harmful ultraviolet radiation.

Excess ozone in the **troposphere,** the layer of the atmosphere about 10–16 km thick nearest Earth, is a threat to health. Yet in the **stratosphere,** the next layer, which extends from the troposphere to an altitude of about 50 km, ozone forms a lifesaving shield that blocks ultraviolet light from the sun. This energetic radiation, which has wavelengths in the range of 220–330 nm, would harm many exposed forms of life, including humans, if much of it could reach Earth's surface. Fortunately, little does, because ozone is a very effective absorber of light at these wavelengths, undergoing photodissociation to oxygen molecules and oxygen atoms in an excited state,

$$O_3 \xrightarrow{\text{220–330 nm light}} O_2 + \underset{\text{excited atomic oxygen}}{O^*}$$

The excited atomic oxygen atoms can lose their energy by colliding with an oxygen molecule and a third body (M), such as another oxygen molecule or nitrogen molecule, forming ozone, or they can react with ozone to produce two oxygen molecules:

$$O^* + O_2 + M \longrightarrow O_3 + M \text{ (with extra energy)}$$

or

$$O^* + O_3 \longrightarrow 2O_2 + \text{energy}$$

Note that the first reaction produces ozone and the second destroys it. Ozone also is destroyed by traces of nitric oxide (•NO), which are present naturally in the stratosphere. Nevertheless, the amount of ozone remains essentially unchanged. In another reaction, molecular oxygen is photodissociated by light of wavelengths below about 242 nm,

$$O_2 \xrightarrow{\text{light} < 242 \text{ nm}} 2O^*$$

This provides additional excited atomic oxygen atoms, which can react with oxygen molecules and a third body (M) to make more ozone. The processes that form ozone and those that destroy it exist in a natural balance, called a **steady state.** In this model, ozone is formed as fast as it is destroyed, and its concentration remains constant.

Ozone is formed and destroyed in a natural balance called a **steady state.**

This natural balance has been disrupted by human activity, however. *Chlorofluorocarbons* (CFCs) are synthetic chemicals that have found widespread use as refrigerants, blowing agents for making plastic foams, propellants for aerosol sprays, and cleaning solvents for electronic parts. The *Halons,* CF_2ClBr and CF_3Br, are used as fire extinguishers, especially in aircraft. These compounds are chemically inert and nontoxic, and were believed to be environmentally benign. Near the surface of the earth they are indeed so. Their use has merely contaminated the troposphere but has not polluted it. In the stratosphere they are another matter. CFCs (and the Halons) are so chemically inert that they do not react with *anything* in the troposphere. Nor do they dissolve well in water. As a consequence, they are not washed from the troposphere by natural processes; so they are gradually mixed upward into the stratosphere, where highly energetic ultraviolet light easily breaks chemical bonds. Even there, they decompose very slowly.

The two most common CFCs are trichlorofluoromethane (CCl_3F, CFC-11) and dichlorodifluoromethane (CCl_2F_2, CFC-12). In both, the C—Cl bond is easiest to break. For example, CCl_2F_2 undergoes photodissociation as follows,

$$CCl_2F_2 \xrightarrow{\text{ultraviolet light}} {}^{\bullet}CClF_2 + {}^{\bullet}Cl$$

This initiation step produces chlorine free radicals (chlorine atoms), which cause the destruction of ozone by way of the following free-radical chain:

$$\begin{aligned}
{}^{\bullet}Cl + O_3 &\longrightarrow {}^{\bullet}ClO + O_2 \\
{}^{\bullet}ClO + O &\longrightarrow {}^{\bullet}Cl + O_2 \\
{}^{\bullet}ClO + O_3 &\longrightarrow {}^{\bullet}ClO_2 + O_2 \\
{}^{\bullet}ClO_2 + O &\longrightarrow {}^{\bullet}ClO + O_2
\end{aligned}$$

Note that not only does the chlorine free radical, $^{\bullet}Cl$, destroy ozone, but one of the products of its reaction with ozone is chlorine monoxide, $^{\bullet}ClO$, which also destroys ozone. In addition, both $^{\bullet}Cl$ and $^{\bullet}ClO$ are regenerated in reactions with atomic oxygen, so each can destroy another ozone molecule in a continuing, repetitive cycle. Calculations have shown that one chlorine free radical can cause the destruction of up to 100,000 molecules of ozone before its reaction chain is terminated. For this reason, the presence of a very few CFC molecules in the stratosphere can cause a drastic reduction in the

Free-radical chain reactions can repeat themselves many times.

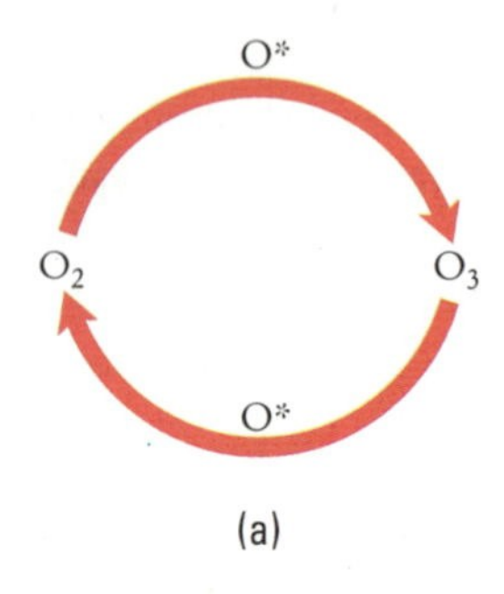

(a)

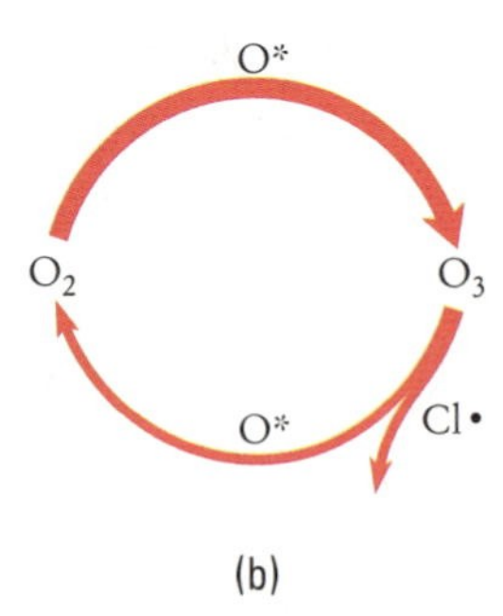

(b)

▶ Figure 13.7
Chlorofluorocarbons Upset the Steady-State Balance of Ozone Formation in the Stratosphere. *(a) Under normal conditions, ozone is formed as fast as it is destroyed, and its concentration remains constant. (b) Chlorine atoms from the photodissociation of CFCs "siphon off" ozone, taking it out of the cycle and lowering its concentration.*

amount of ozone, upsetting the natural balance of its production and destruction (▶ Figure 13.7).

The *ozone hole,* a depletion of the ozone layer over polar Antarctica, was first recognized in 1982. This phenomenon, which is caused by the presence of CFCs in the stratosphere, has increased in intensity and area in recent years; and now another, though smaller, hole is being observed over the polar Arctic. Interestingly, and surprisingly, the greatest depletion of ozone occurs during late polar winters and early spring when the stratosphere is coldest, below −70 °C, and sunlight begins to illuminate the pole. At these extremely low temperatures, *polar stratospheric clouds* are formed that are composed of tiny crystals of ice. Trapped in these ice crystals are photochemically reactive molecules, such as CFCs, chlorine nitrate ($ClONO_2$, from $^{\cdot}ClO + {}^{\cdot}NO_2$), and HCl. These molecules do not react in the dark of the polar winter, but the spring sunlight photodissociates them *on the surface of the solid ice crystals,* rapidly forming chlorine radicals, $^{\cdot}Cl$, which escape to destroy large quantities of ozone by the radical chain reaction previously described. As the stratosphere warms during the summer, the polar stratospheric clouds disappear, and the photochemical reactions cease, allowing the ozone concentrations to build up again and "repair" the hole. There is now disconcerting evidence that the hole is not completely repaired before the cycle resumes the following spring.

Ozone and CFCs can be beneficial or harmful depending on their location in the atmosphere. In the troposphere, close to the surface of the earth, ozone is a pollutant, but in the stratosphere it protects life in the biosphere from harmful ultraviolet radiation. CFCs are very useful substances that are harmless in the troposphere. However, in the stratosphere they upset the balance of ozone production and destruction and serve to endanger life below.

Using computer models based on what we know about the atmosphere and its chemistry, researchers have attempted to predict what might happen

13.4 Identify the different kinds of steps (initiation, propagation, termination) in the overall free-radical chain reaction by which CFC-11 depletes ozone in the stratosphere:

$$CCl_3F \xrightarrow{\text{ultraviolet light}} {}^{\cdot}CCl_2F + {}^{\cdot}Cl$$
$${}^{\cdot}Cl + O_3 \longrightarrow {}^{\cdot}ClO + O_2$$
$${}^{\cdot}ClO + O \longrightarrow {}^{\cdot}Cl + O_2$$
$${}^{\cdot}Cl + O_3 \longrightarrow {}^{\cdot}ClO + O_2$$
$${}^{\cdot}ClO + O \longrightarrow {}^{\cdot}Cl + O_2$$
$${}^{\cdot}Cl + {}^{\cdot}Cl \longrightarrow Cl_2$$

to the ozone layer in the future. Though different scientists have based their studies on somewhat different assumptions, most of them come to a similar conclusion: if we discontinue the use of all chlorofluorocarbons immediately and place no more of them in the atmosphere, those already present will remain in the stratosphere to deplete the ozone layer for as long as a century. There is not much we can do to remedy the situation, except to stop using CFCs and allow natural processes to gradually cleanse the stratosphere of chlorine atoms. How badly the ozone will be depleted in the meantime is uncertain, but assuming that we stop using CFCs now, the amount of chlorine in the stratosphere will probably double before it starts to decline.

Compounds that have the useful properties of CFCs but contain no chlorine, or that decompose in the atmosphere (before they can reach the stratosphere), are actively being examined. The hydrofluorocarbon HFC-134a (CF_3CH_2F) is already replacing CFC-12 (CF_2Cl_2) in automobile air conditioners and refrigerators, and the hydrochlorofluorocarbon HCFC-22 (CHF_2Cl) is now used in air conditioners and as a blowing agent for making styrofoam. Substitution of HFCs and HCFCs for CFCs will be expensive, however, because both refrigeration units and manufacturing facilities must be modified to use the new materials. Production of CFCs in the United States will cease at the end of 1995 and, by international agreement, in the world by the year 2000. All refrigerators and air conditioners presently using CFCs must be recharged with *recycled* CFCs or retrofitted to use the newer refrigerants.

Substitutes that have the useful properties of CFCs and do not deplete ozone are available.

The Greenhouse Effect

Carbon dioxide is a natural component of the atmosphere. In preindustrial times its atmospheric concentration was 270 to 280 ppm (parts per million, or about 0.027 to 0.028%). By 1992, this value had increased to 356 ppm, and today the concentration of CO_2 is increasing at a rate of about 1.5 ppm per year, about double that of 0.8 ppm per year in the 1960s. If present trends continue, the amount of CO_2 in the atmosphere will be double that of preindustrial levels by about 2050.

Most of the new atmospheric carbon dioxide is introduced by human activity, primarily by our burning of enormous quantities of fossil fuels and our massive destruction of tropical forests. The oceans serve as a sink for about half of the carbon dioxide released, and the rest goes into the atmosphere, much of it being used by green plants in photosynthesis. However, because deforestation and changes in the use of land have reduced the amount of vegetation that can carry out photosynthesis, the amount of carbon dioxide that can be removed by this process has decreased. This is expected to contribute substantially to the increase in the amount of CO_2 entering the atmosphere.

Carbon dioxide is not very photoreactive, nor is it toxic in the low concentrations that are found in the atmosphere. (In high concentrations, it could smother someone by excluding oxygen.) So why is its increase a cause for concern? Earth's surface temperature is governed primarily by the balance between the energy absorbed from sunlight and the energy reradiated back into space. Like the steady-state production of ozone in the stratosphere, it is believed that this balance also has been upset.

In the atmosphere, carbon dioxide acts like a pane of glass in a greenhouse.

Carbon dioxide in the atmosphere behaves much like a pane of glass in a greenhouse. It allows visible light from the sun to pass through and reach the surface of the earth. Lighter-hued substances on the surface tend to reflect the light, but the darker ones absorb it and become warm (like the fur of a black cat in the sun on a cold winter day). This warmth is reradiated as infrared light, which has longer wavelengths than those of visible light. Although carbon dioxide is transparent to the visible light that enters the atmosphere, it does absorb the infrared radiation that is trying to leave (much like greenhouse glass traps heat inside), causing the atmosphere to warm. The more carbon dioxide the atmosphere contains, the more heat is absorbed (▶ Figure 13.8.) Nitrogen and oxygen, which make up most of the atmosphere, are transparent to infrared radiation, so they do not contribute to this **greenhouse effect.**

Like carbon dioxide, methane also is a **greenhouse gas.** Since preindustrial times its concentration in the atmosphere has more than doubled, from about 800 ppb (parts per billion) in 1850 to 1738 ppb in 1992. This has been attributed to increased numbers of cattle and to more rice farming, both of which generate methane by **anaerobic** (without oxygen) decomposition of organic matter. Methane is also formed naturally by this process in such places as swamps, forest floors, soils, and the Arctic tundra. Ozone-destroying chlorofluorocarbons also contribute to the greenhouse effect.

Climate is quite sensitive to global temperature changes.

Most climatologists agree that average global temperatures have increased by about 0.3 to 0.6 °C during this century. Whether this is a natural trend or **anthropogenic** (caused by humans) is uncertain, but global temperatures have varied over a range of only 2 °C since the last glaciation ended about 10,000 years ago. A warming of only 1 °C from the average made it possible to farm in Greenland (hence the name) from the late tenth to the early thirteenth centuries, and a cooling of only 0.5° caused the Little Ice Age, a period from about 1500 to 1850, when it snowed in New England in midsummer and some mountain glaciers advanced to lower elevations than they had reached during the last ice age.

Various models of the greenhouse effect have been used to predict that, if today's trends continue, global temperatures could rise by 1.5 to 4.5 °C by the middle of the next century. What environmental changes this may cause is a subject of intense debate. They could be catastrophic, or they might be relatively benign. Some might even be beneficial. No one knows for sure. However, it is clear that we are carrying out an experiment on a global scale that could have irreversible and disastrous consequences for our future. Although

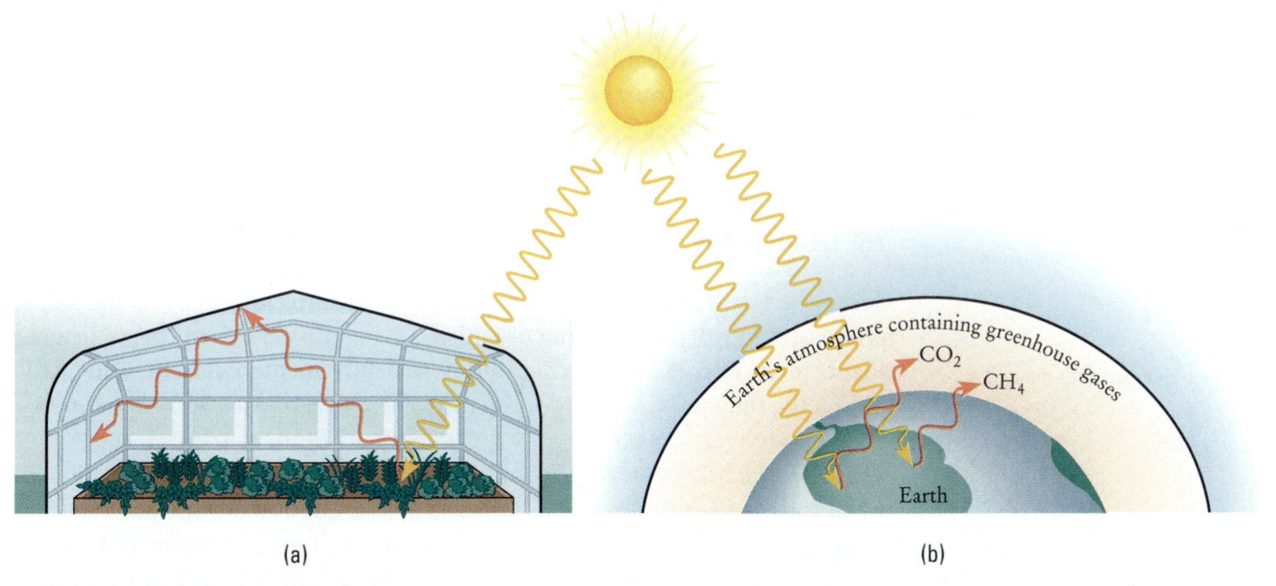

▶ Figure 13.8

The Greenhouse Effect. *(a) Sunlight of short wavelengths passes through glass in a greenhouse and is absorbed by the objects inside. This energy is reradiated as infrared light of long wavelengths, which cannot escape through the glass and is trapped, warming the inside of the greenhouse. (b) Short wavelengths of sunlight also pass through the gases in the atmosphere, including CO_2 and CH_4, which are transparent to visible light. Objects on the earth radiate this energy back toward space as infrared light, which is absorbed by CO_2 and CH_4 molecules, warming the atmosphere.*

we probably cannot stop adding greenhouse gases to the atmosphere altogether, it seems prudent that we should seek ways to keep their emissions to a minimum.

Because much of the energy we use worldwide is obtained from the combustion of fossil fuels, reducing the amount of CO_2 released to the atmosphere will be difficult. This is particularly true in developing nations where energy consumption has not "matured" and is increasing more rapidly than in developed countries. (And it is probably unfair to ask other people to restrict their energy uses when we consume so much.) Clearly, conservation of energy and other natural resources (which require energy use) will be needed. Development of technologies that capitalize on renewable energy sources should be given high priority. Nuclear energy can be substituted for fossil energy, but its use poses other environmental problems.

Proposals for removing excess CO_2 from the atmosphere involve trapping it at the source and either converting it to fuels or disposing of it deep in the oceans, where extreme pressures would keep it in the liquid state. Both of these schemes would require a wasteful amount of energy. To convert CO_2 into hydrocarbons would require 1.5 times the energy the resulting fuel would provide. Probably the best strategy is to reforest areas that have been deforested and encourage growth of vegetative cover on barren lands, because photosynthesis in green plants removes large amounts of carbon dioxide from the atmosphere, using renewable sunshine.

THE LITHOSPHERE

We depend on soils for most of our food.

From the standpoint of life, soils are the most important part of the lithosphere, and most land-dwelling organisms depend on them for their existence. Plants grow in soils, taking their water and nutrients from them. Directly or indirectly, they provide most of the food that we and other terrestrial animals need to survive.

Soils, which are heterogeneous mixtures having widely variable composition, consist of humus, silicate minerals, water, and air. Soils are also very busy places. Full of all kinds of life, they contain fungi, bacteria, algae, roots, termites, ants, nematodes, burrowing animals and, of course, earthworms, up to one million per acre. This biological activity contributes greatly to the fertility of soils by breaking down organic matter to provide nutrients for plants. Decaying animal and vegetable matter make up most of the **humus,** the organic portion of soils, and clay minerals formed from weathered rocks comprise most of the inorganic part. A typical productive soil contains about 5% organic matter, and it may contain up to 35% by volume of air. Because of the decay of organic matter, which uses oxygen and produces carbon dioxide, the air in soil usually contains less oxygen than atmospheric air (as low as 15% vs. 21%) and much more carbon dioxide (up to 5% vs. 0.035%).

Most soils exist in discrete layers called **horizons,** which vary in composition, texture, color, and structure with increasing depth. These are designated O, A, B, and C, starting at the top (► Figure 13.9). The *O horizon* is a thin layer a few centimeters in thickness that consists of the remains of plants,

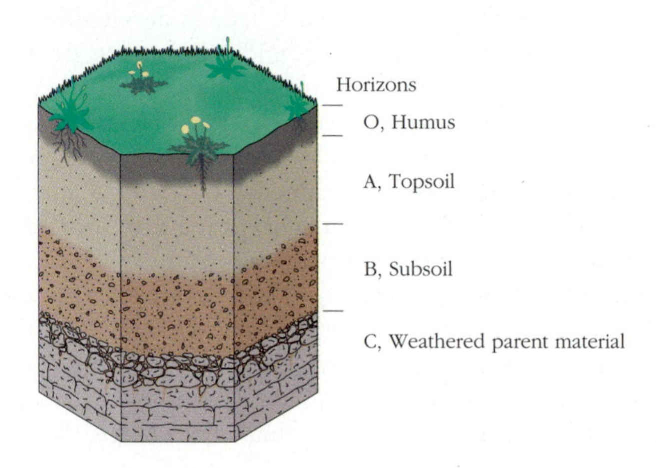

► Figure 13.9
Soil Horizons in a Typical Fully Developed Soil.

such as sticks, leaves, and grass, and animal remains. Its lower part is made up of humus. The *A horizon,* called **topsoil,** contains most of the organic matter and is the place of intense biological activity. It also contains inorganic matter, commonly clays and other silicates such as sand and quartz (Chapter 12). Usually a few inches thick, it can be as deep as several feet in some regions. In cultivated soils, the O and A horizons become mixed together and are often collectively called "topsoil." The *B horizon* is the **subsoil,** made up of inorganic material from weathered parent rocks (which lie below it), along with a little organic matter, ionic substances (salts), and clay particles that have leached (washed) from the topsoil. This layer is usually several feet thick and rests on the *C horizon* of weathered **parent rocks,** from which the soils above have originated.

Although soils are susceptible to many forms of pollution—from the use of pesticides and herbicides to the release of industrial chemicals, radioactive materials, trash, and even human waste—the primary threat to their viability is *erosion.* Plants usually grow best in topsoil because this horizon is richest in nutrients. However, the cultivation of soils to grow crops and the removal of vegetation by domestic grazing animals or by deforestation have made this horizon vulnerable to serious loss by erosion, primarily by the action of water but also of wind. Plants grow poorly in the subsoil that is exposed by these processes. It has been estimated that in the United States about one-third of the topsoil has been lost to erosion since this country was settled and farming began. This has reduced the productivity of the remaining soils, and the loss must be compensated for by the application of more chemical fertilizers. In many parts of the world, such as Greece and Madagascar, almost all of the topsoil has been lost to erosion, and crops there do poorly.

Plants grow poorly in **subsoil**.

Abuse of land along the margins of deserts, where rainfall is barely adequate, has caused **desertification,** a process by which deserts expand into what was once productive farmland. This expansion, estimated at 70,000 km^2 per year (27,000 mi^2), has caused considerable human suffering, particularly along the south rim of the Sahara, where the desert is advancing southward rapidly enough to contribute to persistent and often widespread starvation in Africa.

It is a vicious cycle. Growing human populations in these areas have forced a more intensive use of lands near the fringes of the desert. Many of these lands are covered with grasses and other vegetation whose roots hold the desert at bay. Such lands are also very fragile and suffer from droughts. Clearing them to grow crops and increasing the numbers of livestock that graze on the native grasses has caused a drastic reduction in vegetation cover, allowing the topsoil to blow away. In periods of drought the vegetation dies completely, and the desert takes over. Starving people use—and abuse—the fringe land that remains even more intensely, destroying it also. During the next dry spell, the desert encroaches even further. From 1968 to 1973 a severe drought persisted in the Sahel of Africa, a belt of land 300 to 1100 km wide south of the Sahara. The desert expanded southward as much as 150 km, and almost 250,000 people and 3.5 million cattle died of starvation. Desertification continues at an alarming rate in Africa (▶ Figure 13.10). Large areas in Australia, Asia, South America, and even the southwestern United States are also at risk today.

Soil-forming processes work slowly.

Soil-forming processes work slowly, *starting on the surface and working downward.* In general, the longer a soil remains undisturbed by cultivation the more fully developed and fertile it becomes. This is one reason that the practice of fallowing croplands and planting vegetative cover on them increases their subsequent productivity. Constant cultivation of land has quite the opposite effect because of the topsoil that is lost to erosion and the nutrients that are leached out. Unless these materials are somehow replenished, the soil eventually loses its productivity. Chemical fertilizers are only partially useful for restoring fertility. Although they can supply ionic plant nutrients, they cannot replace the lost organic matter on which soil fertility ultimately depends.

How long does it take for soils to form? That depends on many factors, including climate, the nature of the parent rock, and the availability of organic material. However, a reasonable average would be about 2.5 cm (about 1 inch) *per century.* In the frame of geologic time (millions of years) this is very rapid, but from a human perspective it is exceedingly slow. This means that our soils are like petroleum, coal, and natural gas, and ores and minerals: they

▶ Figure 13.10
Desertification in Niger, Africa. *As the goats eat the remaining bushes, the encroaching dune continues to advance. The pasture in the foreground will soon be lost.*

are nonrenewable natural resources (Chapter 10). We must be good stewards of our soils because without them we will not eat.

Many practices can be employed to conserve soils and even enrich them (▶ Figure 13.11). Erosion can be minimized by *terracing* steep slopes, *contour plowing* (cultivating around a hill rather than up and down it), practicing *minimum tillage* agriculture (seeding and fertilizing in one operation without plowing after the previous crop), using grass waterways and catch basins to retain eroding soil, planting grasses and trees on barren soils, and using windbreaks. Soil fertility can be restored by returning organic matter to fields, *strip cropping* (planting different crops in strips in the same field), crop rotation, planting crops appropriate to soil conditions, *fallowing* (allowing soils to rest), and planting grasses and *legumes* (nitrogen-fixing plants). Because too much irrigation water leaches nutrients from soils, proper irrigation practices preserve soil fertility and also conserve precious water. Some soils can be reclaimed. Though it is expensive, many salinized soils (see later in this chapter) can be restored by proper draining and flooding. Unfortunately, other soils have been damaged beyond repair. Worldwide, an area the size of India and China combined is no longer suitable for growing food.

▶ Figure 13.11
Soil Conservation Practices. *Three soil conservation practices are evident on this farm. Contour plowing, in which fields are plowed along the contours of a hill, is combined with strip cropping and grass waterways. The waterways can be seen in low areas, running downhill across the contours.*

THE HYDROSPHERE

Although 71% of the surface of Earth is covered with water, most of the water in the hydrosphere is unfit for human consumption, or even for agriculture or industrial use. Most of it (97.2%) is in the oceans and another 2%

NU-KLEE-AR CHEMISTRY

"Whatever you do in your lives, whatever you think of nuclear energy, do me one favor . . . especially if you become politicians or news anchors." Rhonda looks beseechingly from face to face, "Don't say *nu-kew-ler!* Look at the word: it's nu-klee-ar."

You started this Saturday afternoon discussion by mentioning your realization that much of what goes on in the world, chemically, is one huge redox reaction and that we humans are tipping the balance toward oxidation:

"We're suffocating the world and I don't see how we can stop. Other people aren't going to stop wanting the good life, and neither are we!"

Carl argued conservation. And Jeanne said of course, but nuclear energy is the solution: "France uses it without difficulty." Charley said trouble is inevitable because some nuclear wastes have half-lives of thousands of years. "Think of that stuff accumulating for hundreds of years! There are tons of it already!" Jeanne said he probably needn't worry about it. Rhonda said Carl's right, we have to stop wasting, Jeanne's right, we have to risk using nuclear energy, Charley's right that long-term use of fission energy would be disastrous. But fission along with conservation might buy time enough for science to find solutions. Charley shakes his head:

"Science got us into this mess!"

No Charley, science is only a method of learning things about the world, and technology is the use of what science learns to make tools. People wanted all the constituents, individually, of "this mess": we wanted longer lives, we wanted our children to stop dying . . . that's not an evil desire . . . and we wanted to be mobile and rich. We started using tools a very long time ago and probably won't stop. The only way for us is onward.

And there's hope. We've been really aware of ecological problems for less than fifty years, at first only that world supplies of fossil fuels are limited. No one thought of carrying camping debris out of the mountains . . . it would vanish. Garbage dumps were necessary, and they were "controlled" by putting them outside of town. Modern technology hasn't suddenly turned us into evil creatures, we've always had this habit of tossing away the remains of our meals. Middens are accumulations of the debris of group living, such as piles of tons of oyster shells and other debris found on the peripheries of ancient encampments. Middens were probably health hazards, but the people who made them didn't know that. So, in a sense, their habit didn't bother them. Now we know and it's bothering us, and we have to change our age-old habits.

And now that we do know, there's hope. We're learning to conserve and recycle, concepts that never occurred to your grandparents unless it saved money. The recycling industry is slowly grinding to a start, and when people can make a good living at it, it will flourish. Chemists are designing less polluting methods of producing commercial chemicals, and ways to reduce pollutants of other industries. Landfills that produce methane for fuel are being designed, and food packaging with edible polymers is on the horizon. For that we'll be our own automatic disposal units.

As for energy, there's long-term hope for nuclear fusion, possibly by 2050. If physicists can learn how to fuse two hydrogen nuclei into one helium nucleus, we'll produce energy the way the sun does. The difficulty is in getting two positive protons close enough together for the powerful but extremely short-range forces that hold nuclear particles together to take over from the electromagnetic force. That takes temperatures of 2×10^8 K, which nothing material can contain. Two methods are being attempted: containing the reactive plasma (*everything* is gaseous at those temperatures and plasma is a gas of separate electrons and nuclei) in "bottles" made of magnetic fields. Such a machine, called Tokamak, has achieved 4.6×10^8 K and generated 9×10^6 watts of power for 240 milliseconds, far from a continuous reaction. Physicists are also experimentally imploding (opposite of explode) tiny pellets of deuterium, which achieves short-lived bursts of the necessary temperatures. It's a start. If science can give us this great gift, we'll have almost unlimited power with much less radioactive waste than fission produces.

"But even if all that works," Rhonda said, "it'll only buy us time. We have to solve our population problem. If we don't, nature will do it for us, and I can guarantee we won't like her methods."

That night you arrive late at an end-of-term party and after you get a soda from a tub, you look around, realizing how your view of the world has changed. The fizz of the soda means something new to you now, and you understand what causes the tartness and sweetness of the drink and the nature of the can that holds it. You know what creates the colors in people's clothes and the nature of the polymers they're made of. You understand what causes the light . . wait! At last you see the person you're looking for and as you move in that direction–trying not to be too obvious–you feel the blood rising in your cheeks. It occurs to you that a course in *this* kind of chemistry would be very helpful.

is frozen in glaciers. This leaves only 0.8% as the *fresh water* that makes up rivers, lakes, swamps, groundwater, and adds moisture to the atmosphere.

The Hydrologic Cycle

The **hydrologic cycle** is driven by solar energy.

Fresh water, though a small fraction of the hydrosphere, is constantly being recycled from the oceans to the atmosphere to the land masses and back to the oceans by means of the **hydrologic cycle** (▶ Figure 3.12). This process is driven by solar energy, which evaporates water from the oceans into the atmosphere. The water vapor rises in the atmosphere and forms clouds, which eventually lose their moisture as precipitation. Most precipitation (about

▶ Figure 13.12
The Hydrologic Cycle.

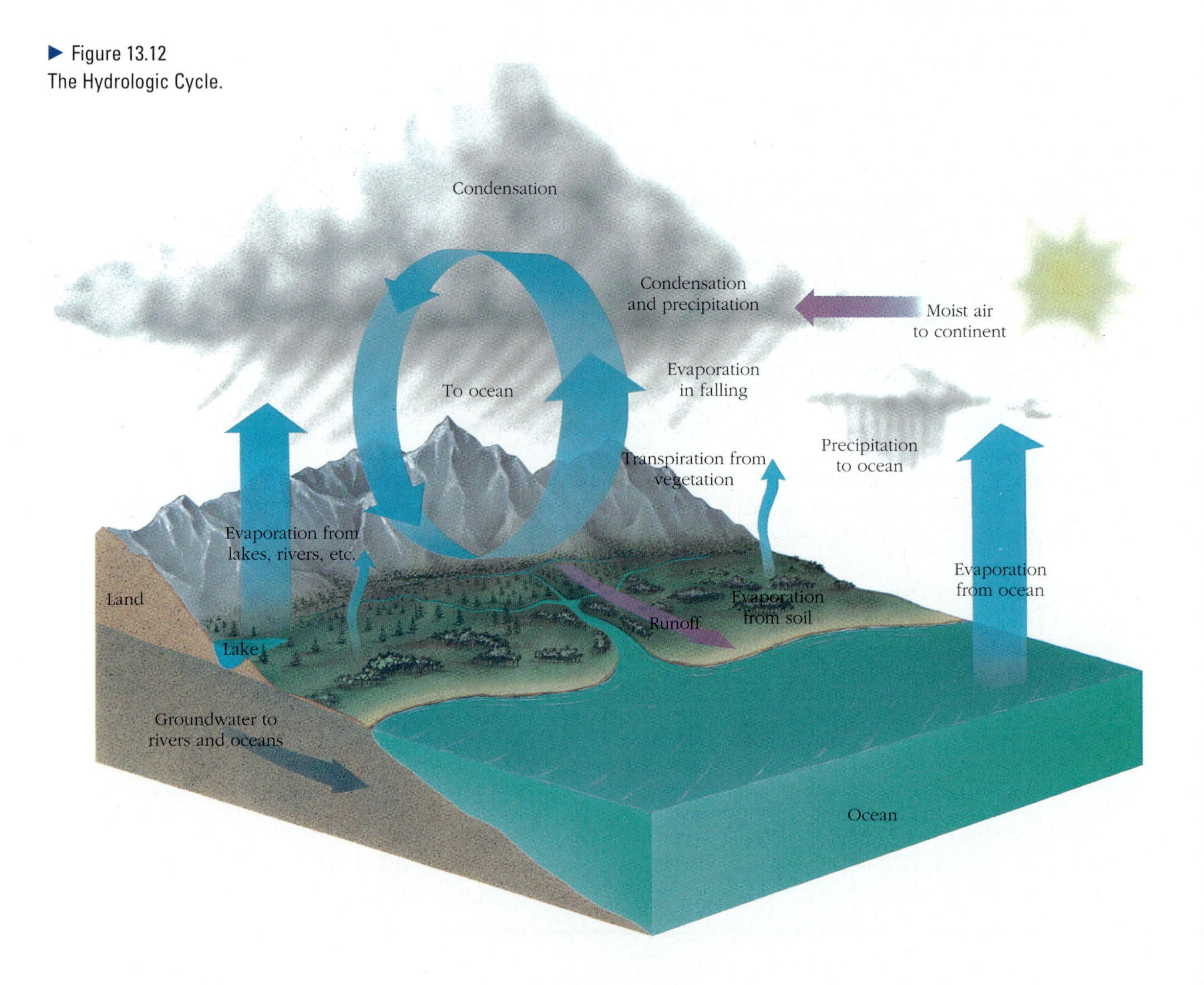

80%) falls back into the oceans, but the rest falls on land as rain or snow. Some of the precipitation evaporates as it falls and reenters the hydrologic cycle. About 15% of the water in the cycle comes from evaporation of water from land.

Once on land, most of the fresh water is temporarily stored in snow fields and glaciers, in lakes and impoundments, or it soaks into the surface to become groundwater. Only a small amount, about 1%, immediately returns to the oceans as runoff; this is still an enormous amount of water, however, about 36,000 km^3 per year.

The Water We Use

We need only one-half gallon (about 2 L) of drinkable water per person each day. Yet in this country we use 80 to 90 gallons (about 300–340 L) of it daily for domestic purposes. About a third of this goes for flushing toilets, another fourth for bathing, and the rest for dishwashing, laundry, and cooking. If we add to this the amount of water we use indirectly—that used in food production, manufacturing, and the production of energy—the amount of water each person uses jumps to over 1700 gallons each day. Agriculture is particularly water intensive (▶ Figure 13.13). For example, to grow 1 kg of dry hay requires over 200 kg of water, and to produce a beefsteak requires 13,000 kg of water. On a global scale, irrigation and other agricultural uses account for two-thirds of the fresh water withdrawn for human use, and only 10% is applied to domestic uses like drinking, bathing, and washing clothes.

It takes 13,000 kg of water to produce a beefsteak.

On average, there is plenty of fresh water to satisfy the needs of everyone in the world. However, not everyone has enough water; and even when water supplies are adequate there is often not enough pure water. One reason for water shortages is that the distribution of water is uneven; where populations have grown or where agriculture has expanded, water consumption often has outgrown the supply. Climate also can cause periodic shortages of water. We shall be more concerned here with the second reason: namely, that human activity has badly polluted much of the supply of fresh water, rendering it unfit for use or too expensive to purify.

Some Common Water Pollutants

About half of the fresh water withdrawn in this country is **surface water,** mostly from lakes, rivers, and reservoirs. The other half is **groundwater,** pumped from **aquifers,** porous layers of underground gravel or rock that are full of groundwater, usually flowing slowly from one place to another. Pollution of these waters comes from numerous sources and takes a number of forms. Some of the more important of these pollutants are heat from industrial processes, pathogens (disease-causing agents), oxygen-consuming wastes, organic chemicals, inorganic chemicals from industry, mining and agriculture, sediments, and oil. These will be discussed briefly in turn.

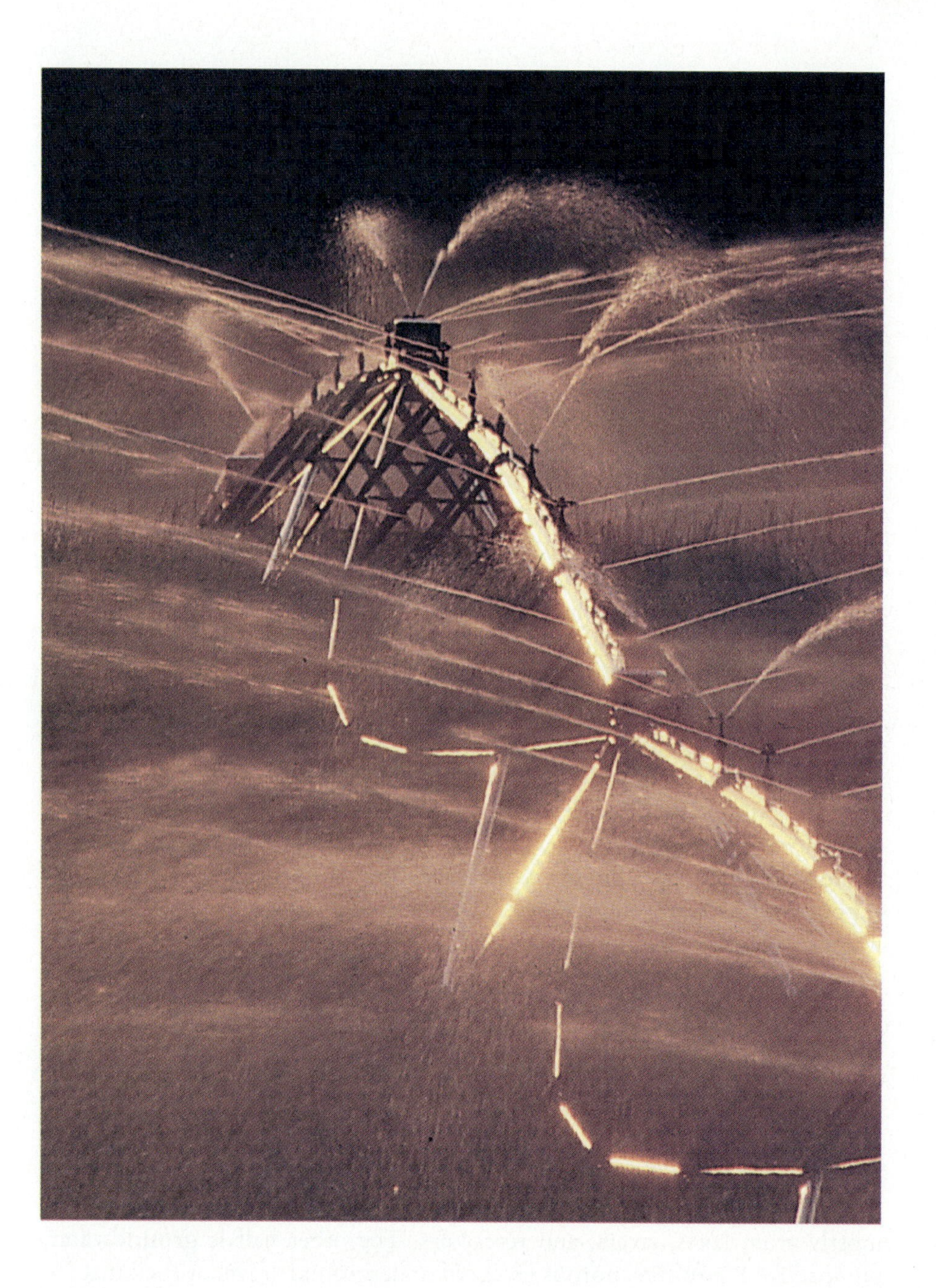

▶ Figure 13.13
The Ogallala Aquifer Irrigates a Nebraska Cornfield. *This center-pivot irrigation system can deliver 4,000 liters per minute (about 1,000 gallons per minute) from a well in this aquifer, 24 hours a day. Globally, agricultural uses account for two-thirds of the fresh water withdrawn for human use.*

Thermal pollution Whenever energy is converted, some is wasted, usually in the form of heat, which escapes into the environment (Chapter 10). Much of this heat is removed from industrial process and power plants by the use of cooling water, which often exits into rivers, lakes, or the oceans and causes **thermal pollution** (▶ Figure 13.14). Unlike the solubility of most solid materials, the solubility of gases *decreases* as the temperature is increased. Raising the temperature of these bodies of water thus drives out dissolved oxygen, making it more difficult for **aerobic** organisms (organisms that require oxygen) to survive. At the same time, their metabolic processes are speeded up by the temperature increase, which increases their need for oxygen. For some aquatic life, this one-two punch can prove fatal. Also, changing the water temperature often causes a change in the kinds of species that can survive there. For example, trout cannot live at temperatures much above 15 °C, but bass do fine.

▶ Figure 13.14
Preventing Thermal Pollution. *Cooling towers such as this are used to recycle industrial cooling water. In the tower, some of the warm water evaporates. This cools the rest of the water, which is recycled.*

13.5 Why do goldfish die when placed in water that has been boiled and then cooled?

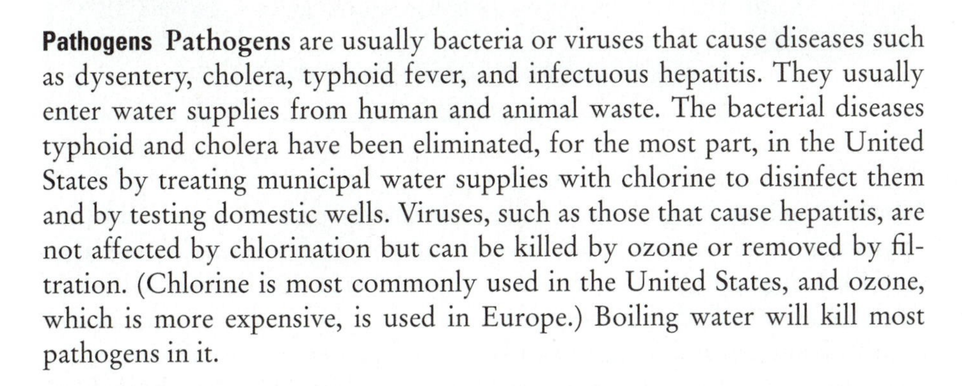

Pathogens **Pathogens** are usually bacteria or viruses that cause diseases such as dysentery, cholera, typhoid fever, and infectuous hepatitis. They usually enter water supplies from human and animal waste. The bacterial diseases typhoid and cholera have been eliminated, for the most part, in the United States by treating municipal water supplies with chlorine to disinfect them and by testing domestic wells. Viruses, such as those that cause hepatitis, are not affected by chlorination but can be killed by ozone or removed by filtration. (Chlorine is most commonly used in the United States, and ozone, which is more expensive, is used in Europe.) Boiling water will kill most pathogens in it.

Oxygen-consuming wastes **Oxygen-consuming wastes** consist primarily of organic matter from domestic sewage, detergents, decaying aquatic plants, animal manure, and effluent from industries such as those that process food, make paper, and refine petroleum. Aerobic bacteria usually degrade these materials if enough oxygen is dissolved in the water,

$$\underset{\text{(containing C, H, N, O, S)}}{\text{organic wastes}} + O_2 \xrightarrow{\text{aerobic bacteria}} CO_2 + H_2O + NO_3^- + SO_4^{2-}$$

The quantity of dissolved oxygen that is needed to oxidize a given amount of this organic matter is called the **biochemical oxygen demand,** or **BOD.** If the BOD is low, the water is reasonably unpolluted with organic matter and the bacteria can process it readily. On the other hand, a high BOD means that

Biochemical oxygen demand(BOD) is a measure of the amount of organic waste in water.

there is a large amount of organic waste, and oxygen may be depleted. If this happens, aerobic oxidation can no longer continue, making way for the anaerobic bacteria present to *reduce* the organic waste, producing flammable gases such as methane and foul, smelly ones such as hydrogen sulfide, amines, and ammonia,

$$\underset{\text{(containing C, H, N, O, S)}}{\text{organic wastes}} \xrightarrow{\text{anaerobic bacteria}} CH_4 + H_2O + NH_3 + \text{amines} + H_2S$$

These odors are commonly associated with marshes and swamps. Most fish and other higher forms of aquatic life cannot live under anaerobic conditions.

13.6 Urea, H_2NCONH_2, is an organic compound present in untreated sewage.

a. In fresh water, what compounds will urea produce if the BOD is low?
b. What compounds will urea form if the BOD is high?

Organic chemicals Organic herbicides, pesticides, preservatives, industrial chemicals, cleaning fluids, and solvents are some of the compounds that constitute **organic chemical waste** (▶ Figure 13.15). Herbicides and pesticides commonly find their way into waterways by means of agricultural runoff or municipal drainage from lawns and gardens. Leftover paints, insecticides, and solvents from homes lie buried in landfills. Also buried are industrial organic wastes, which sometimes are also discharged into watercourses. At many places where such chemicals were buried in the past, little consideration was given to factors that determine the ease with which these pollutants can enter the environment, such as the local geology, kinds of underlying soil, and the climate. Nor was much thought been given to construction of the dump site in ways that would prevent pollutants from escaping. Typically, buried pollutants enter the environment by **leaching** through underlying sediments and into an aquifer below. If sufficient surface water percolates through buried substances, they are often transported with it (▶ Figure 13.16).

Even in semiarid places such as Pocatello, Idaho, leaching of organic pollutants can be a problem. There are two areas near the city where cleaning solvents such as TCE (1,1,2-trichloroethene) and PERC (1,1,2,2-tetrachloroethene) (Chapter 11) have contaminated several city water wells and a

▶ Figure 13.15
The Careless Dumping of Hazardous Waste. *This kind of irresponsible activity pollutes both land and water, creates health hazards to humans and other species, and requires costly remediation.*

number of domestic water supplies. Suspected sources of this pollution are the present county landfill and an old industrial complex. Near another landfill, which has been closed for years, other organic pollutants have contaminated private wells. The areas affected so far are small, and the matter is being taken very seriously by local government. To the north, the water supply of Missoula, Montana, is threatened with contamination from creosote. Spilled on the ground by the ton at a facility that once treated railroad ties to preserve them, the creosote has soaked into the soil over a large area located directly above a shallow aquifer (one lying near the surface). Scenarios of this kind are common throughout the country.

Inorganic chemicals **Inorganic chemical wastes** enter surface water from sources such as mines, ore processing, electroplating and other industrial processes, domestic water softening, salting of highways in winter, and agricultural runoff. Like organic chemicals, they also can enter groundwater by leaching from landfills and other burial sites. These common pollutants include mine drainage, cyanides, heavy metals, acid rain (Chapter 8), salt, and fertilizers. Acid mine drainage is often formed in abandoned coal mines and tailings from copper mines, which are rich in pyrite (FeS_2) and other sulfur minerals (▶ Figure 13.17). Pyrite oxidizes in the presence of air and moisture to form sulfuric acid, which drains into streams and other bodies of water.

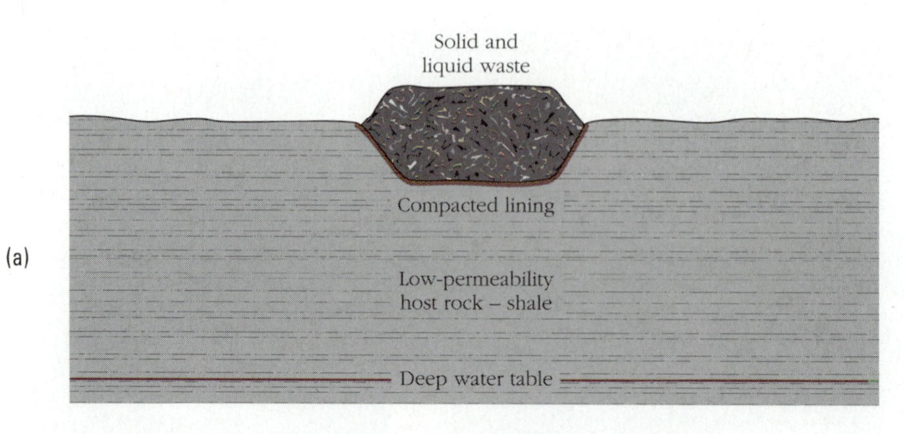

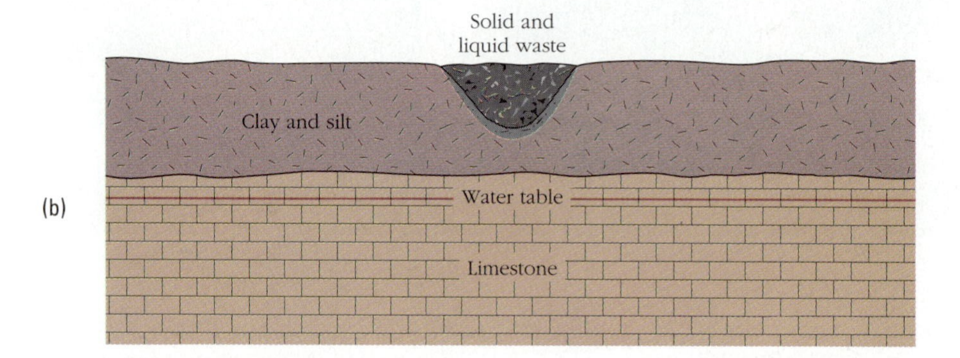

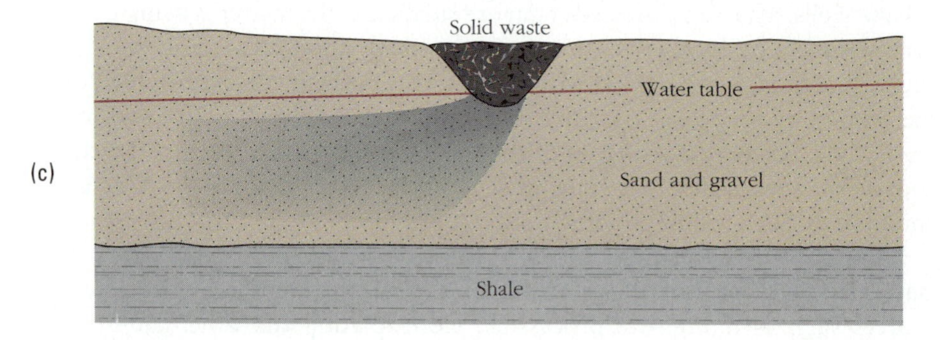

► Figure 13.16
Three Kinds of Landfills. *(a) A Class I landfill gives maximum protection to underground water. (b) A Class II landfill provides a moderate amount of protection. (c) A Class III landfill affords almost no protection to underground water. Note that leaching of pollutants is most severe from a Class III landfill.*

More than 11,000 km of streams in the United States are contaminated by drainage of sulfuric acid from abandoned coal mines. This pollution has an effect on aquatic life similar to that of acid rain.

The use of cyanide ion, CN^-, is an effective and economical way to recover gold and silver from their ores. Sodium cyanide is widely used in industry for such purposes as electroplating, mineral processing and recovery, and metal

▶ Figure 13.17
A Stream Near the Open-Pit Copper Mine at Morenci, Arizona. *The blue color of the water is due to a high concentration of copper ions leached from the mine wastes.*

cleaning. Cyanides are also very poisonous. In the past, it was common for gold mining operations to consume large quantities of cyanide and simply dump their effluent into a nearby stream. It is said that at one time the Uncompahgre River near Ridgeway in western Colorado was barren of life, and that trees would not grow near its banks. Upstream, near Ouray and Telluride, cyanide was used to process thousands of tons of ore. Today, cyanide still is used in the *heap-leaching process* (▶ Figure 13.18). A large pile (heap) of the crushed ore is sprayed with a very dilute solution of cyanide, which trickles down through the ore, removing (leaching) the gold or silver on its way to the bottom. There the solution is collected for removal of the metal and the cyanide is recycled. Numerous safeguards are made to ensure that the cyanide is contained. Though not without hazard, heap-leaching is environmentally much safer than previous methods.

Some other inorganic chemicals that are serious water pollutants are mercury and organomercury compounds, cadmium, lead, and arsenic. Mercury, for example, is used industrially as an electrode in the electrolytic production of chlorine gas and caustic (sodium hydroxide), and in various kinds of pressure gauges and electrical switches. Other sources of mercury in the environment are laboratory chemicals, pharmaceuticals, batteries, and broken thermometers. Compounds of mercury are found in coal in trace amounts (up to 100 ppb or higher). When this fuel is burned in large quantities, the mercury can pollute not only water, but air and land as well. Organic mercury compounds have been widely used as fungicides. Phenyl mercuric dimethyldithiocarbamate has been used as a slimicide and mold retardant in the

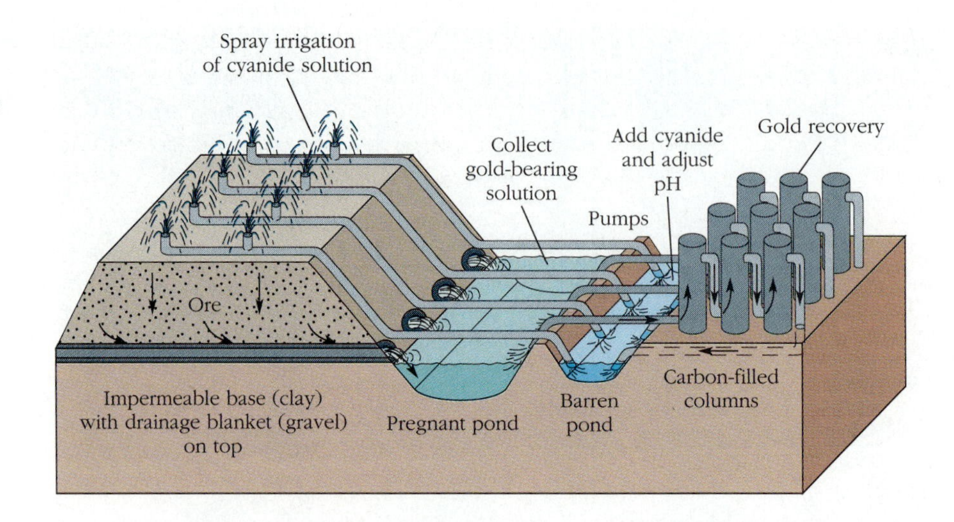

► Figure 13.18
A Cyanide Heap-Leaching Process for Extracting Gold from its Ore. *This method is environmentally much safer than older mining processes that used cyanide.*

paper industry, and ethylmercuric chloride at one time was used as a fungicide on seeds used for planting.

$$C_6H_5\text{—Hg—S—}\overset{\overset{\displaystyle S}{\|}}{C}\text{—}N(CH_3)_2 \qquad CH_3CH_2HgCl$$

phenyl mercuric dimethyldithiocarbamate ethylmercuric chloride

Under anaerobic conditions in water, mercury and its salts, such as $HgCl_2$, can be converted to the monomethylmercury ion, CH_3Hg^+, and dimethylmercury, $(CH_3)_2Hg$. Both of these serve to concentrate mercury in the fatty tissues of fish and aquatic birds, such as ducks, in sufficient amounts to cause mercury poisoning in persons eating them.

Cadmium is used in nickel–cadmium batteries, in metal plating for corrosion prevention, and in paint and plastics. It is present in mining wastes, in effluent from burning coal and municipal waste, in tobacco smoke, and even in emissions from the combustion of transportation fuels. It eventually finds its way into fresh water from these sources as Cd^{2+} ion. Note that cadmium and zinc appear in the same family in the periodic table. Zinc (as Zn^{2+} ion) is a dietary trace element necessary for human health. Because the chemistry of cadmium is similar to that of zinc, cadmium can replace zinc in the body. However, cadmium compounds are quite toxic to humans and can cause kidney damage, high blood pressure, and destruction of red blood cells.

Lead pollution of water is not as prevalent as it once was. In the past, lead commonly found its way into water from such sources as smelter effluent, emissions from burning leaded gasoline, and lead plumbing. Still, lead in drinking water today can come from some kinds of solder and pipe-joint compounds used in domestic plumbing. Water which stands for long periods in pipes joined with these materials can accumulate surprisingly high concentrations of lead, copper, zinc, and cadmium, particularly if the pH is low. Under these circumstances, it is wise to run the water for a few minutes before drinking it.

The metalloid arsenic is not only poisonous, it is suspected of causing cancer. It enters water from the combustion of coal and other fossil fuels, from phosphate mineral processing (again, note the family resemblance), and from mine tailings. In water, arsenic behaves like mercury and is converted by bacteria into methyl arsenic compounds, which are more toxic and more mobile in the environment than arsenic itself.

Salinity in water has been increased by human introduction of soluble salts, such as sodium chloride, into fresh water. Sources include mines, salt used for water softening and highway de-icing, industrial wastes, and fertilizers. Some of the effects of fertilizers as pollutants have been discussed previously (Chapter 6). Fertilizers also contribute to soil **salinization** of irrigated farmlands in arid regions. As irrigation water flows over and through the soil, it dissolves small amounts of salts, commonly sodium, magnesium, and calcium sulfates and salts derived from fertilizers. The water that collects in low areas of fields evaporates in the heat, leaving these salts behind. Where there is insufficient rain water to flush these ionic materials from the soil, particularly if groundwater is just below the surface, they collect over time and render the soil infertile. Soil salinization is a major problem in the western United States, southwest Asia, North Africa, and the Middle East.

Sediments Soil erosion is the major source of suspended particles, called **sediments,** in water. On a global basis these cause the greatest amount of water pollution. For example, the Mississippi River carries about 750 million tons of sediments each year (▶ Figure 13.19). Because they prevent sunlight from penetrating water as deeply, these pollutants reduce the amount of photosynthesis that can be carried out by aquatic plants and make it harder for many organisms to see to find food. Certain freshwater fish, such as trout and pike, cannot tolerate turbid water for very long. Sediments also absorb many substances and therefore spread other pollutants by carrying them along. Their removal from municipal water supplies can be accomplished by filtration, settling, or coagulation (adding a chemical that causes the sediment particles to clump together).

Oil Oil spills from offshore drilling platforms, such as the one caused by sabotage during the Persian Gulf War, or from broken tankers like the Exxon Valdez, which poured 11 million gallons (42 million L) of crude oil into Prince William Sound in Alaska in 1989, attract international attention in the

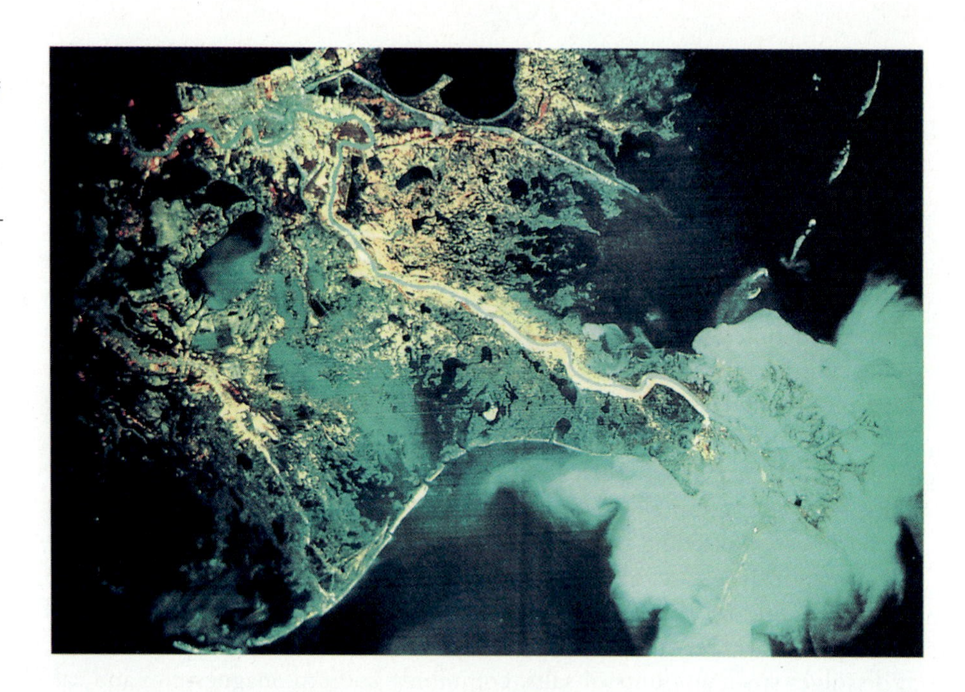

▶ Figure 13.19
A Color-Enhanced Satellite Image of the Mississippi River Delta. *The Mississippi River flows past New Orleans (upper center) and into the Gulf of Mexico, where its fresh, sediment-laden water appears a cloudy light-green as it mixes with salt water. Each year, the Mississippi River carries 750 million tons of sediments, most of which are deposited here.*

news. These tragic events wreak considerable havoc on marine mammals, sea birds, fish, and shellfish. Less spectacular, but no less damaging, is the pollution of fresh water with motor oil and transportation fuels. Used engine oil contains not only hydrocarbons but also additives, traces of toxic metals, and products of decomposition that make it a potentially serious water pollutant. A huge volume of motor oil is used in this country each year, about 1.4 billion gallons (5.3 billion L), and most of this is handled responsibly. However, do-it-yourself oil changes account for about one-seventh of this amount, and some of this oil is dumped into sewers or onto the ground, sometimes to settle the dust on unpaved roads, where it eventually finds its way into surface and groundwater.

Used motor oil is a potentially serious water pollutant.

Even more serious is the threat to groundwater posed by **leaking underground storage tanks,** or **LUST.** Many storage tanks for gasoline and diesel fuel at service stations have been buried for long periods of time, and a large number of older underground tanks have been abandoned. In many places gasoline has entered the groundwater near service stations (or the place they once were located), contaminating water wells nearby. It is now of great concern that many of these underground storage tanks, possibly several hundred thousand of them, may have begun to leak and that more groundwater may become contaminated. Not even remote small towns such as Rockland, Idaho, (population 283) have been immune. Several years ago a LUST there contaminated a shallow aquifer with gasoline. The soil was quite porous, and the basements of several homes nearby filled with gasoline fumes that had drifted upward from the aquifer, forcing the residents to leave their homes. Domestic wells were also contaminated with gasoline.

Protecting our water It is much easier to prevent the pollution of water than to purify it after it has been polluted. (Recall the discussion of entropy in Chapter 10.) Our industries, utilities, businesses, farms, and government agencies are becoming aware of this fact. For example, many chemical producers are recycling waste chemicals for reuse or for sale as specialty products. Chemical wastes that cannot be used are properly stored or destroyed. Considerable research is being done to invent benign technologies for making industrial chemicals. These processes, which are based on chemical reactions that require fewer (or no) hazardous chemicals as intermediates in manufacture, will greatly decrease the likelihood that unwanted pollutants may escape into the environment.

Proper construction of landfills and waste storage facilities can do much to prevent pollutants from leaching into surface water and groundwater (▶ Figure 13.20). Public awareness and cooperation can ensure that pesticides, in-

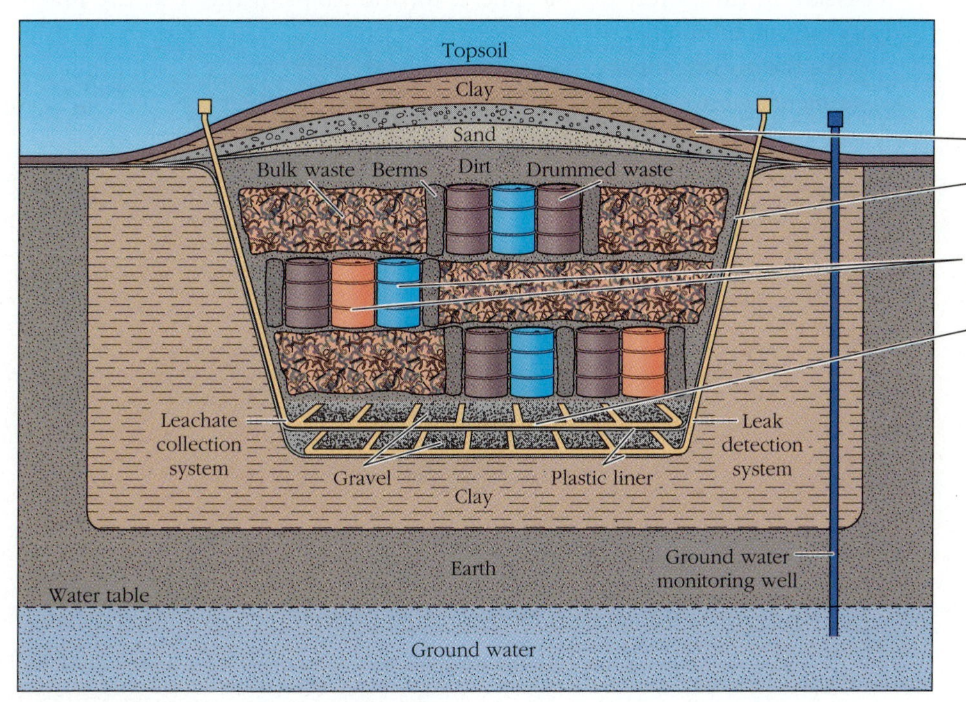

What could go wrong?

- Burrowing gopher attacks cover.
- Freezing temperatures shrink and tear liner.
- Mineral acids and solvents mix, triggering a chemical fire.
- Chemicals corrode waste collection pipes.

▶ Figure 13.20

A Secure Toxic Waste Storage Facility. *Its basic safety features are liners, a leachate collection system, a leak detection system, and a monitoring well. Note that precautions have been taken should something go wrong.*

secticides, fertilizers, motor oil, and solvents are properly disposed of or recycled. Conscientious agricultural practices can prevent soil losses (and hence sedimentation) and decrease the amounts of fertilizers, pesticides, and herbicides that pollute water supplies. And everyone must conserve water: most of us use far more than we need.

IMPLICATIONS FOR THE BIOSPHERE

The biosphere exists at the interface of the atmosphere, the lithosphere, and the hydrosphere, and all realms are interrelated and interconnected. A change that affects one affects all of the others. Pollution of the atmosphere with sulfur and nitrogen oxides causes acidification of the hydrosphere and destruction of life in the biosphere. Salinization in the lithosphere results from misuse of the hydrosphere. Improper treatment of solid and liquid waste in the lithosphere pollutes the hydrosphere and poisons the biosphere. Removal of vegetation in the biosphere contributes to erosion of the lithosphere, sedimentation of the hydrosphere, and changes in the atmosphere. These all in turn affect the biosphere. Relationships of this kind are practically infinite in number.

Like the holistic ideas expressed in Taoist thought, we are a part of our environment and it is a part of us. The way we affect it affects us.

DID YOU LEARN THIS?

For each of the statements given, fill in the blank with the most appropriate word or phrase from the following list. You may use a word or phrase more than once.

lithosphere	free radicals	hydrologic cycle
hydrosphere	CO_2	glaciers
biosphere	BOD	lakes
atmosphere	oxidation	rivers
·NO	reduction	aquifer
$\cdot NO_2$	horizons	leaching
steady state	erosion	sedimentation
photodissociation	desertification	LUST
SO_2	salinization	contaminant
troposphere	oceans	pollutant
stratosphere		

a. _______ The kind chemical degradation of organic matter that takes place in water under anaerobic conditions
b. _______ Replenishes the fresh water on land
c. _______ The name for molecules that have unpaired electrons
d. _______ Acid precursor(s)
e. _______ Expansion of deserts into productive farmland
f. _______ An underground reservoir of water
g. _______ Gives photochemical oxidative smog its brown color
h. _______ Discrete layers of soil
i. _______ A condition in which a substance is consumed and formed at the same rate
j. _______ The liquid environment
k. _______ Changes the normal composition of the environment but does not seem to harm it
l. _______ The region of the atmosphere in which ozone is beneficial to life
m. _______ The transport of pollutants through soil by the action of water
n. _______ Contain almost all of the water on the earth
o. _______ The greatest threat to the fertility of soils

EXERCISES

1. List some factors that are often considered when assessing the possible risks and benefits of an endeavor.

2. Distinguish between a pollutant and a contaminant. List several substances that are poisons in large amounts but are harmless in small quantities.
3. Describe the four basic realms of our physical environment. How are they related to one another? What is a closed system? What happens to pollutants when they are released into any one of the four realms of the environment?
4. Give an example that shows the interrelationship of two or three realms of the environment.
5. List several reasons that human activity is causing more environmental damage today that it did, say, 200 years ago.
6. What is the primary source of pollution in the atmosphere? What general kinds of atmospheric pollution does it cause?
7. In general, what is photochemical oxidative smog? List several compounds that are emitted from various human activities that contribute to its formation. Why does it form only during daylight hours and not at night? What molecule is present in photochemical smog that gives it its red-brown color?
8. What are free radicals? How do they differ from most ions and molecules? How do they usually behave chemically?
9. Write the chemical equation for the reaction that occurs between nitrogen and oxygen at the high temperatures generated in the cylinders of a gasoline engine.
10. What are PAN? What compounds emitted by human activity are necessary for their formation? How do they serve to export air pollution to unpolluted areas?
11. What is a photochemical reaction? Give an example of one.
12. Ozone is a pollutant in the troposphere, yet in the stratosphere it protects life on the surface of the earth. Explain how ozone can have this dual character.
13. Ozone is continually being created and destroyed in the stratosphere, but normally its concentration remains nearly constant. What is this condition called? What has human activity done to change it?
14. What property of CFCs makes it possible for them to reach the stratosphere where they can cause destruction of the ozone that is there? What chemical species do they form in the stratosphere that causes depletion of the ozone? What is it about the chemical reaction it causes that makes it so effective in destroying ozone?
15. Under what conditions and where on earth does an ozone hole form? What role do polar stratospheric clouds play in the destruction of ozone? How do CFCs enter into this process? Why are CFCs merely a contaminant in the troposphere?

16. What is the primary source of anthropogenic (human caused) CO_2 in the atmosphere? Go back to Chapter 8 and review how the oceans act as a sink for CO_2. Write the equations for the process. Try to figure out why the oceans can only absorb only a limited amount of CO_2. What other way is CO_2 removed from the atmosphere? How is this process affected by human activity?
17. Besides CO_2, list two other greenhouse gases. In general, what property do these compounds have (that N_2 and O_2 do not) to make them act to cause warming in the atmosphere? Briefly describe the process by which the greenhouse effect works.
18. Why are soils the most important part of the lithosphere when it comes to sustaining life on Earth? In general, what are soils composed of?
19. What are the layers in soils called? Briefly describe each and tell what its composition is. Which layer is the most productive for growing food?
20. Why is biological activity in soils necessary to keep them fertile and productive?
21. Is the amount of CO_2 in soils greater or less than that in the atmosphere? Is the amount of O_2 in soils greater or less than that in the atmosphere? Name the process that goes on in soils that causes these differences. What does it do to change the amounts of CO_2 and O_2?
22. What is the primary human-caused environmental threat to the productivity of soils?
23. Under what kinds of conditions is the desertification of soils most likely to happen? Describe briefly how the process of desertification occurs.
24. On average, how long does it usually take to produce an inch (2.5 cm) of new topsoil? What effect does constant cultivation have upon soils? How does fallowing of soils and planting cover crops on them increase their fertility? Why are chemical fertilizers only partially effective in restoring soil fertility? Do you suppose desertified soils usually can be reclaimed?
25. Explain why soils should be treated as nonrenewable natural resources, like petroleum and metal ores.
26. What makes most of the water on the surface of Earth unfit for human consumption? How is most of the fresh water provided to sustain life on land?
27. Describe how the hydrologic cycle works. What powers it?
28. Considering all of the fresh water drawn in the world for human consumption, how is most of it used?
29. In this country, what household activities use the most water?

30. List several factors that limit the amount of fresh water that can be available for human use.
31. How does thermal pollution affect the amount of oxygen that will dissolve in water? How does raising the temperature of water affect the metabolism of aquatic organisms? How can this affect the survival of some organisms that need oxygen to live?
32. What are pathogens? How do they enter drinking water supplies? Why are water-borne diseases not prevalent in the United States today?
33. What do the terms aerobic and anaerobic mean? What kinds of wastes consume dissolved oxygen in water? What does the BOD tell about the amount of organic matter in a body of water? If the BOD in water becomes too high, what changes occur in the kinds of chemical reactions that occur? Write a general, overall chemical reaction for the decomposition of organic matter (containing C, H, N, O, and S) when the BOD is low. What kind of chemical process is this? Do the same for the situation in which the BOD is high. What kind of chemical process is this?
34. What is an aquifer?
35. List several ways that toxic organic compounds can enter watercourses. What is the term used to describe the transport of pollutants from burial sites into sources of fresh water by surface water, such as from rainfall?
36. How do toxic inorganic substances enter fresh water?
37. List two ways that acids can enter watercourses.
38. Describe how mercury and arsenic as water pollutants can cause poisoning in humans. What kinds of compounds do they form in water that makes them behave this way?
39. What do cadmium and zinc have in common? How does this relationship make cadmium pollution a threat to human health?
40. Explain how irrigation of desert soils causes salinization. How does fertilization accelerate this process?
41. Why do you suppose that sedimentation is the major cause of water pollution in the world? List several ways that sediments pollute water.
42. List several human sources of oil pollution in salt water and fresh water. What is a LUST? Why are they of such concern? How do they threaten our supply of fresh water?
43. How is the biosphere affected by changes in the atmosphere, the lithosphere, and the hydrosphere? Can you think of an example that shows it is possible to pollute one realm without affecting any of the others?

THINKING IT THROUGH

Crops for Fuel?

In Chapter 10 it was stated that using modern agricultural methods for growing crops to make gasohol is not efficient energywise. There is another important reason that this is not a wise practice. What might it be?

Pollutants 1,000 Years Ago

By today's standards, would sediments have been considered to be pollutants in this country's fresh waters 1,000 years ago? Explain.

Living on a Remote Island

You have been exiled to a remote island for the rest of your life and must live entirely on the natural resources that exist there. The island is uninhabited now, but it has been studied well, and you have access to all that is known about it before you depart. You also have access to all of modern technology and may take anything that you choose, so long as all of it fits into *one* back pack having a volume of 1 cubic foot. If you wish, one other person may live with you on the island, but must share the contents of your pack. Once you are on the island, no one will visit and you will be unable to leave.

What kinds of items would you take with you in your pack and why? (How would you use them to help you survive?) How would you prioritize them in case some would not fit into the pack? Would you have another person accompany you? Explain.

Energy and a Safe Environment

We have become accustomed to having an abundant, inexpensive source of energy, and we also want a clean, safe environment. Can we have both? Explain.

ANSWERS TO "DID YOU LEARN THIS?"

a. reduction
b. hydrologic cycle
c. free radicals
d. $\cdot NO_2$, SO_2
e. desertification
f. aquifer
g. $\cdot NO_2$
h. horizons
i. steady state
j. hydrosphere
k. contaminant
l. stratosphere
m. leaching
n. oceans
o. erosion

CHEMISTRY IS EVERYWHERE

Miriam C. Nagel

Freelance Science Writer

B. S. Chemical Education, Boston University

M. S. Chemical Education, Simmons College

As I stood near the lake formed by the Tasman glacier on the South Island of New Zealand, I felt like the luckiest person in the world. I had an assignment to write an article on the use of ethylene to control the ripening of fruit when I heard of a group trip to New Zealand. Kiwifruit, which is exported from New Zealand, is one of the fruits treated with ethylene. What better excuse to take the trip?

In New Zealand, I spent a day visiting a kiwifruit producer and exporter. Like bananas and many other fruits, kiwifruit is shipped firm and unripe. Just before marketing, the fruit is exposed to controlled amounts of ethylene to accelerate ripening. Consequently, the customer gets the fruit in prime condition.

The significance of ethylene in the maturation of plants was discovered during years of effort to find out why a commercial grower suffered economic disaster one cold night. His greenhouses were maintained at an even temperature by natural-gas heaters. On that evening, a chrysanthemum crop was almost ready for market, but by the next morning, the plants had lost their leaves and were nearly dead. Eventually, scientists learned that ethylene released into the air by an improperly regulated gas heater had caused the plants to age overnight.

Many fruits produce ethylene in the late stages of maturation or when fruit is damaged. When they do, the maturation of neighboring fruits is accelerated. A rotten apple does spoil the barrel.

Chemistry, especially applied chemistry, has interested me since I heard the Du Pont slogan "Better things for better living through chemistry" when I was in high school. Today Du Pont has dropped "through chemistry" from its ad. Too bad—I am not sure the new message is as inspirational.

I was lucky. My high school chemistry class was enriched by a teacher who told anecdotes about how the theories and chemicals related to everyday life. In one lab experiment, we precipitated a white solid. He commented that the calcium carbonate (our precipitate) was a chemical in most toothpastes.

Freshman chemistry in college was even better. Again I was fortunate in having teachers who made the subject come alive. As a treat, the last lab was a redox reaction using silver nitrate. The unexpected reward was a beautiful mirror. At the start of my third year, I decided to major in chemical education because academic chemistry had to be an exciting career.

Teaching was challenging. The syllabus, text, and lab book prescribed for the first-level chemistry course seemed sterile. Students were not enthusiastic. At first, I was at a loss to provide inspiration. Then one day an art major in my lab was working late. He did not usually seem to care much for the course, but this day he was excited about a bright yellow zinc chromate(VI) hydroxide precipitate. "Why can't we make paint?" he asked enthusiastically. Why not?

Encouraged by the change in this student's morale, I worked paint making into a basic chemistry lesson that so intrigued the students that they wanted another project package. Most amazing of all, they put many extra hours into lab work. Among other inducements, I used the redox lesson that culminated in a silvered mirror. I was working long hours because the required material had to be rearranged and rewritten to accommodate my new approach.

Word got around that this course was different. I was invited to share what we were doing with other teachers and was soon writing and rewriting information on all aspects of the course. Enrollments increased. More writing and more conferences followed. I was honored with an award from the Chemical Manufacturers Association. Then came some commercial writing opportunities.

Writing was not a career I had ever considered. But while I was researching background information I needed to make the laboratory projects as complete as possible, I found that I enjoyed reading and writing about chemistry almost as much as teaching it. And writing had a very distinct advantage—I would be free from the restricting routine.

Kiwifruit growing.

Until I received the assignment to research the use of chemical ripening, I never gave a thought to how bananas from Central America arrive ready to eat at the supermarket. I have learned many new and interesting things and have even had a few extra excuses to travel. Opportunity is nearly unlimited if you look for it—chemistry is everywhere.

Appendix A
Useful Conversion Relationships

Length
1 foot (ft) = 12 inches (in.)
1 yard (yd) = 3 feet (ft) = 0.914 m
1 mile (mi) = 5280 ft = 1760 yd = 1.609 km
1 km = 0.6214 mi
1 in. = 2.54 cm
1 cm = 0.394 in

Mass
1 pound (lb) = 16 ounces (oz) = 454 g
1 kg = 2.20 lb
1 oz = 28.4 g
1 g = 0.035 oz
1 ton = 2000 lb = 907.2 kg
1 metric ton = 1000 kg = 1.103 tons

Volume
1 quart (qt) = 2 pints (pt) = 32 fluid ounces (fl oz) = 0.946 L
1 gallon (gal) = 4 qt = 8 pt = 128 fl oz = 3.705 L
1 L = 1.06 qt
1 fl oz = 29.6 mL
1 mL = 1 cm^3

Time
1 minute (min) = 60 seconds (s)
1 hour (h) = 60 min
1 day = 24 h
1 year = 365 days

Energy
1 cal = 4.18 joules (J)
1 kcal = 1 Cal = 3.97 British thermal units (Btu)

Appendix B
Exponential Notation and Significant Figures

MULTIPLYING AND DIVIDING EXPONENTIAL NUMBERS

When we multiply exponential numbers we simply add the exponents of the exponential terms, thus,

$$(2.0 \times 10^{3})(4.0 \times 10^{6}) = (2.0 \times 4.0)(10^{3} \times 10^{6}) = 8.0 \times 10^{3+6} = 8.0 \times 10^{9}$$

and

$$(3.0 \times 10^{5})(1.5 \times 10^{-7}) = (3.0 \times 1.5)(10^{5} \times 10^{-7}) = 4.5 \times 10^{5+(-7)} = 4.5 \times 10^{-2}$$

In dividing exponential numbers we subtract the exponents of the exponential terms in the denominator from the exponents of those in the numerator, or,

$$\frac{3.0 \times 10^{4}}{1.5 \times 10^{2}} = \frac{3.0}{1.5} \times \frac{10^{4}}{10^{2}} = 2.0 \times 10^{4-2} = 2.0 \times 10^{2}$$

and

$$\frac{7.5 \times 10^{-3}}{5.0 \times 10^{-4}} = \frac{7.5}{5.0} \times \frac{10^{-3}}{10^{-4}} = 1.5 \times 10^{-3-(-4)} = 1.5 \times 10^{1}$$

Problem B.1
Carry out the following multiplications and divisions of exponential terms.

a. $(1.5 \times 10^{6})(3.0 \times 10^{2})$
b. $(6.0 \times 10^{-2})(1.0 \times 10^{9})$
c. $(2.5 \times 10^{-4})(2.0 \times 10^{-3})$
d. $(9.0 \times 10^{5})/(2.0 \times 10^{3})$
e. $(5.0 \times 10^{7})/(2.5 \times 10^{-2})$
f. $(8.0 \times 10^{-3})/(4.0 \times 10^{-5})$
g. $(7.0 \times 10^{-3})/(2.0 \times 10^{8})$

SCIENTIFIC NOTATION

A number such as 174,932 can be expressed in exponential notation several ways: namely, 174.932×10^{3}, 17493.2×10^{1}, 17.4932×10^{4}, and so forth. A convention for expressing exponential numbers, called *scientific notation,* has been adopted in science to make it easier to compare numbers of different magnitudes and, as we shall see shortly, to clearly indicate their precision. In scientific notation, an exponential number is always expressed as some number between one and 10, this number being multiplied by 10 raised to the appropriate exponent. In scientific notation, then, the number just given would be written 1.74932×10^5. There are a few simple rules for writing numbers in scientific notation:

1. Move the decimal point of the number until you have a number between one and 10:

2. Note in which direction and by how many places you have moved the decimal point:

8.9764	000003.109
⟵	⟶
three places to the left	six places to the right

3. For each digit you moved the decimal point to the left, multiply the number between one and 10 by that many positive powers of 10. Because in our example we moved the decimal point three digits to the left, we multiply 8.9764 by 10^3 (three positive powers of 10) and write

$$8.9764 \times 10^3$$

4. For each digit you moved the decimal point to the right, multiply the number between one and 10 by that many negative powers of 10. In our example we moved the decimal point six places to the right, so we multiply 3.109 by 10^{-6} (six negative powers of 10) and write

$$3.109 \times 10^{-6}$$

There is an easy way to remember steps 3 and 4. Note that when the decimal point is moved to the left to make a number between one and 10, as in step 3, the original number (the number you started with) gets smaller. This means that the exponential term that you multiply by must correspondingly be made larger. The only way it can increase is to give it an exponent that is more positive. Likewise, moving the decimal point to the right, as in step 4, causes the original number to become larger when the number between one and 10 is produced. This means that the exponential term you multiply by must correspondingly be made to decrease, and to do this the exponent must be made more negative.

Problem B.2

Express the following numbers in scientific notation:

a. 1094.2	g. 262
b. 0.00759	h. 0.8317
c. 93.2633	i. 236×10^{3}
d. 0.000005972	j. 197×10^{-3}
e. 473215	k. 0.0373×10^{7}
f. 0.00000000013975	l. 0.000764×10^{-2}

SIGNIFICANT FIGURES: INDICATING THE PRECISION OF A MEASUREMENT

If you want to know how much change you have on hand to feed a photocopy machine, you can count it exactly to the nearest cent. Likewise, you can determine the number of students in your classroom on a certain day, or the number of eggs you have left in the refrigerator. These quantities, which you can count to the nearest item, are called exact numbers.

Numbers resulting from measurement do not share this exactness. Let's consider an example from a study question in Chapter 2 in which we wanted to know how precisely we could tell the time using a clock whose face was marked at 3, 6, 9, and 12 as compared with one marked at every hour. Both kinds of clocks are illustrated in ▶ Figure B.1, and both are set at the same time.

To determine the hour indicated on the clock whose dial is divided into quarters (▶ Figure B.1a), we note that the hour hand is positioned about halfway between 9 and 12, and our best guess is that it lies somewhere

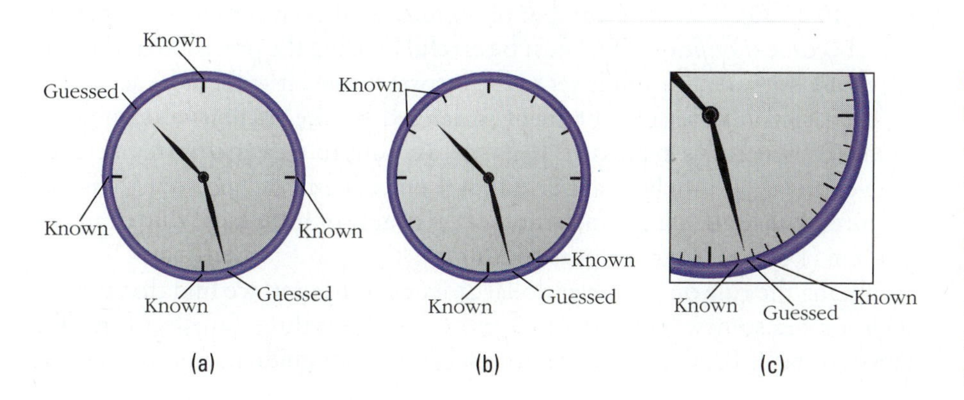

▶ Figure B.1
Clock Faces Showing Time to Different Degrees of Precision. *(a) Clock face marked at 3, 6, 9, and 12 makes precise estimation of the time difficult. (b) Face divided into 12 hours permits estimation of the time to the nearest minute. (c) Face divided into 60 minutes gives the time to the nearest minute and permits its estimation to a few seconds.*

between 10 and 11. Because the hour hand only has to indicate which hour it is, this estimate is not too bad; and it probably is 10-something o'clock. Because we have no graduations between 9 and 12 to guide us, however, we cannot be certain that this is true. Similarly, the minute hand on this clock is located somewhere between 3 and 6, about two-thirds of the way toward the 6, or about 5 on the dial. A good guess is that it is about 25 min after the hour, but that is only a guess. A better, and more honest, estimate would be that the minute hand is somewhere between 20 min and 30 min after the hour. Thus, we could estimate the time by this clock to be 10:25 ± 5 min (plus-or-minus five minutes), and the hour is an estimate.

Finding the time on the clock whose dial is divided into twelve parts is much easier (▶ Figure B.1b). Now we can see that the hour hand indeed lies between 10 and 11, so the 10 o'clock hour is now certain. Also, the minute hand is pointed somewhere between 5 and 6, but a little closer to the 5. We can now be certain that the minute hand is between 25 and 30 min after the hour, and guessing the last minute, we probably would say that it is about 27 min past. Because we have no graduations to go by, however, the 7 in the 27 is still a guess, and a better estimate would be that the time is somewhere between 26 and 28 min after 10 o'clock. We then would write the time as 10:27 ± 1 min.

The third clock dial (▶Figure B.1c) is marked in 60-min divisions, and now it is easy to see that the minute hand lies almost exactly between 27 and 28 min past the hour. We can now determine the number of minutes exactly and even estimate the fraction of a minute, approximately one-half minute (30 sec) in this case, or 10:27:30. (The 30 sec is probably good to ±10 sec or so.) If we wanted to obtain even more precision, to see how many seconds actually had elapsed, we could employ a clock with a second hand.

The precision of the measurement of the time has been successively improved in each example. The clock with the face divided into quarters permitted us to only estimate the hour and be certain of the number of the minutes past to within a range of 10 min or so. The best we can say about the time obtained from this clock is that it is about 10:25. On the other extreme, the clock whose dial is divided into minutes permits us to know the hour and the number of minutes with certainty, and gives us an estimate of the number of seconds (10:27:30). Thus, the number of minutes, and even seconds to some degree, become *significant.* (We must be careful in using the precision of the better clock, however, for if it is set to the incorrect time, it will not be accurate.)

The precision of a measurement is defined by the number of significant figures in which it is expressed. Let's say we want to measure the diameter of a silver dollar, and we have at our disposal a metric rule having 1-cm divisions (Figure ▶B.2a) and a set of metric calipers capable of measuring to 0.1 mm, or 0.01 cm (1 cm = 10 mm) (Figure ▶B.2b).

Placing the rule on the silver dollar at its widest point, we find that this dimension lies somewhere between 3 and 4 cm, about three-fourths of the distance to the 4. Because we have no smaller scale to guide us (a 0.1-cm scale,

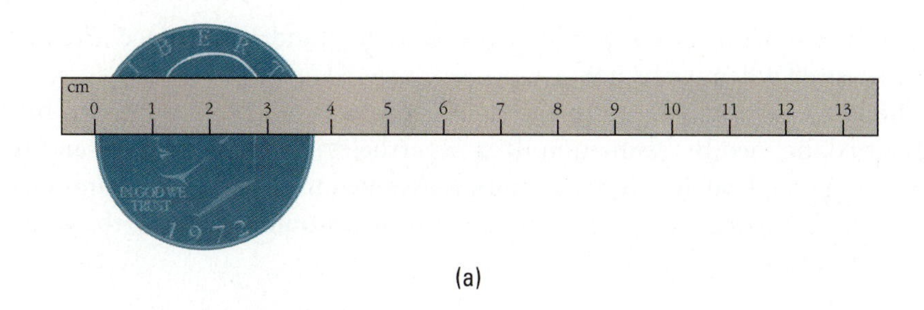

(a)

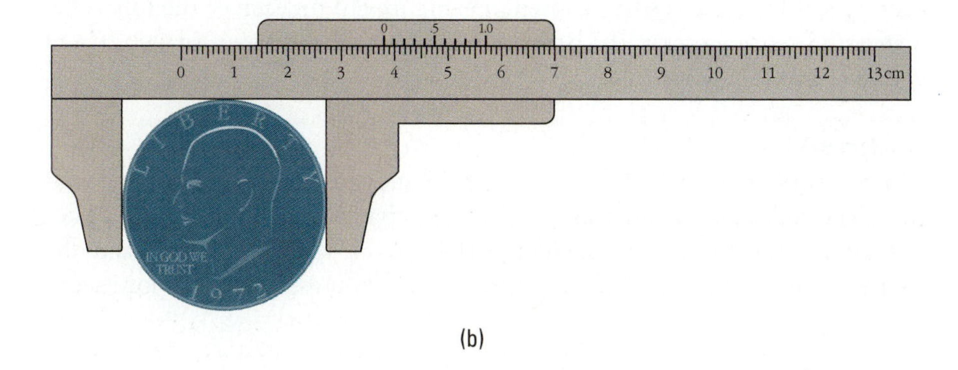

(b)

▶ Figure B.2
Measuring A Silver Dollar With a Simple Metric Rule and With Calipers. *(a) Metric rule divided only into centimeters allows estimation of the diameter of the silver dollar to the nearest 0.1 cm. (b) Vernier calipers afford greater precision, allowing measurement of the diameter of the silver dollar to 0.1 cm and estimation to 0.01 cm.*

for example), we must guess this distance to the nearest 0.1 cm (1 mm). A good estimate is about 0.7 cm, so we can record the diameter of the silver dollar measured this way as being 3.7 cm. We are confident of the units place, or the 3, but the tenths place is uncertain. It is reasonable to expect, however, that our diameter of 3.7 cm is somewhere between 3.6 cm and 3.8 cm, or an uncertainty of about ±0.1 cm.

Measuring the silver dollar with the calipers, we find its diameter to be 3.80 cm. It is possible for us to clearly distinguish 0.1 cm, so this time we have confidence in the units and tenths places, or 3.8. We must now estimate the hundredths place and can do this to the nearest 0.01 cm. This gives us a value of 3.80 ± 0.01 cm. Note that both measurements agree within their limits of precision: 3.7 ± 0.1 cm (3.6 to 3.8 cm) and 3.80 ± 0.01 cm (3.79 to 3.81 cm).

The diameter of the silver dollar measured with the metric rule is expressed to two significant figures, whereas that measured with the calipers is expressed to three. Note the zero in the hundredths place of the three-significant-figure value, 3.80 cm. Since this zero comes after the decimal place and is placed at the end of the number, it has meaning—namely, that the measurement was made with a precision of 0.01 cm and that the hundredths-place number actually was measured to be zero. The value of 3.7 cm obtained from the rule graduated in centimeters has no zero at the end because its precision is only to the nearest 0.1 cm. To record 3.70 cm for this measurement is inappropriate because this would imply that the hundredths place actually was measured when it was not.

The last significant figure to the right is always somewhat uncertain because it is obtained by estimation. It is nevertheless a measured value and is considered real. Usually, its uncertainty is assumed to be within one unit unless otherwise stated. A volume written 35.47 mL would mean that the value was between 35.46 mL and 35.48 mL.

Problem B.3

Record the value of the measurement showing the proper number of significant figures for scale reading shown in ▶ Figure B.3 on each of the following devices: (a) speedometer, (b) bathroom scale, (c) tachometer, (d) graduated cylinder, and (e) metric rule.

Problem B.4

What factors might introduce error into the measurement of the diameter of the silver dollar used as an example above, so that the values might not be accurate? For example, how could the rule give a value of 3.7 cm and the calipers one of 3.80 cm? Which of these would you expect to be more accurate?

THE CONVENTION FOR EXPRESSING SIGNIFICANT FIGURES

Properly recorded measurements must be expressed to the correct number of significant figures. To ensure consistency when communicating significant figures, a convention for expressing significant figures is used within the scientific community. To determine the number of significant figures in a number, follow these simple rules:

1. For a number having a decimal point or ending in a non-zero digit, start at the left. Reading to the right, count the first nonzero digit and all other digits that follow, including zeros. Thus, the number 30.09 contains four significant figures and the number 46702 contains five. The number 0.00296 has three significant figures and the number 0.0000001357728 has seven. (Start counting at the first nonzero digit on the left: 2 and 1, respectively.) The number 3.210 has four significant figures and the number 0.00795000 has six. (The zeros on the right are counted because they follow a nonzero digit in a number that has a decimal point.)
2. In numbers written without a decimal point, zeros that follow (are to the right of) all nonzero digits may or may not be significant. Thus, the number 1700 may have two, three, or four significant figures depending on the precision to which it was measured.

▶ Figure B.3
Illustrations for Problem B.3

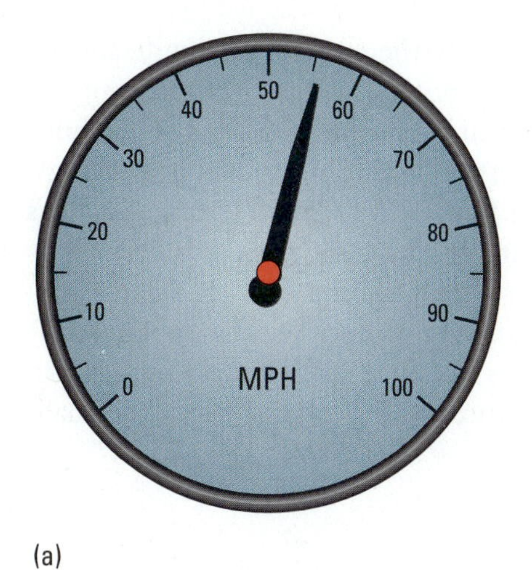

(a)

(b)

(c)

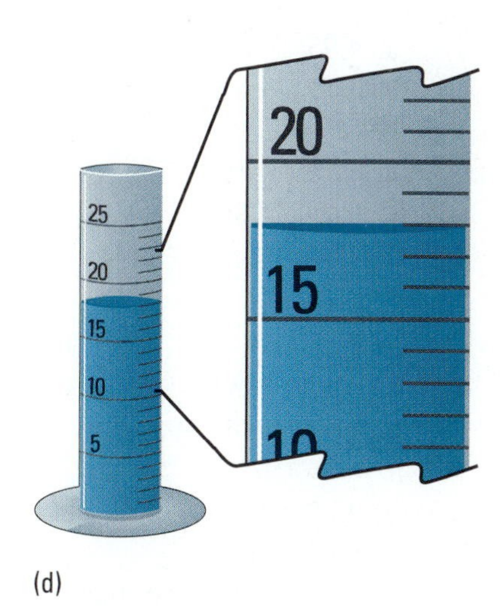

(d)

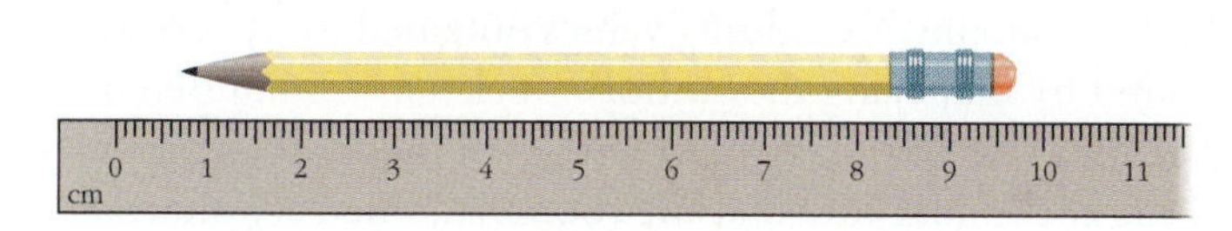

(e)

If all zeros were measured and are significant, the decimal point can be placed after the last zero, or 1700. (*not* 1700.0), to indicate that the two zeros are real. However, if only one, or neither, of the zeros was actually the result of a measurement, the situation is ambiguous. To indicate clearly the number of significant figures in such a number, scientific notation is used, because any digit that follows the decimal point is considered significant. Thus, the number can be expressed to two significant figures by writing it as 1.7×10^3, three significant figures as 1.70×10^3 (one zero has been measured and is significant), and four significant figures as 1.700×10^3 (both zeros measured and significant). Scientific notation, then, provides an unambiguous means of expressing the correct number of significant figures obtained from a measurement, as well as being a useful way for expressing very large and very small numbers.

Problem B.5

Indicate the number of significant figures in each of the following measurements:

a. 4.95×10^2 m
b. 3.701 years
c. 8.720 L
d. 21,500. trout
e. 0.05009 kg
f. 7000 km
g. 45.00 μg
h. 45 mg
i. 0.000000009 m
j. 6.903 g
k. 1.300×10^{-4} cm

Problem B.6

Express the following measurements to the correct number of significant figures using scientific notation:

a. 43.0 mL
b. 0.0298 kg
c. 79,000 cm (3 sig. figs.)
d. 0.000500 s
e. 300,000,000 m/s

Problem B.7

The following sums were obtained by adding values obtained by observation. "Net worth" was found by emptying the author's pockets. "Circumference" (of a very irregular room) was found by finding the length of all of the sides of the room and taking the total. Note that the room had one very short side which had to be measured with calipers.

Net worth:	Change	$.84
	Bills	$35.00
	Checkbook balance	$254.38
	Total	$290.22

Circumference:	Long wall (meter stick)	1042.3 cm
	Long wall (meter stick)	779.8 cm
	Very short wall (calipers)	1.245 cm
	Long wall (meter stick)	167.6 cm
	Circumference of room	1990.945 cm

To how many significant figures can the total net worth be expressed? The circumference of the room cannot be expressed to seven significant figures. To how many should it be expressed? What are the two differences between the total net worth, a sum of what was in the author's pockets, and the circumference of the room, a sum of the lengths of the sides of the room? What can happen to the number of significant figures when measurements are added? (Examine the number of significant figures in each measurement and compare this to the number of significant figures in the total.) What do you suppose could happen to the number of significant figures when measurements are subtracted? Why is it possible to count all of the decimal places (to the cent) in the author's net worth as significant, yet it is not proper to do this with all of the decimal numbers in the circumference?

Problem B.8

A sample of red brass weighed 27.521 g. Its volume, by finding its displacement of water in a graduated cylinder, was determined to be 3.2 mL. The density of the brass then was calculated from these measurements to be 27.521 g / 3.2 mL = 8.6003125 g/mL on a pocket calculator. This calculator was not programmed to express the proper number of significant figures. Using your response to problem B.7 as a point of departure, try to determine the correct number of significant figures to which the density should be reported.

Problem B.9

Problems B.7 and B.8 illustrate questions that occur when calculations are carried out with measurements having different numbers of significant figures. There is a general, and fairly simple, "rule of thumb" that can be used to determine the precision of an answer arrived at by mathematical manipulation of measured values. It begins, "The precision of a value arrived at by mathematical manipulation of measured values can be no better than ____." Complete the rule.

ANSWERS TO SELECTED PROBLEMS

B.1

a. 4.5×10^{8}
b. 6.0×10^{7}
c. 5.0×10^{-7}
d. 4.5×10^{2}
e. 2.0×10^{9}
f. 2.0×10^{2}
g. 3.5×10^{-11}

B.2

a. 1.0942×10^{3}
b. 7.59×10^{-3}
c. 9.32633×10^{1}
d. 5.972×10^{-6}
e. 4.73215×10^{5}
f. 1.3975×10^{-10}
g. 2.62×10^{2}
h. 8.317×10^{-1}
i. 2.36×10^{5}
j. 1.97×10^{-1}
k. 3.73×10^{5}
l. 7.64×10^{-6}

B.5

a. 3
b. 4
c. 4
d. 5
e. 4
f. uncertain (1 to 4)
g. 4
h. 2
i. 1
j. 4
k. 4

B.6

a. 4.30×10^{1} mL
b. 2.98×10^{-2} kg
c. 7.90×10^{4} cm
d. 5.00×10^{-4} s
e. 3×10^{8} m/s

Appendix C
Answers to You Try It Boxes

CHAPTER 2

2.1 a. surveyor's chain; b. postal scale; c. one-quart container; d. six-inch dollar bill; e. by the hour; f. timing it; g. thin lines

2.2 a. washed potatoes; b. stocking feet; c. barn door; d. basketball; e. 20 yards; f. ballpoint pen; g. odometer; h. thin lines

2.3 Setups only given. a. 210 s $\times$ 1 min/60 s; b. 55 gal $\times$ 4 qt/1 gal; c. years $\times$ 12 mo/1 year; d. 72 blocks $\times$ 1 mile/16 blocks; e. 3.5 hr $\times$ 50 mi/1 hr; f. 5 miles $\times$ 1 hr/50 miles $\times$ 60 min/1 hr

2.4 a. .001 (liter); b. 1000 (gram); c. .1 (meter); d. .000000000001 (second); e. .01 (meter); f. .000001 (gram); g. .001 *gram;* h. 1000 *meter;* i. .000001 *second;* j. .000000001 *meter;* k. .01 *gram;* l. 1,000,000 *liter*

2.5 a. 7.73×10^{-2} m; b. 1.2×10^{-3} L; c. 8.27×10^{3} g; d. 3.56×10^{-6} s; e. 5.36×10^{-1} m; f. 4.44×10^{-3} g; g. 9.3×10^{6} L

CHAPTER 3

3.1 a. energy; b. energy; c. matter; d. energy; e. matter; f. energy; g. matter

3.2 a. solid; b. gas; c. solid; d. solid; e. gas; f. liquid

3.3 a. mixture; b. pure substance; c. mixture; d. pure substance; e. pure substance

3.4 a. heterogeneous; b. heterogeneous; c. homogeneous; d. homogeneous; e. heterogeneous

3.5 a. physical; b. chemical; c. chemical; d. physical; e. chemical

3.6 a. on bottom shelf; b. the river below; c. the bottom of the run; d. ashes; e. dead one

CHAPTER 4

4.1 The correctly spelled names and symbols for the elements are given inside the back cover of this book.

4.2 a. $CaCl_2$, 1 atom Ca, 2 atoms Cl; b. N_2H_4, 2 atoms N, 4 atoms H; c. OF_2, 1 atom O, 2 atoms F; d. CH_2Br_2, 1 atom C, 2 atoms H, 2 atoms Br; e. SiH_4, 1 atom Si, 4 atoms H; f. BF_3, 1 atom B, 3 atoms F; g. $(NH_4)_2SO_4$, 2 atoms N, 8 atoms H, 1 atom S, 4 atoms O: h. $Al_2(SO_4)_3$, 2 atoms Al, 3 atoms S, 12 atoms O; i. NH_3, 1 atom N, 3 atoms H; j. Na_2S, 2 atoms Na, 1 atom S; k. C_2H_6O, 2 atoms C, 6 atoms H, 1 atom O; l. Al_2O_3, 2 atoms Al, 3 atoms O; m. K_2SO_4, 2 atoms K, 1 atom S, 4 atoms O; n. HNO_3, 1 atom H, 1 atom N, 3 atoms O; o. $Ca(NO_3)_2$, 1 atom Ca, 2 atoms N, 6 atoms O; p. $Cu_2(CN)_2$ 2 atoms Cu, 2 atoms C, 2 atoms N.

4.3 a. $H_2 + Cl_2 \longrightarrow 2HCl$
b. $S + O_2 \longrightarrow SO_2$ Balanced
c. $2S + 3O_2 \longrightarrow 2SO_3$
d. $SO_3 + H_2O \longrightarrow H_2SO_4$ Balanced
e. $I_2 + Cl_2 \longrightarrow 2ICl$
f. $2Na + Br_2 \longrightarrow 2NaBr$
g. $4K + O_2 \longrightarrow 2K_2O$
h. $Ca + S \longrightarrow CaS$ Balanced
i. $H_2 + S \longrightarrow H_2S$ Balanced

4.4 a. $4Al + 3O_2 \longrightarrow 2Al_2O_3$
b. $2Al + 3Cl_2 \longrightarrow 2AlCl_3$
c. $2S + 3O_2 \longrightarrow 2SO_3$
d. $2P + 3Cl_2 \longrightarrow 2PCl_3$ Balanced
e. $3Mg + N_2 \longrightarrow Mg_3N_2$
f. $4P + 5O_2 \longrightarrow 2P_2O_5$

CHAPTER 5

5.1 a. small; b. large; c. large; d. low; e. large; f. *c*, constant

5.2 a. greater; b. low; c. high; d. high; e. low; f. long, low; g. short, high, high; h. lower, less

5.3 a. *d*; b. *s*; c. *f*; d. *s*; e. *p*; f. *p*; g. *p*; h. *d*

5.4 a. 12; b. 49; c. 33; d. 52; e. 14; f. 35; g. 26; h. 80; i. 19; j. 25

CHAPTER 6

6.1 a. Li^+, lithium ion; b. Al^{3+}, aluminum ion; c. Ba^{2+}, barium ion; d. K^+, potassium ion

6.2 a. O^{2-}, oxide ion; b. N^{3-}, nitride ion; c. I^-, iodide ion; d. Cl^-, chloride ion; e. S^{2-}, sulfide ion

6.3 a. LiCl, lithium chloride; b. AlF_3, aluminum fluoride; c. CaS, calcium sulfide; d. MgI_2, magnesium iodide; e. K_2O, potassium oxide; f. Na_3P, sodium phosphide

6.4 a. NaI; b. CaO; c. Al_2O_3; d. K_2S; e. $MgCl_2$; f. Li_3N

6.5 a. sodium nitrate; b. potassim hydroxide; c. sodium hydrogen carbonate (sodium bicarbonate); d. calcium sulfate; e. magnesium hydroxide; f. ammonium chloride

6.6 a. $KHCO_3$; b. NH_4Br; c. Li_2CO_3; d. $Ca(NO_3)_2$; e. $Mg_3(PO_4)_2$; f. Na_2SO_4

6.7 The Lewis structures are not the same for each, since these compounds are made up of hydrogen and elements all in the same group (7A).

H:F:

6.8 All of these compounds are made up of hydrogen and elements of Group 6A. The same is true for those made up of hydrogen and elements of Group 5A. Likewise, for Group 4A. Within a family of elements, a series of compounds made from these elements and another nonmetal element will have the same Lewis structures.

H:Se:H   H:P:H (H)   H:Si:H (H, H)

6.9 They have the same electron-dot formulas, since carbon is bonded in each case to Group 7A elements.

6.10 a. As; b. S; c. N; d. P; e. Te; f. N; g. C; h. O

6.11 One of the electron pairs in each NF_3 and PCl_3 is not bonded to a second atom. That is, each of these molecules has a nonbonded, or unshared (or "lone") electron pair. In OF_2 and SCl_2 there are two nonbonded pairs of electrons.

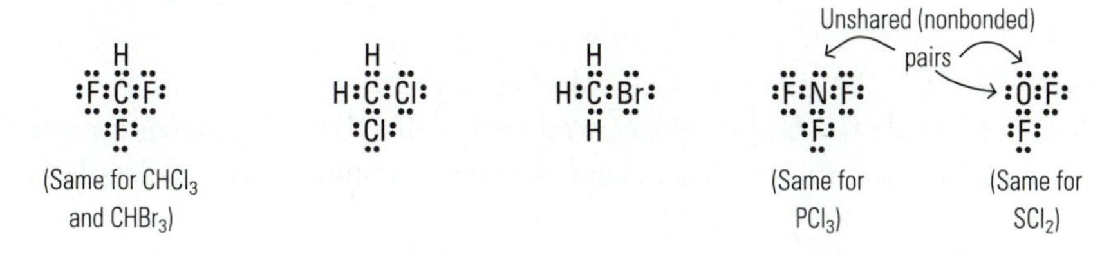

6.12 S::C::S H₂C::O H₂C::S

6.13 :C:::O H:C:::C:H

6.14 H:O:N::O (with :O: on N) H:O:C:O:H (with :O: on C) H:O:P:O:H (with :O: above P and :O:H below P)

6.15 :O:N::O ($^-$) H:O:C:O: ($^-$) :O:C:O: ($^{2-}$) :O:P:O: with :O: above and below P ($^{3-}$)

6.16 All are tetrahedral.

6.17 Simple molecules in a series having central atoms that are elements in the same family usually have geometries that are the same. For example, O, S, and Se are in the same family and when these elements make up the central atom in molecules such as OF_2, H_2S, and H_2Se the molecules all have the same arrangement (geometry) of electron pairs (tetrahedral) and the same molecular geometry (bent, or V-shaped). Likewise, CF_4, CH_2Cl_2, SiH_4, and SiF_4 have a tetrahedral arrangement of electron pairs and a tetrahedral shape, and NCl_3, PH_3, AsH_3, and $AsCl_3$ have a tetrahedral arrangement of electron pairs and a trigonal pyramidal (or pyramidal) shape.

6.18 H_2CS is triangular, SO_3 is triangular, $CO_3{}^{2-}$ is triangular, and $NO_3{}^-$ is triangular. The arrangement of the electron pairs in the SO_2 molecule is triangular (trigonal planar), but since it has a nonbonded pair of electrons on the sulfur atom, the molecule itself has a bent, or U-shaped geometry.

H₂C=S O₂S=O O₂C=O ($2-$) O₂C=O ($-$)

O–S=O (with lone pair on S)

6.19 All are linear.

H:N:::C: S::C::S O::C::S H:C:::C:H

Geometry at *each* carbon atom is linear.

6.20 a. C; b. Br; c. O; d. Cl; e. N; f. O; g. F

6.21 The rule for finding the central atom when drawing Lewis structures states that it is the least electronegative element in the molecule.

6.22 a. carbon monoxide; b. nitrogen dioxide; c. oxygen dichloride; d. disulfur dichloride; e. diphosphorus trioxide; f. tricarbon dioxide; g. dinitrogen trioxide; h. phosphorus pentachloride; i. sulfur tetrafluoride; j. nitrogen trifluoride

6.23 a. molecular; b. ionic; c. molecular; d. ionic; e. molecular; f. ionic; g. molecular; h. molecular; i. ionic; j. ionic

6.24 a. ionic; b. ionic; c. molecular; d. ionic; e. molecular; f. molecular; g. ionic; h. molecular; i. ionic; j. molecular; k. molecular; l. ionic; m. molecular

CHAPTER 7

7.1 a) salt, sugar, baking soda (sodium hydrogen carbonate), epsom salts (magnesium sulfate), boric acid, etc., dissolved in water, pine pitch dissolved in turpentine, iodine dissolved in alcohol (tincture of iodine), grease dissolved in kerosene; b) rubbing alcohol (70% isopropyl alcohol, 30% water), gasoline and kerosene (mixtures of liquid hydrocarbons), antifreeze in water, fingernail polish (acetone, ethyl acetate, and other components), paint thinner (mixture of liquid hydrocarbons); c) many alloys, some kinds of glass; d) carbonated water, ammonia solution, hydrogen chloride solution (muriatic acid and some toilet bowel cleaners), oxygen in water (so fish can breathe); e) the air (oxygen dissolved in nitrogen with a little carbon dioxide and other gases in minor amounts), oxygen dissolved in helium for diving to great depths underwater.

7.2 Freezing water rupturing a water pail, the buckling of ice on lakes, etc.

7.3 a. $KF(s) \xrightarrow{\text{water}} K^+ + F^-$ "Solid potassium fluoride dissolved in water yields potassium ions and fluoride ions in the ratio of 1:1."

b. $CaBr_2(s) \xrightarrow{\text{water}} Ca^{2+} + 2Br^-$ "Solid calcium bromide dissolved in water yields calcium ions and bromide ions in the ratio of 1:2."

c. $NaHCO_3(s) \xrightarrow{\text{water}} Na^+ + HCO_3^-$ "Solid sodium hydrogen carbonate dissolves in water to produce sodium ions and hydrogen carbonate ions in the ratio of 1:1."

d. $MgSO_4(s) \xrightarrow{\text{water}} Mg^{2+} + SO_4^{2-}$ "Solid magnesium sulfate when dissolved in water produces magnesium ions and sulfate ions in the ratio of 1:1."

e. $NH_4Cl(s) \xrightarrow{\text{water}} NH_4^+ + Cl^-$ "Solid ammonium chloride produces ammonium ions and chloride ions in the ratio of 1:1 when dissolved in water."

f. $Al(NO_3)_3(s) \xrightarrow{\text{water}} Al^{3+} + 3NO_3^-$ "Solid aluminum nitrate when dissolved in water yields aluminum ions and nitrate ions in the ratio of 1:3."

7.4 Many sugars are so soluble in water that very little water is needed to dissolve a large amount of sugar, and often the percentage of sugar in the syrup is greater than 50%. This means that the solid is the substance that will be in greater abundance.

7.5
a. 14.0 g/200. g × 100% = 7.00%
b. 250. g/1250 g × 100% = 20.0%
c. 1.8 g/90.0g × 100% = 2.0% (2 sig figs here)
d. Total amount of solution is 22.0 g + 178 g = 200. g. Then, 22.0 g/ 200. g × 100% = 11.0%
e. Total amount of solution is 1.4 g + 138.6 g = 140.0 g. Then, 1.4 g/ 140.0 g × 100% = 1.0%

7.6
a. 100. g soln × 10.0 g KNO_3/100. g soln = 10.0 g KNO_3
b. 750. g soln × 20.0 g HCl/100. g soln = 150. g HCl
c. 2.5 kg soln × 10^3 g soln/1 kg soln × 0.10 g $HC_2H_3O_2$/100. g soln = 2.5 g $HC_2H_3O_2$

7.7
a. N_2, 28.0 g; b. Br_2 159.8 g; c. Mg, 24.3 g; d. Zn, 65.4 g;
e. NH_3, 17.0 g; f. CaO, 56.1 g; g. KNO_3, 101.1 g;
h. C_3H_8O, 72.1 g; i. H_2SO_4, 98.1 g; j. Al_2S_3, 150.3 g;
k. C_5H_5N, 79.0 g; l. Na_3PO_4, 164.0 g; m. $Mg(OH)_2$, 58.3 g;
n. IBR, 207 g; o. CCl_4, 154.0 g

7.8 Since molecules are very tiny and are impossible to weigh individually or in small numbers on ordinary laboratory balances (even very sensitive ones), using molecules as a unit in the laboratory would be impossible.

7.9
a. $H_2 + S \longrightarrow H_2S$ (balanced as is) "One mole of hydrogen reacts with one mole of sulfur to produce one mole of H_2S (hydrogen sulfide)."
b. $P_4 + 5O_2 \longrightarrow P_4O_{10}$ "One mole of phosphorus (as P_4) reacts with five moles of oxygen to produce one mole of P_4O_{10} (tetraphosphorus decaoxide)."
c. $3H_2 + N_2 \longrightarrow 2NH_3$ "Three moles of hydrogen react with one mole of nitrogen to produce two moles of ammonia."
d. $4Na + O_2 \longrightarrow 2Na_2O$ "Four moles of sodium (metal) react with one mole of oxygen to yield two moles of sodium oxide."
e. $3Mg + N_2 \longrightarrow Mg_3N_2$ "Three moles of magnesium (metal) react with one mole of nitrogen to produce one mole of magnesium nitride."

f. $C_2H_4 + 3O_2 \longrightarrow 2CO_2 + 2H_2O$ "One mole of C_2H_4 (ethene) reacts with three moles of oxygen to produce two moles of carbon dioxide and two moles of water."

g. $2C_5H_{10} + 15O_2 \longrightarrow 10CO_2 + 10H_2O$ "Two moles of C_5H_{10} react with fifteen moles of oxygen to yield ten moles of carbon dioxide and ten moles of water."

7.10 a. 2.0 mol I_2 × 254 g I_2/mol I_2 = 510 g I_2
b. 0.30 mol Li × 6.94 g Li/mol Li = 2.1 g Li
c. 0.75 mol SO_2 × 64.1 g SO_2/mol SO_2 = 48 g SO_2
d. 5.0 mol NaCl × 78.5 g NaCl/mol NaCl = 390 g NaCl
e. 0.40 mol C_7H_8O × 108 g C_7H_8O/mol C_7H_8O = 43 g C_7H_8O

7.11 a. 4.5 g Be × 1 mol Be/9.01 g Be = 0.50 mol Be
b. 56.0 g N_2 × 1 mol N_2/28.0 g N_2 = 2.00 mol N_2
c. 4.4 g CO_2 × 1 mol CO_2/44.0 g CO_2 = 0.10 mol CO_2
d. 6.35 mg Cu × 1 g/10^3 mg × 1 mol Cu/63.5 g Cu = 1.00×10^{-4} mol Cu
e. 39.0 g C_6H_6 × 1 mol C_6H_6/78.1 g C_6H_6 = 0.499 mol C_6H_6
f. 299 g Li_2O × 1 mol Li_2O/29.9 g Li_2O = 10.0 mol Li_2O

7.12 a. 4.0 mol/2.0 L = 2.0 mol/L = 2.0 *M*
b. 0.25 mol/.750 L = 0.333 mol/L = 0.333 *M* (Note change of mL to L.)
c. 1.0 mol/0.400 L = 2.5 mol/L = 2.5 *M* (Note change of mL to L.)
d. 0.75 mol/3.0 L = 0.25 mol/L = 0.25 *M*
e. 1.25 mol/0.25 L = 5.0 mol/L = 5.0 *M*

7.13 a. 11.9 g KBr × 1 mol KBr/119 g KBr = 0.100 mol KBr
0.100 mol/0.50 L = 0.20 mol/L = 0.20 *M*
b. 32.5 g HNO_3 × 1 mol HNO_3/63.0 g HNO_3 = 0.500 mol HNO_3
0.500 mol/2.0 L = 0.25 mol/L = 0.25 *M*
c. 1.68 g $NaHCO_3$ × 1 mol $NaHCO_3$/84.0 g $NaHCO_3$ = 0.0200 mol $NaHCO_3$
0.0200 mol/0.100L = 0.200 mol/L = 0.200 *M* (Note change of mL to L.)

CHAPTER 8

8.1 $HBr(g) \xrightarrow{\text{water}} H^+ + Br^-$

$HI(g) \xrightarrow{\text{water}} H^+ + I^-$

$HNO_3(l) \xrightarrow{\text{water}} H^+ + NO_3^-$

8.2 $KOH(s) \xrightarrow{\text{water}} K^+ + OH^-$

$LiOH(s) \xrightarrow{\text{water}} Li^+ + OH^-$

8.3 $Al(OH)_3(s) + 3H^+ + 3Cl^- \longrightarrow Al^{3+} + 3H_2O + 3Cl^-$

8.4 $C_2H_3O_2^- + H_2O \rightleftharpoons HC_2H_3O_2 + OH^-$

$SO_4^{2-} + H_2O \rightleftharpoons HSO_4^- + OH^-$

$PO_4^{3-} + H_2O \rightleftharpoons HPO_4^{2-} + OH^-$

8.5 a. $HC_2H_3O_2(l) \overset{\text{water}}{\rightleftharpoons} H^+ + C_2H_3O_2^-$

b. $HC_3H_5O_3(l) \overset{\text{water}}{\rightleftharpoons} H^+ + C_3H_5O_3^-$

c. $HC_6H_7O_7(s) \overset{\text{water}}{\rightleftharpoons} H^+ + C_6H_7O_7^-$

d. $HC_4H_7O_2(l) \overset{\text{water}}{\rightleftharpoons} H^+ + C_4H_7O_2^-$

e. $HC_6H_7O_6(s) \overset{\text{water}}{\rightleftharpoons} H^+ + C_6H_7O_6^-$

f. $HC_2HO_4(s) \overset{\text{water}}{\rightleftharpoons} H^+ + C_2HO_4^-$

g. $HCN(g) \overset{\text{water}}{\rightleftharpoons} H^+ + CN^-$

CHAPTER 9

9.1 The N_2H_4 is oxidized. It lost hydrogen.

9.2 a. CH_4 b. Zn c. CH_3CH_2OH d. Cu e. Si f. H_2O

9.3 a. SO_3 b. HNO_3 c. NO_2 d. P_2O_5 e. H_2O_2

9.4 The SO_2 is reduced. In SO_2, the sulfur atom is reduced. (The sulfur atom in elemental sulfur is less oxidized than the one in SO_2.) In H_2S, the sulfur atom is oxidized. (The sulfur atom in elemental sulfur is more oxidized than the one in H_2S.)

9.5 and 9.6

a.	H_2	O_2
	oxidized	reduced
	reducing agent	oxidizing agent
b.	Ca	H_2O
	oxidized	reduced
	reducing agent	oxidizing agent
c.	SnO_2	C
	reduced	oxidized
	oxidizing agent	reducing agent
d.	H_2S	O_2
	oxidized	reduced
	reducing agent	oxidizing agent

e. Na — oxidized, reducing agent; Cl_2 — reduced, oxidizing agent
f. Br^- — oxidized, reducing agent; Cl_2 — reduced, oxidizing agent
g. CuO — reduced, oxidizing agent; CH_4 — oxidized, reducing agent
h. Mg — oxidized, reducing agent; N_2 — reduced, oxidizing agent

CHAPTER 10

10.1
a. chemical
b. electrical, mechanical, thermal
c. nuclear, thermal
d. chemical
e. electrical
f. chemical, electrical
g. mechanical
h. nuclear, radiant, thermal
i. radiant, thermal
j. mechanical, thermal, chemical
k. thermal, radiant
l. mechanical

10.2
a. kinetic
b. potential, kinetic
c. kinetic
d. potential (if in use, kinetic)
e. potential
f. kinetic
g. potential (if lighted, kinetic)
h. potential
i. kinetic
j. potential
k. potential
l. kinetic

10.3
a. steam radiator
b. bucket full of cool water
c. kilogram of molten lead
d. gas burner flame

10.4 There is a difference in the strengths of the attractive forces holding their particles (ions, molecules) together. The water molecules in the ice are held to one another by fairly strong hydrogen bonds, the gas molecules in the air are attracted to one another by very weak London forces, and the attractive forces between particles in the cold pack are somewhere in between.

10.5
a. 212 °F
b. 140 °F
c. –40 °F
d. 41 °F

10.6
a. 25 °C
b. –25 °C
c. 23 °C
d. 100 °C
e. 15 °C

10.7
a. chemical to electrical
potential to kinetic
b. chemical to electrical to mechanical
potential to kinetic to kinetic
c. chemical to thermal
potential to kinetic
d. chemical to thermal to mechanical
potential to kinetic to kinetic
e. mechanical to mechanical to electrical
potential to kinetic to kinetic
f. electrical to mechanical to mechanical
kinetic to kinetic to potential
g. mechanical to electrical to chemical
kinetic to kinetic to potential
h. radiant to electrical
kinetic to kinetic
i. radiant to thermal
kinetic to kinetic
j. mechanical to mechanical to mechanical
kinetic to potential (full draw) to kinetic

10.8
a. on
b. by
c. by
d. on
e. by
f. by
g. by
h. on
i. on
j. by
k. by
l. on
m. on

10.9
a. heat
b. heat, sound
c. heat, sound, gases out lungs
d. heat, sound, vibration
e. heat, sound, vibration
f. heat, sound, moving air out of way, gases out exhaust
g. heat
h. sound, gases out barrel, heat, recoil moves cannon
i. waste hot water, sound, heat

10.10
a. loose in room
b. salvage yard
c. in the air
d. ore
e. bolt of cloth
f. polluted water
g. messy one
h. hot water and ashes
i. water vapor
j. landfill

10.11 Several factors. Energy is cheap right now. Other precious metals in the ore can be reclaimed to defray cost. Price of gold is not too low.

CHAPTER 11

11.1 $CH_3CH_2CH_2CH_3$

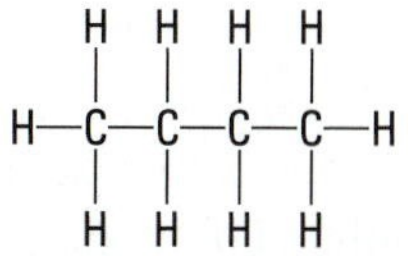

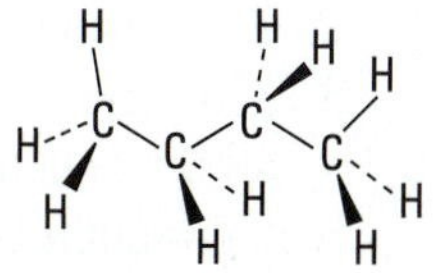

```
   H  H  H  H
   |  |  |  |
H—C—C—C—C—H
   |  |  |  |
   H  H  H  H
```

```
  H        H  H
  |        |/    H
H--C      C     /
  /  \  /   \  C
 H    C       \  H
      | \ H    H
      H
```

11.2 a. branched-chain
b. branched-chain
c. straight-chain
d. straight-chain
e. branched-chain
f. branched-chain

11.3 $CH_3CH_2CH_2CH_2CH_2CH_3$

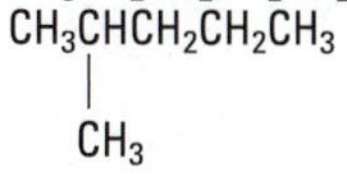

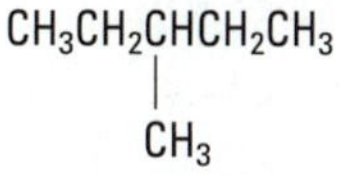

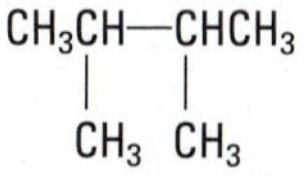

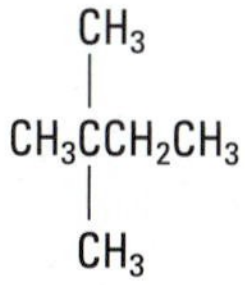

```
CH3CHCH2CH2CH3
   |
   CH3
```

```
CH3CH2CHCH2CH3
      |
      CH3
```

```
CH3CH—CHCH3
   |   |
 CH3  CH3
```

```
     CH3
     |
CH3CCH2CH3
     |
     CH3
```

11.4 a. 2-methylbutane
b. 3-methylheptane
c. 5-isopropylnonane
d. 3-ethylhexane
e. 4-ethyloctane
f. 3-ethyloctane

11.5 a. isopropylcyclopentane
b. ethylcyclohexane

11.6 a. 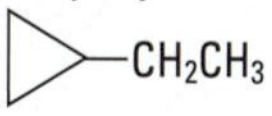 (cyclopropane ring)—CH_2CH_3

b. 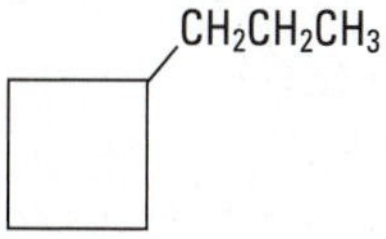 (cyclobutane ring)—$CH_2CH_2CH_3$

11.7 There are three electron pairs to orient as far apart as possible around each carbon atom in ethylene. (The double bond counts as one pair of electrons.) This makes a triangular (trigonal planar) arrangement around each carbon atom and a bond angle of about 120°.

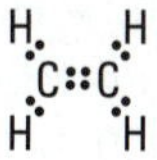

```
H     H
 .C::C.
H     H
```

11.8 There are two electron pairs to orient as far apart as possible around each carbon atom in ethyne. (The triple bond counts as one pair of electrons.) This makes a linear arrangement around each carbon atom and a bond angle of about 180°.

H:C:::C:H

11.9 a. 2-butene
b. 2-heptyne
c. 1-propyne
d. 3-octene

11.10 a. $CH_3CH_2C{\equiv}CCH_2CH_3$
b. $CH_2{=}CHCH_2CH_2CH_2CH_2CH_2CH_2CH_3$
c. $CH_3CH{=}CHCH_2CH_3$
d.

11.11 a. H_2N–(benzene ring)–SO_2NH_2
b. OCH_3, CH_3O, CH_3O, O, C–N, O
c. O–CH_2–C(=O)–NH, S, CH_3, CH_3, N, O, CO_2
d.
e. N, $NHCH_3$, Cl, N, O

11.12 a. $C_5H_{12} + 8O_2 \longrightarrow 5CO_2 + 6H_2O$
b. $2C_3H_6 + 9O_2 \longrightarrow 6CO_2 + 6H_2O$
c. $C_7H_8 + 9O_2 \longrightarrow 7CO_2 + 4H_2O$
d. $C_{10}H_8 + 12O_2 \longrightarrow 10CO_2 + 4H_2O$
e. $2C_8H_{18} + 25O_2 \longrightarrow 16CO_2 + 18H_2O$
f. $C_5H_{12} + 8O_2 \longrightarrow 5CO_2 + 6H_2O$
g. $2C_4H_6 + 11O_2 \longrightarrow 8CO_2 + 6H_2O$
h. $C_6H_{12} + 9O_2 \longrightarrow 6CO_2 + 6H_2O$
i. $2C_8H_{18} + 25O_2 \longrightarrow 16CO_2 + 18H_2O$
j. $2C_6H_{10} + 17O_2 \longrightarrow 12CO_2 + 10H_2O$

11.13 Consider what structural feature of fats is primarily responsible for their being liquid or solid.

11.14 a. $CH_3CH_2CHCH_3$ (with OH on the third carbon)
b. $CH_3CH_2CH_2CH_2CH_3$
e. Br (on benzene ring)
f. SO_3H (on benzene ring)

c. $CH_3CH_2CH_2CH_3$

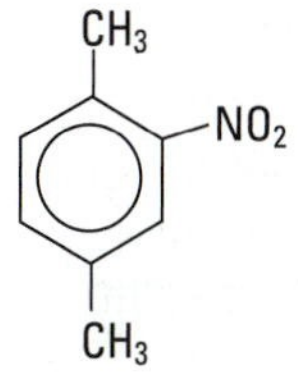

d. 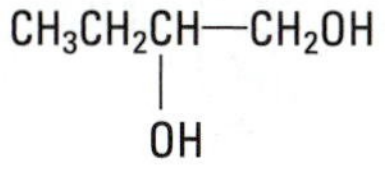 $CH_3CH_2CH(OH)—CH_2OH$

g. (benzene ring with CH_3, NO_2 ortho, and CH_3 para)

h. 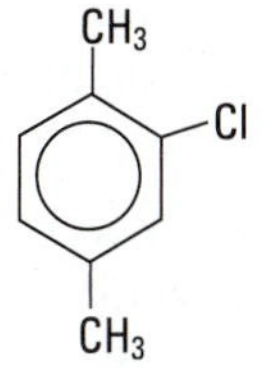 (benzene ring with CH_3, Cl ortho, and CH_3 para)

11.15 a. 2-chloropentane
b. 1-iodobutane
c. 4-bromooctane
d. 1-chloropropane
e. 3-fluorohexane
f. chlorocyclopentane

11.16 a. 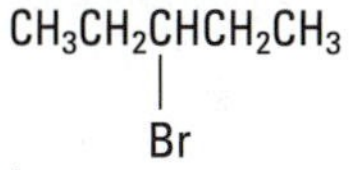 $CH_3CH_2CH(Br)CH_2CH_3$
b. 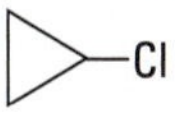 cyclopropane—Cl
c. 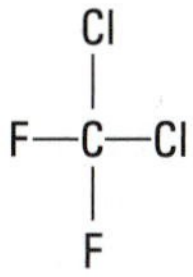 CCl_2F_2 (F—C(Cl)(F)—Cl)
d. $CH_3CH_2CH_2CH_2CH_2CH_2CH_2F$
e. 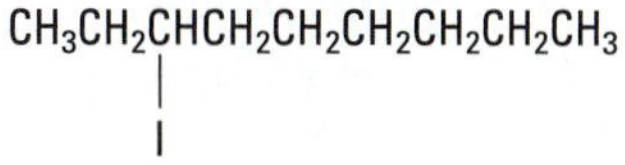 $CH_3CH_2CH(I)CH_2CH_2CH_2CH_2CH_2CH_3$

11.17 Coffee is now decaffeinated using carbon dioxide. There are three more uses, which are very common, that you are familiar with.

11.18 a. 1-butanol
b. 1-propanol
c. 3-pentanol
d. 2-heptanol
e. cyclobutanol
f. 1-nonanol

11.19 a. 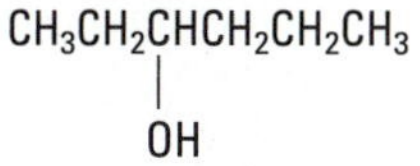 $CH_3CH_2CH(OH)CH_2CH_2CH_3$
b. 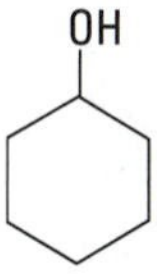 cyclohexane—OH
c. 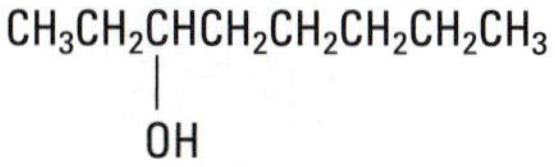 $CH_3CH_2CH(OH)CH_2CH_2CH_2CH_2CH_3$
d. 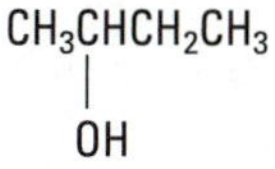 $CH_3CH(OH)CH_2CH_3$
e. $CH_3CH_2CH_2OH$
f. 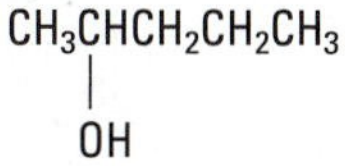 $CH_3CH(OH)CH_2CH_2CH_3$

11.20 $CH_3(CH_2)_{10}CH_2O(CH_2CH_2O)_8CH_2CH_2OH$

hydrocarbon end hydrophobic, soluble in fats or grease
ether bonds polar, dipole-dipole attractions with water molecules
hydroxyl group hydrogen bonds with water molecules

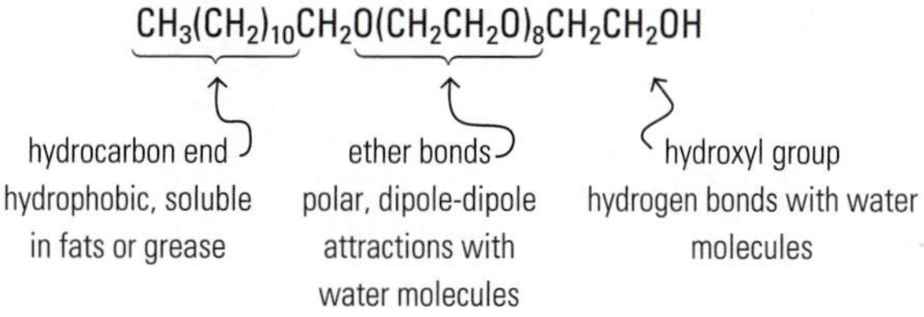

11.21 a. propanal
b. hexanal
c. 3-pentanone
d. 2-heptanone

11.22 a. $CH_3CH_2CH_2CH_2CHO$
b. $CH_3CH_2CH_2CH_2CH_2CH_2CH_2CH_2CHO$
c. $CH_3CH_2C(=O)CH_2CH_2CH_2CH_2CH_3$
d. cyclopentanone (five-membered ring with =O)

11.23 a. propanoic acid
b. methanoic acid
c. pentanoic acid
d. heptanoic acid

11.24 a. $CH_3CH_2CH_2CO_2H$
b. $CH_3CH_2CH_2CH_2CH_2CH_2CH_2CH_2CH_2CO_2H$
c. CH_3CO_2H
d. $CH_3CH_2CH_2CH_2CH_2CH_2CH_2CO_2H$

11.25 a. dimethylamine
b. ethyldimethylamine
c. propylamine
d. methylisopropylamine

11.26 a. $CH_3CH(CH_3)—NH_2$
b. $CH_3—N(CH_2CH_3)—CH_2CH_2CH_3$
c. $CH_3CH_2NHCH_2CH_3$
d. NH_2 on a benzene ring

CHAPTER 12

12.1 a. $CH_2{=}CHNO_2$
b. $CH_2{=}CCl_2$
c. $CFCl{=}CF_2$
d. $CH_2{=}CHOH$

12.2 a.

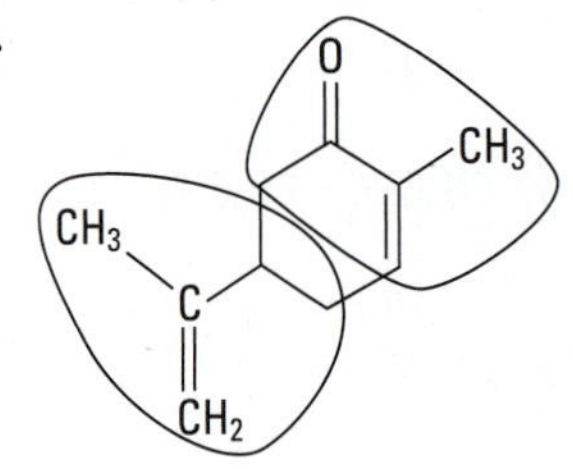

b.

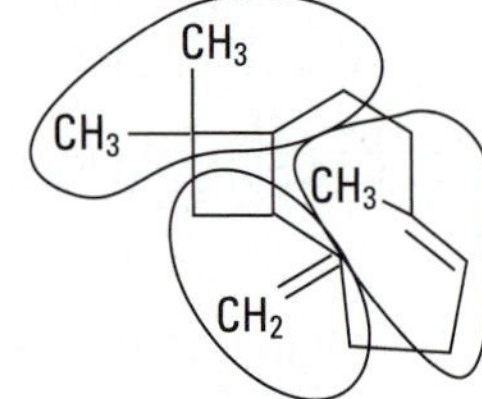

c.

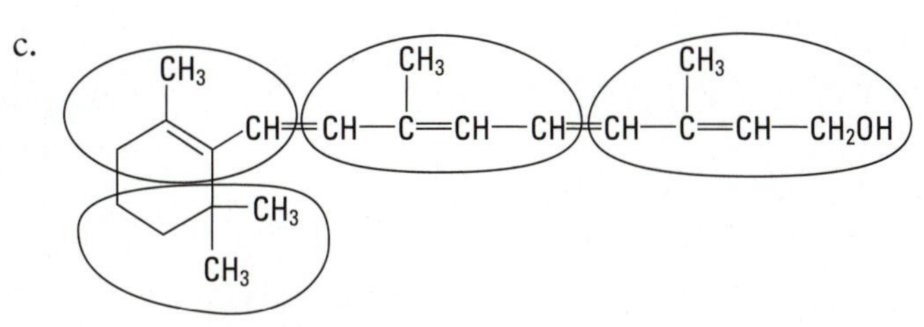

12.3 a. $CH_2{=}CHCl$ and $CH_2{=}CCl_2$
b. $CH_2{=}CHC(CH_3){=}CH_2$ (isoprene) and $CH_2{=}C(CH_3)_2$

12.4 a. $\text{+}CH_2CH_2CH_2\overset{O}{\overset{\|}{C}}NH\text{+}_n$ polyamide
b. $\text{+}CH_2CH_2O\overset{O}{\overset{\|}{C}}CH_2\overset{O}{\overset{\|}{C}}O\text{+}_n$ polyester
c. $\text{+}CH_2CH_2CH_2O\text{+}_n$ polyether

CHAPTER 13

13.1 Probably not. Situations come about which, for example, are desperate; and any action, risky or not, may be worth trying. For example, at this writing no effective treatment of cure for AIDS is on the horizon, and some people feel that using drugs (such as AZT) of unknown effectiveness and safety in an attempt to slow the disease is preferable to not treating the disease at all.

13.2 Salt kills many microorganisms, essentially by taking water from their cells (dehydrating them).

13.3 b, c, e, i, j

13.4 Initiation step is: $CCl_3F \xrightarrow{\text{ultraviolet light}} {\cdot}CCl_2F + {\cdot}Cl$
Propagation steps are:
$${\cdot}Cl + O_3 \longrightarrow {\cdot}ClO + O_2$$
$${\cdot}ClO + O \longrightarrow {\cdot}Cl + O_2$$
$${\cdot}Cl + O_3 \longrightarrow {\cdot}ClO + O_2$$
$${\cdot}ClO + O \longrightarrow {\cdot}Cl + O_2$$
Termination step is: ${\cdot}Cl + {\cdot}Cl \longrightarrow Cl_2$

13.5 Boiling the water drives all of the air (and hence the oxygen) out of it.

13.6 a. CO_2, H_2O, NO_3^-
b. CH_4, H_2O, NH_3, amines

Appendix D
Answers to Selected End-of-Chapter Exercises

CHAPTER 1

1. An operational definition is a statement of a test, and results that should be obtained from that test, that can be applied (by a person) to something to determine whether it fits into a particular category or not. A descriptive definition simply tells what something looks like.

4. a. scientific fact: a scientific observation that has been accepted by the scientific community. c. scientific law: a statement or generalization that summarizes a large number of scientific facts and appears to be always true.

8. a. theory c. law e. hypothesis g. fact i. theory

9. Science has a lot of trouble when experimental control and objectivity cannot be established. For example, it is impossible to know or predict exactly what a given person will do in a certain situation at a certain time. Religious and political beliefs are pretty much outside the realm of science, as are events that cannot be replicated, such as historical occurrences, "visions," and stories of extraterrestrial visits. Sometimes, the experiment disturbs the object or phenomenon being observed to such an extent that objectivity cannot be maintained (see the discussion on the Heisenberg uncertainty principle in Chapter 5).

13. Scientific laws *describe* nature. Civil laws *prescribe* what is acceptable behavior in society.

CHAPTER 2

1. In the Newtonian view of reality, the observer is separate from that which is being observed and does not affect the experiment in any way. Reality is "out there," or is what it is regardless of what the observer does.

3. The assumptions of science are that the world exists, that we can know the world through observation of nature, that there is an orderliness in nature, and that the acquisition of this knowledge is a worthwhile enterprise. They imply that reality can be observed, that a consensus of ob-

servations can be reached, that orderliness allows for classifications, and that such work leads to useful knowledge such as technology.

5. Science seeks to understand nature, while technology seeks to control it. The intent of technology is to use scientific knowledge to predict, control, or change our environment.

9. Units of measurement should be standardized in order to facilitate scientific communication, reproducibility, and verifiability. The latter two in particular are virtually impossible to attain without calibration of instruments to unit standards.

11. All measurements have some built-in error. The object of making measurements is to minimize this error by refining techniques, by taking great care, and by paying attention to detail. A precise measurement may or may not be error minimal, depending upon whether the instrument used to make it was calibrated and operated properly. That is, it could be inaccurate. Error-minimal values require a great degree of both precision and accuracy in making a measurement.

14. A compound unit is made by taking a ratio of two measurements. Compound units are useful because the denominator of these units is always unity, and it is easy to make direct comparisons of measured values having the same compound unit.

16. a. $16 \text{ in.} \times \frac{2.5 \text{ cm}}{1 \text{ in.}}$ c. $5 \text{ oz} \times \frac{30 \text{ mL}}{1 \text{ oz}}$

e. $100 \text{ mL} \times \frac{1 \text{ tsp}}{5 \text{ mL}} \times \frac{1 \text{ dose}}{2 \text{ tsp}}$

g. $10.0\ \mu\text{L} \times \frac{1 \text{ mL}}{10^3\ \mu\text{L}} \times \frac{0.998 \text{ g}}{1 \text{ mL}}$

$10.0\ \mu\text{L} \times \frac{1 \text{ mL}}{10^3\ \mu\text{L}} \times \frac{0.998 \text{ g}}{1 \text{ mL}} \times \frac{10^6\ \mu\text{g}}{1 \text{ g}}$

17. a. 6.52×10^{-1} m c. 8.637×10^{-3} kg e. 9.909×10^{-8} m
g. 4.05×10 L i. $1.25 \times 10^6\ \mu$g k. 3.0×10^2 cm/s
m. 4.390×10^4 nm $= 4.390 \times 10^{-8}$ km

CHAPTER 3

1. Chemistry is the study of matter, its properties, what its structure is, how it is classified, and how it undergoes change. Although other nat-

ural science disciplines deal with matter in some way, their interest is not in matter itself but rather in matter as it relates to something else, namely its processes, energy, motion, etc.

3. a. energy c. matter e. energy g. energy i. energy

5. a. liquid c. solid e. solid g. solid i. solid k. liquid

7. Much of the frost on the freezing compartment walls is ice that has sublimed from the frozen food, becoming water vapor and deposited as frost. Some of the frost also is formed from water vapor already in the air (humidity) that gets into the compartment each time the door is opened.

9. a. homogeneous c. either e. heterogeneous g. heterogeneous i. either k. either m. either o. homogeneous

11. a. physical change c. physical change e. chemical change g. chemical change i. chemical change

13. a. pure substance b. pure substance c. mixture

16. Energy either must, like a reactant, be put in to make a chemical reaction go or, like a product, it is given off during a chemical reaction.

19. a. neon c. phosphorus e. fluorine g. beryllium

20. a. Al c. Si e. Fe g. Br i. Au k. Sn

CHAPTER 4

3. Qualitative analysis tells us what a substance is made of, while quantitative analysis gives us the amounts of constituents present.

5. The 9.56-g sample of copper sulfide contained 3.21 g of sulfur. Since sulfur was the only other element present in the compound, 9.56 g – 6.35 g = 3.21 g of sulfur. The law of conservation of mass describes this behavior.

7. a. helium c. carbon e. neon g. silicon i. argon k. copper m. tin o. lead q. lithium s. nitrogen u. sodium w. phosphorus y. potassium aa. zinc cc. iodine

8. a. group, metals c. period, metalloids e. period, metals g. Be, Ca, Sr, or Ra, metals i. metal, nonmetal, period

11. a. C_5H_{12} c. $C_{12}H_{22}O_{11}$ e. $Al_2(SO_4)_3$

13. a. 2 atoms Na, 2 atoms O, 2 atoms H c. 6 atoms Na, 3 atoms S, 12 atoms O e. 10 atoms Al, 15 atoms C, 45 atoms O

15. a. $C_3H_8 + 5O_2 \longrightarrow 3CO_2 + 4H_2O$ c. $P_4 + 6Cl_2 \longrightarrow 4PCl_3$ e. $P_2O_5 + 3H_2O \longrightarrow 2H_3PO_4$

17. The relative combining masses of the elements in a compound depend upon the relative atomic masses of the elements and the number of each kind of atom in the compound. The relative atomic masses of the elements are simply the average relative masses of their individual atoms.

CHAPTER 5

1. Many scientists in the late-1800s believed that just about everything was known to science, and that most of what was left to do in science was to simply refine measurements. The belief was common that most of the new discoveries in science had already been made.

3. J.J. Thomson showed that the positive part of the atom makes up most of its mass.

7. Light radiating from a source may undergo interference, reinforcement, refraction, and diffraction. The most conclusive demonstration that light is wavelike is that its properties can be described by the same equations that apply to other wavelike phenomena such as sound. Light is different from sound, earthquakes, and water waves in that it does not need to be transmitted through a medium. It is easily transmitted through a vacuum (nothing or the absence of matter).

9. The photoelectric effect is the instantaneous production of moving electrons (an electric current, usually) when light strikes a semiconductor or some metals, such as potassium or zinc. The energy of the electrons emitted in the photoelectric effect has an energy proportional to the frequency (energy) of the incident light. The energy of the emitted electrons is not related to the amount of light but only to its wavelength (frequency).

   Bohr reasoned that since light of discrete energies was emitted when electrons fell from excited states to lower energy states in atoms, this emission of certain quantities of energy (quanta) could be accommodated nicely by Einstein's idea of the photon. Each event of an electron falling from a higher state to a lower one produces photons having a unique energy corresponding to the amount of energy lost when the electron fell. Each line in a line spectrum consists of a photon of light of a particular energy (color).

11. Something in the ground state is in a state of lowest energy, and it is most stable. Excited states are of higher energy than the ground state, and something in this unstable state has extra energy. Excited states are less stable than the ground state. Bohr observed that the energies of the excited state atoms must lie at discrete energy levels, or steps. These energy levels are *quantized.*

13. It only worked well in explaining the line spectra of the hydrogen atom and even when modified did not account well for the line spectra of larger atoms.

16. The harmonics (overtones) of a vibrating string or column of air (as in an organ pipe) show a coalescing pattern of frequencies starting from the fundamental and going up in frequency. The energies (thus frequencies) of the excited states of atoms show this same coalescing pattern of energies, starting at the ground state. Since vibrating strings obeyed the equations of the harmonics of sound and were wavelike, Schrödinger reasoned that the electrons in atoms were too.

19. There are four different kinds of orbitals: *s*, *p*, *d*, and *f*.

20. Valence electrons are those electrons in an atom that are involved in chemical bonding. They are usually those electrons that occupy the outermost orbitals, farthest from the nucleus where they are easiest to remove.

23. The Heisenberg uncertainty principle tells us that there are limits to our measurements. All measurements have a built-in tolerance. Measurements can never be exact, and thus there are limits to our ability to perceive reality. This is especially true for very small objects. For the electron, for example, both position and momentum cannot be known simultaneously because measurement of one precludes measurement of the other, since the first measurement disturbs the system.

25. a. Cl c. Mo e. Li g. Pd

27. a. 2 c. 128 e. 75

CHAPTER 6

1. a. 2+ b. 3+ c. 1– c. 3–

3. a. F^-, fluoride ion c. As^{3-}, arsenide ion e. Se^{2-}, selenide ion

5. a. S^{2-} c. Al^{3+} e. F^- g. P^{3-}

6. a. $BeCl_2$ c. $BaBr_2$ e. Ba_3N_2 g. MgO

7. a. beryllium chloride c. barium bromide e. barium nitride g. magnesium oxide

9. a. OH^-, anion c. $PO_4{}^{3-}$, anion e. $NO_3{}^-$, anion g. $SO_4{}^{2-}$, anion

10. a. aluminum phosphate c. calcium hydroxide e. lithium acetate g. lithium phosphate i. potassium sulfate

11. a. $BaSO_4$ c. $Al(NO_3)_3$ e. NH_4NO_3 g. $KC_2H_3O_2$

15. A diatomic molecule or ion has two atoms, a triatomic molecule or ion has three, and a polyatomic molecule or ion has "many."

17. Electron pairs bonded to a covalent atom will repel each other and will get as far away from one another as possible to minimize these repulsive forces. These repulsive forces are often better minimized when the orbitals containing the electron pairs (both bonded and nonbonded) can arrange themselves in three dimensions instead of two. This is the reason that many molecules have three-dimensional arrangements of their atoms and must be looked at as three-dimensional objects, just as most things in our everyday environment are three dimensional.

20. A molecule with polar bonds can be nonpolar if the arrangement of the polar bonds within the molecule causes the bond dipoles to cancel one another out. Usually, this means that if the molecule is viewed from the outside, there will be no apparent positive end or negative end. Or, it will not appear to be "lopsided."

21. a. molecular c. molecular e. molecular g. ionic i. molecular

22. a. ionic c. molecular e. molecular g. ionic i. molecular

23. a. dinitrogen tetraoxide (dinitrogen tetroxide) c. silicon dioxide
    e. oxygen dichloride g. selenium dioxide

CHAPTER 7

1. A solvent is the component of a solution in which other substances are dissolved. A solute is the substance that is dissolved in the solvent. A solution is a homogeneous mixture of solvent and solute. Generally, the substance in greater amounts in a solution is considered to be the solvent, and the substance in lesser amount the solute.

3. Dipole-dipole attractive forces hold polar molecules together in the liquid and solid phases. London dispersion forces hold nonpolar molecules together, and hydrogen bonding is the primary force that holds associated molecules together. Hydrogen bonding forces are the strongest, and London dispersion forces are the weakest. Nonpolar substances would have the lowest boiling point, and associated substances would have the highest boiling and melting points, assuming they had similar molecular masses.

6. A molecule of dimethyl ether is bent, just like a water molecule. This gives the molecule a lopsidedness, similar to the water molecule. The trimethylamine molecule has pyramidal geometry around the central N atom, similar to the ammonia molecule.

8. Water is a highly polar substance and solvates ions primarily by ion-dipole attractive forces, which, for water, are quite strong. The negative end of the water molecule is attracted to positive ions, and its positive end is attracted to negative ions, so that the water molecules arrange themselves around the ions and solvate (surround) them. Water solvates polar substances by dipole-dipole attractions. That is, its negative end is attracted to the positive end of the dipole (partial charge) of a polar molecule, and its positive end is attracted to the negative end of the dipole. Water molecules solvate polar molecules by surrounding them and isolating them from other polar molecules in solution. Hydrogen bonding forces are the most important when water solvates associated substances. Instead of hydrogen bonding just with itself, water also hydrogen bonds with substances that have —OH and —NH bonds and nonbonded pairs of electrons on the O and N atoms.

9. Dissociation is the process of separating or becoming separated. In the case of ions dissociating in solution, the solid substance, in which the

positive and negative ions are closely packed in a lattice structure, is torn apart as the ions are attracted to solvent molecules (usually water) and solvated to become essentially independent of one another.

a. $KBr(s) \xrightarrow{\text{water}} K^+ + Br^-$ c. $NaOH(s) \xrightarrow{\text{water}} Na^+ + OH^-$

e. $K_2SO_4(s) \xrightarrow{\text{water}} 2K^+ + SO_4^{2-}$

g. $Ca(NO_3)_2(s) \xrightarrow{\text{water}} Ca^2 + 2NO_3^-$

11. Although the intermolecular forces attracting nonpolar molecules to one another are relatively weak (London dispersion forces), there is little attraction of water molecules (solvent) for the nonpolar molecules either. In addition, the strong attractive forces (hydrogen bonds) between water molecules in liquid water (the solvent) result in a high degree of structure in the water itself. Introduction of a nonpolar molecule into this structure breaks hydrogen bonds, but no new strong bonds between water molecules and the nonpolar molecule can form. Breaking strong bonds while forming weak ones requires an input of energy, and the process tends not to be spontaneous. The nonpolar substances are *immiscible* with water.

13. Polar solvents usually will dissolve polar solutes, but often will dissolve some ionic substances and/or nonpolar solutes as well. This depends upon the degree of polarity of the solvent. Very polar solvents, like acetone and chloroform ($CHCl_3$), will dissolve some ionic substances as well as some substances of low polarity. Less polar substances, like ether and dichloromethane (CH_2Cl_2), dissolve nonpolar substances very well but are generally poor solvents for ions. Since polar solvents have permanent dipoles, they solvate ions primarily by ion-dipole interactions and polar substances by dipole-dipole interactions. Polar molecules can solvate nonpolar molecules to some extent by inducing dipoles in them, resulting in permanent dipole-induced dipole attractions.

16. Soap and detergent molecules both have a long nonpolar hydrocarbon tail (R) and an anionic end. In soaps, the anionic end is a carboxylate group ($R—CO_2^-$), and in detergents is a sulfonate group ($R—SO_3^-$). Detergents remain soluble in hard water, which contains calcium and magnesium ions, usually as their carbonates or hydrogen carbonates. Soaps react with the calcium and magnesium ions in hard water to form a precipitate or solid that falls out of solution. The solid product is called *soap scum.* This removes the soap from solution and renders it unavailable for cleaning. Since detergents stay in solution and do not precipitate in the presence of calcium and magnesium ions, they are effective cleaning agents in hard water.

$$Ca^{2+} + 2R{-}CO_2^- \longrightarrow Ca(R{-}CO_2)_2$$

ion precipitate

21. The mole is a counting unit, like a dozen, a ream, etc., which is used to express the number of molecules, atoms, or formula units of a substance being measured. The mole is based upon the number of particles (atoms, ions, molecules, etc.) of a substance and thus is related directly to the number of chemical species present in a chemical equation. In fact, chemical equations are expressed in moles and balanced in moles. Units of weight or volume cannot be used for this purpose, since the same weight or volume of two different substances will likely contain a different number of particles of each. Avogadro's number corresponds to the number of particles in the formula weight of any substance in grams. Thus, it is directly related to the mole, which is used in chemical equations. It is large because a large number of atoms, molecules, etc., is required to make enough material to weigh readily in the laboratory.

22. Molarity is a unit of concentration expressed as a ratio of the number of moles of a solute contained in one liter of a solution. It is written as mol/L, or *M*. Molar solutions are convenient for doing solution chemistry because they are expressed on a chemical basis instead of a weight or volume basis. Solutions of the same molarity will always contain the same number of moles of solute for a given volume regardless of solute composition. Molar solutions are convenient in the laboratory because a certain number of moles of solute can quickly and accurately be measured by volume.

CHAPTER 8

1. The chemical species that produces acidic properties in water is H^+, the hydrogen ion. Hydroxide ion, OH^-, produces basic properties in aqueous solutions. Acids turn blue litmus red, and bases turn red litmus blue.

3. $H_3PO_4 \overset{\text{water}}{\rightleftharpoons} H^+ + H_2PO_4^-$

6. a. $H^+ + Cl^- + Li^+ + OH^- \longrightarrow H_2O + Li^+ + Cl^-$
   H^+ is the acid, OH^- is the base, and Li^+ and Cl^- are spectator ions.
   c. $2H^+ + 2Br^- + Ca(OH)_2(s) \longrightarrow 2H_2O + Ca^{2+} + 2Br^-$
   H^+ is the acid, OH^- is the base, and Ca^{2+} and Br^- are spectator ions.
   e. $CaCO_3\ (s) + 2H^+ + 2Cl^- \longrightarrow 2H_2O + Ca^{2+} + 2Cl^- + CO_2\ (g)$
   H^+ is the acid, OH^- is the base, and Cl^- is a spectator ion. (OH^- is formed from a small amount of $CO_3{}^{2-}$ in solution.)

7. a. LiCl c. $CaBr_2$ e. $CaCl_2$

9. $Ca^{2+} + 2HCO_3^- \xrightarrow{\text{heat}} CO_2\,(g) + CaCO_3 + H_2O$

11. In a neutral solution there are equal numbers of hydrogen ions and hydroxide ions. In an acidic solution there is an excess of hydrogen ions, and in a basic solution there is an excess of hydroxide ions.

13. A change of 1.0 on the pH scale is a tenfold change in H^+ concentration. a. 10^2 times more acidic b. 10^4 times more basic c. 10^6 times more acidic

CHAPTER 9

2. a. Na is oxidized, H_2O is reduced, H_2O is the oxidizing agent, Na is the reducing agent c. CH_4 is oxidized, Cl_2 is reduced, Cl_2 is the oxidizing agent, CH_4 is the reducing agent e. Zn is oxidized, S is reduced, S is the oxidizing agent, Zn is the reducing agent g. C is oxidized, P_4O_{10} is reduced, P_4O_{10} is the oxidizing agent, C is the reducing agent

3. CO, carbon monoxide, forms if there is not enough oxygen.
   a. $2C_2H_6 + 7O_2 \longrightarrow 4CO_2 + 6H_2O$
   c. $C_5H_{12} + 8O_2 \longrightarrow 5CO_2 + 6H_2O$

5. The electrons from oxidation must go somewhere, to a species that must accept the electrons and be reduced. The redox term that is used implies the simultaneity of the *red*uction and *ox*idation processes.

7. Photosynthesis uses sunlight as a source of energy to produce carbohydrates and oxygen gas from carbon dioxide and water. Carbohydrates are a source of food for plants and animals, and oxygen is necessary for the survival of many forms of life, including humans.

9. a. HCl(*g*), b. NaOH(*s*), d. baking soda, f. HNO_3

11. A dry cell is not dry but instead contains a little water to make an electrolyte solution that will allow ions to move through the cell. If the cell were really dry, the ions could not migrate inside and the cell could not produce an electric current outside, between its terminals.

13. A fuel cell converts a fuel and an oxidant directly into electricity. It produces an electric current, as does a voltaic cell, but it does so as long as fuel is fed into it.

CHAPTER 10

1. Temperature is a measure of the intensity of heat. More specifically, it is the measure of the average kinetic energy of the particles in a substance. The higher the temperature, the faster the particles are in motion. The amount of heat depends not only upon the temperature, but also upon the amount of matter that is heated.

3. Conduction of heat depends upon the contact of the hotter body with the colder one, so that the more vigorous motion of the particles of the hotter body can be transmitted by collisions of the particles with the particles of the colder body. These collisions cause the particles of the colder body to move more vigorously, raising the temperature of the colder body. Radiant heat is electromagnetic radiation of long-wavelength light and is transmitted from the hotter body to the colder one through space. This radiation couples with the particles in the cooler body, giving up its energy to them and causing them to move faster. This raises their temperature.

6. Absolute zero, in principle, is the temperature to which any substance can be cooled to make the motion of its particles completely cease. If these particles actually stopped moving at absolute zero, then we would know their position and momentum exactly, at the same time. However, since these particles are very small objects, we can only know how "stopped" they are within the limits of error prescribed by the uncertainty principle. We can calculate the kinetic energy associated with this uncertainty. This is the zero-point energy, or the energy of residual motion.

9. The amount of heat contained by a substance depends upon its temperature and its mass. The third factor is the ability of the substance to hold heat. (This is called the specific heat.)

11. Whenever energy is converted from one form to another, some of the energy becomes incapable of doing useful work: it is wasted. Other kinds of energy are electrical (a form of kinetic energy), kinetic (motion of large objects, water, wind, sound, etc.), chemical, nuclear, and electromagnetic (light, radio waves, etc.).

13. *Endergonic* means that work is done *on* a system, such as a bag of cement being lifted onto a truck. *Exergonic* means that work is done *by* the system, such as the person who lifts the bag of cement. (Properly, *exergonic* and *endergonic* refer to *all* forms of energy, work being one kind.) The person does the work (exergonic) on the cement (endergonic). The cement will be at a higher energy level (less stable) when the process is finished, and the person will be at a lower energy level (more stable). *En-*

dothermic refers to the flow of heat *into* a system, such as in the boiling of water. Heat flows into the water from the outside (say, from a hot stove), and when the water has vaporized, the vapor contains more heat than the water originally did. *Exothermic* refers to the flow of heat *from* one system to some other system: in this example, from the hot vapor to its surroundings (the air) when the vapor condenses back to the liquid state. The air in the room will be warmer when the vapor has condensed because it has taken up heat from the water vapor as the vapor condenses. The flow of heat from the hot vapor to the room air tends to be spontaneous, since this process gives off heat. Vaporizing the water on the burner tends not to be spontaneous, since it must be made to occur by the input of heat.

14. Work is the use of energy to move mass. Power is the rate (how fast) work is done. The greater the power, the faster work can be done, so the production of greater power is often desirable. Some units of power are the horsepower, watt, kilowatt, Btu/hr, and therm/day.

17. In an energy conversion process, some energy does useful work. The remainder is converted to low-potential, low-temperature heat that cannot do useful work.

CHAPTER 11

1. Vitalism is the theory that suggested that organic compounds originated from living organisms. It was supported by the fact that all organic compounds then known were associated with life processes. It was refuted when cyanate and ammonium chloride (inorganic compounds) were used to generate urea, an organic compound.

3. Carbon can bond extensively with itself, forming four bonds to each carbon atom. This allows long chains, branched chains, or rings of carbon atoms to form, creating a large variety of compounds with a wide range of structural features.

5. A functional group is an atom or group of atoms present in a molecule that exhibits a particular kind of chemical behavior.

7. Aromatic hydrocarbons have carbon rings with alternating double and single bonds between the carbons. The electrons in the double bonds are delocalized around the ring, making them more stable and less reactive. Aromatic compounds tend to undergo substitution reactions rather than the addition reactions of alkenes and alkynes.

10. a. hexane b. $CH_3(CH_2)_{13}CH_3$ c. 1-butanol d. 2-hexanone

12. Hydrocarbons are obtained from petroleum, natural gas, and coal. Natural gas is the primary source of methane. Petroleum is the primary source of alkanes, cycloalkanes, and alkenes. Heating of coal in the absence of air or high-temperature reactions of petroleum fractions produce aromatic hydrocarbons.

14. Catalytic cracking is used to convert other petroleum fractions into the gasoline fraction.

$$C_{18}H_{38} \xrightarrow[\text{catalyst}]{\text{zeolite}} \text{a mixture of alkanes and alkenes like } C_9H_{20} + C_9H_{18}$$

16. Incomplete combustion produces some carbon monoxide in addition to carbon dioxide and water.

$$\text{Complete combustion: fuel} + O_2 \longrightarrow CO_2 + H_2O + \text{heat}$$
$$\text{Incomplete combustion: fuel} + O_2 \longrightarrow CO_2 + CO + H_2O + \text{heat}$$

This is dangerous in an enclosed space because the carbon monoxide interferes with oxygen transport by your blood.

18. Alcohols can be used for the production of any other kind of hydrocarbon derivative. They are starting materials for a range of products, including perfumes, flavorings, medicines, cosmetics, antifreeze, and solvents. Carbohydrates are alcohols that provide us with metabolic energy, clothing, and shelter.

21. An alcohol has the —OH group attached to a hydrocarbon structure, whereas a phenol has the —OH group attached to an aromatic hydrocarbon structure. Phenols: a, d, f; alcohols: b, e, g; ethers: d, f, h

23. Formaldehyde is sometimes used in resins in building materials and in some insulations that could make it an indoor pollutant.

25. Baking soda is a weak base, so it neutralizes acids that might have a sour smell. Acetic, butyric, and lactic acids are often found in refrigerators:

$$CH_3CH_2CH_2CO_2H + HCO_3^- \longrightarrow CH_3CH_2CH_2CO_2^- + CO_2 + H_2O$$

26. Dimethylamine is a basic substance, making a water solution feel slippery when touched and making the solution turn red litmus blue:

$$(CH_3)_2NH + H_2O \rightleftharpoons (CH_3)_2NH_2^+ + OH^-$$

29. The glyceryl tripalmitate is a saturated fat; it lacks any double or triple bonds.

$$\text{glyceryl tripalmitate} + 3Na^+ + 3OH^- \xrightarrow{\text{heat}} \text{glycerol} + 3CH_3(CH_2)_{14}CO_2^- + 3Na^+$$

31. a. amine c. alcohol e. aldehyde, alcohol g. alkyne i. ester k. ester

32. a. acetone c. chloroform e. ethyl alcohol (grain alcohol)

33. a. $CH_3CH(OH)CH_3$ c. $CH_3C(=O)CH_2CH_3$ e. $H_2C=O$
g. $HO—CH_2CH_2—OH$

CHAPTER 12

1. A polymer is made from the repetitious combination of smaller molecules called monomers. A polymer is thus a very large molecule and therefore a macromolecule.

4. Properties of addition polymers can be modified by changing the substituents off the backbone. Thus, polypropylene is different from polysterene, which is different from poly (vinyl chloride), etc.

5. A homopolymer is made from the repeated combination of one monomer, whereas a copolymer is made from the repeated combination of two or more monomers in a roughly alternating sequence. The polymer shown is made from the monomers $CH(CH_3)=CH_2$ and $CH_2=CH_2$.

9. a. polybutadiene: tires, rubber belts, hoses b. polystyrene: housings for large appliances, auto instrument panels, clear cups and food containers, foam cups and packaging c. polyacrylonitrile: carpets and knitwear d. HDPE: rigid containers, toys, housewares e. poly (vinyl chloride): pipe, siding, floor tile, raincoats, shower curtains f. SBR: tires

11. Monomers in condensation polymers must be bifunctional so that the condensation reaction can occur on both ends of each monomer molecule to form a long polymer chain, much like joining hands to make a chain of people. With only one functional group, only a dimer could

form. If the monomer is trifunctional, reactions can occur at three sites and a space-network polymer is formed.

13. Space-network polymers spread out in three directions, whereas linear polymers spread out in two directions. The space-network polymers thus extend in all directions, connecting the monomer units in a three-dimensional network and forming very rigid structures.

15. Polymers that make good fibers are linear, highly ordered, and have strong intermolecular forces. Polymers that are good elastomers are linear with weak intermolecular forces and tangled chains. HDPE would not make a good fiber because it has weak intermolecular forces.

18. Thermosetting resins are not plastic because they usually have a rigid three-dimensional structure. Heating and remolding is not feasible with such interconnected molecular structures because strong covalent bonds would have to be broken.

21. Cellulose is a polymer of glucose that has its glycoside linkages "up," whereas starch is a polymer of glucose that has it glycoside linkages "down." Cellulose is formed in plants for support material (in trees, for example). Starch is produced in plants as stored energy. We can metabolize the starch but not the cellulose.

23. Cellulose acetate is preferred over cellulose nitrate because cellulose nitrate is very flammable.

25. The silicon atoms form strong single bonds to oxygen atoms in a tetrahedral geometry. In quartz, the silicon atoms are crosslinked by the oxygen atoms into a space-network polymer. Silicones are similar in that they have a repeating O—Si—O backbone, but the other two oxygen positions are replaced with hydrocarbon substituents like $—CH_3$ to make oils or greases. Silicone oils, greases, and rubber are very stable and are useful in high-temperature applications and under other harsh conditions, such as in strong sunlight and in outer space.

CHAPTER 13

1. When assessing risks and benefits, one should examine not only the immediate personal issues, but also the impact on other population groups and the effects over time. High risk/low benefits do not necessarily indicate that the risk should not be taken. Development of nuclear energy

and exploration of space were certainly high-risk endeavors, but one could argue that both have the potential for benefits.

3. Our physical environment can be divided into the following realms: atmosphere, lithosphere, hydrosphere, and biosphere. A closed system is one in which nothing enters or leaves. Pollutants in any one of the four realms of the system can transfer to other realms of the system.

5. We are causing such environmental damage for several reasons, including population growth, high expectations for standard of living, industrialization, and "disposable" societal attitudes.

7. Photochemical oxidative smog is the atmospheric pollution that forms from gasoline combustion products in a warm, sunny environment. The combustion and evaporation of gasoline produces hydrocarbons. Nitrogen burns in engines to form nitric oxide, NO. The nitric oxide reacts with the sunlight and oxygen to generate free radicals, atomic oxygen, and nitrogen dioxide, NO_2, which gives smog its brown color. These react with other substances to produce ozone, hydroxyl radicals, and hydrocarbon products like aldehydes, ketones, and PAN.

9. $N_2 + O_2 \longrightarrow 2NO$

11. A photochemical reaction is one that is catalyzed by the energy provided by light. For example,

$$\cdot NO_2 \xrightarrow{\text{sunlight}} \cdot NO + O$$

13. This is a *steady-state condition,* which we have disrupted by introduction of CFCs.

15. The ozone hole forms over Antarctica in the polar spring. The ice crystals in the clouds there concentrate CFCs, chlorine nitrate, and HCl, so that they react with spring sunlight to form chlorine radicals that begin the ozone-depleting chain reactions. The CFCs are a contaminant in the troposphere because they do no harm there; they are inert in the troposphere.

19. The layers of soil include the O, A, B, and C horizons. The O horizon is the top few centimeters with remains of plants, etc.; the A horizon is the topsoil with organic matter; the B horizon is the subsoil with inorganic material from weathered parent rocks, salts, and clay; and the C horizon is the weathered parent rocks. The topsoil is most productive for growing food.

21. The amount of CO_2 is higher and the amount of O_2 is lower in soils because the bacteria are degrading the organic matter. These reactions use O_2 and make CO_2.

24. The rate of rebuilding of topsoil is about 1 in./century. Constant cultivation increases erosion and removes nutrients from the soil. Fallowing increases production of topsoil and fertility by adding to the humus. Chemical fertilizers only partially restore soil fertility because organic matter is also necessary. Desertified soils would be very difficult to reclaim because plant growth and moisture needed for fallowing would be lacking.

27. The hydrologic cycle is a process by which sunlight evaporates water from the oceans into the atmosphere, where the water accumulates into clouds and then precipitates back to earth in rain, snow, etc. Once precipitated, it becomes surface water, flows through the ground, or is stored in glaciers or snow fields. Eventually it reaches the oceans and the process starts all over again.

30. Fresh water availability includes such factors as location of surface water and aquifers with respect to population centers and farming regions, climatic situations, and pollution of available water sources.

33. Aerobic and anaerobic indicate processes that occur in oxygen and without oxygen, respectively. The BOD indicates the amount of oxygen needed to degrade organic matter, with high BOD indicating a high organic content. If the BOD is too high, the system becomes anaerobic and reduction reactions instead of oxidation reactions occur.

$$\text{organic wastes} + O_2 \xrightarrow[]{\text{aerobic bacteria}} CO_2 + H_2O + NO_3^- + SO_4^{2-}$$

$$\text{organic wastes} \xrightarrow[]{\text{anaerobic bacteria}} CH_4 + H_2O + NH_3 + \text{amines} + H_2S$$

35. Organic compounds enter watercourses by leaching out of the soils and burial sites, by runoff from land surfaces, and by direct emptying from sewers and industry. The process of transport of pollutants from burial sites into fresh water is called leaching.

37. Acids enter watercourses by acid mine drainage and from acid rain.

39. Cadmium and zinc are in the same family in the Periodic Table; thus, when cadmium enters the body, it is treated like zinc and displaces the

zinc. Disruption of the natural roles of zinc leads to high blood pressure, anemia, and kidney failure.

42. Oil pollution results from large oil spills during transport, from improper household oil disposal, and from leakage of underground storage tanks (LUST). Since leakage from these old storage tanks can place these hydrocarbons into the aquifers, it endangers our water supply.

Glossary

A

α-particle See alpha particle

absolute zero The temperature at which, in principle, the motion of particles of a substance would cease, –273.16 °C (0 K)

absorption spectrum A continuous spectrum having dark spaces or lines in it, caused by absorption of certain wavelengths of light by a substance

accuracy How closely a measurement agrees with the actual dimensions of an object

acid A chemical species that produces hydrogen ions in water solution

addition reactions Reactions in which a double or triple bond becomes saturated

addition polymers Polymers formed by the successive addition of monomer molecules to one another to form a long chain

aerobic Biological processes that require the presence of oxygen

alcohols Hydrocarbon derivatives that contain the hydroxyl, or —OH, group

aldehydes Carbonyl compounds in which the carbonyl carbon atom is bonded to at least one hydrogen atom: $C{-}\overset{\overset{O}{\|}}{C}{-}H$ and $H{-}\overset{\overset{O}{\|}}{C}{-}H$

alkanes Hydrocarbons with the molecular formula C_nH_{2n+2}

alkenes Hydrocarbons that contain carbon–carbon double bonds and have the molecular formula C_nH_{2n}

alkynes Hydrocarbons that contain carbon–carbon triple bonds and have the molecular formula C_nH_{2n-2}

alpha particle A helium ion, He^{2+}, produced by radioactive decay

amides Derivatives of carboxylic acids containing the amide group $-\overset{\overset{O}{\|}}{C}-\underset{|}{N}-$ and having the general formulas $R{-}\overset{\overset{O}{\|}}{C}{-}NH_2$, $R{-}\overset{\overset{O}{\|}}{C}{-}NHR$, and $R{-}\overset{\overset{O}{\|}}{C}{-}NR_2$

amines Derivatives of ammonia, in which one or more hydrogen atoms on the ammonia molecule have been replaced by an organic group R Weak bases with general formulas RNH_2, R_2NH, and R_3N

anaerobic Biological processes that occur in the absence of oxygen

anion Any negative ion

anode The electrode in an electrochemical cell at which oxidation takes place

anthropogenic Caused or introduced by humans

aquifers Porous layers of underground gravel or rock that are full of groundwater, which is usually flowing from one place to another

aromatic Describes compounds that possess an extra-stable, cyclic system of alternating double and single bonds

Arrhenius model A model of acids and bases in water solution in which acids furnish hydrogen ions and bases furnish hydroxide ions

associated Describes molecules that are hydrogen bonded to each other

atmosphere The blanket of gases that surrounds the earth Our gaseous environment

atomic theory Proposed in 1808 by John Dalton: that all matter is composed of small pieces called atoms; that atoms of each element are different; that atoms combine in whole-number ratios to form compounds; that atoms are merely joined, rearranged, or separated; and that atoms retain their identity in chemical reactions

atomic number The number of protons in the nucleus of an atom of an element, or the number of electrons contained in its uncombined neutral atom

atoms The smallest piece of an element See atomic theory

Avogadro's number 6.02×10^{23}

B

β-particle See beta particle

base A chemical species that produces hydroxide ions in water solution

battery One or more voltaic cells in a package used to produce and store electrical energy

beta particle A particle of 1– charge, identical to the electron, that is produced by radioactive decay

bifunctional A molecule containing two functional groups

binary compound A compound composed of two different elements

binder An organic substance that polymerizes to form the protective film formed by paints and varnishes

biochemical oxygen demand (BOD) The quantity of oxygen dissolved in water that is required to oxidize a given amount of organic matter A measure of the amount of organic waste in water

biosphere The interface between the atmosphere, the lithosphere, and the hydrosphere The part of our environment that supports life

BOD See biochemical oxygen demand

boiling point The temperature at which a liquid boils Same as the condensation point for pure substances

boiling The condition of a liquid when bubbles rise rapidly from beneath its surface and escape as vapor as the liquid is heated

C

calibration A process of comparison with a standard by which instruments are made to be accurate

calorie cal, the amount of heat required to increase the temperature of 1 g of water by 1 °C

carbohydrate Molecular compounds that contain carbon, hydrogen, and oxygen in roughly the ratio $C_n(H_2O)_n$, such as starch, cellulose, and sugar

carbonyl compounds Hydrocarbon derivatives that contain the carbonyl group, $-\overset{O}{\overset{\|}{C}}-$, as their functional group Aldehydes and ketones

carbonyl group The functional group $-\overset{O}{\overset{\|}{C}}-$

carboxyl group The functional group $-\overset{O}{\overset{\|}{C}}-O-H$

carboxylic acids Organic weak acids that contain the carboxyl group

catalyst A substance that alters the rate of a chemical reaction but is not itself consumed in the process

catalytic cracking Breaking large hydrocarbon molecules into smaller ones by the use of a catalyst and heat

catenation The bonding of carbon atoms with each other to form long chains, rings, and many other structures

cathode The electrode in an electrochemical cell at which reduction takes place

cation Any positive ion

chemical symbol A unique one-, two-, or three-letter abbreviation for the name of an element

chemical formula Shows what elements are present in a compound and in what proportions

chemical energy Energy produced in chemical reactions when stronger bonds are formed at the expense of weaker ones

chemical change A process by which the composition of matter is changed

chemical equation A means of describing a chemical reaction with abbreviations instead of words

chemical equilibrium A dynamic, reversible chemical process in which reactants are forming products and products are forming reactants at the same rate

chemistry The natural science that deals with the categories, properties, structure, and transformations of matter

classical view of reality (also called the Newtonian view) Holds that the objects we observe have an existence independent of our observation and that we can observe them objectively. Theories represent some element of actual truth.

closed system A system into which nothing can enter and from which nothing can leave

combustion The rapid reaction of fuels with oxygen

compound unit A dimension that is expressed as a ratio of two kinds of measurements, such as density (mass/volume)

compounds Substances that arise from the chemical combination of two or more elements

concentrated Describes a solution in which a relatively large amount of solute is dissolved in a given amount of solvent

concentration The ratio of the amount of solute to a given amount of solvent or solution

conceptual model A mental construct based on analogy and inference that describes things that cannot be observed directly or entirely

condensation The formation of a liquid from a gas as it is cooled

condensation Describes reactions in which two molecules are joined to form a larger molecule, and a small molecule such as water is lost See condensation polymers

condensation point The temperature at which a gas condenses to a liquid Same as the boiling point for pure substances

condensation polymers Polymers formed from reactions that join bifunctional monomer molecules, with the loss of a small molecule such as water See condensation

conduction The transfer of heat by the collision of the rapidly moving particles of one substance with the slower moving particles of another

contaminant A substance placed in the environment, in greater than natural amounts, that changes the makeup of that environment but does not appear to harm it

continuous spectrum A spectrum of light of all colors A rainbow Produced by glowing solids and liquids

Copenhagen interpretation Holds that the observer is a part of an experiment and affects its outcome. Theories permit correlation of observations and permit predictions, but do not necessarily represent actual truth

copolymer An addition polymer made up of more than one kind of monomer unit

corrosion The unwanted oxidation of metals in the environment

covalent bond A bond in which a pair of electrons is shared between the two atoms that are joined by it A molecular bond

crosslinking Joining separate chains in a polymer together with bridges of atoms, such as sulfur or carbon, to produce a more rigid structure

crystal lattice An orderly, three-dimensional array of ions, atoms, or molecules

cycloalkanes Saturated hydrocarbons that exist in rings of carbon atoms and have the molecular formula C_nH_{2n}

D

de Broglie's equation $h\nu = mc^2$ (wave/particle duality)

delocalized Describes electrons that are not bound to a particular atom but are free to move from one atom to another

density A compound unit that is a ratio of mass to volume of a substance

deposition Conversion of a gas to the solid state, bypassing the liquid state The reverse of sublimation

desertification Expansion of deserts into productive farmland at desert margins

dilute Describes a solution in which a relatively small amount of solute is dissolved in a given amount of solvent

dimensional analysis (also called factor-label method) A method for solving problems that uses units associated with measurements or quantities as a guide to setting up calculations and as a means for checking answers

dipole–dipole attraction Forces of attraction between polar molecules in which a (partially) positive end of the dipole of one molecule is attracted to a (partially) negative end of the dipole of another

dissociation The separation of ions from one another when an ionic substance is dissolved

drying oils Naturally occurring unsaturated and polyunsaturated fatty acid esters of glycerol

E

efficiency The amount of work done by an energy conversion process divided by the amount of energy converted, multiplied by 100 percent

Einstein's equation $E = mc^2$ (matter/energy duality)

elastomers Polymers that can be stretched considerably and then return to their original shapes

electrical energy Energy carried by moving electrons, such as electricity

electrodes Conductors in an electrochemical cell at which oxidation and reduction take place

electrolysis The decomposition of compounds by the use of electricity

electrolytes Substances that produce mobile ions when dissolved in water

electrolytic cell An electrochemical cell in which nonspontaneous chemical reactions are made to proceed by the application of electrical energy

electron A small particle having a 1– charge with a mass about 1/1837 that of the hydrogen atom

electron sea model A three-dimensional lattice of positive metal ions immersed in a "sea" of valence electrons The metallic bond

electronegativity The ability of an atom in a covalent bond to attract electrons to itself

electrostatic forces The attraction of positive charges for negative charges

elements Pure substances that cannot be broken down into other pure substances by chemical means

emission spectrum A spectrum produced by glowing gases characterized by bright lines of color separated by dark spaces Also called a line spectrum

emulsion Finely divided droplets of an immiscible liquid suspended in another

endergonic Describes processes that absorb, or take up, any form of energy

endothermic Describes processes that absorb, or take up, heat

energy That part of the natural world that makes things move

enthalpy The heat available to be used in an energy conversion process

entropic Describes changes in which a state of greater disorder is created Tend to be spontaneous

entropy A measure of disorder in a system

esters Derivatives of carboxylic acids containing the ester group $-\overset{\displaystyle O}{\overset{\|}{C}}-O-$ and having the general formula $R-\overset{\displaystyle O}{\overset{\|}{C}}-O-R$

ethers Derivatives of alcohols and phenols containing the ether linkage, C—O—C (two hydrocarbon groups bonded to an oxygen atom)

excited state The condition of an atom having extra energy An unstable state

exergonic Describes processes that evolve, or give off, any form of energy

exothermic Describes processes that evolve, or give off, heat

exponent The power to which the number 10 is raised

exponential number 10^n, where n is any number

F

fibers Fine threads formed by extruding (squeezing) a molten or dissolved polymer through a die with small holes in it (spinneret)

first law of thermodynamics Energy is neither created or destroyed (it is conserved) in energy conversion processes

formula mass The sum of the atomic masses of all of the atoms in the chemical formula of a substance expressed in grams

formula unit The collection of atoms contained in the chemical formula of a compound

fractionation Also called fractional distillation A means of separating volatile compounds on the basis of differences in their boiling points

free energy Actual work done in an energy conversion process

free radicals Chemical species (atoms, molecules, fragments of molecules) that contain unpaired electrons

freezing The change from liquid to solid state as the temperature of a substance is lowered

freezing point The temperature at which a substance freezes Same as the melting point for pure substances

frequency The number of wave crests that pass a given point in a second

fuel cells Electrochemical cells that produce an electric current as long as fuel and oxidant are fed into them

functional group An atom or group of atoms present in a molecule that exhibits a particular kind of chemical behavior

G

galvanic cell See voltaic cell

galvanizing Coating iron with a thin layer of zinc to prevent rusting

gas A state of matter that has no shape or volume but expands to completely fill its container

greenhouse effect The absorption of solar energy reradiated from the surface of the earth, which would normally escape into space, by gases such as carbon dioxide and methane in the atmosphere

greenhouse gas Gases in the atmosphere, such as carbon dioxide, methane, and chlorofluorocarbons, that absorb long wavelength radiation

ground state The condition of an atom in its lowest, most stable, energy state

groundwater Water underground that is stored in aquifers

groups Vertical rows of elements in the Periodic Table which bear similarities to one another in their chemical and physical properties

H

halogenated hydrocarbons Hydrocarbons in which one or more hydrogen atoms have been replaced by fluorine, chlorine, bromine, or iodine atoms (the halogens)

halogens The Group 7A elements: F, Cl, Br, and I

hard water Water containing high concentrations of calcium and magnesium ions

heat Energy contained in the motion of ions, atoms, or molecules that make up a substance Thermal energy

Heisenberg uncertainty principle States that if we know with any precision where a very small particle is going, we will have almost no idea where it may be, and conversely

heterogeneous Not having the same composition, or being nonuniform, throughout

homogeneous Having the same composition, or being uniform, throughout

homopolymer A polymer made up of one kind of monomer unit

horizons Discrete layers in soils that vary in composition, texture, color, and structure

humus Decaying animal and vegetable matter that makes up most of the organic portion of soils

hydrocarbon derivatives Organic compounds that, in addition to carbon and hydrogen, also contain one or more nonmetallic elements, such as oxygen, nitrogen, and chlorine

hydrocarbons Compounds containing only carbon and hydrogen

hydrogen bond A partial bond formed between a hydrogen atom on an —OH, —N(—)—H, or —NH_2 group and a nonbonded pair of electrons on another group of similar kind The strongest of the intermolecular forces

hydrogenation The reaction of hydrogen gas, in the presence of a catalyst, with compounds containing double and triple bonds

hydrologic cycle The process, driven by solar energy, by which the water in the oceans is recycled to the land masses through the atmosphere

hydrolysis Reaction with water

hydrophilic Water loving

hydrophobic Repelled by water

hydrosphere The system of waters that covers the earth Our liquid environment

hygroscopic Having a great affinity for water

hypothesis A scientific proposition to explain certain facts and which must be tested for validity by experiment

I

incubation period See lag time

inorganic chemical waste Pollutants such as mine drainage, agricultural fertilizer runoff, salt from domestic water softening and highway deicing, and effluent from electroplating and ore processing industries

instruments Devices used to extend our senses and/or make measurements

intensive properties Properties that do not depend on the amount of substance present

intermolecular forces Attractive forces that hold molecules together in the solid or liquid state

ion An atom, or small group of atoms bound together, that carries a positive or negative charge

ion–dipole attraction Forces of attraction between the dipoles of polar molecules and the ions they solvate

ionic substances Compounds composed of ions, often of a metal cation and a nonmetal anion or a polyatomic ion

ionization The formation of ions from a neutral chemical species

isomers Compounds that have the same molecular formula but differ in the way their atoms are arranged

IUPAC system An international system for the naming of chemical compounds

J

joule The SI unit of energy, J (1 cal = 4.18 J)

K

ketones Carbonyl compounds in which the carbonyl carbon atom is bonded to two carbon atoms, C—C(=O)—C

kinetic energy Energy of motion

L

lag time The time between making a basic discovery and applying it for widespread use

law of conservation of mass States that matter is neither created nor destroyed in a chemical process

law of definite proportions States that in a pure compound the elements are present in fixed proportions by mass

law of multiple proportions States that if two elements combine to form two or more compounds, the masses of one element that can combine with a fixed mass of the other can be reduced to a ratio of small whole numbers

Le Châtelier's principle When a system in equilibrium is subjected to some kind of stress, the position of equilibrium will shift to eliminate that stress

leaching The transport of substances through sediments, soils, and rock caused by the movement of water

leaking underground storage tanks (LUST) Old and abandoned or forgotten underground gasoline and fuel storage tanks that have corroded and begun to leak their contents into the underground environment

light A form of energy Electromagnetic radiation

line spectrum See emission spectrum

liquid A state of matter that has a definite volume for a given mass but has an indefinite shape

lithosphere The thin outer shell of the solid earth The solid environment we live on

London dispersion forces Forces of attraction between nonpolar molecules in which a temporary dipole on one molecule interacts with another temporary dipole present or induced in another molecule The weakest of the intermolecular forces

M

macromolecules Very large molecules See polymers

mass The resistance of matter to being accelerated by a force

mathematical model A description of a process or relationship that uses mathematical equations or other mathematical concepts

matter Anything that has mass and occupies space

mechanical energy Energy produced by moving objects, such as a falling weight, flowing water, and the wind

mechanical model A physical copy or reproduction of something on a smaller or larger scale

melting The change from solid to liquid state as the temperature of a substance is increased

melting point The temperature at which a solid melts Same as the freezing point for pure substances

metabolism The oxidation of carbohydrates in the mitochondria of living cells to produce energy

metalloids Elements that have properties of both metals and nonmetals and are found along the Zintl border in the Periodic Table

metals Elements that possess metallic luster, are malleable and ductile, are good conductors of heat and electricity, and tend to lose valence electrons in chemical reactions

metric system A system of units that uses a basic unit (meter, liter, gram, etc.) and a system of prefixes that multiply the basic unit by a given power of 10

micelle A microscopic globule of a hydrophobic substance with a charged or highly polar surface

miscibility The solubility of liquids in one another

mixtures Two or more pure substances together in various proportions Have no unique set of properties

molarity The number of moles of solute contained in a liter of solution, expressed as either mol/L or *M*

mole The formula mass of an element or compound in grams Equivalent to 6.02×10^{23} atoms, ions, or molecules

molecular bond A bond in which a pair of electrons is shared between the two atoms that are bound together A covalent bond

molecular compounds Compounds formed from the combination of nonmetals with one another

molecular orbital An orbital, containing a pair of electrons, that envelopes two atoms such that the electrons are shared between the atoms

molecule A chemical entity composed of the number and kinds of atoms contained in its chemical formula

monomers Small molecules that are combined to form the repeating unit(s) in polymers

monounsaturated Describes compounds containing one carbon–carbon double (or triple) bond

N

negentropic Describes changes in which a state of greater order (less disorder) is created These changes tend not to be spontaneous

net ionic equation A chemical equation in which spectator ions are omitted

neutralization The reaction of an acid with a base to produce water and a salt

noble gases Helium, neon, argon, krypton, and xenon

noble metals Copper, silver, gold, and platinum

node A place in an orbital where there is zero probability of finding an electron

nomenclature A system of labeling or naming things that implies they fall into a particular order

nonelectrolytes Substances that do not produce mobile ions when dissolved in water

nonmetals Elements that lack metallic luster, are brittle solids or liquids or gases, are poor conductors of heat and electricity, and tend to gain valence electrons in chemical reactions

nonrenewable energy source A source of energy found on the earth that is not rapidly being replenished naturally, such as coal, oil, natural gas, and uranium

nuclear energy Energy released in nuclear reactions when matter is converted to energy

nucleus The small, dense, positively charged core of an atom that contains almost all of the atom's mass

O

observations In science, the results of observations must be repeatable and verifiable to be considered valid

octet rule Most nonmetal atoms tend to form covalent bonds so as to have a total of eight valence electrons around them

orbital A volume in space around the nucleus in which there is a high probability of finding an electron

organic chemical waste Pollutants such as herbicides, pesticides, solvents, and industrial organic chemicals

organic chemistry The chemistry of the compounds of carbon

oxidation The loss of electrons from a chemical species Also, the gain of oxygen by or the loss of hydrogen from a chemical species

oxidizing agent A substance that causes some other species to be oxidized and is itself reduced in the process

oxygen-consuming wastes Organic matter from many sources including domestic sewage, detergents, animal manure, decaying aquatic plants, and effluents from the food processing, paper, and petroleum refining industries

P

PAN See peroxyacylnitrates

parent rocks The bottom horizon of most soils, consisting mostly of weathered underlying rock from which the soils originate

passivation Formation of a metal-oxide coating on the surface of a metal that protects the metal from further oxidation

pathogens Bacteria or viruses that cause disease

periodic law States that the properties of the elements are periodic functions of atomic number

Periodic Table A chart showing the arrangement of the elements based on the periodic repetition of similar properties as a function of atomic number

periods Rows of elements that run from left to right in the Periodic Table

peroxyacylnitrates (PAN) Oxidizing agents formed in photochemical oxidative smog that have the formula $R{-}\overset{\overset{\large O}{\|}}{C}{-}O{-}O{-}NO_2$

pH acidity scale A quantitative, logarithmic scale of acidity based on the concentration of hydrogen ion in water solution

phenols Aromatic compounds having a hydroxyl (—OH) group bonded to a carbon atom of the aromatic ring

phosphor A substance, such as zinc sulfide, that produces flashes of light (scintillations) when struck by energetic particles

photochemical oxidative smog An airborne mixture of oxidizing agents, such as NO, NO_2, O_3, and peroxyacylnitrates (PAN), formed by simultaneous free radical reactions of anthropogenic air pollutants in the presence of sunlight

photodissociation Dissociation caused by light

photoelectric effect The production of moving electrons when light strikes a semiconductor or a metal such as potassium or zinc

photon A massless particle of light

photosynthesis The production of carbohydrates and oxygen from carbon dioxide and water in the leaves of green plants by the use of sunlight and the chlorophyll in their leaves

physical change A process that does not alter the composition of a substance

pictorial model A picture or drawing used to show and describe things that are complex or not easily visualized

pigment An inorganic solid that provides opaqueness and color to paints

polar bond A covalent bond in which the electron pair is unequally shared between the two atoms that it joins, creating a dipole

pollutant A substance placed in the environment by human activity, in greater than natural amounts, that has a detrimental effect on that environment or something of value within it

polyatomic ions Ions that contain two or more atoms in a single unit

polymers Very large molecules, containing as many as 10^6 atoms, that are made by the repetitious combination of small molecules called monomers Macromolecules

polyunsaturated Describes compounds containing two or more carbon–carbon double (or triple) bonds

potential energy Something that has the capacity to become kinetic energy Stored energy

power The rate at which energy is converted

precipitate (noun) A solid that forms from a reaction in solution

precipitate (verb) A process in which a solid forms from a reaction in solution

precision The amount of fineness and reproducibility in a measurement

primary source An energy resource found on the earth in an unconverted state, such as coal, petroleum, natural gas, and solar energy

products Substances that are the result of a chemical reaction (are produced) and are written on the right-hand side of a chemical equation

properties Attributes of a substance such as color, odor, physical state, density, mass, and volume

proton A particle having a 1+ charge that is found in the nuclei of all atoms (a nucleon) Also, a hydrogen ion

pure substance A form of matter having definite, constant composition and a unique set of properties

Q

qualitative analysis Determining what a substance is made of

quanta Small bundles, or packets, of energy

quantitative analysis Determining how much of each constituent a substance contains

quantized energy levels Energy levels in atoms that have discrete relationships to one another

R

radiant energy Electromagnetic radiation such as light, X-rays, and microwaves

radical chain reaction A repetitive series of reactions in which a free radical reacts with a molecule to produce new free radicals that continue to react in a similar manner until the original free radical is regenerated and begins the cycle again

radioactivity A process in which large atoms spontaneously fall apart into smaller atoms and emit energetic rays such as alpha particles, beta particles, and gamma rays

reactants Substances that enter into a chemical reaction and are written on the left-hand side of a chemical equation

recrystallization A commonly used means of purifying solid substances that takes advantage of the fact that the solubility of most substances in a solvent increases with increasing temperature

redox Oxidation–reduction

reducing agent A substance that causes some other species to be reduced and is itself oxidized in the process

reduction The gain of electrons by a chemical species Also, the loss of oxygen from or the gain of hydrogen by a chemical species

renewable energy source A source of energy that ultimately comes from the sun, such as water power, wood, and solar energy

representative elements Elements in Groups 1A through 8A (A-Group elements)

risk–benefit analysis Weighing the negative effects of a technology against its positive attributes to determine whether using that technology is worthwhile

S

salinization The collection of ionic materials in poorly drained, irrigated soils in arid regions

salt An ionic compound

saponification Cleaving a fat (triglyceride) with hot lye (NaOH) solution to make a soap and glycerol

saturated The condition of a solution in which no more solute can be dissolved at a given temperature

saturated hydrocarbons Hydrocarbons that contain only single bonds

scientific fact Something that exists, discernible by observation, that has the consensus of most observers

scientific law A description of a natural regularity or pattern that summarizes a large number of scientific facts

scientific theory An explanation that accounts for all of the facts currently known about a phenomenon and is an attempt to provide an understanding of its underlying structure

scientific truth A scientific observation

scintillations Flashes of light

second law of thermodynamics States that in any energy conversion process, some of the energy converted cannot do useful work (is wasted)

secondary source An energy source obtained by conversion of a primary energy source to another form, such as electricity, a battery, or hydrogen fuel

sediments Suspended particulate matter in water

SI system The system of units based on the metric system presently used by the international scientific community

soft water Hard water in which calcium and magnesium ions have been replaced by sodium ions

solid A state of matter that has a definite shape and volume

solute The component that is present in the lesser amount in a solution

solution A homogeneous mixture

solvated Describes a solvent particle (ion or molecule) that is surrounded by solvent molecules

solvent The component that is present in the greater amount in a solution

space-network polymer A three-dimensional, crosslinked polymer that is essentially one very large, rigid molecule

spectator ions Ions that do not participate in chemical reactions in solution

spinneret A die with small holes in it through which a molten or dissolved polymer is extruded (squeezed) to make fibers

stability Relative levels (states) of energy, a lower level of energy being more stable than a higher level of energy

standard An object used for comparison in calibrating instruments See calibration

states of matter Solid, liquid, and gas

steady state A condition in which a substance is formed and consumed at the same rate, causing its concentration to remain essentially constant

stratosphere The layer of the atmosphere that extends from the upper troposphere (16 km) to an altitude of about 50 km

strong acids Acids that ionize completely in water to produce a large number of hydrogen ions in solution

strong bases Bases that react completely in water to furnish a large number of hydroxide ions in solution

sublimation Conversion of a solid directly into the gaseous state, bypassing the liquid state

subsoil The intermediate horizon of most soils, composed mainly of inorganic matter and some organic matter

substitution reactions Reactions in which an atom or group of atoms on a molecule is replaced by another atom or group of atoms

supersaturated Describes a solution that contains more solute than it would ordinarily hold at a given temperature

surface water Fresh water in lakes, reservoirs, and rivers

systematic errors A combination of errors in a measurement caused by such things as operator mistakes and built-in instrument inaccuracies

T

technology The social use of science, also called applied science

temperature A measure of the average kinetic energy of the particles that make up a substance

thermal pollution Increasing the temperature of surface waters by placing waste heat into them

thermodynamics The study of the movement of heat

thermoplastic A polymer that is a solid at room temperature and can be shaped by heating and molding

thermosetting resins Polymers that solidify into a rigid mass when heated and molded

tincture A solution of a medicine in alcohol

tolerance Built-in inaccuracy in measurements

topsoil The top horizon of most soils, which contains most of the organic matter and supports most of the biological activity

transition elements The B-Group elements

triglycerides Esters of long-chain fatty acids with glycerol

troposphere The layer of the atmosphere, about 10–16 km thick, nearest Earth

U

units Labels that tell what a measurement is and give it a size

unsaturated Describes a solution that contains less than the amount of solute required to make it saturated

V

vacuum The absence of all matter

valence electrons Outermost electrons in atoms that are involved in forming chemical bonds

valence-shell electron pair repulsion The principle of minimizing the repulsive forces (hence maximizing the distance) between electron pairs in covalent bonds around a central atom in a molecule Abbreviated VSEPR

voltaic cell An electrochemical cell that produces an electric current A galvanic cell

vulcanization Heating natural rubber with sulfur to produce an elastomer that is stronger, more durable, and more elastic

W

wave model Treats the electron(s) in an atom as both a particle and a standing wave, with the result that a every electron in an atom has a unique energy associated with it, much like the frequency associated with a particular overtone of a vibrating string

wavelength The distance between crests (or troughs) in a wave

weak acids Acids that ionize slightly in water to produce a small number of hydrogen ions in solution

weak bases Bases that react slightly in water to furnish a small number of hydroxide ions in solution

weight The attraction of gravity for an object

work The expenditure of energy to move mass with a force through a distance

Z

zero-point energy The residual motion of particles at absolute zero

Zintl border A zigzag line in the Periodic Table that separates the metals from the nonmetals

Index

D

T

Figure Credits

Chapter 1: 1.1 Richard Megna 1990/Fundamental Photographs. **1.2** Radel/Navidi: *Chemistry,* 2/e, © 1994 by West Publishing Company. Adapted with permission. **Fig. 1** © 1987 Paul Silverman/Fundamental Photographs. **1.3** E. R. Degginger.

Chapter 2: 2.1a E. R. Degginger. **2.2** © Paul Shambroom/Science Source, Photo Researchers, Inc. **2.3** Courtesy of Digital Instruments, Inc. **Fig. 1** Phil Degginger. **2.4a** Courtesy of the National Institute of Standards and Technology. **2.4b** © Royal Greenwich Observatory/Science Photo Library. **2.5** Courtesy of the National Institute of Standards and Technology. **2.6ab** Radel/Navidi: *Chemistry,* 2/e, © 1994 by West Publishing Company. Adapted with permission. **2.7** Notebook courtesy of Renee Bunde. **2.8** Phil Degginger. **2.9** © Lori Adamski/Tony Stone Worldwide.

Chapter 3: 3.2 © Tony Freeman, PhotoEdit. **3.3** Science VU/Visuals Unlimited. **3.4a-b** Spencer Seager. **3.4c** © Dr. E. R. Degginger. **3.6** Seager/Slabaugh: *Chemistry for Today: General, Organic, and Biochemistry,* 2/e, © 1994 by West Publishing Company. Adapted with permission. **3.7a-c** © 1986 Joel Gordon. **3.8a-d** © 1988 Joel Gordon. **3.9** Science VU/Visuals Unlimited. **Fig. 1** The Bettmann Archive. **3.13a-b** © 1988 Joel Gordon. **3.13c-d** E. R. Degginger. **3.13e** Glenn Olivet/Visuals Unlimited.

Chapter 4: 4.1 Monroe/Wicander: *Physical Geology: Exploring the Earth,* © 1992 by West Publishing Company. Reprinted with permission. **Fig. 1** Courtesy of P. K. Link, Idaho State University. **4.2a** © Dan McCoy/Rainbow. **4.2b** © 1990 James Prince/Photo Researchers. **4.3a-b** Phil Degginger. **4.6** Dewey: *Understanding Chemistry: An Introduction,* © 1994 by West Publishing Company. Adapted with permission. **4.9** © Dr. E. R. Degginger. **4.10** E. F. Smith Collection, Dept. of Special Collections, Van Pelt-Dietrich Library, University of Pennsylvania. **4.13** Lester V. Bergman, NY.

Chapter 5: 5.3 Dewey: *Understanding Chemistry: An Introduction,* © 1994 by West Publishing Company. Reprinted with permission. **5.4** Dewey: *Understanding Chemistry: An Introduction,* © 1994 by West Publishing Company. Adapted with permission. **5.5** Radel/Navidi: *Chemistry,* 2/e, © 1994 by West Publishing Company. Adapted with permission. **5.7a-b** Bob Firth/Firth Photobank. **5.8** Runk/Schoenberger from Grant Heilman. **5.9** Radel/Navidi: *Chemistry,* 2/e, © 1994 by West Publishing Company. Reprinted with permission. **5.10a** © Martin Dohrn/Science Photo Library. **5.10b** © Yoav Levy/Phototake NYC. **5.11** Umland: *General Chemistry,* © 1993 by West Publishing Company. Adapted with permission. **5.12** Radel/Navidi: *Chemistry,* 2/e, © 1994 by West Publishing Company. Adapted with permission. **5.15** Lester U. Bergmann and Associates, Inc. **5.14** Radel/Navidi: *Chemistry,* 2/e, © 1994 by West Publishing Company. Adapted with permission. **5.22** Radel/Navidi: *Chemistry,* 2/e, © 1994 by West Publishing Company. Adapted with permission. **Fig. 1** Umland: *General Chemistry,* © 1993 by West Publishing Company. Reprinted with permission. **Fig. 2** Radel/Navidi: *Chemistry,* 2/e, © 1994 by West Publishing Company. Reprinted with permission. **Fig. 3** Vases designed by Gunnar Ander, Lindshammar Glassworks. American Swedish Institute Collection. Photo credit: Sue Hartley. **5.24** Umland: *General Chemistry,* © 1993 by West Publishing Company. Adapted with permission. **5.26** Umland: *General Chemistry,* © 1993 by West Publishing Company. Adapted with permission. **5.27** PSSC PHYSICS, 2nd Ed., 1965; D. C. Health and Company with Education Development Center, Inc., Newton, MA. **Fig. 1** Umland: *General Chemistry,* © 1993 by West Publishing Company. Adapted with permission. **Fig. UN** Umland: *General Chemistry,* © 1993 by West Publishing Company. Adapted with permission. **Fig. 2** National Optical Astronomy Observatories.

Chapter 6: 6.1a-b E. R. Degginger. **6.2a** Paul Silverman/Fundamental Photographs. **6.2b-c** E. R. Degginger. **6.2d** Phil Degginger. **6.2e** E. R. Degginger. **6.6** Dewey: *Understanding Chemistry: An Introduction,* © 1994 by West Publishing Company. Adapted with permission. **Fig. 1** © Tony Freeman. All rights reserved/PhotoEdit. **Fig. 2** E. R. Degginger. **6.11** Seager/Slabaugh: *Chemistry for Today: General, Organic, and Biochemistry,* 2/e © 1994 by West Publishing Company. Adapted with permission. **6.12** © Phil Degginger. **6.18a** © Phil Degginger. **6.19a** © Phil Degginger. **6.20** Umland: *General Chemistry,* © 1993 by West Publishing Company. Reprinted with permission. **6.25** Umland: *General Chemistry,* © 1993 by West Publishing Company. Adapted with permission. **Photo p. 201** Monroe/Wicander: *Physical Geology: Exploring the Earth,* © 1992 by West Publishing Company. Adapted with permission.

Chapter 7: 7.2 Monroe/Wicander: *Physical Geology: Exploring the Earth,* © 1992 by West Publishing Company. Adapted with permission. **7.1a** E. R. Degginger. **7.1b** Phil Degginger. **7.1c** E. R. Degginger. **7.5a-b** Joel Gordon. **7.9** Tom Pantages. **7.11** Dewey: *Understanding Chemistry: An Introduction,* © 1994 by West Publishing Company. Adapted with permission. **Fig. 1** Courtesy of R. V. Dietrich. **7.16** Umland: *General Chemistry,* © 1993 by West Publishing Company. Adapted with permission. **7.17** © 1987 Joel Gordon. **7.19** Adapted and reprinted with the permission of Macmillan College Publishing Company from *Chemistry: A Science for Today* by Stephen H. Stoker. Copyright © 1989 by Macmillan College Publishing Company, Inc. **7.20a-b** © 1993 Joel Gordon. **7.21** © 1993 Joel Gordon. **7.22a-c** © E. R. Degginger. **7.23a-d** © 1988 Joel Gordon.

Chapter 8: 8.1 Frank Kujawa, University of Central Florida, GeoPhoto Publishing Co. **8.2a-b** E. R. Degginger. **8.5** © 1990 Robert Mathena/Fundamental Photographs **8.11** Virginia Runk/Schoenberger/Grant Heilman. **Fig. 1** Richard Megna/Fundamental Photographs. **8.15** Paul Silverman/Fundamental Photographs. **Fig. 1** Richard Megna/Fundamental Photographs. **Fig. 2** Seager/Slabaugh: *Chemistry for Today: General, Organic, and Biochemistry,* 2/e, © 1994 by West Publishing Company. Adapted with permission.

Chapter 9: 9.1 E. R. Degginger. **9.2** © Vulcain/Explorer/Photo Researchers. **Fig. 1** © Bill Horsman/Stock Boston. **9.5** E. R. Degginger. **9.6** © Felicia Martinez/PhotoEdit. **9.7** Runk/Schoenberger from Grant Heilman. **9.8** Richard Megna/Fundamental Photographs. **9.9ab** Dewey: *Understanding Chemistry: An Introduction,* © 1994 by West Publishing Company. Adapted with permission. **9.9** Photos by E. R. Degginger. **9.10** Radel/Navidi: *Chemistry,* 2/e, © 1994 by West Publishing Company. Adapted with permission. **9.12** Seager/Slabaugh: *Chemistry for Today: General, Organic and Biochemistry,* 2/e, © 1994 by West Publishing Company. Adapted with permission. **9.13** Radel/Navidi: *Chemistry,* 2/e, © 1994 by West Publishing Company. Adapted with permission. **9.15** Tom Pantages. **9.17** Radel/Navidi: *Chemistry,* 2/e, © 1994 by West Publishing Company. Adapted with permission. **9.18** Mason Morfit/FPG. **9.19** Dewey: *Understanding Chemistry: An Introduction,* © 1994 by West Publishing Company. Adapted with permission. **9.20** Adapted with permission from the publisher of Bockris, J. O'M.; Reddy, A. K. N. *Modern Electrochemistry;* Plenum:

New York, 1970, p. 1388. **9.21** Bethlehem Steel Corporation. **10.1** © Pat and Tom Leeson/Photo Researchers.

Chapter 10: 10.2 Gendel: *Basic Chemistry: A Problem Solving Approach,* © 1993 by West Publishing Company. Reprinted with permission. **10.3** Phil Degginger. **10.4** Radel/Navidi: *Chemistry,* 2/e, © 1994 by West Publishing Company. Reprinted with permission. **10.5** Phil Degginger. **10.6** Seager/Slabaugh: *Chemistry for Today: General, Organic, and Biochemistry,* 2/e, © 1994 by West Publishing Company. Reprinted by Permission. **10.8** Camerique Stock Photography. **10.10** G. Prance/Visuals Unlimited. **10.11** E. R. Degginger. **10.12a-b** © Joel Gordon. **10.13** Science VU/Visuals Unlimited. **10.14** © Mark Greenberg/VISIONS. **Fig. 1** Smithsonian Institution. **10.16** © Hank Morgan/Science Photo Library. **10.17** © Grant Heilman Photography. **Fig. 1** © Dan McCoy/Rainbow. **Fig. 2** Dewey: *Understanding Chemistry: An Introduction,* © 1994 by West Publishing Company. Adapted with permission. **10.25** T. Havill/Visuals Unlimited. **10.27** Dr. Jeremy Burgess/Science Photo Library.

Chapter 11: 11.3 © 1988 Joel Gordon. **11.4** Seager/Slabaugh: *Chemistry for Today: General, Organic, and Biochemistry,* 2/e, © 1994 by West Publishing Company. Reprinted with permission. **11.5** Seager/Slabaugh: *Chemistry for Today: General, Organic, and Biochemistry,* 2/e, © 1994 by West Publishing Company. Reprinted with permission. **11.8** Seager/Slabaugh: *Chemistry for Today: General, Organic, and Biochemistry,* 2/e, © 1994 by West Publishing Company. Reprinted with permission. **11.9** Phil Degginger. **11.10** Diagram redrawn, with permission from a diagram, provided courtesy of Exxon Company, U.S.A. **11.11** Joseph P. Sinnot/Fundamental Photographs. **11.13** Dr. E. R. Degginger. **11.14** Photo direction by Fran Sizer and photo taken by Quest Photographic, Inc. **11.15** © John Zoiner/International Stock. **11.16** Phil Degginger. **11.18** David S. Addison/Visuals Unlimited. **Fig. 1** Gendel: *Basic Chemistry: A Problem Solving Approach,* © 1993 by West Publishing Company. Reprinted with permission. **11.19** © 1992 B. Kramer/Custom Medical Stock Photo. All rights reserved. **11.20a-c** © 1988 Joel Gordon. **11.23** Walt Anderson/Visuals Unlimited.

Chapter 12: 12.4 © Bill Pogue/Tony Stone Worldwide. **12.5** © Frank Cezus/Tony Stone Worldwide. **12.6** Phil Degginger. **12.9** DuPont deNemours Company. **12.10** E. R. Degginger. **12.12** © Dr. E. R. Degginger. **12.13** John Blaustein, Woodfin Camp and Associates. **12.14** Courtesy of Frankl & Thomas Incorporated, Greenville, South Carolina. **Fig. 4** Michael Long/Visuals Unlimited. **12.16** E. R. Degginger. **12.20** Courtesy of the Monsanto Company. **12.21** E. R. Degginger. **12.22** Dewey: *Understanding Chemistry: An Introduction,* © 1994 by West Publishing Company. Reprinted with permission. **Photo, p. 467** Courtesy of Mile Schmitt, Soil Scientist, University of Minnesota.

Chapter 13: Fig. 1 © Rand Raraku/International Stock. **13.1** Reflective art rendered by Darwen and Vally Hennings for Chiras: *Biology: The Web of Life,* © 1993 by West Publishing Company. Reprinted with permission. **13.3** © A. J. Copley/Visuals Unlimited. **Fig. 1** Dewey: *Understanding Chemistry: An Introduction,* © 1994 by West Publishing Company. Adapted with permission. **13.4** Phil Degginger. **Fig. 2** Peticolas/Megna of Fundamental Photographs. **13.6** Dewey: *Understanding Chemistry: An Introduction,* © 1994 by West Publishing Company. Reprinted with permission. **13.9** Monroe/Wicander: *Physical Geology: Exploring the Earth,* © 1992 by West Publishing Company. Adapted with permission. **13.10** Steve McCurry/Magnum. **13.11** Science VU/Visuals Unlimited. **13.12** Monroe/Wicander: *Physical Geology: Exploring the Earth,* 2/e, © 1995 by West Publishing Company. Reprinted with permission. **13.13** Jim Richardson. **13.14** Philip Wright/Visuals Unlimited. **13.15** © Dana Richter/Visuals Unlimited. **13.16a-c** Pipkin: *Geology and the Environment,* © 1994 by West Publishing Company. Reprinted with permission. **13.17** D. D. Trent. **13.18** From Kenneth A. Cole and Ann Kirkpatrick, "Cyanide Leaching in California," *California Geology.* Also from Pipkin: *Geology and the Environment,* © 1994 by West Publishing Company. Reprinted with permission. **13.19** Visuals Unlimited. **13.20** Pipkin: *Geology and the Environment,* © 1994 by West Publishing Company. Reprinted with permission. **Photo, p. 517** Courtesy of the New Zealand Tourism Board.

SYMBOLS, ATOMIC NUMBERS, AND ATOMIC MASSES FOR THE ELEMENTS

Element	Symbol	Atomic Number	Atomic Mass
Actinium	Ac	89	*
Aluminum	Al	13	27.0
Americium	Am	95	
Antimony	Sb	51	122
Argon	Ar	18	40.0
Arsenic	As	33	74.9
Astatine	At	85	
Barium	Ba	56	137
Berkelium	Bk	97	
Beryllium	Be	4	9.01
Bismuth	Bi	83	209
Boron	B	5	10.8
Bromine	Br	35	79.9
Cadmium	Cd	48	112
Calcium	Ca	20	40.1
Californium	Cf	98	
Carbon	C	6	12.0
Cerium	Ce	58	140
Cesium	Cs	55	133
Chlorine	Cl	17	35.5
Chromium	Cr	24	52.0
Cobalt	Co	27	58.9
Copper	Cu	29	63.5
Curium	Cm	96	
Dysprosium	Dy	66	162
Einsteinium	Es	99	
Erbium	Er	68	167
Europium	Eu	63	152
Fermium	Fm	100	
Fluorine	F	9	19.0
Francium	Fr	87	
Gadolinium	Gd	64	157
Gallium	Ga	31	69.7
Germanium	Ge	32	72.6
Gold	Au	79	197
Hafnium	Hf	72	178
Helium	He	2	4.00
Holmium	Ho	67	165
Hydrogen	H	1	1.01
Indium	In	49	115
Iodine	I	53	127
Iridium	Ir	77	192
Iron	Fe	26	55.8
Krypton	Kr	36	83.8
Lanthanum	La	57	139
Lawrencium	Lr	103	
Lead	Pb	82	207
Lithium	Li	3	6.94
Lutetium	Lu	71	175
Magnesium	Mg	12	24.3
Manganese	Mn	25	54.9
Mendelevium	Md	101	
Mercury	Hg	80	201
Molybdenum	Mo	42	95.9
Neodymium	Nd	60	144
Neon	Ne	10	20.2
Neptunium	Np	93	
Nickel	Ni	28	58.7
Niobium	Nb	41	92.9
Nitrogen	N	7	14.0
Nobelium	No	102	
Osmium	Os	76	190
Oxygen	O	8	16.0
Palladium	Pd	46	106
Phosphorus	P	15	31.0
Platinum	Pt	78	195
Plutonium	Pu	94	
Polonium	Po	84	
Potassium	K	19	39.1
Praseodymium	Pr	59	141
Promethium	Pm	61	
Protactinium	Pa	91	
Radium	Ra	88	
Radon	Rn	86	
Rhenium	Re	75	186
Rhodium	Rh	45	103
Rubidium	Rb	37	85.5
Ruthenium	Ru	44	101
Samarium	Sm	62	150
Scandium	Sc	21	45.0
Selenium	Se	34	79.0
Silicon	Si	14	28.1
Silver	Ag	47	108
Sodium	Na	11	23.0
Strontium	Sr	38	87.6
Sulfur	S	16	32.1
Tantalum	Ta	73	181
Technetium	Tc	43	
Tellurium	Te	52	128
Terbium	Tb	65	159
Thallium	Tl	81	204
Thorium	Th	90	232
Thulium	Tm	69	169
Tin	Sn	50	119
Titanium	Ti	22	47.9
Tungsten	W	74	184
Uranium	U	92	238
Vanadium	V	23	50.9
Xenon	Xe	54	131
Ytterbium	Yb	70	173
Yttrium	Y	39	88.9
Zinc	Zn	30	65.4
Zirconium	Zr	40	91.2

*Elements for which no atomic mass is listed have no known stable isotopes.